W9-DDO-542

WIRELESS BROADBAND NETWORKS

WIRELESS BROADBAND NETWORKS

DAVID TUNG CHONG WONG
PENG-YONG KONG
YING-CHANG LIANG
KEE CHAING CHUA
JON W. MARK

WILEY
A JOHN WILEY & SONS, INC., PUBLICATION

Published by John Wiley & Sons, Inc., Hoboken, New Jersey.
Published simultaneously in Canada.

For general information on our other products and services or for technical support, please contact our Customer Care Department within the United States at (800) 762-2974, outside the United States at (317) 572-3993 or fax (317) 572-4002.

Wiley also publishes its books in a variety of electronic formats. Some content that appears in print may not be available in electronic formats. For more information about Wiley products, visit our web site at www.wiley.com.

Library of Congress Cataloging-in-Publication Data:

Wireless broadband networks / David Tung Chong Wong . . . [et al.].
p. cm.
Includes bibliographical references and index.
ISBN 978-0-470-18177-5 (cloth)
1. Wireless communication system. 2. Broadband communication systems. I. Wong, David Tung Chong.
TK5103.2.W557 2009
621.384–dc22

2008053469

Printed in the United States of America

10 9 8 7 6 5 4 3 2 1

To my family
David Tung Chong Wong

To my parents and wife
Peng-Yong Kong

To my wife, Shuo, and my kids, Paul and Wendy
Ying-Chang Liang

To Nancy, Daryl, and Kevin
Kee Chaing Chua

To my wife, Betty
Jon W. Mark

CONTENTS

PREFACE

This book is divided into two main parts. The first part covers the enabling technologies for *wireless broadband networks*, and the second part covers the various systems for *wireless broadband networks*. The enabling technologies are clearly explained, with illustrations to provide readers with the necessary knowledge to better understand the rationale for the design of advanced practical systems, which are presented in detail in the second part of the book. The important enabling technologies for wireless broadband networks include OFDM, MIMO, UWB, MAC, mobility resource management, radio resource management, routing, and quality of service for multimedia services. The advanced systems that are covered for wireless broadband networks include 3.9G long-term evolution (LTE) cellular systems, WiMax, WLAN (IEEE 802.11e and 802.11n), WUSB, and WiMedia. The 3.9G LTE cellular system and IEEE 802.11n WLAN are still currently under standardization, but the latest information on them is provided in the book. The treatment is such that essential wireless broadband networks are covered together with a thorough explanation of the theories and rationales in the design of these advanced practical networks. The objective of the book is to provide a good foundation in theories and to apply some of these to advanced practical systems in wireless broadband networks, embodying the physical layer to the network layer in the ISO/OSI model. Applications of these systems are also presented. Extensive references are available for those readers who want to explore the theories or advanced practical systems in greater depth. Our approach is to couple theories with advanced practical systems for wireless broadband networks. Thus, the book is unique in these aspects and differentiates itself from other books in the marketplace.

Part I consists of eight chapters. Chapter 1 is devoted to orthogonal frequency-division multiplexing (OFDM) and other block-based transmissions. A brief introduction to the basics of wireless communications systems is provided, and various multiple-access schemes, such as OFDM, single-carrier cyclic prefix, orthogonal frequency-division multiple access, interleaved frequency-division multiple access, cyclic prefix-based-division multiple access, and multicarrier code-division multiple access, are presented. Linear and iterative equalizers are also reviewed for the general channel.

Chapter 2 deals with multiple-input, multiple-output (MIMO) antenna systems. The MIMO system model, channel capacity, diversity gain, and relationship between spatial diversity gain and spatial multiplexing gain are introduced at the beginning of the chapter. SIMO systems, MISO systems, and space–time coding are explained together with the signal-to-noise ratio expressions. MIMO transceiver design is also presented in this chapter. In the final part of the chapter, SVD-based eigen-beamforming, MIMO and transmit diversity for frequency-selective fading channel, and cyclic delay diversity are explained in detail.

In Chapter 3 we describe and analyze the performance of time-hopping and direct-sequence ultrawideband (UWB) systems. Both single and multiple traffic classes, both with and without multipath channels, are considered. Other types of UWB systems, such as transmitted reference UWB, chirp UWB, multicarrier UWB, and MIMO UWB are also presented.

In Chapter 4 we describe and provide analytical frameworks for medium access control (MAC) protocols. The MAC protocols include slotted Aloha, carrier-sense multiple access with collision avoidance, polling, reservation, energy efficient, multichannel, directional, time-division multiple access, frequency-division multiple access, and code-division multiple access.

In Chapter 5 we categorize the types of horizontal and vertical handoffs as well as the types of handoff strategies. Channel assignment schemes for single and multiple traffic classes are presented with analytical models. The channel assignment schemes for single traffic classes include nonprioritized, prioritized (guard channels), limited fractional guard channel, fractional guard channels, guard channel with queue, and two-level guard channels. The channel assignment schemes for multiple traffic classes include complete partitioning, complete sharing, and virtual partitioning. Link-layer resource allocation schemes are presented and analyzed for both single and multiple traffic classes. Location management is also presented briefly. Finally, mobile IP, cellular IP, and HAWAII for mobility handling are presented.

Chapter 6 covers routing protocols for multihop wireless broadband networks. The routing metrics are also classified in this chapter. Furthermore, six types of classification for routing protocols are listed: topology-based versus position-based, proactive versus reactive, distance vector versus link state, hop-by-hop routing versus source routing, flat versus hierarchical, and single-path versus multipath. Existing routing protocols such as ad hoc on-demand distance vector, optimized link state routing, and dynamic source routing are also discussed in detail.

Chapter 7 deals with radio resource management for wireless broadband networks. Two important aspects of radio resource management discussed and analyzed in this chapter are packet scheduling and admission control. The packet-scheduling schemes include channel error avoidance scheduling for fair bandwidth sharing and channel error-avoidance scheduling with quality-of-service differentiation. Model-, measurement-, and resource-based admission controls are also discussed in this chapter.

In Chapter 8 we deal with various traffic models and quality of service in wireless systems. The traffic models include voice, video, data, web browsing, and network gaming. The wireless systems include universal mobile telecommunications systems, WiMax, IEEE 802.11 wireless local area network (WLAN), and WiMedia wireless personal area network (WPAN).

Part II consists of five chapters. In Chapter 9 we introduce the latest 3.9G LTE cellular system, which is still under standardization. This chapter covers the architecture, physical layer, radio link control, packet data convergence protocol, and radio resource control of LTE cellular networks. Mobility management, radio resource management, and quality of service in LTE cellular networks are also described and discussed. Applications in this wireless broadband network are also described.

In Chapter 10 we introduce WiMAX and its competing technologies. Different modes of operations in WiMAX are also described and discussed in detail. These modes of operations include PMP, mesh, and multihop relay.

In Chapter 11 we describe and discuss IEEE 802.11 WLAN and its architectures, physical layer (IEEE 802.11n), and medium access control protocols (IEEE 802.11, 802.11e, 802.11n, and 802.11s). The focus of the physical layer and medium access control is on IEEE 802.11n, which has a data rate of up to 600 Mbps. An analytical model is presented for IEEE 802.11n MAC and 802.11e MAC. Mobility resource management of IEEE 802.11 WLAN, and quality of service and applications of IEEE 802.11n WLAN are also described and discussed.

In Chapter 12 we introduce WiMedia WPAN and its architectures, physical layer, and medium access control. WiMedia has a data rate of up to 480 Mbps. An approximate analytical model for the WiMedia MAC is presented in this chapter. Wireless universal serial bus (WUSB) is also described briefly. Mobility resource management, quality of service, and applications of WiMedia WPAN are described and discussed.

Chapter 13 looks at an envisaged vision of a future convergence of networks with WPANs, WLANs, WiMax, and cellular networks. The issues arising from the interworkings of these networks are also discussed. Six 3GPP/WLAN interworking scenarios are presented. IEEE 802.11u for interworking with external networks is also described briefly. IEEE 802.21 media-independent handoff is described in this chapter. Finally, an analytical model for cellular/WLAN interworking is presented.

For completeness, an appendix that presents a concise review of the basics of probability, random variables, exponential random process, birth–death processes, and simple queueing systems is included.

ACKNOWLEDGMENTS

There are many people that we want to thank. First and foremost, we are deeply indebted to the series editors, Dr. Vincent Lau and Dr. Russell Hsing, who invited us to write this book. We sincerely thank Dr. Francois Chin, Dr. Sumei Sun, and Dr.

Chen Khong Tham for supporting this project. We would like to thank Sumei for providing the latest IEEE 802.11n draft documents and sharing her tutorial material on IEEE 802.11n, and Higuchi-san for sharing his seminar material on LTE. We would also like to express our gratitude to Serene, Jianxin, Cheng Heng, Shajan, Sai Ho, Winston, Zhiwei, Ananth, Lijuan, Lokesh, and The Hanh for proofreading some parts of the book. Thanks are due to Mr. Paul Petralia, Ms. Whitney Lesch, Ms. Anastasia Wasko, Mr. Michael Christian, and Ms. Angioline Loredo for their assistance and professional advice. We would also like to thank all the people who have helped in the preparation and production of this book in one way or the other. Those contributions notwithstanding, this book has been devised and written by us alone, and we remain responsible for any errors in the final version of the book. Last but not least, we would like to thank our family and friends, who provided love and encouragement throughout this project.

DAVID TUNG CHONG WONG
PENG-YONG KONG
YING-CHANG LIANG
KEE CHAING CHUA
JON W. MARK

PART I

ENABLING TECHNOLOGIES FOR WIRELESS BROADBAND NETWORKS

CHAPTER 1

ORTHOGONAL FREQUENCY-DIVISION MULTIPLEXING AND OTHER BLOCK-BASED TRANSMISSIONS

1.1 INTRODUCTION

In this chapter we first provide a brief introduction to the basics of wireless communication systems. Then we focus on reviewing various block-based transmission schemes that play important roles in the physical-layer design of wireless broadband networks. These schemes include orthogonal frequency-division multiplexing (OFDM), single-carrier cyclic prefix (SCCP), orthogonal frequency-division multiple access (OFDMA), interleaved frequency-division multiple access (IFDMA), single-carrier frequency-division multiple access (SC-FDMA), cyclic prefix-based code-division multiple access (CP-CDMA), and multicarrier code-division multiple access (MC-CDMA). From these schemes, we also establish a generic input–output model and review linear and nonlinear equalizers that can be used to recover the transmitted signals.

1.2 WIRELESS COMMUNICATION SYSTEMS

From a physical-layer perspective, the block diagram of a wireless communication system shown in Figure 1.1 consists of three key components: the transmitter, the wireless channel, and the receiver. On the transmitter side, the design objective is to transform the information bits into a signal format suitable for transmission over

Wireless Broadband Networks, By David Tung Chong Wong, Peng-Yong Kong, Ying-Chang Liang, Kee Chaing Chua, and Jon W. Mark

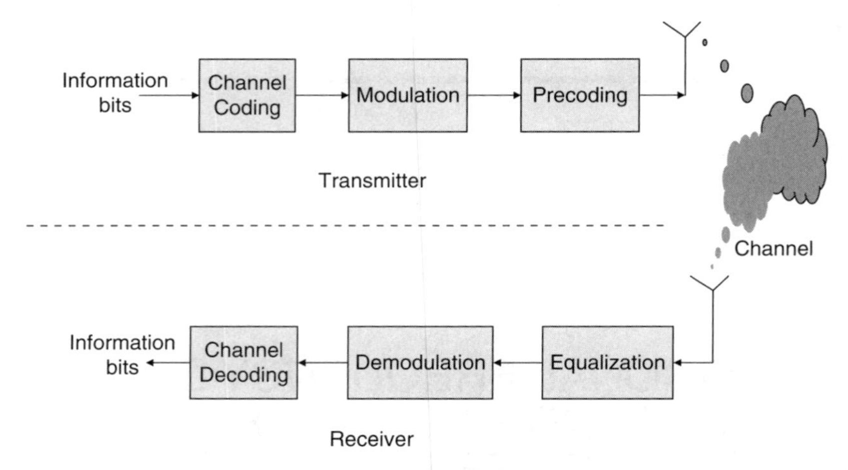

FIGURE 1.1 Block diagram of a wireless communication system.

the wireless channel. The major elements in the transmitter include channel coding, modulation, and linear or nonlinear precoding.

When the signal passes through the wireless channel, the signal will be attenuated due to propagation loss, shadowing, and multipath fading, and the waveform of the signal received will be different from the one transmitted, due to multipath delay, the time/frequency selectivity of the channel, and the addition of noise and unwanted interference. Finally, at the receiver side, the information bits transmitted are to be recovered through the operations of equalization, demodulation, and channel decoding.

With channel coding, the information bits are converted into coded bits with redundancy so that the effect of channel noise and multipath fading can be minimized. The modulation operation transforms the coded bits into modulated symbols for the purpose of achieving efficient transmission of the signal over the channel with a given bandwidth. The objective of the precoding operation is to provide robustness over the fading channel with multipath delay, or to compensate for the unwanted interference.

The equalization operation estimates the modulated symbols by removing the effect of the channel. Through proper design of the precoding operation, equalization sometimes becomes very simple. The demodulation operation converts the estimated symbols into a bit format, which is then used to recover the information bits through the channel decoding operation.

1.2.1 Frequency-Selective Fading Channels

In a wireless propagation environment, the signal transmitted arrives at the receiver with multiple delayed and attenuated versions, and these versions are added up and received by the receiver. The difference in traveling time, τ, between the shortest and longest paths is called *excess delay spread*. When the excess delay spread is much smaller than the symbol period, T_s, the channel can be described by a single delay tap. With this single delay tap, in the frequency domain, the channel responses are flat within the channel bandwidth; thus, the channel is said to be a *flat fading channel*. If the excess delay spread is relatively large compared to the symbol period, the channel can be described by multiple delay taps, and in the frequency domain, the channel responses are no longer flat for all frequencies of interest; thus, this channel is called a *frequency-selective fading channel*.

Suppose that we have a sequence of modulated symbols $\{x(n)\}$ transmitted at the symbol rate of $1/T_s$, through a frequency-selective fading channel. At the receiver, after sampling at the symbol rate, we receive a sequence of received samples $\{y(n)\}$. The relationship between $\{y(n)\}$ and $\{x(n)\}$ is given by

$$y(n) = \sum_{l} h_l(n)x(n-l) + \tilde{u}(n), \tag{1.1}$$

where $\tilde{u}(n)$ is the additive noise and $h_l(n)$ is the lth delay tap of the channel at time n.

We can further characterize the channel as a fast fading or slow fading channel, based on the relationship between the bandwidth of the transmitted signal and the Doppler shift of the wireless channel. When the Doppler shift is relatively significant compared to the signal bandwidth, the channel is called a *fast-fading channel* and $h_l(n)$ changes with time n; otherwise, the channel is referred to as a *slow-fading channel* and $h_l(n)$ is invariant to the time instant n. When slow fading is considered, for description brevity, we drop the time variable in the channel coefficients.

1.2.2 Receiver Equalization

It is seen from equation (1.1) that for a frequency-selective fading channel, the signal received at a time instant is the superposition of weighted and delayed versions of the symbols transmitted. This results in introducing *intersymbol interference* (ISI). Let N be the number of transmitted symbols and L be the number of channel taps spaced at a symbol interval; then the receiver collects $N + L - 1$ samples, which are related to the entire number of symbols transmitted. Equalizers have to be designed at the receiver to compensate for ISI and to recover the symbols transmitted.

The optimal equalizer involves maximum likelihood (ML) detection, which requires very high computational complexity. Suboptimal equalizers have thus been proposed which can be implemented in either linear or nonlinear fashion and require complexity much reduced from ML detection. The performance of these suboptimal receivers is, however, usually far away from the performance bound achieved by ML detection.

1.3 BLOCK-BASED TRANSMISSIONS

To reduce the computational complexity of equalization, block-based transmissions have been proposed. Specifically, in a *block-based transmission*, the entire sequence of modulated symbols is first divided into multiple blocks, each is preprocessed further using linear transforms, and guard symbols are inserted between two consecutive blocks. If the length of the guard symbols is longer than the channel memory, two consecutive blocks will not interfere with each other; thus, each block can be equalized separately.

Two types of guard symbols are applicable in block-based transmissions. One is *zero padding*, which inserts zeros between two consecutive blocks; the other is *cyclic prefix* (CP), which is the copy of the last portion of the signal block. In the following, we describe the properties of CP-based block transmissions.

1.3.1 Use of a Cyclic Prefix

The block diagram of a CP-based block transmission system is shown in Figure 1.2. Let N be the length of one signal block and denote the signal block to be transmitted through a frequency-selective fading channel as follows:

$$\boldsymbol{x} = [x(0) \quad x(1) \quad \cdots \quad x(N-1)]^T. \tag{1.2}$$

The channel is characterized by a channel impulse response (CIR) $\boldsymbol{h} = [h_0 \quad h_1 \quad \cdots \quad h_{L-1}]^T$, which contains L equally spaced time-domain taps (spaced at symbol intervals T_s).

Instead of transmitting block $\boldsymbol{x}$ directly, a new block, $\tilde{\boldsymbol{x}}$, is generated and transmitted through the channel. The new block is formed by appending the last P symbols of $\boldsymbol{x}$ to the head of itself. The portion of the first P symbols in the new block $\tilde{\boldsymbol{x}}$ is the

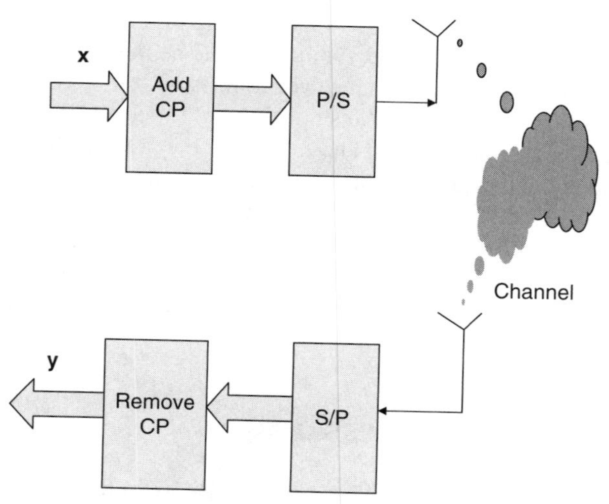

FIGURE 1.2 Block diagram of a CP-based block transmission system.

cyclic prefix (CP). Then the new block $\tilde{\boldsymbol{x}}$ can be represented by

$$\begin{aligned}\tilde{\boldsymbol{x}} &= [\tilde{x}(0) \ \cdots \ \tilde{x}(P-1) \ \tilde{x}(P) \ \tilde{x}(P+1) \ \tilde{x}(P+2) \ \cdots \ \tilde{x}(P+N-1)]^T \\ &= \Big[x(N-P) \ \cdots \ x(N-1) \ \underbrace{x(0) \ x(1) \ \cdots \ x(N-P) \ \cdots \ x(N-1)}_{x^T} \Big]^T. \end{aligned} \tag{1.3}$$

With the transmitted signal $\tilde{\boldsymbol{x}}$, the received signal becomes

$$\tilde{y}(v) = \sum_{l=0}^{L-1} h_l \tilde{x}(v-l) + \tilde{u}(v), \qquad v = 0, 1, \ldots, P+N+L-1, \tag{1.4}$$

where $\tilde{u}(v)$ is additive complex Gaussian random variable with zero mean and variance $E\left\{|\tilde{u}(v)|^2\right\} = N_0$.

The P received signal samples from $\tilde{y}(0)$ to $\tilde{y}(P-1)$ associated with the CP portion are discarded, and we are interested in the received signal samples from $\tilde{y}(P)$ to $\tilde{y}(P+N-1)$, which are associated with the data block. From (1.4), when $P \geq L-1$, we can write the following equations:

$$\begin{aligned}\tilde{y}(P) &= h_0 x(0) + h_1 x(N-1) + \cdots + h_{L-1} x(N-L+1) + \tilde{u}(P), \\ \tilde{y}(P+1) &= h_0 x(1) + h_1 x(0) + \cdots + h_{L-1} x(N-L+2) + \tilde{u}(P+1), \\ &\vdots \\ \tilde{y}(P+N-1) &= h_0 x(N-1) + h_1 x(N-2) + \cdots + h_{L-1} x(N-L) \\ &\quad + \tilde{u}(P+N-1). \end{aligned} \tag{1.5}$$

If we collect the N received signal samples in (1.5) to form a vector $\boldsymbol{y} = [\tilde{y}(P) \quad \tilde{y}(P+1) \quad \cdots \quad \tilde{y}(P+N-1)]^T$, this vector can be written as

$$\boldsymbol{y} = \boldsymbol{H}\boldsymbol{x} + \tilde{\boldsymbol{u}}, \tag{1.6}$$

where $\tilde{\boldsymbol{u}} = [\tilde{u}(P) \quad \tilde{u}(P+1) \quad \cdots \quad \tilde{u}(P+N-1)]^T$, and thanks to the addition of CP, $\boldsymbol{H}$ is now a circular matrix of size $N \times N$ given by

$$\boldsymbol{H} = \begin{bmatrix} h_0 & 0 & \cdots & 0 & h_{L-1} & h_{L-2} & \cdots & h_1 \\ h_1 & h_0 & 0 & \cdots & 0 & h_{L-1} & \cdots & h_2 \\ \vdots & \ddots & \ddots & \ddots & \ddots & \ddots & \ddots & \vdots \\ 0 & \cdots & 0 & h_{L-1} & h_{L-2} & \cdots & h_1 & h_0 \end{bmatrix}. \tag{1.7}$$

Note that what we have developed so far is for the transmission of a single block. To transmit multiple blocks consecutively, Figure 1.3 shows the structure of continuous transmission with CP. From this figure it is clear that some signals received at the beginning of a block are affected by symbols transmitted from the previous block. This

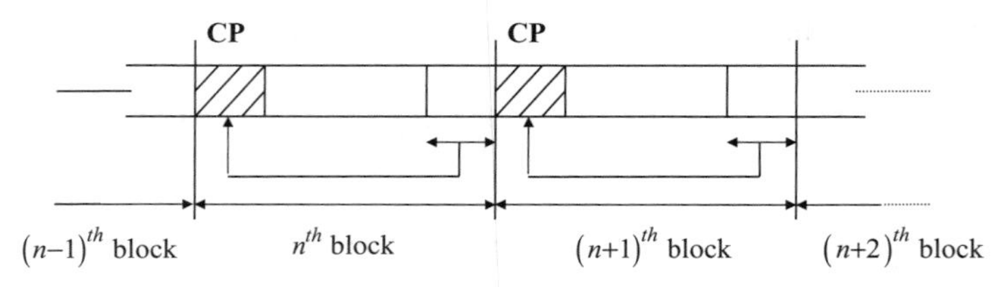

FIGURE 1.3 Structure of continuous transmission with CP.

phenomenon is called *interblock interference* (IBI). Again, if $P \geq L - 1$, inserting CP and discarding the received signals associated with CP help to eliminate the IBI.

1.3.2 Relation Between Vectors *x* and *y*

The circular matrix $\boldsymbol{H}$ in (1.7) can be decomposed into the form

$$\boldsymbol{H} = \boldsymbol{W}_N^H \boldsymbol{\Lambda} \boldsymbol{W}_N, \tag{1.8}$$

where:

- $\boldsymbol{W}_N \in \mathbb{C}^{N\times N}$ is the $N \times N$ discrete Fourier transform (DFT) matrix, given by

$$\boldsymbol{W} = \frac{1}{\sqrt{N}} \begin{bmatrix} 1 & 1 & \cdots & 1 \\ 1 & e^{-j2\pi/N} & \cdots & e^{-j2\pi(N-1)/N} \\ \vdots & \ddots & \ddots & \vdots \\ 1 & e^{-j2\pi(N-1)/N} & \cdots & e^{-j2\pi(N-1)(N-1)/N} \end{bmatrix} \tag{1.9}$$

  Note that $\boldsymbol{W}_N^H \boldsymbol{W}_N = \boldsymbol{I}_N$.
- $\boldsymbol{\Lambda} = \mathrm{diag}\{H_0, H_1, \ldots, H_{N-1}\} \in \mathbb{C}^{N\times N}$ is a diagonal matrix with diagonal elements defined by frequency responses, H_k, of the channel; that is, $H_k = \sum_{l=0}^{L-1} h_l e^{-j2\pi kl/N}$ for $k = 0, 1, \ldots, N-1$.

The proof of (1.8) is given in the appendix at the end of the chapter. From (1.6) and (1.8), we have the following relation between $\boldsymbol{x}$ and $\boldsymbol{y}$:

$$\boldsymbol{y} = \boldsymbol{W}_N^H \boldsymbol{\Lambda} \boldsymbol{W}_N \boldsymbol{x} + \tilde{\boldsymbol{u}}. \tag{1.10}$$

1.3.3 Overview of Block-Based Transmissions

By proper design of the transmitted signal vector $\boldsymbol{x}$ in (1.10), various block-based transmission schemes can be developed, including but not limited to the following:

- Orthogonal frequency-division multiplexing (OFDM) system
- Single-carrier cyclic prefix (SCCP) system
- Orthogonal frequency-division multiple access (OFDMA)

- Interleaved frequency-division multiple access (IFDMA)
- Single-carrier frequency-division multiple access (SC-FDMA)
- Cyclic prefix–based code-division multiple access (CP-CDMA)
- Multicarrier code-division multiple access (MC-CDMA)

OFDMA, IFDMA, SC-FDMA, CP-CDMA, and MC-CDMA are designed to support multiple users to share the same radio resource simultaneously. OFDM and SCCP systems, however, are designed to support single-user communication only. Thus, to support multiple users in sharing the same radio resource, they have to be used in conjunction with other multiple-access schemes, such as time-division multiple access (TDMA) or frequency-division multiple access (FDMA). In TDMA, the time resource is divided into time slots and each user is allowed to use the entire frequency band when it is allocated to use the time slot. In FDMA, the frequency resource is divided into frequency subbands, and each user is allowed to use the entire time resource at the frequency subband allocated to the user. In the following sections, the details of each scheme are provided.

1.4 ORTHOGONAL FREQUENCY-DIVISION MULTIPLEXING SYSTEMS

Orthogonal frequency-division multiplexing (OFDM) is a discrete Fourier transform (DFT)–based multicarrier modulation (MCM) scheme [1,2]. The basic idea of OFDM is to transform a frequency-selective fading channel into several parallel frequency flat fading subchannels on which modulated symbols are transmitted. OFDM has been widely adopted in various communications systems, including the digital audio broadcast (DAB) [3] and digital terrestrial video broadcast (DVB-T) [4] standards in Europe and Japan, the IEEE 802.11a/11n wireless local area network (WLAN) [5], and the asymmetric digital subscriber loop (ADSL) [6].

1.4.1 System Description

The block diagram of an OFDM system is depicted in Figure 1.4. A sequence of data symbols $\{s(n;v)\}_{v=0}^{N-1}$ is first serial-to-parallel (S/P)-converted to form the nth data block $\boldsymbol{s}(n) = [s(n;0) \quad s(n;1) \quad \cdots \quad s(n;N-1)]^T$. This block is transformed by the inverse DFT (IDFT) operation. The output of the IDFT is the transmitted signal block $\boldsymbol{x}(n)$. This block is added with the CP portion as shown in Section 1.2, and the resulting block is parallel-to-serial (P/S)-converted for transmission. The relation between the data block and the transmitted signal block is given by

$$\boldsymbol{x}(n) = \boldsymbol{W}_N^H \boldsymbol{s}(n). \tag{1.11}$$

At the receiver side, the receiver first discards the received signal samples associated with the CP portion. The received signal samples associated with the data block are then fed to the DFT operation. The output of the DFT is passed through the

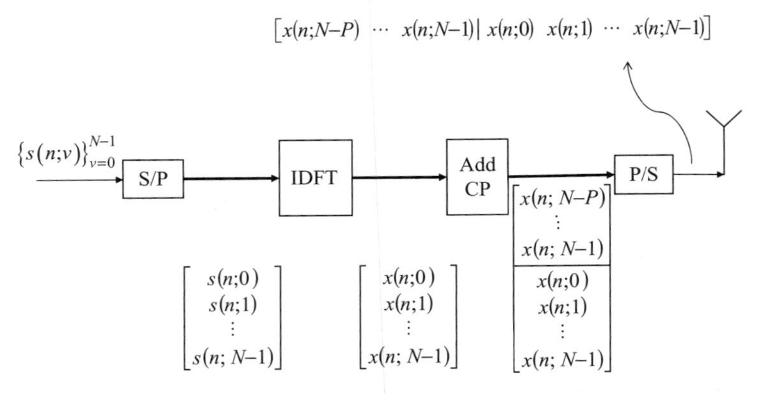

FIGURE 1.4 Block diagram of an OFDM transmitter.

equalizer to recover the data symbols. These operations are illustrated in Figure 1.5. Based on (1.10) and (1.11), we have

$$\boldsymbol{y}(n) = \boldsymbol{W}_N^H \boldsymbol{\Lambda} \boldsymbol{s}(n) + \tilde{\boldsymbol{u}}(n). \tag{1.12}$$

After the DFT, we obtain

$$\boldsymbol{z}(n) = \boldsymbol{W}_N \boldsymbol{y}(n) = \boldsymbol{\Lambda} \boldsymbol{s}(n) + \boldsymbol{u}(n), \tag{1.13}$$

where $\boldsymbol{u} = \boldsymbol{W}_N \tilde{\boldsymbol{u}}$.

It is clear from (1.13) that OFDM transforms a frequency-selective fading channel into a number of frequency flat channels which are called *subchannels*. The output at the kth subcarrier is given by

$$z(n;k) = H_k s(n;k) + u(n;k), \qquad k = 0, 1, \ldots, N-1. \tag{1.14}$$

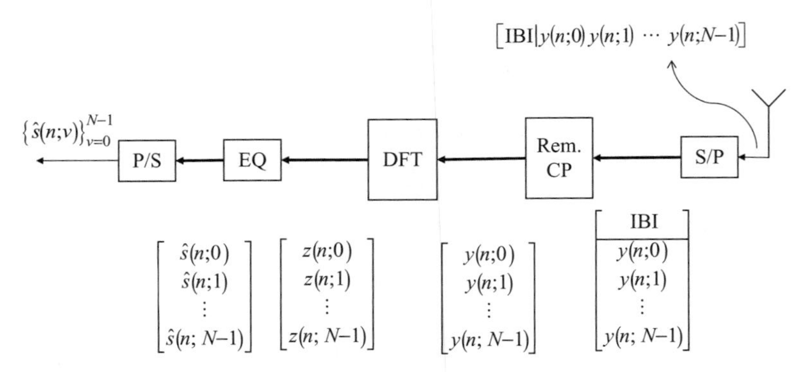

FIGURE 1.5 Block diagram of an OFDM receiver.

At the receiver, thanks to (1.14), a one-tap equalizer can be applied to each of the subcarrier outputs to recover the transmitted symbol delivered over that subcarrier. More specifically, the coefficient of the zero-forcing (ZF) equalizer is

$$C(n;k) = \frac{1}{H_k}, \qquad k = 0, 1, \ldots, N-1, \tag{1.15}$$

while that of the minimum mean square error (MMSE) equalizer becomes

$$C(n;k) = \frac{H_k^*}{|H_k|^2 + N_0/E_s}, \qquad k = 0, 1, \ldots, N-1, \tag{1.16}$$

where E_s is the average energy on every modulated symbol. The recovered symbol on the kth subcarrier is obtained by rounding

$$\bar{s}(n;k) = C(n;k)z(n;k), \qquad k = 0, 1, \ldots, N-1, \tag{1.17}$$

to the closest element of the signal constellation in use. This process is also referred to as *slicing*.

1.4.2 Advantages and Disadvantages of OFDM Systems

One of the attractive features offered by OFDM is that it provides relatively simple one-tap frequency-domain equalization over the complex time-domain equalization used in conventional single-carrier systems. Furthermore, since OFDM has decoupled the frequency-selective fading channel into a parallel set of flat fading channels over the subcarriers, a more fascinating advantage of OFDM is that it allows power and bit loading over the subcarriers, and by doing so, for a given power budget the channel capacity can be maximized.

OFDM suffers from some drawbacks, however. For example, timing synchronization error results in the IBI. Carrier frequency offset destroys the orthogonality among the subcarriers and introduces intercarrier interference (ICI). The presence of IBI and/or ICI degrade the system's performance dramatically [7, Chap. 2]. Another shortcoming of OFDM is its high peak-to-average power ratio (PAPR). When OFDM signal with high PAPR passes through a nonlinear device, high peak signals may be clipped. The distortions caused by this clipping will affect the orthogonality of subcarriers.

1.5 SINGLE-CARRIER CYCLIC PREFIX SYSTEMS [8]

As OFDM suffers from high PAPR, a single-carrier duo of OFDM has been proposed, and this system is called single-carrier cyclic prefix (SCCP) system. The block diagram of a SCCP system is depicted in Figure 1.6. The symbols transmitted are first grouped into blocks. Unlike OFDM, where these blocks are transformed using IDFT

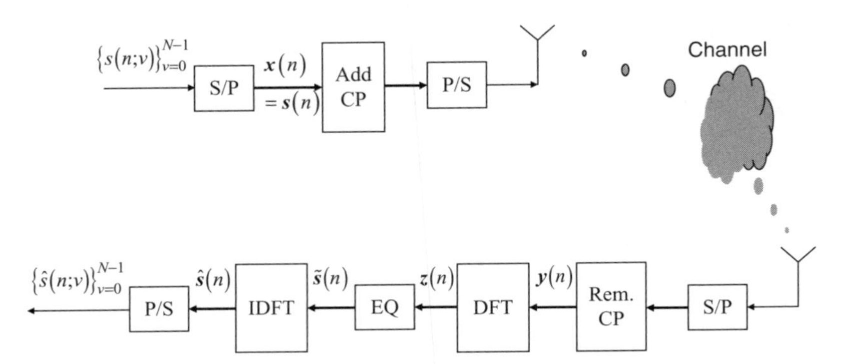

FIGURE 1.6 Block diagrams of SCCP transmitter and receiver.

before CPs are appended, in SCCP, CPs are inserted directly to these blocks. After that, the resulting blocks are parallel-to-serial-converted and delivered to the transmitting antenna for transmission. At the receiver, the received signal samples associated with the CP portion are removed and the received signal block is transformed to the frequency domain to perform equalization. Finally, the estimated block symbols are transformed back to the time domain for symbol detection.

From Figure 1.6, if we choose $\boldsymbol{x}(n) = \boldsymbol{s}(n)$ and $\boldsymbol{y}(n)$ is filtered as $\boldsymbol{z}(n) = \boldsymbol{W}_N\, \boldsymbol{y}(n)$, we have

$$\boldsymbol{z}(n) = \boldsymbol{\Lambda} \boldsymbol{W}_N \boldsymbol{s}(n) + \boldsymbol{u}(n). \tag{1.18}$$

Based on (1.18), a frequency-domain equalizer can be deployed to recover the symbols transmitted. More specifically, let $\tilde{\boldsymbol{s}}(n) = \boldsymbol{W}_N \boldsymbol{s}(n)$; then (1.18) reads

$$\boldsymbol{z}(n) = \boldsymbol{\Lambda} \tilde{\boldsymbol{s}}(n) + \boldsymbol{w}(n). \tag{1.19}$$

A ZF or MMSE equalizer can be applied in (1.19) to estimate $\tilde{\boldsymbol{s}}(n)$. After that, the transmitted symbols are recovered as

$$\hat{\boldsymbol{s}}(n) = \boldsymbol{W}_N^H \tilde{\boldsymbol{s}}(n). \tag{1.20}$$

SCCP has been adopted as a part of the IEEE 802.16 standards for wireless metropolitan area networks (WMANs).

1.6 ORTHOGONAL FREQUENCY-DIVISION MULTIPLE ACCESS

With the increasing demand on high-data-rate applications and more users supported in a geographical area, orthogonal frequency-division multiple access (OFDMA), which is a combination of OFDM and FDMA, has been proposed to support multiple users simultaneously. First adopted for cable TV (CATV) networks [9], OFDMA has been used in the uplink of the interaction channel for digital terrestrial television (DVB-RCT) [10] and in the IEEE 802.16 standards for WMAN [11]. OFDMA has also been used in satellite communication [12] and third-generation cellular system long-term evolution (3G-LTE).

1.6.1 Subcarrier Allocation

In an OFDMA system, a number of users transmit their information data simultaneously on a number of available subcarriers. Each user is assigned to a set of subcarriers called a *subchannel*. Different subchannels are mutually exclusive. More specifically, suppose that there are N subcarriers and U users in the system. N subcarriers are divided into S subchannels, in which one subchannel consists of $P = N/S$ subcarriers. It is obvious that the system can, at maximum, support only $U \leq S$ users simultaneously.

In a subcarrier assignment scheme [13], each user's subchannel occupies a group of P adjacent subcarriers. This scheme is called *localized subcarrier allocation*. For this scheme, frequency diversity offered by a multipath channel is not obtained because a deep fade can occur over a large number of subcarriers assigned to a given user.

To overcome the drawback of localized subcarrier allocation, distributed subcarrier allocation was proposed [14]. In this scheme, subcarriers belonging to a given user are uniformly distributed over the entire set of N subcarriers. This allocation method reduces the probability that a substantial number of carriers of a user experience a deep fade at the same time. Hence, the frequency diversity can be exploited fully. In a more flexible way, each user can select the best available subcarriers (i.e., those available subcarriers having the highest signal-to-noise ratios for that particular user). By doing so, the sum rate of the system can be maximized.

1.7 INTERLEAVED FREQUENCY-DIVISION MULTIPLE ACCESS

Interleaved frequency-division multiple access (IFDMA) is a multiple-access scheme combining the advantages of spread-spectrum and multicarrier transmission. By assigning each user a set of orthogonal subcarriers, no multiple-access interference (MAI) arises even in a severe frequency-selective fading channel. At the receiver side, user discrimination is done using FDMA. Selecting the subcarriers for a particular user from the set of interleaved subcarriers, IFDMA is by nature a single-carrier-based system.

IFDMA has the following advantages over other multiple-access schemes. In comparison with TDMA, IFDMA uses continuous transmission. Compared to CDMA, no MAI is present. With respect to traditional FDMA, IFDMA is capable of achieving better frequency diversity. Moreover, IFDMA overcomes the large power backoff problems associated with its competitor OFDM/OFDMA by reducing the peak/average power ratio, since it employs a single-carrier modulation [16]. The IFDMA symbols transmitted can be generated in two ways, in either the time [15] or the frequency domain [16].

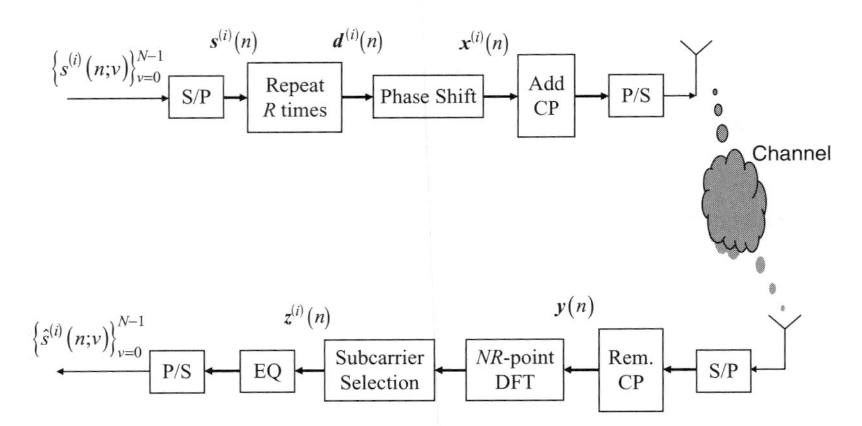

FIGURE 1.7 Block diagram of an IFDMA system with the transmitter implemented in the time domain.

1.7.1 Time-Domain Implementation

Consider a CP-based block transmission scheme in which the nth data block of length N belonging to the ith user is denoted as

$$s^{(i)}(n) = [s^{(i)}(n;0) \quad s^{(i)}(n;1) \quad \cdots \quad s^{(i)}(n;N-1)]^T.$$

It is obtained from the S/P operation on a data sequence $\{s^{(i)}(n;v)\}_{v=0}^{N-1}$. Denote R as the repetition times, or alternatively, the spreading factor in the frequency-domain implementation, which has to satisfy the condition to avoid overloading the system. The block diagram for time-domain implementation of an IFDMA transmitter is depicted in Figure 1.7.

Next, the nth data block is compressed and repeated R times. The resulting nth IFDMA symbol is given by

$$\boldsymbol{d}^{(i)}(n) = \frac{1}{\sqrt{R}} \left[\underbrace{s^{(i)}(n;0) \quad \cdots \quad s^{(i)}(n;N-1) \quad \cdots \quad s^{(i)}(n;0) \quad \cdots \quad s^{(i)}(n;N-1)}_{R \text{ times}} \right]^T. \tag{1.21}$$

This IFDMA symbol for the ith user is then modified by a phase vector $\boldsymbol{p}^{(i)}$ of size NR in which the lth component is expressed as

$$p^{(i)}(l) = \exp(-jl\varphi^{(i)}), \qquad l = 0, 1, \ldots, NR-1, \tag{1.22}$$

where $\varphi^{(i)} = i(2\pi/NR)$ is the user-dependent phase shift. The elementwise multiplication of $\boldsymbol{d}^{(i)}(n)$ and $\boldsymbol{p}^{(i)}$ assures orthogonality among different users [15]. The resulting signals for the ith user are given as

$$x^{(i)}(n)=\left[d^{(i)}(n;0) \quad d^{(i)}(n;1)e^{-j\varphi^{(i)}} \quad \cdots \quad d^{(i)}(n;NR-1)e^{-j(NR-1)\varphi^{(i)}}\right]^T. \tag{1.23}$$

With the user-dependent phase shift, each user is assigned a set of orthogonal frequencies and such operation facilitates easy user separation at the receiver side. Before transmission, CP is inserted to the front of each block $x^{(i)}(n)$ to eliminate IBI.

Assume that the channel for the ith user is a frequency-selective fading channel with L equally spaced time-domain taps, $h^{(i)}=[h_0^{(i)} \quad h_1^{(i)} \quad \cdots \quad h_{L-1}^{(i)}]^T$. The signal samples received from user i after the removal of CP can be written as

$$y^{(i)}(n)=W_{NR}^H \Lambda_{NR}^{(i)} W_{NR} x^{(i)}(n)+\tilde{u}^{(i)}(n), \tag{1.24}$$

where W_{NR} is the $NR \times NR$ DFT matrix, $\Lambda_{NR}^{(i)}=\text{diag}\{\lambda_0^{(i)}, \lambda_1^{(i)}, \ldots, \lambda_{NR-1}^{(i)}\}$, in which $\lambda_k^{(i)}=\sum_{l=0}^{L-1} h_l^{(i)} e^{-j(2\pi/NR)kl}$ denotes the frequency response of the kth subcarrier of the channel and $\tilde{u}^{(i)}(n)$ is the addictive noise. At the receiver, the signals received contain signal samples from all users:

$$y(n)=\sum_{i=1}^{U} y^{(i)}(n), \tag{1.25}$$

where U is the total number of users. If we choose

$$U \le R, \tag{1.26}$$

after NR-point DFT, the orthogonality among different users allows us to separate the users by selecting the subcarriers allocated to the ith user. The relation between the resulting signal $z^{(i)}(n)$ for the ith user with respect to the data block $s^{(i)}(n)$ is given as

$$z^{(i)}(n)=\Lambda_N^{(i)} W_N s^{(i)}(n)+u^{(i)}(n), \tag{1.27}$$

where

$$z^{(i)}(n)=[y^{(i)}(n;i) \quad y^{(i)}(n;i+R) \quad \cdots \quad y^{(i)}(n;i+(N-1)R)]^T$$

and $\Lambda_N^{(i)}=\text{diag}\{\lambda_i^{(i)}, \lambda_{i+R}^{(i)}, \ldots, \lambda_{i+(N-1)R}^{(i)}\}$, and $u^{(i)}$ is a column vector containing the elements of $\tilde{u}^{(i)}$ at positions $\{i, i+R, \ldots, i+(N-1)R\}$. Equation (1.27) is the same as the SCCP model given in (1.18). Thus, although no MAI is present for IFDMA system, intersymbol interference still exists. Various equalization techniques have to be employed to recover the signals transmitted.

1.7.2 Frequency-Domain Implementation

Equivalently, the IFDMA transmission signal can be constructed in the frequency domain as illustrated in Figure 1.8. First, N-point DFT is performed on each data block $\boldsymbol{s}^{(i)}(n)$. Then the frequency-domain symbols $\boldsymbol{b}^{(i)}(n) = \boldsymbol{W}_N \boldsymbol{s}^{(i)}(n)$ are interleaved in such a way that each user occupies an orthogonal set of subcarriers and the resulting expression of the kth subcarrier ($k = 0, 1, \ldots, NR - 1$) from the ith user is given as

$$c^{(i)}(n;k) = \begin{cases} b^{(i)}(n;k), & k = k'R + i \quad (k' = 0, 1, \ldots, N-1) \\ 0, & \text{otherwise,} \end{cases} \tag{1.28}$$

where R represents the frequency spacing between adjacent subcarriers (sometimes called the *spreading factor*) [16]. Finally, the time-domain symbols $\boldsymbol{x}^{(i)}(n)$ obtained after NR-point IDFT is inserted with a CP portion before transmission. The resulting transmitted signal implemented in the frequency domain is exactly the same as the one generated in the time domain. Hence, the receiver structure and analysis remain the same as in the previous case.

1.8 SINGLE-CARRIER FREQUENCY-DIVISION MULTIPLE ACCESS

For IFDMA, each user is allocated with interleaved subcarriers; thus, frequency diversity can be achieved for all users. The frequency-domain implementation structure of IFDMA provides other ways to allocate subcarriers to the users. For example, in Figure 1.8, each user can be allocated a different set of N consecutive subcarriers, called *localized subcarrier allocation*. With localized subcarrier allocation, the system is called *single-carrier frequency-division multiple access* (SC-FDMA) [17, 18]. Similar to OFDMA systems, SC-FDMA can achieve multiuser diversity.

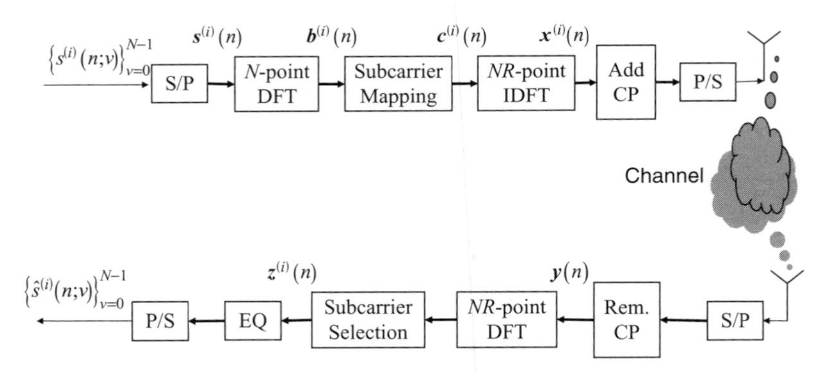

FIGURE 1.8 Block diagram of an IFDMA system with the transmitter implemented in the frequency domain.

1.9 CP-BASED CODE DIVISION MULTIPLE ACCESS

In this section we review CDMA-based block transmissions. In particular, we consider CP-CDMA and MC-CDMA systems. We assume that there are U active users in both systems. Each user has the same processing gain of G. The short code for the ith user is denoted as $\boldsymbol{c}_i$, where $\boldsymbol{c}_i = [c_i(0) \quad c_i(1) \quad \cdots \quad c_i(G-1)]^T$. All U short codes make up a set of U orthonormal basis vectors (i.e., $\boldsymbol{c}_i^H \boldsymbol{c}_j = 1$ for $i = j$ and $\boldsymbol{c}_i^H \boldsymbol{c}_j = 0$ for $i \neq j$). The long scrambing code for the nth data block is denoted by the diagonal matrix $\boldsymbol{D}(n)$.

1.9.1 CP-CDMA [19]

In a CP-CDMA system, each user transmits $Q = N/G$ symbols in one data block where N is the size of the data block. The Q symbols of one user are first spread out with the user's specific spreading code. After that, all the chip sequences of all users are added up. The total chip sequence, which has the length of N, is then passed through the CP inserter. At the receiver, the received signal samples associated with the CP portion are removed, and then the DFT transform is performed on the remaining signals associated with the data block.

We have the following equation to model the CP-CDMA system:

$$z(n) = \boldsymbol{\Lambda W D}(n)\boldsymbol{C s}(n) + \boldsymbol{w}(n), \tag{1.29}$$

where:

- $\boldsymbol{s}(n) = [\bar{\boldsymbol{s}}_1^T(n) \ \bar{\boldsymbol{s}}_2^T(n) \ \cdots \ \bar{\boldsymbol{s}}_Q^T(n)]^T; \quad \bar{\boldsymbol{s}}_v(n) = [s_1(n; v) \ s_2(n; v) \ \cdots \ s_T(n; v)]^T$ for $v = 1, 2, \ldots, Q$, and $s_i(n; v)$ is the vth symbol of the ith user transmitted on the nth data block.
- $\boldsymbol{C} = \text{diag}\left\{\underbrace{\bar{\boldsymbol{C}}, \ldots, \bar{\boldsymbol{C}}}_{Q \text{ times}}\right\}; \ \bar{\boldsymbol{C}} = [\boldsymbol{c}_1 \quad \boldsymbol{c}_2 \quad \cdots \quad \boldsymbol{c}_U].$

1.9.2 MC-CDMA [20]

In an MC-CDMA system, each of the Q symbols from a user is spread in the frequency domain on some subcarriers. The total number of subcarriers in the system is $N = GQ$. On the receiver side, the nth block received after DFT can be written as

$$z(n) = \boldsymbol{\Lambda \Pi D}(n)\boldsymbol{C}\,\boldsymbol{s}(n) + \boldsymbol{w}(n), \tag{1.30}$$

where $z(n)$, $\boldsymbol{C}$, $\boldsymbol{s}(n)$, and $\boldsymbol{w}(n)$ are defined as in the CP-CDMA system. The matrix $\boldsymbol{\Pi}$ is the interleaver matrix, which is used to allocate to nonconsecutive subcarriers chips belonging to a symbol.

1.10 RECEIVER DESIGN

The fundamental objective of the receiver design in communication systems is to recover the information bits dedicated to a particular receiver. From the derivations above for various multiple-access schemes, we arrive at the following generic input–output model:

$$z(n) = \boldsymbol{H}\boldsymbol{s}(n) + \boldsymbol{u}(n), \tag{1.31}$$

where the channel matrix depends on the specific scheme of interest.

To describe the linear receivers, we make the following assumptions:

- $\boldsymbol{s}(n) = [s(n;1) \quad s(n;2) \quad \cdots \quad s(n;K-1)]^T$ is the signal vector transmitted. We assume that the symbols transmitted, $s(n;k)$' s, are zero mean, independent, identically distributed (i.i.d.) with power $E\{|s(n;k)|^2\} = E_S$.
- $\boldsymbol{u}(n) \in \mathbb{C}^{N\times 1}$ is the noise vector, which is assumed to be a zero-mean complex Gaussian random vector with a covariance matrix of $E\{\boldsymbol{u}(n)\boldsymbol{u}^H(n)\} = N_0\boldsymbol{I}_N$.
- The symbols transmitted are statistically independent of the noises.

To facilitate subsequent derivations, we define the signal-to-noise ratio (SNR) as $\Gamma = E_S/N_0$. Our objective is to recover the transmitted signal vector $\boldsymbol{s}(n)$ based on the received signal vector $z(n)$. The ML estimate of $\boldsymbol{s}(n)$ is given by

$$\breve{\boldsymbol{s}}_{\mathrm{ML}}(n) = \arg \min_{\boldsymbol{s}(n)\in C^K} \|z(n) - \boldsymbol{H}\boldsymbol{s}(n)\|^2, \tag{1.32}$$

where $C = \{c_0 \quad c_1 \quad \cdots \quad c_{|C|-1}\}$ is the set containing all constellation points of the modulation scheme in use and $|C|$ is the size of the constellation. Equation (1.32) requires an exhaustive search of all $|C|^K$ possible vectors to find the ML estimate of $\boldsymbol{s}(n)$. This number should be very large when we use higher modulation schemes and/or when the signal dimension K is large. Hence, the ML receiver involves an exponential complexity, and low-complexity suboptimal receivers are desired in practical systems.

In this section we give an overview of two types of suboptimal receivers: linear and nonlinear. Among the linear receivers, zero-forcing and minimum mean square error receivers are chosen due to their simplicity and popularity. However, these receivers provide performance that is far from that of the ML receiver. Hence, nonlinear receivers are also reviewed. We cover the MMSE receiver with soft interference cancellation and a block-iterative generalized decision feedback equalizer.

1.10.1 Linear Receivers

The model of linear receivers is represented by matrix $\boldsymbol{C}^H$ in Figure 1.9, where $\boldsymbol{C} = [\boldsymbol{c}_0 \quad \boldsymbol{c}_1 \quad \cdots \quad \boldsymbol{c}_{K-1}]$. A linear receiver uses a weighting vector $\boldsymbol{c}_k$ of dimension $N \times 1$ to decode the kth symbol, $s(n;k)$, based on the received signal vector. More

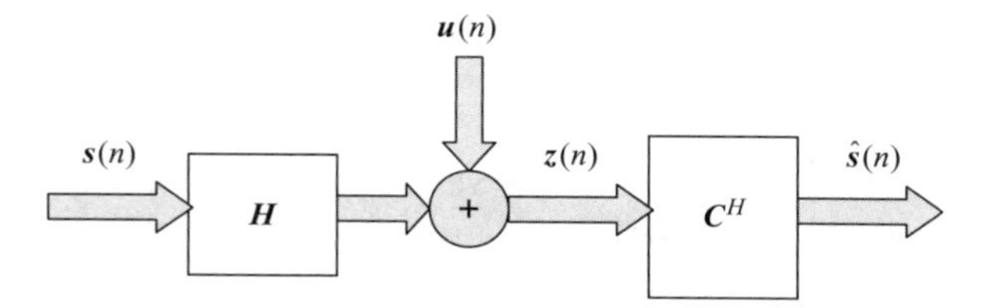

FIGURE 1.9 Block diagram of linear receivers.

specifically, we calculate the quantity

$$\breve{s}(n;k) = \boldsymbol{c}_k^H \boldsymbol{z}(n) = \boldsymbol{c}_k^H \boldsymbol{h}_k s(n;k) + \sum_{i \neq k} \boldsymbol{c}_k^H \boldsymbol{h}_i s(n;i) + \boldsymbol{c}_k^H \boldsymbol{u}(n). \tag{1.33}$$

The signal component in $\breve{s}(n;k)$ is $\boldsymbol{c}_k^H \boldsymbol{h}_k s(n;k)$, and its power is given by

$$P_S = \left|\boldsymbol{c}_k^H \boldsymbol{h}_k\right|^2 E_S. \tag{1.34}$$

The interference-plus-noise component is $\sum_{i \neq k} \boldsymbol{c}_k^H \boldsymbol{h}_i s(n;i) + \boldsymbol{c}_k^H \boldsymbol{u}(n)$, and its variance is determined by

$$\sigma_k^2 = \sum_{i \neq k} \left|\boldsymbol{c}_k^H \boldsymbol{h}_i\right|^2 E_S + \|\boldsymbol{c}_k\|^2 N_0. \tag{1.35}$$

Hence, the signal-to-interference-plus-noise ratio (SINR) for determining the symbol $s(n;k)$ is given by

$$\text{SINR}_k = \frac{\left|\boldsymbol{c}_k^H \boldsymbol{h}_k\right|^2}{\sum_{i \neq k} \left|\boldsymbol{c}_k^H \boldsymbol{h}_i\right|^2 + (1/\Gamma) \|\boldsymbol{c}_k\|^2 N_0}. \tag{1.36}$$

1.10.1.1 ZF Receiver For the ZF receiver, $\boldsymbol{C}_{\text{ZF}}^H$ is determined by the Moore–Penrose pseudoinverse of $\boldsymbol{H}$. When $\boldsymbol{H}$ is a square matrix and is invertible, $\boldsymbol{C}_{\text{ZF}}^H$ reduces to $\boldsymbol{H}^{-1}$. When $N \geq K$, the Moore–Penrose pseudoinverse of $\boldsymbol{H}$ becomes [21]

$$\boldsymbol{C}_{\text{ZF}}^H = (\boldsymbol{H}^H \boldsymbol{H})^{-1} \boldsymbol{H}^H. \tag{1.37}$$

In this case, the output of the ZF receiver is given by

$$\begin{aligned} \breve{s}(n) &= (\boldsymbol{H}^H \boldsymbol{H})^{-1} \boldsymbol{H}^H (\boldsymbol{H} \boldsymbol{s}(n) + \boldsymbol{w}(n)) \\ &= \boldsymbol{s}(n) + (\boldsymbol{H}^H \boldsymbol{H})^{-1} \boldsymbol{H}^H \boldsymbol{w}(n). \end{aligned} \tag{1.38}$$

It is observed that the ZF receiver tries to null out the interferences from other symbols, and it usually suffers from noise enhancement because of the nulling purpose.

1.10.1.2 MMSE Receiver The MMSE receiver is the optimal linear receiver that maximizes the SINR at the equalization output. To derive this receiver, we define the mean square error (MSE) between the transmitted symbol $s(n;k)$ and the equalization output $\breve{s}(n;k)$ as follows:

$$\begin{aligned} J(\boldsymbol{c}_k) &= E\left\{\left|s(n;k) - \boldsymbol{c}_k^H z(n)\right|^2\right\} \\ &= E\left\{\left|s(n;k)\right|^2\right\} - \boldsymbol{r}_k^H \boldsymbol{c}_k - \boldsymbol{c}_k^H \boldsymbol{r}_k + \boldsymbol{c}_k^H \boldsymbol{R}_z \boldsymbol{c}_k, \end{aligned} \tag{1.39}$$

where $\boldsymbol{R}_z = E\{z(n)z^H(n)\}$ and $\boldsymbol{r}_k = E\{z(n)s^*(n;k)\}$. Differentiating (1.39) with respect to $\boldsymbol{c}_k^*$ [22, App. B], we obtain

$$\frac{\partial J(\boldsymbol{c}_k)}{\partial \boldsymbol{c}_k^*} = \boldsymbol{R}_z \boldsymbol{c}_k - \boldsymbol{r}_k. \tag{1.40}$$

Letting $\partial J(\boldsymbol{c}_k)/\partial \boldsymbol{c}_k^* = \boldsymbol{0}$ yields

$$\boldsymbol{c}_k = \boldsymbol{R}_z^{-1} \boldsymbol{r}_k. \tag{1.41}$$

From (1.31) we have

$$\boldsymbol{R}_z = E\left\{z(n)z^H(n)\right\} = \boldsymbol{H}\boldsymbol{H}^H E_S + N_0 \boldsymbol{I}_N \tag{1.42}$$

and

$$\boldsymbol{r}_k = E\{z(n)s^*(n;k)\} = \boldsymbol{h}_k E_S. \tag{1.43}$$

Therefore, (1.41) can be written as

$$\boldsymbol{c}_k = (\boldsymbol{H}\boldsymbol{H}^H + (1/\Gamma)\boldsymbol{I}_N)^{-1}\boldsymbol{h}_k = \left(\boldsymbol{h}_k\boldsymbol{h}_k^H + \sum_{i \neq k} \boldsymbol{h}_i\boldsymbol{h}_i^H + (1/\Gamma)\boldsymbol{I}_N\right)^{-1}\boldsymbol{h}_k. \tag{1.44}$$

Equivalently, the MMSE receiver for all symbols transmitted can be represented as

$$\boldsymbol{C}_{\text{MMSE}} = (\boldsymbol{H}\boldsymbol{H}^H + (1/\Gamma)\boldsymbol{I}_N)^{-1}\boldsymbol{H}. \tag{1.45}$$

Some parameters of the MMSE receiver that we are interested in are:

- The output SINR for the symbol $s(n;k)$: The generic SINR formula in (1.36) can be written as

$$\mathrm{SINR}_k = \frac{\left|\boldsymbol{c}_k^H \boldsymbol{h}_k\right|^2}{\boldsymbol{c}_k^H\left(\boldsymbol{H}_k \boldsymbol{H}_k^H + (1/\Gamma)\boldsymbol{I}_N\right)\boldsymbol{c}_k}, \tag{1.46}$$

  where $\boldsymbol{H}_k$ is the matrix $\boldsymbol{H}$ in which the kth column is removed. Using the matrix inversion lemma† with $\boldsymbol{A} = \boldsymbol{H}_k \boldsymbol{H}_k^H + (1/\Gamma)\boldsymbol{I}_N$, $\boldsymbol{B} = \boldsymbol{h}_k$, $\boldsymbol{C} = \boldsymbol{h}_k^H$, and $\boldsymbol{D} = -1$, we have

$$\left(\boldsymbol{H}\boldsymbol{H}^H + (1/\Gamma)\boldsymbol{I}_N\right)^{-1} = \boldsymbol{A}^{-1} - \frac{\boldsymbol{A}^{-1}\boldsymbol{h}_k \boldsymbol{h}_k^H \boldsymbol{A}^{-1}}{1 + \boldsymbol{h}_k^H \boldsymbol{A}^{-1}\boldsymbol{h}_k}. \tag{1.47}$$

  Thus, the MMSE weight vector $\boldsymbol{c}_k$ in (1.44) is given by

$$\begin{aligned} \boldsymbol{c}_k &= \left(\boldsymbol{H}\boldsymbol{H}^H + (1/\Gamma)\boldsymbol{I}_N\right)^{-1}\boldsymbol{h}_k \\ &= \frac{1}{1 + \boldsymbol{h}_k^H \boldsymbol{A}^{-1}\boldsymbol{h}_k}\boldsymbol{A}^{-1}\boldsymbol{h}_k = \frac{1}{1+\beta_k}\boldsymbol{A}^{-1}\boldsymbol{h}_k, \end{aligned} \tag{1.48}$$

  where $\beta_k = 1 + \boldsymbol{h}_k^H \boldsymbol{A}^{-1}\boldsymbol{h}_k$. Substituting (1.48) in (1.46), we obtain

$$\mathrm{SINR}_k = \beta_k = \boldsymbol{h}_k^H\left(\boldsymbol{H}_k \boldsymbol{H}_k^H + (1/\Gamma)\boldsymbol{I}_N\right)^{-1}\boldsymbol{h}_k. \tag{1.49}$$

- If we define $\alpha_k = \boldsymbol{c}_k^H \boldsymbol{h}_k$, we can easily obtain

$$\alpha_k = \frac{\beta_k}{1+\beta_k}. \tag{1.50}$$

- After placing (1.48) in the general form of variance of interference plus noise in (1.35), we obtain

$$\sigma_k^2 = \boldsymbol{c}_k^H\left(\boldsymbol{H}_k \boldsymbol{H}_k^H E_s + N_0 \boldsymbol{I}_N\right)\boldsymbol{c}_k = \frac{\beta_k}{(1+\beta_k)^2}. \tag{1.51}$$

In the MMSE receiver, for a given $s(n;k)$, when the signal dimension K becomes large, the interference plus noise can be modeled as a zero-mean complex Gaussian random variable. From (1.33) we have

$$E\{\breve{s}(n;k)\} = \alpha_k s(n;k) \neq s(n;k). \tag{1.52}$$

†$(\boldsymbol{A} - \boldsymbol{B}\boldsymbol{D}^{-1}\boldsymbol{C}) = \boldsymbol{A}^{-1} + \boldsymbol{A}^{-1}\boldsymbol{B}(\boldsymbol{D} - \boldsymbol{C}\boldsymbol{A}^{-1}\boldsymbol{B})^{-1}\boldsymbol{C}\boldsymbol{A}^{-1}$ [23].

This result implies that the MMSE receiver produces a biased estimate of $s(n;k)$. For phase-shift-keying (PSK) modulation, if hard decisions are made on $\breve{s}(n;k)$, the BER performance will not be affected by this biased estimate. However, for quadrature amplitude modulation (QAM), this bias does affect the BER performance. Therefore, we need to obtain an unbiased estimate of $s(n;k)$. We can achieve this estimate by multiplying $\breve{s}(n;k)$ with a scaling factor of $1/\alpha_k$, which yields

$$\widehat{s}(n;k) = \frac{1}{\alpha_k}\breve{s}(n;k). \tag{1.53}$$

It is easy to see that $E\{\widehat{s}(n;k)\} = s(n;k)$; hence, $\widehat{s}(n;k)$ is called an *unbiased estimate* of $s(n;k)$.

1.10.2 Iterative Receivers

In the preceding section, we studied linear equalizers. In this section, two nonlinear iterative receivers, MMSE-SIC and BI-GDFE, are presented. These receivers have near-ML performance and much reduced complexity.

1.10.2.1 MMSE-SIC [24] We first present the MMSE-SIC receiver to estimate the transmitted symbols in (1.31). Let $\boldsymbol{c}_{k,l}$ be the MMSE weighting vector for the kth symbol $s(n;k)$ at the lth iteration. At the first iteration, the conventional MMSE is used. Hence, from (1.44), $\boldsymbol{c}_{k,1}$ is determined as

$$\boldsymbol{c}_{k,1} = \left(\boldsymbol{h}_k\boldsymbol{h}_k^H + \sum_{i\neq k}\boldsymbol{h}_i\boldsymbol{h}_i^H + (1/\Gamma)\boldsymbol{I}_N\right)^{-1}\boldsymbol{h}_k. \tag{1.54}$$

From (1.53), the unbiased output of this MMSE receiver for the kth symbol is written as

$$\widehat{s}_1(n;k) = \frac{1}{\alpha_{k,1}}\boldsymbol{c}_{k,1}^H\boldsymbol{z}(n) = s(n;k) + r_1(n;k), \tag{1.55}$$

where $\alpha_{k,1} = \boldsymbol{c}_{k,1}^H\boldsymbol{h}_k$ and $r_1(n;k)$ is the residual interference and noise after the bias removal, which is given by

$$r_1(n;k) = \frac{1}{\alpha_{k,1}}\left(\sum_{i\neq k}\boldsymbol{c}_{k,1}^H\boldsymbol{h}_i + \boldsymbol{c}_{k,1}^H\boldsymbol{u}(n)\right). \tag{1.56}$$

With the help of (1.51), we can derive the variance of $r_1(n;k)$ as

$$\sigma_{k,1}^2 = \frac{\boldsymbol{c}_{k,1}^H\left(\boldsymbol{H}_k\boldsymbol{H}_k^H E_s + N_0\boldsymbol{I}_N\right)\boldsymbol{c}_{k,1}}{|\alpha_{k,1}|^2}. \tag{1.57}$$

Hence, the SINR of the output of the kth symbol can easily be obtained:

$$\mathrm{SINR}_{k,1} = \frac{\left|\boldsymbol{c}_{k,1}^H \boldsymbol{h}_k\right|^2}{\boldsymbol{c}_{k,1}^H \left(\boldsymbol{H}_k \boldsymbol{H}_k^H + (1/\Gamma)\boldsymbol{I}_N\right) \boldsymbol{c}_{k,1}}. \tag{1.58}$$

The soft estimate of $s(n;k)$ is then calculated as

$$\tilde{s}_1(n;k) = E\left\{s(n;k)|\widehat{s}_1(n;k)\right\} = \frac{\sum_{v=0}^{|C|-1} c_v f(\widehat{s}_1(n;k)|s(n;k)=c_v)}{\sum_{v=0}^{|C|-1} f(\widehat{s}_1(n;k)|s(n;k)=c_v)}, \tag{1.59}$$

where $f(a|b)$ is the probability density function of a given b. When the signal dimension K becomes large, the residual $r_1(n;k)$ can be modeled approximately as a zero-mean complex Gaussian random variable with variance of $\sigma_{k,1}^2$; hence,

$$f(\widehat{s}_1(n;k)|s(n;k)=c_v) = \frac{1}{\sqrt{2\pi}\,\sigma_{k,1}^2} \exp\left\{-\frac{|\widehat{s}_1(n;k) - c_v|^2}{2\sigma_{k,1}^2}\right\}. \tag{1.60}$$

Repeating those calculations for $k = 0, 1, \ldots, K-1$, we obtain the soft information for all symbols after the first iteration. Suppose that we have the soft information for all symbols after the lth iteration. At the $(l+1)$th iteration, for the kth symbol, the MMSE-SIC receiver performs soft cancellation of the interference to produce $z_{k,l+1}(n)$ to estimate $s(n;k)$ at the $(l+1)$th iteration as

$$\begin{aligned} z_{k,l+1}(n) &= z(n) - \sum_{i\neq k} \boldsymbol{h}_i \tilde{s}_l(n;i) \\ &= \boldsymbol{h}_k s(n;k) - \sum_{i\neq k} \boldsymbol{h}_i (s(n;i) - \tilde{s}_l(n;i)) + \boldsymbol{u}(n). \end{aligned} \tag{1.61}$$

Based on (1.61), a new MMSE weighting vector, $\boldsymbol{c}_{k,l+1}$, is obtained for the kth symbol as follows:

$$\boldsymbol{c}_{k,l+1} = \left(\boldsymbol{h}_k \boldsymbol{h}_k^H + \boldsymbol{H}_k \boldsymbol{D}_{k,l} \boldsymbol{H}_k^H + (1/\Gamma)\boldsymbol{I}_N\right)^{-1} \boldsymbol{h}_k, \tag{1.62}$$

where $\boldsymbol{D}_{k,l} = \mathrm{diag}\{d_{0,l} \quad \cdots \quad d_{k-1,l} \quad d_{k+1,l} \quad \cdots \quad d_{k-1,l}\}$ with

$$\begin{aligned} d_{i,k} &= E\{|s(n;i) - \tilde{s}_l(n;i)|^2 \big| \widehat{s}_l(n;l)\} \\ &= \frac{1}{E_s}(E\{|s(n;i)|^2 \big| \widehat{s}_l(n;i)\} - |\tilde{s}_l(n;i)|^2) \end{aligned} \tag{1.63}$$

and

$$E\{|s(n;i)|^2 \Big| \widehat{s}(n;i).\} = \sum_{v=0}^{|C|-1} |c_v|^2 p(s(n;i) = c_v \Big| \widehat{s}_l(n;i)). \tag{1.64}$$

From the new weighting vector in (1.62) and (1.55), the new unbiased output of the kth symbol is determined. Hence, soft interference cancellation is carried out for every iteration. Note that the weighting vectors are different from one symbol interval to another; thus, the computational complexity for MMSE-SIC is still very high.

1.10.2.2 BI-GDFE [25] The block diagram of a BI-GDFE receiver is given in Figure 1.10. This receiver is an iterative receiver in which the decisions obtained from the previous iteration are used to reconstruct the ISI, which is then canceled out from the received signal vector for the purpose of improving the detection performance in later iterations. At the lth iteration, the received signal vector $z(n)$ is passed through the feed forward equalizer (FFE) $\boldsymbol{K}_l$. At the same time, the hard decisions from the previous iteration $\hat{\boldsymbol{s}}_{l-1}(n)$ are filtered by the feed-backward equalizer (FBE) $\boldsymbol{D}_l$. The output from the FFE then subtracts the output from the FBE to generate $\hat{\boldsymbol{z}}_l(n)$, which is exploited further to obtain the hard decision $\hat{\boldsymbol{s}}_l(n)$.

The optimal values of $\boldsymbol{K}_l$ and $\boldsymbol{D}_l$ that maximize the SINR at the lth iteration are given by [25]

$$\boldsymbol{K}_l = \left[\left(1 - \rho_{l-1}^2\right) \boldsymbol{H}\boldsymbol{H}^H + (1/\Gamma)\boldsymbol{I}_N\right]^{-1} \boldsymbol{H} \tag{1.65}$$

and

$$\boldsymbol{D}_l = \rho_{l-1}\left(\boldsymbol{K}_l^H \boldsymbol{H} - \boldsymbol{A}_l\right), \tag{1.66}$$

where $\boldsymbol{A}_l$ is a diagonal matrix whose diagonal elements are equal to those of $\boldsymbol{K}_l^H \boldsymbol{H}$; ρ_{l-1} is a coefficient that indicates the statistical reliability between the hard decision $\hat{\boldsymbol{s}}_{l-1}(n)$ and the transmitted signal vector $\boldsymbol{s}(n)$, and $E\{\boldsymbol{s}(n)\hat{\boldsymbol{s}}_l^H(n)\} = \rho_{l-1} E_s \boldsymbol{I}_K$.

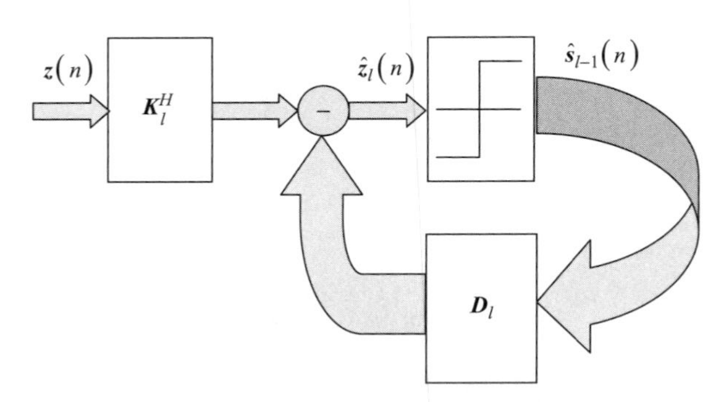

FIGURE 1.10 Block diagram of a BI-GDFE receiver.

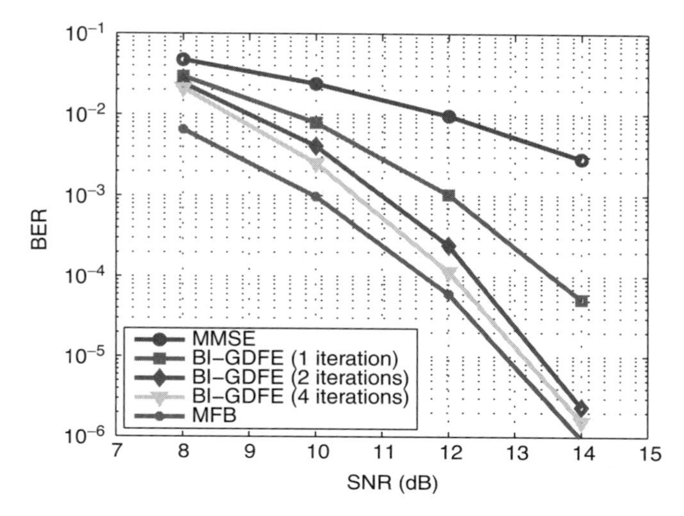

FIGURE 1.11 Performance of MMSE and BI-GDFE receivers for IFDMA.

Methods for determining the statistical reliability coefficients are studied in [25] for QPSK modulation and in [26] for high-order QAM.

For the first iteration, ρ_0 is chosen to be zero and thus the BI-GEFE receiver functions as the conventional MMSE receiver. If the channel matrix $\boldsymbol{H}$ is static over some symbol intervals (block fading channel), the values of $\boldsymbol{K}_l$ and $\boldsymbol{D}_l$ need to be determined once and can be applied to the entire block. This helps to reduce the complexity of a BI-GDFE receiver compared to a MMSE-SIC receiver.

In [25], computer simulations have shown that the BI-GDFE receiver is capable of achieving the single-user matched-filter bound (MFB) for spatial multiplexing based on large random multiple-input, multiple-output (MIMO) channels when the received SNR is high enough. That is, the BI-GDFE receiver is very effective in suppressing the interference between the data streams for MIMO systems.

The use of BI-GDFE in CP-CDMA and MC-CDMA can also be found [27,28], while a reconfigurable BI-GDFE has been proposed [29] for various CP-based block transmissions. As an example, in Figure 1.11 we plot the performance curves of MMSE and BI-GDFE receivers for IFDMA systems with 256 subcarriers, each user being allocated 64 subcarriers and with a channel length of 64. Here, for IFDMA, discontinuous subcarrier mapping is used. From the simulation results it is clear that a BI-GDFE outperforms an MMSE and can achieve a performance close to that of an MFB.

SUMMARY

Multiple-access schemes play important roles in designing the physical-layer air interfaces of broadband wireless networks. In this chapter, various multiple-access schemes have been reviewed in detail from transmitter and receiver perspectives. We have also reviewed linear and nonlinear equalizers for generic transmission models. Recommendations for further reading are as follows:

- Equivalence and relationship of MMSE-SIC and BI-GDFE [26]
- Resource allocation for multiuser OFDM systems [30–32]
- Resource allocation for OFDMA [33]

APPENDIX: PROOF OF (1.8)

We first prove the following theorem.

Theorem 1.1 Consider the following two sequences:

- *Sequence 1:* $[x(0) \quad x(1) \quad \cdots \quad x(N-1)]$.
- *Sequence 2:* $[\tilde{x}(0) \quad \tilde{x}(1) \quad \cdots \quad \tilde{x}(N-1)]$.

Their corresponding Fourier transforms are $[X(0) \quad X(1) \quad \cdots \quad X(N-1)]$ and $[\tilde{X}(0) \quad \tilde{X}(1) \quad \cdots \quad \tilde{X}(N-1)]$, respectively. If sequence 2 is a circular shift version of sequence 1 by right-shifting the elements to v positions [i.e., $\tilde{x}(n)=x((n-v)_N)$, where $(\cdot)_N$ denotes the modulo operation over period of N], then

$$\tilde{X}(k)=X(k)e^{-j(2\pi kv/N)}, \qquad k=0,1,\ldots,N-1. \tag{1.67}$$

Proof: For any value of k, we have

$$\begin{aligned}
\tilde{X}(k) &= \sum_{n=0}^{N-1} \tilde{x}(n)e^{-j(2\pi kn/N)} \\
&= \sum_{n=0}^{N-1} x((n-v)N)e^{-j(2\pi kn/N)} \\
&= \sum_{n=0}^{v-1} x(N-v+n)e^{-j(2\pi kn/N)} + \sum_{n=v}^{v-1} x(n-v)e^{-j(2\pi kn/N)} \\
&= \sum_{n'=N-v}^{N-1} x(n')e^{-j[2\pi k(n'-N+v)]/N} + \sum_{n'=0}^{N-v-1} x(n')e^{-j[2\pi k(n'+v)]/N} \\
&= e^{-j(2\pi kv/N)} \sum_{n=0}^{N-1} x(n)e^{-j(2\pi kn/N)} = X(k)e^{-j(2\pi kv/N)}.
\end{aligned} \tag{1.68}$$

To prove (1.8), let $\boldsymbol{A} = \boldsymbol{\Lambda W}$ and $\tilde{\boldsymbol{H}} = \boldsymbol{W}^H \boldsymbol{A}$. Notice that

$$\boldsymbol{A} = \boldsymbol{\Lambda W} = \frac{1}{\sqrt{N}} \begin{bmatrix} H_0 & H_0 & \cdots & H_0 \\ H_1 & H_1 e^{-j2\pi/N} & \cdots & H_1 e^{-[j2\pi(N-1)]/N} \\ \vdots & \ddots & \ddots & \vdots \\ H_{N-1} & H_{N-1} e^{-j2\pi(N-1)/N} & \cdots & H_{N-1} e^{-[j2\pi(N-1)(N-1)]/N} \end{bmatrix}. \tag{1.69}$$

Using MATLAB notation‡ we have $\tilde{\boldsymbol{H}}(:, m) = \boldsymbol{W}^H \boldsymbol{A}(:, m)$, and we only need to prove that $\tilde{\boldsymbol{H}}(:, m) = \boldsymbol{H}(:, m)$ for $m = 1, 2, \cdots, N$.

From the definition of H_k, $H_k = \sum_{l=0}^{L-1} h_l e^{-j(2\pi kl/N)}$, we have

$$\begin{bmatrix} H_0 \\ H_1 \\ \vdots \\ H_{N-1} \end{bmatrix} = \sqrt{N}\ \boldsymbol{W} \begin{bmatrix} h_0 \\ h_1 \\ \vdots \\ h_{L-1} \\ 0 \\ \vdots \\ 0 \end{bmatrix} = \sqrt{N}\ \boldsymbol{WH}(:, 1). \tag{1.70}$$

Thus, $\boldsymbol{A}(:, 1) = \boldsymbol{WH}(:, 1)$ or $\tilde{\boldsymbol{H}}(:, 1) = \boldsymbol{W}^H \boldsymbol{A}(:, 1) = \boldsymbol{W}^H \boldsymbol{W} \boldsymbol{H}(:, 1) = \boldsymbol{H}(:, 1)$. From (1.69) we observe that the mth column of $\tilde{\boldsymbol{H}}$ is just the circular shift version of the first column of $\tilde{\boldsymbol{H}}$ by down-shifting the elements $m - 1$ positions. Furthermore, from (1.7), the mth column of $\boldsymbol{H}$ is just the circular shift version of the first column of $\boldsymbol{H}$ by down-shifting the elements $m - 1$ positions. Besides, $\tilde{\boldsymbol{H}}(:, 1) = \boldsymbol{H}(:, 1)$. Hence, we conclude that $\tilde{\boldsymbol{H}} = \boldsymbol{H}$.

REFERENCES

[1] S. B. Weinstein and P. M. Ebert, "Data transmission by frequency division multiplexing using the discrete Fourier transform," *IEEE Trans. Commun. Technology*, vol. COM-19, pp. 628–634, Oct. 1971.

[2] J. A. C. Bingham, "Multicarrier modulation for data transmission: an idea whose time has come," *IEEE Commun. Mag.*, vol. 28, pp. 5–14, May 1990.

[3] "Radio broadcasting systems: digital audio broadcasting to mobile, portable and fixed receivers," *European Telecommunication Standard ETS 300 401*, ETSI, Sophia Antipolis, France, 1995.

‡For an arbitrary matrix $\boldsymbol{A}$, the mth column of $\boldsymbol{A}$ is denoted by $\boldsymbol{A}(:, m)$. Here, we also count the columns from number 1 onward according to MATLAB.

[4] "Digital video broadcasting (DVB-T): frame structure, channel coding, modulation for digital terrestrial television," *European Telecommunicaton Standard ETS 300 744*, ETSI, Sophia Antipolis, France 1997.

[5] "Part 11: Wireless LAN medium access control (MAC) and physical layer (PHY) specifications, higher-speed physical layer extension in the 5 GHz band," *IEEE802.11a*, 1999.

[6] J. S. Chow, J. C. Tu, and J. M. Cioffi, "A discrete multitone transceiver system for HDSL applications," *IEEE J. Sel. Areas Commun.*, vol. 9, pp. 895–908, Aug. 1991.

[7] Y. G. Li and G. Stüber, *OFDM for Wireless Communications*, Springer, Boston, 2006.

[8] D. Falconer, S. L. Ariyavisitakul, A. Benyamin-Seeyar, and B. Edison, "Frequency domain equalization for single-carrier broadband wireless systems," *IEEE Commun. Mag.*, vol. 40, pp. 58–66, Apr. 2002.

[9] H. Sari and G. Karam, "Orthogonal frequency-division multiple access and its application to CATV networks," *Eur. Trans. Telecommun.*, vol. 9, pp. 507–516, Nov.–Dec. 1998.

[10] "Interaction channel for digital terrestrial television (RCT) incorporating multiple access OFDM," *ETSI DVB RCT*, ETSI, Sophia Antipolis, France, Mar. 2001.

[11] Draft amendament to IEEE standard for local and metropolitan area networks, "Part 16: Air interface for fixed broadband wireless access system–amendament 2: Medium access control modifications and additional physical layer specifications for 2–11 GHz," *IEEE P802.16a/D3-2001*, Mar. 2002.

[12] L. Wei and C. Schlegel, "Synchronization requirements for multi-user OFDM on satellite mobile and two-path Rayleigh fading channels," *IEEE Trans. Commun.*, vol. 43, pp. 887–895, Feb.–Apr. 1995.

[13] S. Barbarossa, M. Pompili, and G. Giannakis, "Channel-independent synchronization of orthogonal frequency division multiple access systems," *IEEE J. Sel. Areas Commun.*, vol. 20, pp. 474–486, Feb. 2002.

[14] Z. Cao, U. Tureli, and Y. D. Yao, "Efficient structure-based carrier frequency offset estimation for interleaved OFDMA uplink," *Proceedings of IEEE ICC 2003*, vol. 5, pp. 3361–3365, May 2003.

[15] U. Sorger, I. D. Broeck, and M. Schnell, "Interleaved FDMA: new spread spectrum multiple-access scheme," *Proceedings of IEEE ICC 1998*, vol. 2, pp. 1013–1017, June 1998.

[16] R. Dinis, D. Falconer, C. T. Lam, and M. Sabbaghian, "A multiple access scheme for the uplink of broadband wireless systems," *Proceedings of IEEE GLOBECOM 2004*, vol. 6, pp. 3808–3812, Nov. 2004.

[17] H. Ekstrom, A. Furusk, J. Karlsson, M. Meyer, S. Parkvall, J. Torsner, and M. Wahlqvist, "Technical solutions for the 3G long-term evolution," *IEEE Commun. Mag.*, vol. 42, pp. 38–45, Mar. 2003.

[18] Y. Ofuj, K. Higuch, and M. Sawahashi, "Frequency domain channel-dependent scheduling employing an adaptive transmission bandwidth for pilot channel in uplink single-carrier-FDMA radio access," *Proceedings of IEEE VTC 2006–Spring*, vol. 1, pp. 334–338 May 2006.

[19] K. L. Baum, T. A. Thomas, F. W. Vook, and V. Nangia, "Cyclic-prefix CDMA: an improved transmission method for broadband DS-CDMA cellular systems," *Proceedings of IEEE WCNC 2002*, vol. 1, pp. 183–188, Mar. 2002.

[20] S. Hara and R. Prasad, "Overview of multicarrier CDMA," *IEEE Commun. Mag.*, vol. 35, pp. 126–133, Dec. 1997.

[21] C. R. Rao and S. K. Mitra, *Generalized Inverse of Matrices and Its Applications*, Wiley, New York, 1971.

[22] S. Haykin, *Adaptive Filtering Theory*, 3rd ed., Prentice Hall, Upper Saddle River, NJ, 1995.

[23] G. H. Golub and C. F. V. Loan, *Matrix Computations*, 3rd ed.; Johns Hopkins University Press, Baltimore, 1996.

[24] X. Wang and H. V. Poor, "Iterative (turbo) soft interference cancellation and decoding for coded CDMA," *IEEE Trans. Commun.*, vol. 47, pp. 1046–1061, July 1999.

[25] Y.-C. Liang, S. Sun, and C. K. Ho, "Block-iterative generalized decision feedback equalizers for large MIMO systems: algorithm design and asymptotic performance analysis," *IEEE Trans. Signal Process.*, vol. 54, pp. 2035–2048, June 2006.

[26] Y.-C. Liang, E. Y. Cheu, L. Bai, and G. Pan, "On the relationship between MMSE-SIC and BI-GDFE receivers for multiple input multiple output channels," *IEEE Trans. Signal Process*, accepted for publication, vol. 56, No. 8, pp. 3627–3637, Aug. 2008.

[27] Y.-C. Liang, Block-iterative GDFE (BI-GDFE) for CP-CDMA and MC-CDMA, *Proceedings of IEEE VTC 2005–Spring*, vol. 5, pp. 3033–3037, June 2005.

[28] Y.-C. Liang, Asymptotic performance of BI-GDFE for large isometric and random precoded systems, *Proceedings of IEEE VTC 2005C–Spring*, vol. 3, pp. 1557–1561, May–June 2005.

[29] L. B. Thiagarajan, S. Attallah, and Y.-C. Liang, "Reconfigurable transceivers for wireless broadband access schemes," *IEEE Wireless Commun. Mag.*, vol. 14, pp. 48–53, June 2007.

[30] C. Y. Wong, R. S. Cheng, K. B. Lataief, and R. D. Murch, "Multiuser OFDM with adaptive subcarrier, bit, and power allocation," *IEEE J. Sel. Areas Commun.*, vol. 17, no. 10, pp. 1747–1758, Oct. 1999.

[31] W. Rhee and J. M. Cioffi, "Increase in capacity of multiuser OFDM system using dynamic subchannel allocation," *Proceedings of IEEE VTC 2000–Spring*, vol. 2, 2000, pp. 1085–1089.

[32] M. Tao, Y.-C. Liang, and F. Zhang, "Resource allocation for delay differentiated traffic in multiuser OFDM systems," *IEEE Trans. Wireless Commun.*, accepted for publication.

[33] D. Kivanc, G. Li, and H. Liu, "Computationally efficient bandwidth allocation and power control for OFDMA," *IEEE Trans. Wireless Commun.*, vol. 2, no. 6, pp. 1150–1158, Nov. 2003.

CHAPTER 2

MULTIPLE-INPUT, MULTIPLE-OUTPUT ANTENNA SYSTEMS

2.1 INTRODUCTION

In this chapter we consider wireless communication systems with multiple transmitting and/or multiple receiving antennas. When both transmitter and receiver sides have multiple antennas, these systems are referred to as *multiple-input, multiple-output* (MIMO) *antenna systems*, or MIMO systems. In [6] and [7], Foschini and Telatar have proven that for a given power budget and a given bandwidth, the ergodic capacity of a MIMO Rayleigh fading channel increases linearly with the minimum number of transmitting and receiving antennas. It is this promising result that makes MIMO an attractive solution for achieving high-speed wireless connections over a limited amount of bandwidth.

The increased data rate for MIMO systems is achieved through spatial multiplexing. MIMO also offers improved transmission reliability through transmit and/or receive diversity. In fact, even when there is only one transmitting antenna, receive diversity can be achieved through the use of multiple receiving antennas. Such a system is called a *single-input, multiple-output* (SIMO) *system*. Equivalently, transmit diversity can be achieved through the use of multiple transmitting antennas. A *multiple-input, single-output* (MISO) *system* has multiple transmitting antennas and a single receiving antenna.

We first study the fundamental capacity limits of MIMO systems; then we look at the transceiver design for MIMO systems with channel state information (CSI)

Wireless Broadband Networks, By David Tung Chong Wong, Peng-Yong Kong, Ying-Chang Liang, Kee Chaing Chua, and Jon W. Mark

known or unknown at the transmitter side. We also consider the design for two special systems, SIMO and MISO systems. For a frequency-selective fading channel, OFDM, SCCP, and IFDMA are designed in conjunction with MIMO to achieve increased data rate transmission or in conjunction with space–time coding to achieve transmit diversity.

2.2 MIMO SYSTEM MODEL

We consider a narrowband MIMO wireless communication system with N_t transmitting antennas and N_r receiving antennas. This system is illustrated in Figure 2.1 and the channel is referred to as an $N_t \times N_r$ *MIMO channel*. Considering the flat fading environment, the MIMO system can be represented by a discrete-time model at time index n as follows:

$$z(n) = \boldsymbol{H}\boldsymbol{x}(n) + \boldsymbol{u}(n), \tag{2.1}$$

where:

- The transmitted signal vector $\boldsymbol{x}(n)$ of size $N_t \times 1$ is drawn from a white Gaussian codebook. In other words, the elements of $\boldsymbol{x}(n)$ are i.i.d. zero-mean Gaussian random variables. The correlation matrix of the signal vector is defined as $\boldsymbol{R_x} = E\{\boldsymbol{x}(n)\boldsymbol{x}^H(n)\}$ with $\text{trace}\{\boldsymbol{R_x}\} = P$, where P is the fixed total transmit power.
- We use $h_{k,l}$ to denote the channel coefficient from the lth transmitting antenna to the kth receiving antenna, where $l = 1, 2, \ldots, N_t$ and $k = 1, 2, \ldots, N_r$. Hence,

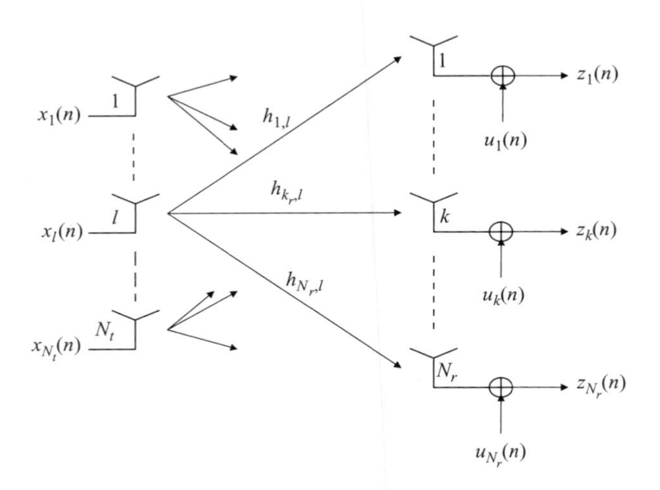

FIGURE 2.1 MIMO antenna system model.

the channel matrix $\boldsymbol{H}$ in (2.1) is written as

$$\boldsymbol{H} = \begin{bmatrix} h_{1,1} & h_{1,2} & \cdots & h_{1,N_t} \\ h_{2,1} & h_{2,2} & \cdots & h_{2,N_t} \\ \vdots & \vdots & \vdots & \vdots \\ h_{N_r,1} & h_{N_r,2} & \cdots & h_{N_r,N_t} \end{bmatrix} \in \mathbb{C}^{N_r \times N_t}. \tag{2.2}$$

Furthermore, we assume that $h_{k,l}$ is a complex Gaussian random variable with zero mean and a variance of $E\{|h_{k,l}|^2\} = 1$.

- The elements of noise vector $\boldsymbol{u}(n)$ of size $N_r \times 1$ are i.i.d. complex Gaussian random variables with zero mean and a variance of N_0. Therefore, the correlation matrix of the noise vector is $\boldsymbol{R_u} = E\{\boldsymbol{u}(n)\boldsymbol{u}^H(n)\} = N_0\boldsymbol{I}_{N_r}$.

Two special cases are:

1. When $N_t = 1$ and $N_r > 1$, the system becomes a single-input, multiple-output (SIMO) system.
2. When $N_t > 1$ and $N_r = 1$, the system becomes a multiple-input, single-output (MISO) system.

2.3 CHANNEL CAPACITY

Channel capacity is a fundamental performance indicator that describes the maximum rate of data transmission that a channel can support with an arbitrarily small probability of error incurred due to channel impairments. The channel capacity for additive white Gaussian noise (AWGN) channels was derived by Claude Shannon in 1948 [1]. For single-input single-output systems, the capacity limits for fading channels have been well documented: for example, in [2–5]. In this section we consider the channel capacity of MIMO systems in a fading channel environment. We first look at the capacity for single-input, single-output (SISO) fading channels.

2.3.1 SISO Channels

Consider the following AWGN channel:

$$x(n) = s(n) + u(n), \tag{2.3}$$

and assume that:

- The transmitted signal $s(n)$ is zero-mean i.i.d. Gaussian with $E\{|s(n)|^2\} = E_s$.
- The noise $u(n)$ is zero-mean i.i.d. Gaussian with $E\{|u(n)|^2\} = N_0$.

We define the signal-to-noise ratio (SNR) as $\Gamma = E_s/N_0$. The capacity of the channel is determined by the mutual information between the input and output, which is given by

$$C = \log_2(1 + \Gamma). \tag{2.4}$$

Here the unit of capacity is bits per second per hertz (bits/s/Hz). In a high-SNR region, the channel capacity increases 1 bit/s/Hz for every 3-dB increase in SNR.

Next, we consider the following SISO fading channel:

$$z(n) = hx(n) + u(n), \tag{2.5}$$

and assume the following:

- The transmitted signal $x(n)$ is zero-mean i.i.d. Gaussian with $E\{|x(n)|^2\} = E_s$.
- The noise $u(n)$ is zero-mean i.i.d. Gaussian with $E\{|u(n)|^2\} = N_0$, and the SNR is $\Gamma = E_s/N_0$.
- The fading state h is a random variable with $E\{|h|^2\} = 1$.

Let us first introduce the concept of a *block fading channel*, which is a slowly fading channel whose coefficient is constant over an interval of time T and which changes to another independent value, again constant over an interval of time T, and so on. The instantaneous mutual information between $z(n)$ and $x(n)$ of the fading channel conditional on channel state h is given by

$$I(x; z|h) = \log_2\left(1 + |h|^2\Gamma\right). \tag{2.6}$$

Since h is a random variable, the instantaneous mutual information is also a random variable. Thus, if the distribution of $|h|^2$ is known, the distribution of $I(x; z|h)$ can be calculated accordingly.

The channel capacity of a fading channel can be quantified either in an ergodic sense or in an outage sense, yielding ergodic capacity and outage capacity. The ergodic capacity of the SISO fading channel in (2.5) is defined as

$$C = E\left\{\log_2\left(1 + |h|^2\Gamma\right)\right\}, \tag{2.7}$$

where the expectation is taken over the channel state variable h. Physically speaking, the ergodic capacity defines the maximum (constant) rate of codes that can be transmitted over the channel and recovered with an arbitrarily small probability of error when the codes are long enough to cover all possible channel states.

In Figure 2.2 we compare the capacities of the AWGN channel and the SISO Rayleigh fading channel with respect to the received SNR. Here, for the fading channel case, we have used the average received SNR. It can be seen that at high SNR, the capacity of the fading channel increases 1 bit/s/Hz for every 3-dB increase

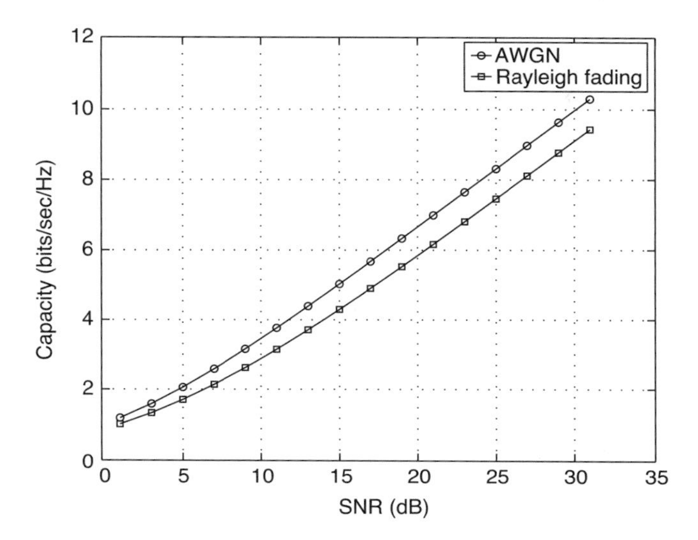

FIGURE 2.2 Capacity comparison for AWGN channels and SISO Rayleigh fading channels.

in SNR, which is the same as the AWGN channel. Since the instantaneous mutual information is a random variable, if a code with constant rate C_0 is transmitted over the fading channel, this code cannot be recovered correctly at the receiver at a fading block whose instantaneous mutual information is lower than the code rate C_0, thus causing an outage event. We define the outage probability as the probability that the instantaneous mutual information is less than the rate of C_0; that is,

$$P_{\text{out}}(C_0) = \Pr(I(x; z|h) < C_0). \tag{2.8}$$

Based on this, the $q\%$ outage capacity $C_{\text{out},q\%}$ is defined as the maximum information rate of codes transmitted over the fading channel for which the outage probability does not exceed $q\%$.

2.3.2 MIMO Channel Capacity for One-Channel Realization

We next investigate the channel capacity for MIMO channel with a given realization of the channel matrix. After that, the ergodic channel capacity of the system under fading channels is presented. We assume that the receiver has the perfect information on channel matrix $\boldsymbol{H}$. Furthermore, we differentiate two cases: One is that the channel matrix is known at the transmitter, referred to as the *CSI-known case*, and the other is when the channel matrix is unknown at the transmitter, referred to as the *CSI-unknown case*.

2.3.3 General Capacity Formula

The *MIMO channel capacity* is defined as [6,7]

$$C = \max_{f(\boldsymbol{x})} I\left(z; \boldsymbol{x}\right) = \max_{f(\boldsymbol{x})} \left(h\left(z\right) - h\left(z \,|\boldsymbol{x}\right)\right), \tag{2.9}$$

where $h(z)$ is the differential entropy of z and $h(z|\boldsymbol{x})$ is the conditional differential entropy of z given $\boldsymbol{x}$. The maximization in (2.9) is over all possible probability distributions of $\boldsymbol{x}$, $f(\boldsymbol{x})$.

We have $h(z|\boldsymbol{x}) = h(\boldsymbol{H}x + \boldsymbol{u}|\boldsymbol{x}) = h(\boldsymbol{u})$, where

$$h(\boldsymbol{u}) = N_r + N_r \log_2(2\pi) + N_r \log_2(N_0) \qquad \text{bits.} \tag{2.10}$$

The correlation matrix of z is given by

$$\boldsymbol{R}_z = E\{zz^H\} = N_0 \boldsymbol{I}_{N_r} + \boldsymbol{H}\boldsymbol{R}_x\boldsymbol{H}^H, \tag{2.11}$$

where $\boldsymbol{R}_x = E\{\boldsymbol{x}\boldsymbol{x}^H\}$. Among all random vectors with covariance matrix $\boldsymbol{R}_z$, the differential entropy $h(z)$ is maximized when z is a Gaussian random vector [8]. This can be obtained if $\boldsymbol{x}$ is also a complex Gaussian random vector (which is one of our assumptions). To this end, the entropy $h(z)$ is

$$h(z) = N_r + N_r \log_2(2\pi) + \log_2(\det(\boldsymbol{R}_z)) \qquad \text{bits.} \tag{2.12}$$

Hence, the mutual information between z and $\boldsymbol{x}$ in (2.9) reduces to

$$\begin{aligned} I\left(z; \boldsymbol{x}\right) &= \log_2 \left\{ \frac{\det\left(N_0 \boldsymbol{I}_{N_r} + \boldsymbol{H}\boldsymbol{R}_x\boldsymbol{H}^H\right)}{N_0^{N_r}} \right\} \\ &= \log_2 \left\{ \det\left(\boldsymbol{I}_{N_r} + \frac{1}{N_0}\boldsymbol{H}\boldsymbol{R}_x\boldsymbol{H}^H \right) \right\} \qquad \text{bits/s/Hz,} \end{aligned} \tag{2.13}$$

and the capacity of the MIMO system is now written as

$$C = \arg \max_{\text{trace}\{\boldsymbol{R}_x\}=P} \log_2 \left\{ \det\left(\boldsymbol{I}_{N_r} + \frac{1}{N_0}\boldsymbol{H}\boldsymbol{R}_x\boldsymbol{H}^H \right) \right\} \qquad \text{bits/s/Hz.} \tag{2.14}$$

Based on (2.14), if the bandwidth allocated to the system is W, the maximum achievable data rate over this bandwidth is CW bits/s.

2.3.3.1 Channel Capacity of the CSI-Known Case

The channel matrix $\boldsymbol{H}$ can be decomposed by using singular-value decomposition (SVD [9]) as follows:

$$\boldsymbol{H} = \boldsymbol{U}\boldsymbol{\Sigma}\boldsymbol{V}^H, \tag{2.15}$$

where:

- $\boldsymbol{U}$ and $\boldsymbol{V}$ are unitary matrices of size $N_r \times N_r$ and $N_t \times N_t$, respectively.
- $\boldsymbol{\Sigma}$ is the $N_r \times N_t$ singular-value matrix of $\boldsymbol{H}$. All elements of $\boldsymbol{\Sigma}$ are zeros except that $(\boldsymbol{\Sigma})_{i,i} = \sigma_i \geq 0$, where $\sigma_1 \geq \sigma_2 \geq \cdots \geq \sigma_M \geq 0$ are singular values of $\boldsymbol{H}$. M is the rank of matrix $\boldsymbol{H}$. Since M cannot exceed the number of rows or columns of $\boldsymbol{H}$, $M \leq \min(N_t, N_r)$. For the case that $\boldsymbol{H}$ is full rank, $M = \min(N_t, N_r)$.

Based on the decomposition in (2.15), we have

$$\boldsymbol{H}\boldsymbol{H}^H = \boldsymbol{U}\boldsymbol{\Sigma}\boldsymbol{\Sigma}^H\boldsymbol{U}^H. \tag{2.16}$$

It is easy to note that $\lambda_i = \sigma_i^2$, $i = 1, 2, \ldots, M$ are the singular values of matrix $\boldsymbol{H}\boldsymbol{H}^H$.

If we precode the transmitted signal, a processing called *transmit eigen-beamforming*, as

$$\boldsymbol{x} = \boldsymbol{V}\boldsymbol{s}, \tag{2.17}$$

the received signal becomes

$$\boldsymbol{z} = \boldsymbol{U}\boldsymbol{\Sigma}\boldsymbol{V}^H\boldsymbol{V}\boldsymbol{s} + \boldsymbol{u}. \tag{2.18}$$

At the receiver side, if we premultiple $\boldsymbol{z}$ with $\boldsymbol{U}^H$, a processing called *receive eigen-beamforming*, we obtain

$$\boldsymbol{y} = \boldsymbol{U}^H\boldsymbol{z} = \boldsymbol{\Sigma}\boldsymbol{s} + \boldsymbol{U}^H\boldsymbol{u} = \boldsymbol{\Sigma}\boldsymbol{s} + \tilde{\boldsymbol{u}}. \tag{2.19}$$

Note that the receive eigen-beamforing processing does not alter the distribution of the noise; that is, $\tilde{\boldsymbol{u}}$ is still a complex zero-mean Gaussian random vector with covariance matrix of $E\{\tilde{\boldsymbol{u}}\tilde{\boldsymbol{u}}^H\} = E\{\boldsymbol{U}^H\boldsymbol{u}\boldsymbol{u}^H\boldsymbol{U}\} = N_0\boldsymbol{I}_{N_r}$.

Equation (2.19) states that the joint transmit and receive eigen-beamforming transforms the MIMO system into M parallel independent SISO systems in which the input, channel coefficient, noise, and output of the ith SISO channel are s_i, σ_i, $\tilde{u}_i$, and y_i, respectively. This decomposition process using joint transmit and receive eigen-beamforming is shown in Figure 2.3, and the equivalent channels are illustrated in Figure 2.4.

Suppose that the transmission power of the ith data stream is $\gamma_i = E\{|s_i|^2\}$; the SNR for this data stream is given by

$$\text{SNR}_i = \frac{\sigma_i^2 E\left\{|s_i|^2\right\}}{E\left\{|\tilde{u}_i|^2\right\}} = \frac{\sigma_i^2\gamma_i}{N_0}. \tag{2.20}$$

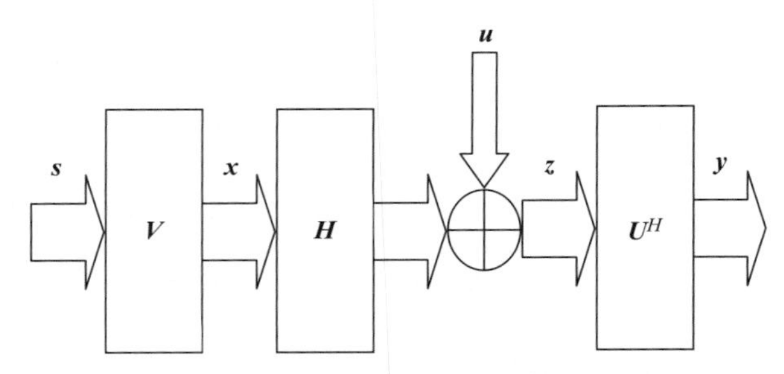

FIGURE 2.3 Block diagram of joint transmitter and receiver eigen-beamforming.

The transmission power values γ_i's must satisfy the power constraint, $\sum_{i=1}^{M} \gamma_i = P$. The channel capacity of the given MIMO system is the sum of the individual SISO channel's capacity. More specifically, we have

$$C = \sum_{i=1}^{M} \log_2 \left(1 + \frac{\sigma_i^2 \gamma_i}{N_0}\right) = \sum_{i=1}^{M} \log_2 \left(1 + \frac{\lambda_i \gamma_i}{N_0}\right) \qquad \text{bits/s/Hz.} \tag{2.21}$$

The capacity in (2.21) can be maximized by allocating different transmission powers to different SISO channels. This maximization problem is solved using the Lagrangian method. More specifically, we define the object function as

$$J = \sum_{i=1}^{M} \log_2 \left(1 + \frac{\lambda_i \gamma_i}{N_0}\right) + \mu \left(\sum_{i=1}^{M} \gamma_i - P\right). \tag{2.22}$$

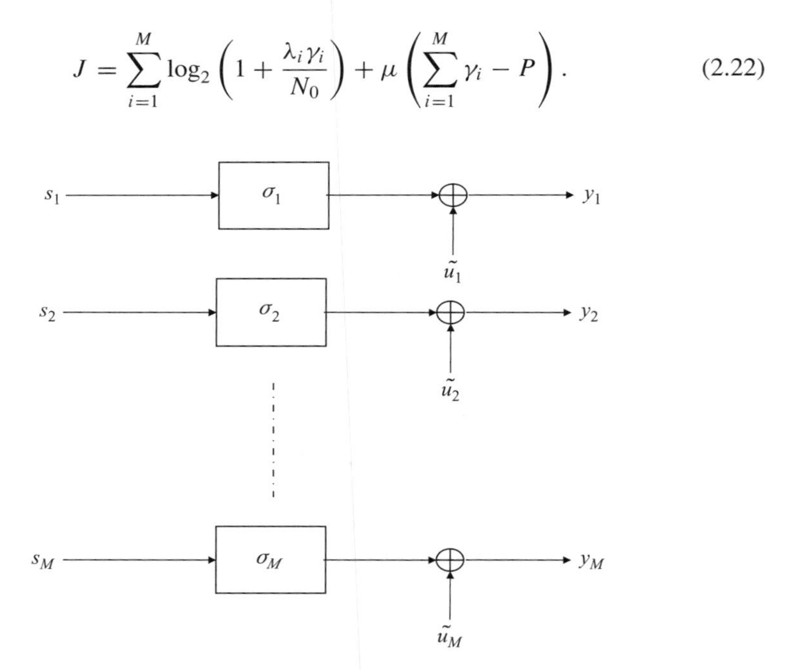

FIGURE 2.4 A MIMO channel is equivalent to a parallel of a SISO channel.

Calculating $\partial J/\partial \gamma_i$ and equating $\partial J/\partial \gamma_i = 0$ for all i's, we obtain the following *water-filling solution:*

$$\overline{\gamma}_i = \left(\mu - \frac{N_0}{\lambda_i}\right)^+, \qquad i = 1, 2, \ldots, M, \tag{2.23}$$

where $(x)^+ = x$ for $x \geq 0$ and $(x)^+ = 0$ for $x < 0$; μ is calculated to satisfy the power constraint. With the power allocation, the channel capacity can then be determined as

$$C = \sum_{i=1}^{M} \log_2\left(1 + \frac{\lambda_i \overline{\gamma}_i}{N_0}\right). \tag{2.24}$$

The water-filling solution states that in the parallel transmission on M SISO channels, we assign higher transmission power to strong SISO channels and lower transmission power to weak SISO channels and even assign zero power to SISO channels that are extremely weak.

2.3.3.2 Channel Capacity of the CSI-Unknown Case When a transmitter does not have access to CSI, it does not have the information to perform the transmitting eigen-beamforming operation as in Section 2.3.3.1. Instead, it divides the available transmission power evenly to the transmitting antennas. If that is the case, we have $\boldsymbol{R}_{\boldsymbol{x}} = (P/N_t)\boldsymbol{I}_{N_t}$. Substituting this correlation matrix to (2.14), in conjunction with the result of (2.16), the capacity is expressed as

$$C = \log_2\left\{\det\left(\boldsymbol{I}_{N_r} + \frac{P}{N_t N_0}\boldsymbol{U}\boldsymbol{\Sigma}\boldsymbol{\Sigma}^H\boldsymbol{U}^H\right)\right\} = \log_2\left\{\det\left(\boldsymbol{I}_{N_r} + \frac{\Gamma}{N_t}\boldsymbol{U}\boldsymbol{\Sigma}\boldsymbol{\Sigma}^H\boldsymbol{U}^H\right)\right\}, \tag{2.25}$$

where $\Gamma = P/N_0$ is the average SNR at one receiving antenna. Using the determinant identity,† we have the channel capacity as follows:

$$C = \log_2\left\{\det\left(\boldsymbol{I}_{N_r} + \frac{\Gamma}{N_t}\boldsymbol{\Sigma}\boldsymbol{\Sigma}^H\right)\right\} = \sum_{i=1}^{M} \log_2\left(1 + \frac{\Gamma}{N_t}\lambda_i\right). \tag{2.26}$$

2.3.4 Channel Capacity for Fading Channels

We now turn to practical scenarios where the channel matrix $\boldsymbol{H}$ evolves with time. We distinguish the following two cases:

1. Matrix $\boldsymbol{H}$ is constant during one symbol period and changes randomly to the next one. This channel model is referred to as a *fast-fading channel.*

†Suppose that $\boldsymbol{A}$ and $\boldsymbol{B}$ are two matrices with sizes of $m \times n$ and $n \times m$, respectively. The determinant identity states that $\det(\boldsymbol{I}_m + \boldsymbol{AB}) = \det(\boldsymbol{I}_n + \boldsymbol{BA})$.

2. Matrix $\boldsymbol{H}$ remains constant for long time of many symbols. This model is called a *slow-fading channel*; it arises when the channel coherence time is much larger than the symbol period.

2.3.4.1 Channel Capacity for Fast-Fading Channels For each realization of channel matrix $\boldsymbol{H}$, the channel has the maximum information rate in (2.24) and (2.26), depending on whether the matrix $\boldsymbol{H}$ is known or unknown at the transmitter. When the channel matrix $\boldsymbol{H}$ is known at the transmitter, the ergodic capacity of the MIMO system is the ensemble average of the capacity in (2.24) when the water-filling solution is applied for each realization of $\boldsymbol{H}$. This ergodic capacity is given by

$$\overline{C} = E\left\{\sum_{i=1}^{M} \log_2\left(1 + \frac{\lambda_i \overline{\gamma}_i}{N_0}\right)\right\}. \tag{2.27}$$

Similarly, when the channel matrix $\boldsymbol{H}$ is unknown at the receiver, the ergodic capacity is

$$\overline{C} = E\left\{\sum_{i=1}^{M} \log_2\left(1 + \frac{\sigma_s^2}{N_0}\lambda_i\right)\right\}. \tag{2.28}$$

The ergodic capacity for the CSI-known case is greater than that for the CSI-unknown case. However, the gap between these two decreases at a sufficiently high SNR region.

The expectation operation of ergodic capacities can be derived based on the eigenvalue distribution [10]. Alternatively, we may quatify the performance gain obtained using multiple antennas by looking at the lower bound of the ergodic capacites. In fact, for the CSI-known case with equal power allocation, a lower bound of the ergodic capacity in (2.28) is given by [11, 30]

$$\overline{C} = \overline{C}(\Gamma) \geq M \log_2\left[1 + \frac{\Gamma}{M}\exp\left(\frac{1}{M}\sum_{j=1}^{M}\sum_{p=1}^{Q-j}\frac{1}{p} - \beta\right)\right], \tag{2.29}$$

where $\beta \approx 0.57721566$ is Euler's constant and $Q = \max(N_t, N_r)$.

Let us define the *spatial multiplexing gain* as

$$r = \lim_{\Gamma\to\infty} \frac{\overline{C}(\Gamma)}{\log_2 \Gamma}. \tag{2.30}$$

From (2.29) we can see that $r = M$. Therefore, in a high-SNR region, the ergodic capacity increases M bits/s/Hz for every 3-dB increase in the average SNR, Γ. However, for SISO AWGN channels and SISO Rayleigh fading channels, the capacity increases only 1 bit/s/Hz for every 3-dB increase of average SNR, Γ, in the high-SNR region. This shows the tremendous capacity gain obtained by using MIMO systems.

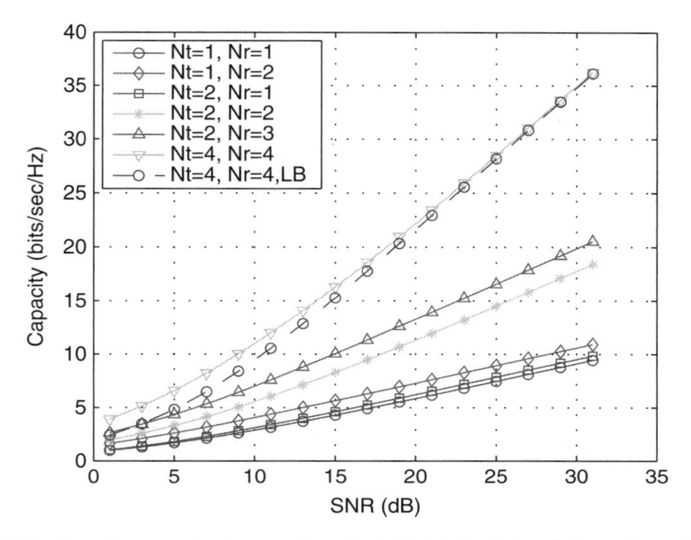

FIGURE 2.5 Ergodic capacity comparison for MIMO Rayleigh fading channels with different numbers of antennas.

Similar to the CSI-known case, we have the following lower bound on the ergodic capacity for the CSI-unknown case as [30]

$$\overline{C}(\Gamma) \geq M \log_2 \left[1 + \frac{\Gamma}{N_t} \exp\left(\frac{1}{M} \sum_{j=1}^{M} \sum_{p=1}^{Q-j} \frac{1}{p} - \beta \right) \right]. \tag{2.31}$$

Thus, a spatial multiplexing gain of M can be achieved for MIMO systems even when the CSI is unavailable at the transmitter.

Figure 2.5 illustrates the ergodic capacity of MIMO fading channels with different antenna configurations for the CSI-unknown case. Here we also plot the lower bound of the ergodic capacity for 4×4 MIMO channels. It is seen that this lower bound is tight when the SNR is greater than 25 dB. Further, for asymmetric antenna configurations, for a given M and in the high-SNR region, there exists a fixed SNR loss when the transmitting antenna number is larger than the receiving antenna number.

2.3.4.2 Channel Capacity for Slow-Fading Channels Suppose that we want to communicate at the rate of R bits/s/Hz using a MIMO system. With the slow-fading assumption, the goal is to find $\boldsymbol{R}_x$ such that

$$\log_2 \left\{ \det \left(\boldsymbol{I}_{N_r} + \frac{1}{N_0} \boldsymbol{H} \boldsymbol{R}_x \boldsymbol{H}^H \right) \right\} > R, \tag{2.32}$$

with the power constraint being satisfied, i.e., $\mathrm{trace}\{\boldsymbol{R}_x\} = P$. When the MIMO system does not satisfy the condition in (2.32), we have an outage. Hence, we must relate the capacity with some level of reliability. Suppose that we require the information rate to be guaranteed for $(100 - q)\%$ of channel matrix realization $\boldsymbol{H}$; that is, if the probability of outage is $q\%$, the $q\%$ outage capacity $C_{\mathrm{out},q\%}$ satisfies

$$\Pr\left\{\log_2\left\{\det\left(\boldsymbol{I}_{N_r} + \frac{1}{N_0}\boldsymbol{H}\boldsymbol{R}_x\boldsymbol{H}^H\right)\right\} \leq C_{\mathrm{out},q\%}\right\} = q\%. \tag{2.33}$$

2.4 DIVERSITY

To guarantee a reliable communication in fading channels where the transmitted signal is distorted, *diversity* techniques are deployed. The key idea of diversity is to provide the receiver with multiple versions of the transmitted signal. If the diversity scheme is well designed, there is a high probability that at least one version of the transmitted signal is not degraded severely by the channel. Diversity can be deployed in three different approaches: time, frequency, and space.

In *time diversity*, the same transmitted signal is sent out in different time slots. The interval between the successive time slots is equal to or greater than the coherence time of the channel. Hence, independently faded versions of the same transmitted signal are available to the receiver.

In *frequency diversity*, the same transmitted signal is carried on a number of different frequencies. The separation between the two adjacent frequencies must be wide enough to ensure independent fading. This separation can be equal or greater than the coherence bandwidth of the channel.

Space diversity is realized by deploying multiple antennas for transmission or (and) reception. The multiple antennas are separated far enough in space so that independent faded paths exist in the system. In practice, the separation of a few wavelengths is enough to guarantee the condition. In space diversity, multiple versions of the same transmitted signal are provided in the space domain; hence, unlike the time and frequency domains, no loss in bandwidth efficiency arises in this type of diversity. Depending on which side of communication link space diversity is deployed, we can classify the space diversity into the following categories:

- *Receive diversity.* This involves the deployment of multiple antennas at the receiver. The signals received from different paths linking the transmitter and receiver are combined according to different criteria. We can list here the four main types of combining methods: selection combining, switched combining, equal-gain combining, and maximal ratio combining [12].
- *Transmit diversity.* In this case, multiple antennas are placed at the transmitter. Unlike receive diversity, transmit diversity is more difficult in implementation. The difficulties include the additional signal processing at both transmitter and receiver to extract the diversity, and the complication of setting up a reliable

feedback link from the receiver to the transmitter to provide the CSI to the latter. Transmit diversity can be further divided into two smaller categories: transmit diversity with feedback and transmit diversity without feedback. The former uses the CSI provided from the receiver to weight the transmitted signals from transmitting antennas before transmission to maximize the received signal power or to maximize the channel capacity. The latter does not have access to the CSI; hence, signal processing at the transmitter is designed so that the receiver can exploit the diversity. One typical example of this type of diversity is the delay diversity scheme [13], where copies of a transmitted signal are transmitted from different transmitting antennas in different time slots. This makes the sequence of received signals associated with the transmitted signal like the output of a frequency-selective fading channel. The diversity is obtained by using a maximum likelihood sequence estimator or an MMSE equalizer. An alternative is to introduce redundancy in the spatial and temporal domains, which correlates the transmitted signals. This coding technique, called *space–time coding*, is presented in Section 2.8.

- *Joint transmit and receive diversity.* For this case, multiple antennas are deployed for both transmission and reception.

2.5 DIVERSITY AND SPATIAL MULTIPLEXING GAIN

We consider only space diversity in a MIMO system. To quantify the gain offered by diversity, a coefficient called *diversity gain* is defined. If the probability of error $P_e(\Gamma)$ as a function of average SNR, Γ, satisfies, for a fixed transmission rate, the diversity gain.

$$d = -\lim_{\Gamma\to\infty} \frac{P_e(\Gamma)}{\log_2 \Gamma}. \tag{2.34}$$

In other words, the slope of the curve $P_e(\Gamma)$ versus Γ at a high-SNR region equals the diversity gain. Equation (2.34) states that at the high-SNR region, the error probability decreases by 2^{-d} for every 3-dB increase in Γ. The diversity gain therefore indicates the reliability of the communication. For a MIMO system with N_t transmitting antennas and N_r receiving antennas, the *maximum* diversity gain that can be obtained is $d_{\max} = N_t N_r$. The spatial multiplexing gain is defined as in (2.30) for a fixed error probability. The largest spatial multiplexing gain of the MIMO system is $r_{\max} = \min(N_t, N_r)$.

We see from (2.30) and (2.34) that an increase of 3 dB in SNR either brings about a decrease of $2^{-d_{\max}}$ in error probability for a fixed transmission rate or provides a rate of $r_{\max}$ bits/s/Hz for a fixed error probability. Hence, there is a trade-off between the diversity gain and the spatial multiplexing gain; higher diversity gain comes at the price of lower spatial multiplexing gain. In [14, 29], a simple trade-off for block fading channels with block length equal or greater than $N_t + N_r - 1$ was shown for a high-SNR region. The optimal diversity gain as a function of the spatial multiplexing gain r is given by

$$d_{\text{opt}}(r) = (N_t - r)(N_r - r), \quad 0 \leq r \leq r_{\max}. \tag{2.35}$$

2.6 SIMO SYSTEMS

Consider a SIMO system with one transmitting antenna and N receiving antennas. Let the signal received at the ith receiving antenna be represented as

$$x_i(n) = h_i s(n) + u_i(n), \tag{2.36}$$

where $s(n)$ is the signal transmitted with mean zero and variance E_s, $u_i(n)$ is the additive white Gaussian noise (AWGN) received at the ith receiving antenna, and h_i is the channel coefficients from the transmitting antenna to the ith receiving antenna with $E\{|h_i|^2\} = 1$. The received noises are zero mean, independent of each other, and have equal variance N_0. It is also assumed that the transmitted signal is independent of the noises.

The optimal strategy to combine the received signals is called *maximum ratio combining* (MRC), the output of which is given by

$$y(n) = \sum_{i=1}^{N} h_i^* x_i(n). \tag{2.37}$$

It can be easily shown that the output SNR in $y(n)$ is

$$\text{SNR}_{\text{MRC}} = \frac{E_s}{N_0} \sum_{i=1}^{N} |h_i|^2. \tag{2.38}$$

If $|h_i|, i = 1, \ldots, N$, follows the Rayleigh distribution, then SNR_{MRC} follows a χ^2 distribution with $2N$ degrees of freedom, which means that MRC achieves a full diversity order of N.

Next, let us introduce the concept of array gain for multiple-antenna systems in a fading environment. Assuming that each antenna has the same average received SNR, the *array gain* is defined as the increment of average SNR after receive processing at the receiver as compared to that in a SISO system. The average received SNR after MRC is given by

$$E\{\text{SNR}_{\text{MRC}}\} = N \frac{E_s}{N_0}. \tag{2.39}$$

As for the case with a single transmitting antenna, the average received SNR is E_s/N_0; thus, the array gain achieved by MRC is $10 \log_{10}(N)$ dB.

2.7 MISO SYSTEMS

We consider the transmitting strategy for MISO systems with CSI known at the transmitter side. Suppose that the MISO system of interest has N transmitting antennas and one receiving antenna, and denote $s(n)$ as the transmitted signal with mean zero and variance E_s, $x_i(n)$ the signal transmitted through the ith transmitting antenna, and h_i the channel response from the ith transmitting antenna to the receiving antenna. The signal received at the receiver is then given by

$$y(n) = \sum_{i=1}^{N} h_i x_i(n) + u(n), \tag{2.40}$$

where $u(n)$ is the additive white Gaussian noise with zero mean and variance N_0. We also assume that $x_i(n)$ is independent of $u(n)$ and that $h_1, h_2, \ldots, h_N$ are i.i.d. zero-mean complex Gaussian random variables with $E\{|h_i|^2\} = 1, \quad \forall i$.

In order to maximize the diversity and array gain, the signals transmitted need to be designed as follows:

$$x_i(n) = \frac{1}{\sqrt{\sum_{i=1}^{N} |h_i|^2}} h_i^* s(n), \quad \forall i. \tag{2.41}$$

This scheme is called *transmit beamforming*. With this design, the total transmitted power is still constrained to E_s, but the received SNR is given by

$$\mathrm{SNR}_{\mathrm{TB}} = \frac{E_s}{N_0} \sum_{i=1}^{N} |h_i|^2. \tag{2.42}$$

Following the same arguments as those used in SIMO systems, the transmit beamforming scheme achieves a diversity order of N and an array gain of $10 \log_{10}(N)$ dB. That is, if the transmitter side has perfect CSI, a MISO system can perform equally well as a SIMO system if both systems have the same amount of total transmitting power and the same total number of antennas.

It is pointed out that the derivation above is under the assumption that the transmitter has perfect CSI. When there is no CSI information at the transmitter side, different transmission schemes, such as space–time coding, need to be designed in order to achieve transmit diversity. Unfortunately, as we can see later, no array gain can be achieved for such a case.

2.8 SPACE–TIME CODING

Space–time coding is an effective solution to achieve transmit diversity for MISO systems [15]. It introduces the *redundancy* in both spatial and temporal domains for the purpose of providing a correlation among transmitted signals from multiple antennas over multiple time slots. Space–time coding can be classified into two types: space–time trellis codes (STTCs) and space–time block codes (STBCs). Here we are interested in the design of STBCs due to its simplicity in transceiver design. More details of STTCs may be found in [15] and [18].

2.8.1 Space–Time Block Code

A space–time block code is characterized by a *transmission matrix* that characterizes the strategy of data transmission over time and space. This transmission matrix consists of three parameters. The first is the number of transmitting antennas, denoted as N_t; the second is the number of symbols that are transmitted using the transmission matrix, and this number is denoted as k; the third, denoted as p, is the number of time intervals needed to transmit the transmission matrix. The rate of the code is defined as the ratio of the number of transmitted symbols to the required number of symbol intervals:

$$r = \frac{k}{p}. \tag{2.43}$$

2.8.2 Alamouti Code

Alamouti code [16] is the first space–time block code that provides full transmit diversity for systems having two transmitting antennas. Denote x_1 and x_2 as two modulated symbols to be transmitted and assume that $E\{x_1\} = E\{x_2\} = 0$ and $E\{|x_1|^2\} = E\{|x_2|^2\} = E_s$. The transmission strategy using Alamouti code is stated as follows:

- In the first symbol interval, the first antenna transmits x_1, while the second antenna transmits x_2.
- In the second symbol interval, the first antenna transmits $-x_2^*$ and the second antenna sends out x_1^*.

Hence, the transmission matrix of Alamouti code is given by

$$\boldsymbol{X} = \underbrace{\frac{1}{\sqrt{2}} \begin{bmatrix} x_1 & -x_2^* \\ x_2 & x_1^* \end{bmatrix}}_{\rightarrow \text{ time}} \downarrow \text{space}, \tag{2.44}$$

where the spatial dimension is from up to down and the temporal dimension is from left to right. Note that the use of factor $1/\sqrt{2}$ guarantees that the total transmission power is the same as that of the single transmitting antenna case. This transmission matrix exhibits the following *orthogonality* in both spatial and temporal dimensions:

$$\boldsymbol{X}\boldsymbol{X}^H = \boldsymbol{X}^H\boldsymbol{X} = \tfrac{1}{2}\left(|x_1|^2 + |x_2|^2\right)\boldsymbol{I}_2. \tag{2.45}$$

Furthermore, Alamouti code uses two symbol intervals to transmit two symbols; thus, the code rate is $r = 1$ (i.e., Alamouti code is a full-rate code).

Suppose that the channel coefficients from the first and second antennas to the receiving antenna are h_1 and h_2, respectively, and they are unchanged over two symbol intervals for the transmission of $\boldsymbol{X}$. Then the signal received over the first symbol interval is

$$z_1 = \frac{h_1}{\sqrt{2}}x_1 + \frac{h_2}{\sqrt{2}}x_2 + u_1, \tag{2.46}$$

and that over the second symbol interval is

$$z_2 = -\frac{h_1}{\sqrt{2}}x_2^* + \frac{h_2}{\sqrt{2}}x_1^* + u_2, \tag{2.47}$$

where u_1 and u_2 are independent complex Gaussian random variables with mean zero and variance N_0. Assume that h_1 and h_2 are independent of each other and that $E\{|h_1|^2\} = E\{|h_2|^2\} = 1$.

If we collect the two received signals and put them into a vector $\boldsymbol{z} = [\, z_1 \quad z_2^* \,]^T$, we have

$$\boldsymbol{z} = \boldsymbol{H}\boldsymbol{x} + \boldsymbol{u}, \tag{2.48}$$

where

$$\boldsymbol{H} = \frac{1}{\sqrt{2}}\begin{bmatrix} h_1 & h_2 \\ h_2^* & -h_1^* \end{bmatrix}, \tag{2.49}$$

$\boldsymbol{x} = [\, x_1 \quad x_2 \,]^T$, and $\boldsymbol{u} = [\, u_1 \quad u_2^* \,]^T$. We observe that the columns of the squared matrix $\boldsymbol{H}$ are orthogonal. Hence, the decoding operation is performed by first left-multiplying $\boldsymbol{z}$ with $\boldsymbol{H}^H$ (here, the channel coefficients are assumed to be perfectly known at the receiver), which yields

$$\boldsymbol{y} = \boldsymbol{H}^H\boldsymbol{z} = \boldsymbol{H}^H\boldsymbol{H}\boldsymbol{x} + \boldsymbol{H}^H\boldsymbol{u} = \frac{|h_1|^2 + |h_2|^2}{2}\boldsymbol{x} + \boldsymbol{H}^H\boldsymbol{u}, \tag{2.50}$$

where $\boldsymbol{y} = [\, y_1 \quad y_2 \,]^T$. Elements of $\boldsymbol{y}$ can be written as

$$y_1 = \frac{|h_1|^2 + |h_2|^2}{2}x_1 + \frac{1}{\sqrt{2}}\left(h_1^* u_1 + h_2 u_2^*\right), \tag{2.51}$$

$$y_2 = \frac{|h_1|^2 + |h_2|^2}{2}x_2 + \frac{1}{\sqrt{2}}\left(-h_1 u_2^* + h_2^* u_1\right). \tag{2.52}$$

From (2.51) and (2.52), it is observed that due to the orthogonality of Alamouti code, the interference of x_2 does not appear in y_1 and the interference of x_1 does not appear in y_2. It is this property that provides Alamouti code a simple maximum-likelihood decoding rule:

$$\hat{x}_1 = \arg \min_{x \in C} d\left(y_1, \frac{|h_1|^2 + |h_2|^2}{2} x\right) \tag{2.53}$$

and

$$\hat{x}_2 = \arg \min_{x \in C} d\left(y_2, \frac{|h_1|^2 + |h_2|^2}{2} x\right), \tag{2.54}$$

where C is the set of constellations used and $d(a, b)$ is defined as $d(a, b) = |a - b|^2$.

2.8.2.1 SNR Performance

It can be easily shown that the SNR of x_1 in y_1 is given by

$$\mathrm{SNR}_1 = \frac{1}{2}\left(|h_1|^2 + |h_2|^2\right) \frac{E_s}{N_0}. \tag{2.55}$$

Similarly, the SNR of x_2 in y_2 is also given by

$$\mathrm{SNR}_2 = \frac{1}{2}\left(|h_1|^2 + |h_2|^2\right) \frac{E_s}{N_0}. \tag{2.56}$$

If both $|h_1|$ and $|h_2|$ follow Rayleigh distribution, and they are independent of each other, then SNR_1 and SNR_2 follow χ^2 distribution with 4 degrees of freedom, which means that Alamouti code achieves a full diversity order of 2.

2.8.2.2 Array Gain

The average received SNR for Alamouti decoding is given by

$$E\{\mathrm{SNR}_1\} = E\{\mathrm{SNR}_2\} = \frac{E_s}{N_0}. \tag{2.57}$$

As for the case with a single transmitting antenna, the average received SNR is also E_s/N_0; thus, the array gain achieved by Alamouti code is 0 dB. This is reasonable since we have not used any channel state information at the transmitting side when designing Alamouti code.

2.8.2.3 Properties of Alamouti Code [31]

We summarize the four properties of Alamouti code as follows:

1. *Unitary code.* Alamouti code is an orthogonal code [cf. (2.45)].
2. *Full-rate complex code.* Alamouti code is the only complex STBC with code rate equal to 1. With any constellation used, Alamouti code achieves full diversity gain (of order two) at full transmission rate.

3. *Linearity*. The transmission matrix of Alamouti code can be written as

$$\boldsymbol{X} = x_1 \boldsymbol{\Gamma}_{11} + x_2 \boldsymbol{\Gamma}_{21} + x_1^* \boldsymbol{\Gamma}_{22} - x_2^* \boldsymbol{\Gamma}_{12}, \tag{2.58}$$

where $\boldsymbol{\Gamma}_{mn}$ is a 2×2 matrix with all zero elements except the element at row m and column n. Hence, Alamouti code is linear in the symbols transmitted.

4. *Capacity achieving code*. For the system with two transmitting antennas and one receiving antenna, the channel capacity is given by

$$C\left(h_1, h_2\right) = \log_2 \left\{1 + \frac{1}{2}\left(|h_1|^2 + |h_2|^2\right) \frac{E_s}{N_0}\right\}. \tag{2.59}$$

On the other hand, the output SNR after Alamouti decoding is $\frac{1}{2}(|h_1|^2 + |h_2|^2)(E_s/N_0)$. Thus, the achievable rate for Alamouti code is also given by (2.59), and Alamouti code achieves the capacity of a 2×1 MISO system.

2.8.3 Generalized Complex Orthogonal STBC

The full transmit diversity and simple decoding provided by Alamouti code motivated the search for complex space–time block codes for systems with more than two transmitting antennas. Tarokh et al. [17] introduced the theory of generalized complex orthogonal designs that were used to construct the nonsquare transmission matrix to accommodate the use of more than two transmitting antennas. Those codes have only fractional code rates.

With a code rate of $r = \frac{1}{2}$, there are space–time block codes that provide full transmit diversity as well as a simple decoding algorithm for an arbitrary number of transmitting antennas. For example, transmission matrices for systems with three and four transmitting antennas have been designed as follows:

$$\boldsymbol{X}_3 = \frac{1}{\sqrt{3}} \begin{bmatrix} x_1 & -x_2 & -x_3 & -x_4 & x_1^* & -x_2^* & -x_3^* & -x_4^* \\ x_2 & x_1 & x_4 & -x_3 & x_2^* & x_1^* & x_4^* & -x_3^* \\ x_3 & x_4 & x_1 & x_2 & x_3^* & -x_4^* & x_1^* & x_2^* \end{bmatrix}, \tag{2.60}$$

$$\boldsymbol{X}_4 = \frac{1}{2} \begin{bmatrix} x_1 & -x_2 & -x_3 & -x_4 & x_1^* & -x_2^* & -x_3^* & -x_4^* \\ x_2 & x_1 & x_4 & x_3 & x_2^* & x_1^* & x_4^* & -x_3^* \\ x_3 & -x_4 & x_1 & x_2 & x_3^* & -x_4^* & x_2^* & x_2^* \\ x_4 & x_3 & -x_2 & x_1 & x_4^* & x_3^* & -x_1^* & x_1^* \end{bmatrix}. \tag{2.61}$$

It can be verified that

$$\boldsymbol{X}_3\boldsymbol{X}_3^H = \alpha_3\boldsymbol{I}_3 \tag{2.62}$$

and

$$\boldsymbol{X}_4\boldsymbol{X}_4^H = \alpha_4\boldsymbol{I}_4, \tag{2.63}$$

where $\alpha_3 = \frac{2}{3}\sum_{i=1}^{4}|x_i|^2$ and $\alpha_4 = \frac{1}{2}\sum_{i=1}^{4}|x_i|^2$. Thus, orthogonality of the transmission matrix exists in the temporal dimension. However, neither $\boldsymbol{X}_3^H\boldsymbol{X}_3$ nor $\boldsymbol{X}_4^H\boldsymbol{X}_4$ is a scalar of an identity matrix. Therefore, the orthogonality of the transmission matrix does not exist in the spatial dimension.

Compared with Alamouti code, the bandwidth efficiency of the codes corresponding to $\boldsymbol{X}_3$ and $\boldsymbol{X}_4$ have been reduced by half. To increase the bandwidth efficiency, codes with code rate $r = \frac{3}{4}$ can be designed. The following two transmission matrices, $\boldsymbol{X}_3'$ and $\boldsymbol{X}_4'$, are codes for systems with three and four transmitting antennas:

$$\boldsymbol{X}_3' = \frac{1}{\sqrt{3}}\begin{bmatrix} x_1 & -x_2^* & \frac{x_3^*}{\sqrt{2}} & \frac{x_3^*}{\sqrt{2}} \\ x_1 & x_1^* & \frac{x_3^*}{\sqrt{2}} & \frac{-x_3^*}{\sqrt{2}} \\ \frac{x_3}{\sqrt{2}} & \frac{x_3}{\sqrt{2}} & \frac{-x_1 - x_1^* + x_2 - x_2^*}{2} & \frac{x_2 + x_2^* + x_1 - x_1^*}{\sqrt{2}} \end{bmatrix}, \tag{2.64}$$

$$\boldsymbol{X}_4' = \frac{1}{2}\begin{bmatrix} x_1 & -x_2^* & \frac{x_3^*}{\sqrt{2}} & \frac{x_3^*}{\sqrt{2}} \\ x_2 & x_1^* & \frac{x_3^*}{\sqrt{2}} & \frac{-x_3^*}{\sqrt{2}} \\ \frac{x_3}{\sqrt{2}} & \frac{x_3}{\sqrt{2}} & \frac{-x_1 - x_1^* + x_2 - x_2^*}{2} & \frac{x_2 + x_2^* + x_1 - x_1^*}{2} \\ \frac{x_3}{\sqrt{2}} & \frac{-x_3}{\sqrt{2}} & \frac{-x_2 - x_2^* + x_1 - x_1^*}{2} & \frac{-x_1 + x_1^* + x_2 - x_2^*}{2} \end{bmatrix}. \tag{2.65}$$

The equations of the decoding rules for those codes may be found, for examples in [18, Chap. 3].

2.9 MIMO TRANSCEIVER DESIGN

2.9.1 Spatial-Multiplexing System

Here, we consider the MIMO system with the CSI-unknown case. If $N_t \leq N_r$, the selection of $\boldsymbol{R}_x = (P/N_t)\boldsymbol{I}_{N_t}$ implies that the transmitted signals are independent and equally powered at the transmitting antennas. Thus, N_t independent data streams can be sent through the N_t transmitting antennas. This scheme, called *spatial multiplexing for the CSI-unknown case*, was invented by Foshini in Bell Labs. If $N_t > N_r$, special care is needed in designing the transmitted signals. For example, two transmitting antennas can be paired to transmit one data stream using Alamouti code. We do not pursue this direction.

The input–output relation for a static MIMO channel can be represented as

$$\boldsymbol{y}(n) = \boldsymbol{H}\boldsymbol{s}(n) + \boldsymbol{u}(n), \tag{2.66}$$

and the following assumptions are made:

- The elements of transmitted signal vector $\boldsymbol{s}(n)$ of size $N_t \times 1$ are i.i.d. zero-mean random variables and with covariance matrix $\boldsymbol{R}_s = E\{\boldsymbol{s}(n)\boldsymbol{s}^H(n)\} = (E_s/N_t)\boldsymbol{I}$.
- $\boldsymbol{H} = [\boldsymbol{h}_1, \boldsymbol{h}_2, \ldots, \boldsymbol{h}_{N_t}]$, where $\boldsymbol{h}_k$ denotes the channel response from the kth transmitting antenna to all receiving antennas.
- The elements of noise vector $\boldsymbol{u}(n)$ of size $N_r \times 1$ are i.i.d. complex Gaussian random variable with mean zero and variance N_0. Therefore, the covariance matrix of the noise vector is $\boldsymbol{R}_{\boldsymbol{u}} = E\{\boldsymbol{u}(n)\boldsymbol{u}^H(n)\} = N_0\boldsymbol{I}_{N_r}$.

At the receiving side, there are two types of receivers: an optimal maximum-likelihood (ML) receiver and a suboptimal receiver. The latter is further divided into linear and nonlinear receivers.

2.9.1.1 Linear Receivers The material developed in Section 1.7 can be applied to design linear receivers. More specifically, we can have two popular linear receivers: a ZF receiver and an MMSE receiver. For the MMSE receiver, the estimated $\hat{\mathbf{s}}$ is determined by

$$\hat{\mathbf{s}}(n) = \mathbf{W}^H \mathbf{y}(n), \tag{2.67}$$

where

$$W = (HH^H + (N_t/\Gamma)I_N)^{-1}H. \tag{2.68}$$

2.9.1.2 Nonlinear Receivers Linear receivers are simple but provide performance that is far from that of the optimal ML detection. To have a near-ML detection, nonlinear receivers must be used. The MIMO spatial multiplexing systems are equivalent to well-known multiuser communication systems. Thus, advanced receivers developed in those areas can be applied. One example is an ordered serial interference cancellation (OSIC) receiver, also called a *V-BLAST receiver*. Using the MMSE criterion as an example, the V-BLAST receiver consists of the following steps.

Weight Generation and Detection Order Determination This is a preprocessing step. For slow-fading channels, the weights and detection order will be applied for the entire packet, containing possibly many blocks.

1. The SINRs for all N_t data streams are calculated and the stream with the largest SINR is chosen as the stream to be detected first.
2. The channel vector of the stream detected is removed from the channel matrix, and the SINRs are calculated for the rest of the streams. The stream with the largest SINR is chosen as the stream to be detected next.
3. Go to step 2 until the weights for all data streams have been determined.

Signal Detection and Interference Cancellation This is the online processing step, and each block needs its own online processing.

1. Detect one stream based on the weights and order we have determined.
2. Cancel out the data stream just detected.
3. Repeat steps 1 and 2 until all data streams have been detected.

Performance of a V-BLAST Receiver A V-BLAST receiver is, in fact, a capacity-achieving scheme if powerful coding is applied [24]. However, V-BLAST receiver suffers from error propagation, which is a severe problem for uncoded systems. In fact, the earlier-detected streams may have some errors, which will affect the detection performance of the later-detected data streams.

2.10 SVD-BASED EIGEN-BEAMFORMING

For an $N_t \times N_r$ MIMO system, when CSI is available at the transmitter and receiver sides, joint transmit and receive beamforming can be used to decouple the MIMO channel into $M = \min(N_t, N_r)$ SISO channels. Thus, M data streams can be transmitted simultaneously. It is noted that for the CSI-known case, the receiver design becomes simple. However, since each decoupled SISO channel has different SNR, adaptive modulation and coding needs to be designed to match each channel.

2.11 MIMO FOR FREQUENCY-SELECTIVE FADING CHANNELS

In this section we consider an MIMO system that has N_t transmitting antennas and N_r receiving antennas, and whose channel between the lth transmitting antenna and the kth receiving antenna is a frequency-selective fading channel characterized by a channel response of length L, $\boldsymbol{h}_{k,l} = [\, h_{k,l}(0) \quad h_{k,l}(1) \quad \cdots \quad h_{k,l}(L-1)\,]^T$. We incorporate OFDM, SCCP, and IFDMA into the MIMO system to take care of the ISI issue. In all cases we assume that the CP is long enough to remove IBI.

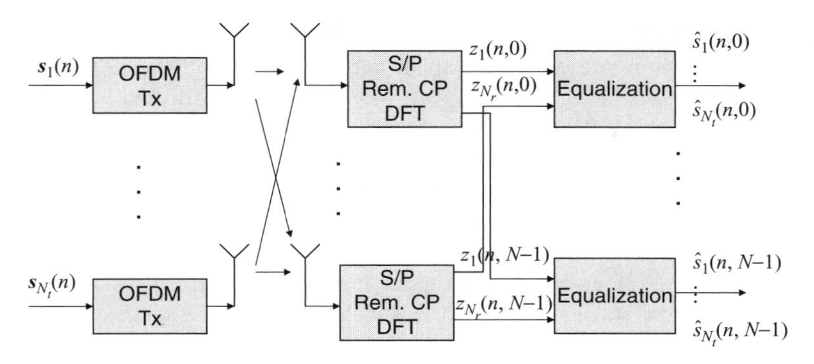

FIGURE 2.6 Block diagram of a MIMO-OFDM transmitter and receiver.

2.11.1 MIMO-OFDM

The block diagram of a MIMO-OFDM transmitter and receiver is shown in Figure 2.6. For MIMO-OFDM, each transmitting antenna transmits an independent OFDM data stream. Specifically, at the nth data block, the lth transmitting antenna transmits a block of symbols $\boldsymbol{s}_l(n) = [\, s_l(n;0) \quad s_l(n;1) \quad \cdots \quad s_l(n;N-1)\,]^T$, where N is the number of subcarriers available. Before transmission, this block is passed to an IDFT operation and added with CP. At the receiver, the block received at any receiving antenna is the superimposition of the signals transmitted from all N_t transmitting antennas. The portion associated with CP is removed, and the resulting block is passed through the DFT operation.

At the mth subcarrier, $m = 0, 1, \ldots, N-1$, the signal received on the kth receiving antenna is given by

$$z_k(n;m) = \sum_{l=1}^{N_t} H_{k,l}(m) s_l(n;m) + u_l(n;m), \tag{2.69}$$

where $H_{k,l}(m)$ is the channel response at the mth subcarrier of the channel between the lth transmitting antenna and the kth receiving antenna; $u_l(n;m)$ is the additive noise. If we collect N_r received signals at the mth subcarrier, $z_k(n;m)$ for $k = 1, 2, \ldots, N_r$, to form a vector $\boldsymbol{z}(n;m) = [\, z_1(n;m) \quad z_2(n;m) \quad \cdots \quad z_{N_r}(n;m)\,]^T$, we have

$$\boldsymbol{z}(n;m) = \boldsymbol{H}(m)\boldsymbol{s}(n;m) + \boldsymbol{u}(n;m), \tag{2.70}$$

where

$$\boldsymbol{H}(m) = \begin{bmatrix} H_{1,1}(m) & H_{1,2}(m) & \cdots & H_{1,N_t}(m) \\ H_{2,1}(m) & H_{2,2}(m) & \cdots & H_{2,N_t}(m) \\ \vdots & \vdots & \vdots & \vdots \\ H_{N_r,1}(m) & H_{N_r,2}(m) & \cdots & H_{N_r,N_t}(m) \end{bmatrix} \in \mathbb{C}^{N_r \times N_t} \tag{2.71}$$

and $\boldsymbol{s}(n;m) = [\,s_1(n;m) \quad s_2(n;m) \quad \cdots \quad s_{N_t}(n;m)\,]^T$ consists of symbols transmitted from N_t transmitting antennas on the mth subcarrier on the nth data block; $\boldsymbol{u}(n;m) = [\,u_1(n;m) \quad u_2(n;m) \quad \cdots \quad u_{N_r}(n;m)\,]^T$ is the additive noise vector.

Based on (2.70), a MIMO OFDM system is decomposed into N MIMO systems corresponding to N subcarriers. This allows data detection to be performed on a subcarrier-to-subcarrier basis. When CSI is unknown at the transmitter, optimal (ML) or suboptimal (ZF, MMSE, V-BLAST) receivers developed for the MIMO flat fading channel can be used for the model in (2.70). When CSI is known at the transmitter, joint transmitter and receiver eigen-beamforming can be performed at each subcarrier, thus generating *NM* subchannels. In [19], subchannel grouping and statistical water filling are proposed to achieve the capacity of MIMO-OFDM channels. In [20], differentiated traffic is supported using decoupled MIMO-OFDM channels.

2.11.2 MIMO-SCCP

The block diagram of a MIMO-SCCP transmitter and receiver is shown in Figure 2.7. The MIMO-SCCP transmitter operates as follows. At the nth data block, the lth transmitting antenna transmits a block of symbols $\boldsymbol{s}_l(n) = [s_l(n;0) \quad s_l(n;1) \quad \cdots \quad s_l(n;N-1)]^T$, where N is the number of available subcarriers. Before transmission, this block is added with CP of length P. At the receiver, the block received at any receiving antenna is the superimposition of the signals transmitted from all N_t transmitting antennas. The portion associated with CP is removed and the resulting block is passed through the DFT operation, yielding at the kth receiving antenna,

$$\boldsymbol{x}_k(n) = \sum_{l=1}^{N_t} \boldsymbol{\Lambda}_{k,l} \boldsymbol{W}_N \boldsymbol{s}_l(n) + \boldsymbol{u}_k(n), \tag{2.72}$$

where the diagonal elements of $\boldsymbol{\Lambda}_{k,l}$ represent the frequency responses in all subcarriers between the kth receiving antenna and the lth transmitting antenna.

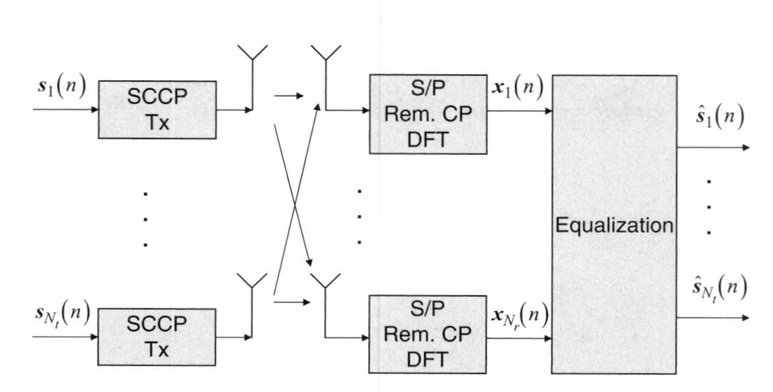

FIGURE 2.7 Block diagram of a MIMO-SCCP transmitter and receiver.

In matrix form, we can write the input–output relationship as follows:

$$x(n) = \tilde{\Lambda}\tilde{W}s(n) + u(n), \tag{2.73}$$

where

$$x(n) = \begin{bmatrix} x_1(n) \\ \vdots \\ x_{N_r}(n) \end{bmatrix}, \quad \tilde{\Lambda} = \begin{bmatrix} \Lambda_{1,1} & \cdots & \Lambda_{1,N_t} \\ \vdots & \ddots & \vdots \\ \Lambda_{N_r,1} & \cdots & \Lambda_{N_r,N_t} \end{bmatrix},$$

$$\tilde{W} = \begin{bmatrix} W_N & \cdots & 0 \\ \vdots & W_N & \vdots \\ 0 & \cdots & W_N \end{bmatrix} \quad s(n) = \begin{bmatrix} s_1(n) \\ \vdots \\ s_{N_t}(n) \end{bmatrix}.$$

From (2.73), linear receivers, such as ZF and MMSE receivers, or iterative receivers, such as the BI-GDFE, can be used to recover the transmitted signal vector s. Because of the special structure of the channel in (2.73), the receivers, in fact, recover the frequency-domain signals of each data stream first, then IDFT is used to convert the frequency-domain signal estimates to time-domain signals. It is pointed that even though the size of $\tilde{\Lambda}$ is $N_rN \times N_tN$, the first step equalization can be implemented in a subcarrier-by-subcarrier basis, because $\Lambda_{k,l}$ is a diagonal matrix for all k and l. By doing so, the computational complexity can be reduced dramatically.

2.11.3 MIMO-IFDMA

The block diagram of a $N_t \times N_r$ $(N_t \leq N_r)$ spatial multiplexed MIMO-IFDMA system is illustrated in Figure 2.8. Each user transmits N_t independent data streams over N_t transmitting antennas. In order not to exceed the total transmission bandwidth, N_t different streams belonging to a single user are multiplexed to occupy the same set of subcarriers. Besides, the transmission power is split and allocated equally to each

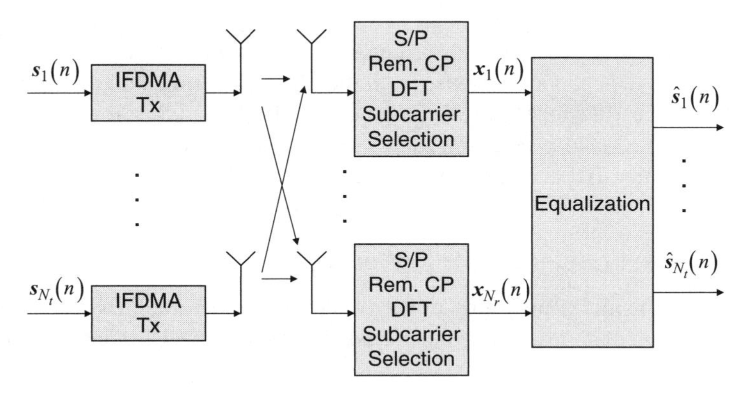

FIGURE 2.8 Block diagram of a MIMO-IFDMA transmitter and receiver.

stream subject to the total power constraint. At the receiver side, the data streams transmitted are superimposed at each receiving antenna.

Because of the user orthogonality, we ignore the user index and consider a user of interest transmitting N_t data streams $\{s_1(n), \ldots, s_{N_t}(n)\}$. Using the equivalence between IFDMA and SCCP, the input–output model for MIMO-IFDMA can also be written as

$$\boldsymbol{x}(n) = \tilde{\boldsymbol{\Lambda}}\tilde{\boldsymbol{W}}\boldsymbol{s}(n) + \boldsymbol{u}(n) \tag{2.74}$$

where

$$\boldsymbol{x}(n) = \begin{bmatrix} \boldsymbol{x}_1(n) \\ \vdots \\ \boldsymbol{x}_{N_r}(n) \end{bmatrix}, \quad \tilde{\boldsymbol{\Lambda}} = \begin{bmatrix} \boldsymbol{\Lambda}_{1,1} & \cdots & \boldsymbol{\Lambda}_{1,N_t} \\ \vdots & \ddots & \vdots \\ \boldsymbol{\Lambda}_{N_r,1} & \cdots & \boldsymbol{\Lambda}_{N_r,N_t} \end{bmatrix},$$

$$\tilde{\boldsymbol{W}} = \begin{bmatrix} \boldsymbol{W}_N & \cdots & \boldsymbol{0} \\ \vdots & \boldsymbol{W}_N & \vdots \\ \boldsymbol{0} & \cdots & \boldsymbol{W}_N \end{bmatrix},$$

and $\boldsymbol{\Lambda}_{k,l}$ is a diagonal matrix with diagonal elements containing the frequency-domain responses in the subcarriers allocated to that user from its lth transmitting antenna to the kth receiving antenna at the base station. Thus, the detection process is the same as that used in MIMO-SCCP systems.

2.12 TRANSMIT DIVERSITY FOR FREQUENCY-SELECTIVE FADING CHANNELS

In this section we consider the transmitter design for two-transmitter, one-receiver systems operating in a frequency-selective fading environment. The objective is to incorporate Alamouti code into the system so that transmit diversity can be achieved. Denote $H_i(k)$ as the channel response from the ith transmitting antenna to the receiving antenna at the kth subcarrier. We incorporate OFDM and SCCP into the transmit diversity system to take care of the ISI issue. In all cases we assume that the CP is long enough so that IBI and ISI do not exist.

2.12.1 Space–Frequency Coding for OFDM

With OFDM transmitter operation at the two transmitting antennas and OFDM receiving operation at the receiver, the signal received at the kth subcarrier is given by

$$z(n;k) = H_1(k)s_1(n;k) + H_2(k)s_2(n;k) + u(n;k). \tag{2.75}$$

Similarly, at the $(k+1)$th subcarrier, the signal received is given by

$$z(n; k+1) = H_1(k+1)s_1(n; k+1) + H_2(k+1)s_2(n; k+1) + u(n; k+1). \tag{2.76}$$

Since $H_1(k) \approx H_1(k+1)$ and $H_2(k) \approx H_2(k+1)$, the transmitted signals $s_1(n; k)$, $s_2(n; k)$, $s_1(n; k+1)$, and $s_2(n; k+1)$ can be designed as the Alamouti code outputs. This scheme is called *space–frequency coding* for OFDM as the coding is done over the two consecutive frequency subcarriers.

2.12.2 Space–Time Coding for OFDM

Suppose that the channel is time invariant, and consider the kth subcarrier's outputs at two consecutive OFDM blocks:

$$z(n; k) = H_1(k)s_1(n; k) + H_2(k)s_2(n; k) + u(n; k), \tag{2.77}$$

$$z(n+1; k) = H_1(k)s_1(n+1; k) + H_2(k)s_2(n+1; k) + u(n+1; k). \tag{2.78}$$

To achieve transmit diversity, we can design $s_1(n; k)$, $s_2(n; k)$, $s_1(n+1; k)$, and $s_2(n+1; k)$ as the Alamouti code outputs. This scheme is called *space–time coding* for OFDM as the coding is done over the two consecutive OFDM blocks.

2.12.3 Generalized Alamouti Code for SCCP

Generalized Alamouti code can be designed for SCCP systems in order to achieve transmit diversity in frequency-selective fading environment. Denote

$$\mathbf{s}_1 = [\, s_1(0) \;\; s_1(1) \;\; \cdots \;\; s_1(N-1)\,]^T \quad \text{and} \quad \mathbf{s}_2 = [\, s_2(0) \;\; s_2(1) \;\; \cdots \;\; s_2(N-1)\,]^T$$

as the two data blocks to be transmitted to the receiver. The principle of generalized Alamouti code works as follows. In the first block interval, antennas 1 and 2 transmit data blocks $\mathbf{s}_{1,1}$ and $\mathbf{s}_{2,1}$, respectively, where

$$\mathbf{s}_{1,1} \triangleq \frac{1}{\sqrt{2}}\mathbf{s}_1, \tag{2.79}$$

$$\mathbf{s}_{2,1} \triangleq \frac{1}{\sqrt{2}}\mathbf{s}_2. \tag{2.80}$$

In the second block interval, antennas 1 and 2 transmit data blocks $\mathbf{s}_{1,2}$ and $\mathbf{s}_{2,2}$, respectively, where

$$\mathbf{s}_{1,2} = \frac{1}{\sqrt{2}}[\, -s_2^*(0) \;\; -s_2^*(N-1) \;\; \cdots \;\; -s_2^*(1)\,]^T, \tag{2.81}$$

$$\mathbf{s}_{2,2} = \frac{1}{\sqrt{2}}[\, s_1^*(0) \;\; s_1^*(N-1) \;\; \cdots \;\; s_1^*(1)\,]^T. \tag{2.82}$$

Note, as in Alamouti code, that the factor $1/\sqrt{2}$ is used to maintain the same transmission power as for a single transmitting antenna. After CP removal, the data block received, corresponding to the first block interval, becomes

$$\boldsymbol{x}_1 = \boldsymbol{W}^H \boldsymbol{\Lambda}_1 \tilde{\boldsymbol{s}}_{1,1} + \boldsymbol{W}^H \boldsymbol{\Lambda}_2 \tilde{\boldsymbol{s}}_{2,1} + \boldsymbol{u}_1, \tag{2.83}$$

where $\tilde{\boldsymbol{s}}_{1,1} = \boldsymbol{W}\boldsymbol{s}_{1,1}$, $\tilde{\boldsymbol{s}}_{2,1} = \boldsymbol{W}\boldsymbol{s}_{2,1}$, $\boldsymbol{\Lambda}_1$, and $\boldsymbol{\Lambda}_2$ are the diagonal matrices containing the frequency-domain responses from the two transmitters to the receiver, and $\boldsymbol{u}_1$ is the noise vector received. Similarly, the data block received, corresponding to the second block interval after CP removal, is

$$\boldsymbol{x}_2 = \boldsymbol{W}^H \boldsymbol{\Lambda}_1 \tilde{\boldsymbol{s}}_{1,2} + \boldsymbol{W}^H \boldsymbol{\Lambda}_2 \tilde{\boldsymbol{s}}_{2,2} + \boldsymbol{u}_2, \tag{2.84}$$

where $\tilde{\boldsymbol{s}}_{1,2} = \boldsymbol{W}\boldsymbol{s}_{1,2}$ and $\tilde{\boldsymbol{s}}_{2,2} = \boldsymbol{W}\boldsymbol{s}_{2,2}$. Let

$$\tilde{\boldsymbol{s}}_{1,1} \triangleq [\,\tilde{s}_{1,1}(0) \quad \tilde{s}_{1,1}(1) \quad \cdots \quad \tilde{s}_{1,1}(N-1)\,]^T = \frac{1}{\sqrt{2}}\tilde{\boldsymbol{s}}_1, \tag{2.85}$$

$$\tilde{\boldsymbol{s}}_{2,1} \triangleq [\,\tilde{s}_{2,1}(0) \quad \tilde{s}_{2,1}(1) \quad \cdots \quad \tilde{s}_{2,1}(N-1)\,]^T = \frac{1}{\sqrt{2}}\tilde{\boldsymbol{s}}_2, \tag{2.86}$$

where $\tilde{\boldsymbol{s}}_1 = \boldsymbol{W}\boldsymbol{s}_1$ and $\tilde{\boldsymbol{s}}_2 = \boldsymbol{W}\boldsymbol{s}_2$. It can easily be verified that

$$\tilde{\boldsymbol{s}}_{1,2} = [\,-\tilde{s}^*_{2,1}(0) \quad -\tilde{s}^*_{2,1}(1) \quad \cdots \quad -\tilde{s}^*_{2,1}(N-1)\,]^T = -\frac{1}{\sqrt{2}}\tilde{\boldsymbol{s}}^*_2, \tag{2.87}$$

$$\tilde{\boldsymbol{s}}_{2,1} = [\,\tilde{s}^*_{1,1}(0) \quad \tilde{s}^*_{1,1}(1) \quad \cdots \quad \tilde{s}^*_{1,1}(N-1)\,]^T = \frac{1}{\sqrt{2}}\tilde{\boldsymbol{s}}^*_1. \tag{2.88}$$

Thus, (2.83) and (2.84) become

$$\boldsymbol{x}_1 = \frac{1}{\sqrt{2}}\boldsymbol{W}^H \boldsymbol{\Lambda}_1 \tilde{\boldsymbol{s}}_1 + \frac{1}{\sqrt{2}}\boldsymbol{W}^H \boldsymbol{\Lambda}_2 \tilde{\boldsymbol{s}}_2 + \boldsymbol{u}_1, \tag{2.89}$$

$$\boldsymbol{x}_2 = -\frac{1}{\sqrt{2}}\boldsymbol{W}^H \boldsymbol{\Lambda}_1 \tilde{\boldsymbol{s}}^*_2 + \frac{1}{\sqrt{2}}\boldsymbol{W}^H \boldsymbol{\Lambda}_2 \tilde{\boldsymbol{s}}^*_1 + \boldsymbol{u}_2, \tag{2.90}$$

respectively. Perform DFT on $\boldsymbol{x}_1$ and $\boldsymbol{x}_2$; we then have $\tilde{\boldsymbol{x}}_1 = \boldsymbol{W}\boldsymbol{x}_1$ and $\tilde{\boldsymbol{x}}_2 = \boldsymbol{W}\boldsymbol{x}_2$, and

$$\begin{bmatrix} \tilde{\boldsymbol{x}}_1 \\ \tilde{\boldsymbol{x}}^*_2 \end{bmatrix} = \frac{1}{\sqrt{2}} \begin{bmatrix} \boldsymbol{\Lambda}_1 & \boldsymbol{\Lambda}_2 \\ \boldsymbol{\Lambda}^*_2 & -\boldsymbol{\Lambda}^*_1 \end{bmatrix} \begin{bmatrix} \tilde{\boldsymbol{s}}_1 \\ \tilde{\boldsymbol{s}}_2 \end{bmatrix} + \begin{bmatrix} \tilde{\boldsymbol{u}}_1 \\ \tilde{\boldsymbol{u}}^*_2 \end{bmatrix}. \tag{2.91}$$

Denote

$$\boldsymbol{H} = \frac{1}{\sqrt{2}} \begin{bmatrix} \boldsymbol{\Lambda}_1 & \boldsymbol{\Lambda}_2 \\ \boldsymbol{\Lambda}^*_2 & -\boldsymbol{\Lambda}^*_1 \end{bmatrix}.$$

Since

$$H^H H = \frac{1}{2}\begin{bmatrix} |\Lambda_1|^2 + |\Lambda_2|^2 & \mathbf{0} \\ \mathbf{0} & |\Lambda_1|^2 + |\Lambda_2|^2 \end{bmatrix},$$

we can easily obtain the estimates of $\tilde{s}_1$ and $\tilde{s}_2$, thus those of s_1 and s_2.

2.13 CYCLIC DELAY DIVERSITY

Using the space–time coding schemes discussed earlier, the receiver has to be informed of the transmission schemes used so that proper decoding can be carried out. Another popular transmit diversity scheme applicable to CP-based block transmissions is called *cyclic delay diversity* (CDD). The main advantage of using CDD is that the receiver structure can be the same for both the CDD case and that of a single transmitting antenna. In the following we consider the operations of CDD for both OFDM and SCCP systems. Since IFDMA is equivalent to SCCP, the application of CDD to IFDMA is also straightforward.

2.13.1 CDD for OFDM

Figure 2.9 is a block diagram of CDD for OFDM using an arbitrary number of transmitting antennas. Let $s(n) = [\, s(n;0) \quad s(n;1) \quad \cdots \quad s(n;N-1)\,]^T$ be the data block to be transmitted to the receiver. Then

$$x_0(n) = \frac{1}{\sqrt{M}} x(n), \tag{2.92}$$

$$x_i(n) = x_0((n-\delta_i)_N), \quad i = 1, 2, \ldots, M-1, \tag{2.93}$$

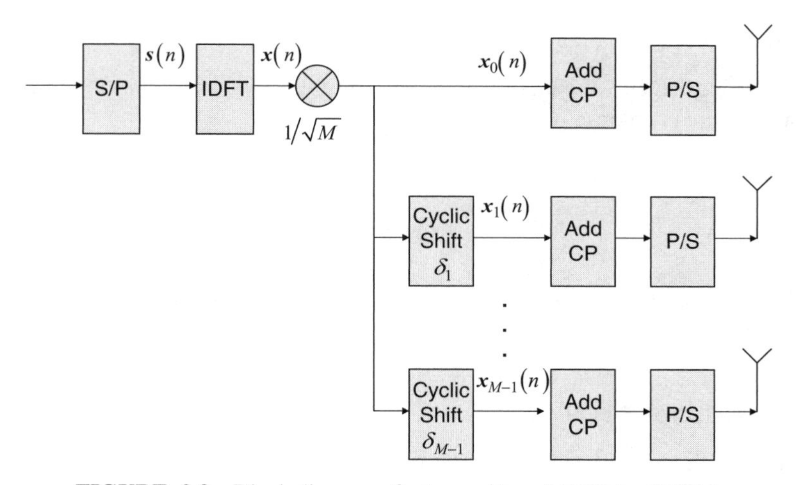

FIGURE 2.9 Block diagram of a transmitter of CDD for OFDM.

where $\boldsymbol{x}(n) = \boldsymbol{W}_N^H \boldsymbol{s}(n)$ and $(\cdot)_N$ denotes the modulo operation over a period of N. Based on Theorem 1.1, for $i = 1, 2, \ldots, M-1$, we have

$$\boldsymbol{W}_N \boldsymbol{x}_i(n) = \boldsymbol{\Delta}_i \boldsymbol{W}_N \boldsymbol{x}_0(n) = \frac{1}{\sqrt{M}} \boldsymbol{\Delta}_i \boldsymbol{s}(n), \tag{2.94}$$

where

$$\boldsymbol{\Delta}_i = \text{diag}\left\{1, \exp\left(-j\frac{2\pi}{N}\delta_i\right), \ldots, \exp\left(-j\frac{2\pi}{N}(N-1)\,\delta_i\right)\right\}. \tag{2.95}$$

At the receiver side, after the part associated with CP is discarded, the received block constructed becomes

$$\boldsymbol{y}(n) = \sum_{i=0}^{M-1} \boldsymbol{W}_N^H \boldsymbol{\Lambda}_i \boldsymbol{W}_N \boldsymbol{x}_i(n) + \boldsymbol{u}(n). \tag{2.96}$$

where $\boldsymbol{\Lambda}_i$ defines the frequency-domain channel responses from the ith transmitting antenna to the receiving antenna. Using (2.94), Equation (2.96) becomes

$$\boldsymbol{y}(n) = \boldsymbol{W}_N^H \left(\frac{1}{\sqrt{M}} \sum_{i=0}^{M-1} \boldsymbol{\Lambda}_i \boldsymbol{\Delta}_i\right) \boldsymbol{s}(n) + \boldsymbol{u}(n). \tag{2.97}$$

Performing DFT on $\boldsymbol{y}(n)$ gives

$$\boldsymbol{z}(n) = \boldsymbol{W}_N \boldsymbol{y}(n) = \left(\frac{1}{\sqrt{M}} \sum_{i=0}^{M-1} \boldsymbol{\Lambda}_i \boldsymbol{\Delta}_i\right) \boldsymbol{s}(n) + \boldsymbol{W}_N \boldsymbol{u}(n). \tag{2.98}$$

That is, using CDD, we generate an equivalent OFDM system, and the frequency-domain channel response becomes

$$\boldsymbol{\Lambda} = \frac{1}{\sqrt{M}} \sum_{i=0}^{M-1} \boldsymbol{\Lambda}_i \boldsymbol{\Delta}_i. \tag{2.99}$$

By choosing the cyclic delays properly and using channel coding, frequency diversity can be achieved.

2.13.2 CDD for SCCP

Figure 2.10 illustrates the block diagram of CDD for SCCP using an arbitrary number of transmitting antennas. Let $\boldsymbol{s}(n) = [\, s(n;0) \quad s(n;1) \quad \cdots \quad s(n;N-1)\,]^T$ be the

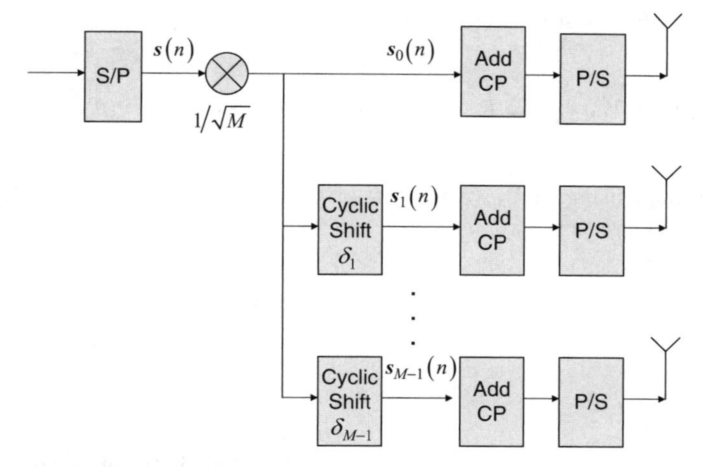

FIGURE 2.10 Block diagram of a transmitter of CDD for SCCP.

data block to be transmitted to the receiver. Then

$$s_0(n) = \frac{1}{\sqrt{M}} s(n), \tag{2.100}$$

$$s_i(n) = s_0((n - \delta_i)_N), \quad i = 1, 2, \ldots, M - 1. \tag{2.101}$$

Based on Theorem 1.1, for $i = 1, 2, \ldots, M - 1$, we have

$$\boldsymbol{W}_N \boldsymbol{s}_i(n) = \boldsymbol{\Delta}_i \boldsymbol{W}_N \boldsymbol{s}_0(n) = \frac{1}{\sqrt{M}} \boldsymbol{\Delta}_i \boldsymbol{W}_N \boldsymbol{s}(n), \tag{2.102}$$

where, again,

$$\boldsymbol{\Delta}_i = \text{diag}\left\{1, \exp\left(-j\frac{2\pi}{N}\delta_i\right), \ldots, \exp\left(-j\frac{2\pi}{N}(N-1)\delta_i\right)\right\}. \tag{2.103}$$

At the receiver side, after the part associated with CP is discarded, the received block constructed becomes

$$\boldsymbol{y}(n) = \sum_{i=0}^{M-1} \boldsymbol{W}_N^H \boldsymbol{\Lambda}_i \boldsymbol{W}_N \boldsymbol{s}_i(n) + \boldsymbol{u}(n), \tag{2.104}$$

where $\boldsymbol{\Lambda}_i$ defines the frequency-domain channel responses of the ith transmitting antenna to the receiving antenna. Using (2.102), (2.104) becomes

$$\boldsymbol{y}(n) = \boldsymbol{W}_N^H \left(\frac{1}{\sqrt{M}} \sum_{i=0}^{M-1} \boldsymbol{\Lambda}_i \boldsymbol{\Delta}_i\right) \boldsymbol{W}_N \boldsymbol{s}(n) + \boldsymbol{u}(n). \tag{2.105}$$

Performing DFT on $\boldsymbol{y}(n)$ gives

$$z(n) = \boldsymbol{W}_N \boldsymbol{y}(n) = \left(\frac{1}{\sqrt{M}} \sum_{i=0}^{M-1} \boldsymbol{\Lambda}_i \boldsymbol{\Delta}_i \right) \boldsymbol{W}_N \boldsymbol{s}(n) + \boldsymbol{W}_N \boldsymbol{u}(n). \qquad (2.106)$$

That is, through using CDD, we generate a SCCP system with equivalent frequency-domain channel response as

$$\boldsymbol{\Lambda} = \frac{1}{\sqrt{M}} \sum_{i=0}^{M-1} \boldsymbol{\Lambda}_i \boldsymbol{\Delta}_i. \qquad (2.107)$$

SUMMARY

In this chapter we have reviewed the fundamental capacity limits of MIMO systems. It is shown that for a given power budget and a given bandwidth, the ergodic capacity of a MIMO Rayleigh fading channel increases linearly with the minimum of the numbers of transmitting and receiving antennas. We have also looked at the transceiver design for MIMO systems with channel state information (CSI) known or unknown at the transmitter side. When CSI is known at the transmitter, joint transmit and receive beamforming can be used to decouple the MIMO channel to a set of parallel SISO channels. Adaptive modulation and coding can be designed to match the achievable rate for each decoupled channel. When CSI is unknown, spatial multiplexing is designed to transmit multiple data streams, thus linear or nonlinear equalizers can be used to recover the data streams transmitted. We have also considered the receiver design for SIMO systems using the maximum ratio combining criterion, and the transmitter design for MISO systems with CSI available at the transmitter. It is seen that these two systems provide the same performance if the total transmission power is the same. When the channel is a frequency-selective fading channel, OFDM, SCCP, and IFDMA are designed in conjunction with MIMO to achieve increased data rate transmission, or in conjunction with space–time coding to achieve transmit diversity. CDD has been analyzed in detail, due to its popularity in CP-based block transmissions. Recommendations for further reading are as follows:

- Guideline for choosing cyclic delays in CDD design [21]
- CDD design for IFDMA [22]
- Low-complexity near-ML detection for MIMO [23, 25–28]

REFERENCES

[1] T. Cover and J. Thomas, *Elements of Information Theory*, Wiley, New York, 1991.

[2] E. Biglieri, C. Caire, and G. Taricco, "Limiting performance of block-fading channels with multiple antennas," *IEEE Trans. Inf. Theory*, vol. 47, pp. 1273–1289, 2001.

[3] E. Biglieri, J. Proakis, and S. S. Shitz, "Fading channels: information-theoretic and communications aspects," *IEEE Trans. Inf. Theory*, vol. 44, pp. 2619–2692, Oct. 1998.

[4] C. Caire, G. Taricco, and E. Biglieri, "Optimal power control over fading channels," *IEEE Trans. Inf. Theory*, vol. 45, pp. 1468–1589, July 1999.

[5] A. J. Doldsmith and P. P. Varaiya, "Capacity of fading channels with channel side information," *IEEE Trans. Inf. Theory*, vol. 43, pp. 1986–1992, Nov. 1997.

[6] G. Foschini, "Layered space-time architecture for wireless communication in fading environment when using multi-element antennas," *Bell Labs Tech. J.*, pp. 41–59, Oct. 1996.

[7] I. Telatar, "Capacity of multi-antenna Gaussian channels," *Eur. Trans. Telecommun.*, vol. 10, pp. 585–596, Nov. 1999.

[8] F. Neeser and J. Massey, "Proper complex random process with application to information theory," *IEEE Trans. Inf. Theory*, vol. 39, pp. 1293–1302, July 1993.

[9] G. H. Golub and C. F. V. Loan, *Matrix Computations*, 3rd ed.; Johns Hopkins University Press, Baltimore, 1996.

[10] Y.-C. Liang, R. Zhang, and J. M. Cioffi, "Subchannel grouping and statistical water-filling for vector block fading channels," *IEEE Trans. Commun.*, vol. 54, pp. 1131–1142, June 2006.

[11] O. Oyman, R. U. Nabar, H. Bolcskei, and A. J. Paulraj, "Tight lower bounds on the ergodic capacity of Rayleigh fading MIMO channels," *Proceedings of IEEE GLOBECOM'02*, pp. 1172–1176, Nov. 2002.

[12] M. K. Simon and M.-S. Alouini, *Digital communication over Fading Channels: A Unified Approach to Performance Analysis*, Wiley, New York, 2000.

[13] N. Seshadri and J. H. Winters, "Two signalling schemes for improving the error performance of FDD transmission systems using transmitter antenna diversity," *Proceedings of IEEE VTC'93*, May 1993, pp. 508–511.

[14] L. Zheng and D. N. Tse, "Diversity and multiplexing: a fundamental tradeoff in multiple antenna channels," *IEEE Trans. Inf. Theory*, vol. 49, pp. 1073–1096, 2003.

[15] V. Tarokh, N. Seshadri, and A. R. Calderbank, "Space-time codes for high data rate wireless communication: performance criterion and code construction," *IEEE Trans. Inf. Theory*, vol. 44, pp. 744–765, Mar. 1998.

[16] S. M. Alamouti, "A simple transmit diversity technique for wireless communications," *IEEE J. Sel. Areas Commun.*, vol. 16, pp. 1451–1458, Oct. 1998.

[17] V. Tarokh, H. Jafarkhani, and A. R. Calderbank, "Space-time block codes from orthogonal designs," *IEEE Trans. Inf. Theory*, vol. 45, pp. 1456–1467, July 1999.

[18] B. Vucetic and J. Yuan, *Space-Time Coding*, Wiley, Hoboken, NJ, 2003.

[19] Y.-C. Liang, R. Zhang and J. M. Cioffi, "Subchannel grouping and statistical water-filling for vector block fading channels," *IEEE Trans. Commun.*, vol. 54, no. 6, pp. 1131–1142, June 2006.

[20] Y.-C. Liang, R. Zhang, and J. M. Cioffi, "Transmit optimization for MIMO-OFDM with mixed delay constrained and no-delay constrained services," *IEEE Trans. Signal Process.*, vol. 54, no. 8, pp. 3190–3199, Aug. 2006.

[21] Y.-C. Liang, W. S. Leon, Y. Zeng and C. L. Xu, "Design of cyclic delay diversity for single carrier cyclic prefix (SCCP) transmissions with block-iterative GDFE (BI-GDFE) receiver," *IEEE Trans. Wireless Commun.*, vol. 7, no. 2, pp. 677–684, Feb. 2008.

[22] J. Xiang, Y. Cai, Y.-C. Liang, K.-H. Li, and K. C. Teh, "Low-complexity iterative receiver for interleaved FDMA (IFDMA) with cyclic delay diversity," *Proceedings of IEEE ICCS'2006*, Nov. 2006.

[23] Y.-C. Liang, S. Sun, and C. Ho, "Block-iterative generalized decision feedback equalizers (BI-GDFE) for large MIMO systems: algorithm design and asymptotic performance analysis," *IEEE Trans. Signal Process.*, vol. 54, no. 6, pp. 2035–2048, June 2006.

[24] R. Zhang, Y.-C. Liang, R. Narasimhan, and J. M. Cioffi, "Approaching MIMO-OFDM capacity with per-antenna power and rate feedback," *IEEE J. Sel. Areas Commun.*, vol. 25, no. 7, pp. 1284–1297, Sept. 2007.

[25] E. Agrell, T. Eriksson, A. Vardy, and K. Zeger, "Closest point search in lattices," *IEEE Trans. Inf. Theory*, vol. 48, no. 8, pp. 2201–2214, Aug. 2002.

[26] M. O. Damen, H. E. Gamal, and G. Caire, "On maximum-likelihood detection and the search for the closest lattice point," *IEEE Trans. Inf. Theory*, vol. 49, no. 10, pp. 2389–2401, Oct. 2003.

[27] H. Vikalo, B. Hassibi, and U. Mitra, "Sphere-constrained ML detection for frequency selective channels," *Proceedings of the IEEE International Conference on Acoustics, Speech, Signal Process. (ICASSP)*, pp. IV-1 to IV-4, 2003.

[28] H. Artes, D. Seetheler, and F. Hlaeatsch, "Efficient detection algorithm for MIMO channels: a geometrical approach to approximate ML detection," *IEEE Trans. Signal Process.*, vol. 51, no. 11, pp. 2808–2820, Nov. 2003.

[29] D. Tse and P. Viswanath, *Fundamentals of Wireless Communication*, Cambridge University Press, Cambridge, UK, 2005.

[30] A. Paulraj, R. Nabar, and D. Gore, *Introduction to Space-Time Wireless Communications*, Cambridge University Press, Cambridge, UK, 2003.

[31] S. Haykin and M. Moher, *Modern Wireless Communications*, Prentice Hall, Upper Saddle River, NJ, 2005.

CHAPTER 3

ULTRAWIDEBAND

3.1 INTRODUCTION

Ultrawideband (UWB) systems have been receiving a lot of attention as they promise a *high data rate* with simple, low-cost, low-power transmission for short-range communications required by wireless personal area networks to ad hoc networks. Such UWB networks are attractive because they can provide portability and support for personal, integrated mobile multimedia services. A wireless mobile multimedia UWB network has to provide a reasonable user-transparent quality of service (QoS) for a variety of service classes. Research works in UWB systems escalated when the Federal Communications Commission (FCC) recently permitted unlicensed operation in the 3.1- to 10.6-GHz band. Thus, low-power unlicensed users can make use of the licensed spectrum on a noninterfering basis.

UWB is a radio technology, and traditional UWB is pulse-based. A transmitted UWB pulse is very narrow in the time domain. Thus, it gives rise to an ultrawideband signal in the frequency domain. Narrow UWB pulses are the building blocks for traditional UWB systems. A basic UWB system uses time-hopping (TH) pulse position modulation (PPM) codes; a slightly more advanced UWB system uses direct-sequence (DS) spreading codes. The codes enable a UWB system to distinguish between its different users. The UWB signal received is correlated with a template signal. Correlation is a mathematical operation that provides a measure of the similarity of the UWB signal received and the template signal. The two signals have

Wireless Broadband Networks, By David Tung Chong Wong, Peng-Yong Kong, Ying-Chang Liang, Kee Chaing Chua, and Jon W. Mark

a strong resemblance if the output value of the correlation function is large. In this way, the signal transmitted can be recovered. It is noted that a DS-UWB system can support more users than can a TH-PPM UWB system.

According to FCC UWB rulings, a signal is considered as UWB if the signal bandwidth is 500 MHz or more, or if the fractional bandwidth has a limit at the minimum of 20%. Fractional bandwidth is defined as $2(f_H - f_L)/(f_H + f_L)$, where f_H and f_L are, respectively, the higher and lower −3-dB points in the spectrum. Another UWB technology that satisfies FCC requirement is multiband orthogonal frequency-division multiplexing (OFDM). Basically, the bandwidth of UWB is divided into bands and OFDM operates in each band of 528 MHz. The signal hops from band to band over a number of bands in a cyclic manner. The main advantages of the multiband technique are lower design complexity and reduction of power consumption over a smaller bandwidth, and spectral flexibility that enables coexistence with existing services. The WiMedia Alliance uses a multiband OFDM technique. WiMedia has standardized a physical layer and a medium access control protocol for a wireless personal area network (WPAN) that operates in a short-range (up to 10 m) environment. This standard is supported by Intel, NXP, Texas Instruments, STMicroelectronics, and many other companies.

There are also other types of UWB systems in the literature. They are transmitted reference (TR) UWB, chirp UWB, multicarrier (MC) UWB, and multiple-input, multiple-output (MIMO) UWB systems. TR UWB transmits a first UWB pulse to serve as the template signal for the second pulse transmitted. The first UWB pulse is unmodulated, and the second UWB pulse is modulated. In this way, no template signal needs to be defined. However, bandwidth efficiency drops by half, as the first UWB pulse does not convey data information. Chirp UWB, which is gaining interest, uses chirp waveforms. A *chirp* is a signal that increases or decreases with frequency in time. The instantaneous frequency in a linear chirp varies linearly with time. Its advantages include easy generation of UWB signals (using a simple circuit), flexibility for meeting the spectrum mask, easy implementation of subband systems and extreme robustness against multipath delay spread. MC UWB has its signals placed at multiple carriers in the frequency domain. MIMO UWB uses multiple transmit and receive antennas to improve the signal-to-interference-noise ratio (SINR). The larger the SINR, the better the signal quality received. With MIMO UWB, the gain in SINR is increased by a factor of the product of the number of transmit and receive antennas.

For consumer communications applications, UWB's advantages include low complexity and lost cost, the ability to operate below the noise floor of traditional narrowband systems, resistantce to severe multipath and jamming, and very good time-domain resolution, suitable for location and tracking applications. *Multipath* is a phenomenon whereby a signal reaches the receive antenna in two or more paths, which can be destructive or constructive. The very narrow pulses of UWB make them less sensitive to the multipath effect. UWB also has very good penetration properties that can be used for imaging.

Applications in UWB systems include connecting a personal computer or laptop to a printer, a storage device, a mobile phone, an MP3/4 player, a television set, and so on. High-data-rate connectivity without wires is a big advantage of WiMedia and wireless universal serial bus (WUSB) based on WiMedia offers a higher rate than Bluetooth and ZigBee, whose data rates are only 1 Mbps and 250 kbps, respectively.

Two of the first important seminal papers on UWB are those of Win and Scholtz [1] and Scholtz [2]. In [1], the performance of a TH-PPM UWB system under an additive white Gaussian noise (AWGN) channel is analyzed. A second derivative Gaussian pulse is used. Due to the narrow pulse in the time domain, an ultrawideband signal is produced in the frequency domain. A single-class traffic with continuous transmission is assumed in the paper. This important work is extended to multiclass traffic with variable-bit-rate transmissions which can represent voice, video, or data traffic in [3]. Closed-form SINR, outage probability, and system capacity are derived in the paper. Jia and Kim [4] analyze the performance of a TH-PPM UWB system under a UWB indoor channel model with single-class traffic with continuous transmission. This channel model is the IEEE 802.15.3a UWB channel model [5]. Extension of this work to multiclass traffic with variable-bit-rate transmissions is analyzed in [6].

An important paper on DS-UWB is that of Boubaker and Letaief [7]. In the paper, the performance of a DS-UWB system under an AWGN channel is analyzed. The results show that DS-UWB can achieve better system capacity than TH-PPM UWB for single-class traffic. The performance under multiclass traffic with variable-bit-rate transmission over an AWGN channel is analyzed in [8] using a multicode DS-UWB system. Extensions of these works [7,8] using the IEEE 802.15.3a UWB channel model for both single-class traffic with continuous transmissions and multiclass traffic with variable-bit-rate transmissions are analyzed in [9].

Lai [10] studied the performance of a multiband OFDM UWB system. Other techniques include TR-UWB in [11], chirp UWB in [12,13], MC-UWB in [14,15] and MIMO UWB.

This chapter is organized as follows. In Section 3.1 we give an overview of the UWB systems to be presented in this chapter. In Sections 3.2 and 3.3 we present performance metrics such as signal-to-interference-noise ratio (SINR), outage probability, and system capacity for TH-PPM-UWB and DS-UWB systems, respectively. Both single class with continuous transmission and multiclass variable-bit-rate transmissions under both idealized multiple access channel and IEEE 802.15.3a indoor UWB channel models are included. In Section 3.4 we present the signal-to-noise ratio (SNR) for a multiband OFDM UWB system; in Section 3.5 we consider signal-to-interference-noise ratio for a TR-UWB system, the MC-UWB system, and the signal-to-interference-noise ratio for a MIMO UWB system. Both TH-PPM UWB and DS-UWB systems can be considered for a simple MIMO architecture. The chapter summary includes extensive references to articles in the literature on analytical formulations of UWB systems.

3.2 TIME-HOPPING ULTRAWIDEBAND

TH-PPM codes are used in most basic UWB systems. These codes enable any user using them to be distinguishable from other users. As mentioned earlier, the UWB signal received is correlated with a template signal. Correlation is a mathematical operation that provides a measure of the similarity of the TH-PPM UWB signal received and the template signal. The value of the correlation function is large when the two signals have a strong resemblance. On the other hand, they have a low correlation when the two signals have weak resemblance. Using correlation, the signal transmitted can be recovered.

3.2.1 Idealized Multiple-Access Channel

3.2.1.1 Single-Class Traffic with Continuous Transmission Here we are concerned with the SINR for a TH-PPM UWB system under an idealized multiple-access channel. That is, we do not consider the multipath phenomenon in UWB. The analysis here is similar to that of Win and Scholtz [1].

The typical transmitted signal of the vth user for TH-PPM UWB can be expressed as

$$s_{\text{tr}}^{(v)}(t) = \sum_{j=-\infty}^{\infty} \omega\big(t - jT_f - c_j^{(v)}T_c - \delta d_{\lfloor j/N_s \rfloor}^{(v)}\big), \tag{3.1}$$

where T_f is the frame time interval, T_c is the chip interval, N_c is the number of chips per frame time interval of T_f such that $N_c = T_f/T_c$, N_s is the number of frames per symbol, $\omega(t)$ is the pulse transmitted, $\{c_j^{(v)}\}$ is the time-hopping value for the jth frame of the vth user depending on the time-hopping code, δ is the modulation index or the time shift associated with the binary PPM modulation, $\{d_{\lfloor j/N_s \rfloor}^{(v)}\}$ is the user-specific data sequence in a data bit for the vth user, $d_{\lfloor j/N_s \rfloor}^{(v)} \in \{0, 1\}$. Note that sometimes it is more convenient to use the unit-energy transmitted pulse, denoted by $w(t)$, instead of the transmitted pulse $\omega(t)$. The signals are encoded by positions of the pulses, each with a certain chip interval. A pulse with a position that has no time shift denotes a '0', and a pulse with a shifted pulse time position of δ denotes a '1'. Figure 3.1 shows the pulse signals of two users using a TH-PPM UWB system.

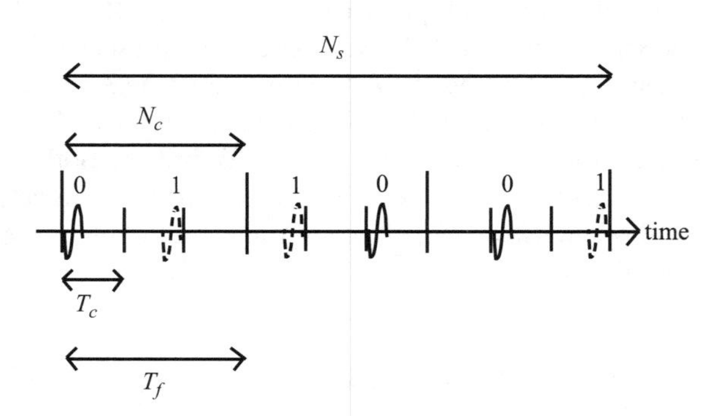

User 1 transmitting bit '0' with time-hopping code {1,3,2}
User 2 transmitting bit '1' with time-hopping code {2,1,3}

FIGURE 3.1 Time-hopping pulse position modulation UWB system.

N_s pulses are used to represent a data bit and there are N_c chips per frame interval, T_f. Each chip time has a time interval of T_c. The solid-line pulses are for user 1; the dashed-line pulses are for user 2. User 1 is transmitting a data bit '0' and thus there is no shift in any of its N_s pulses with respect to their chip intervals. User 1 is also using a time-hopping code of $\{1,3,2\}$. In the first frame, user 1 transmits a '0' pulse in chip 1 of the frame; in the second frame, user 1 transmits a '0' pulse in chip 3 of the frame; and in the third frame, user 1 transmits a '0' pulse in chip 2 of the frame. On the other hand, user 2 is transmitting a data bit of '1' and thus there are time shifts of δ in all of its N_s pulses with respect to their chip intervals. User 2 is using a time-hopping code of $\{2,1,3\}$. Thus, in the first frame, user 2 transmits a '1' pulse in chip 2 of the frame; in the second frame, user 2 transmits a '1' pulse in chip 1 of the frame; and in the third frame, user 2 transmits a '1' pulse in chip 3 of the frame. In this example, each user uses a single time-hopping code. A user can also use multiple time-hopping codes in multicode TH-PPM UWB systems described in Sections 3.2.1.2 and 3.2.2.3, respectively.

A correlation receiver is used for simplicity. This receiver is optimum if there is only one user. Assuming that a '0' is transmitted by the nth user, the received signal is given by

$$r(t) = \sum_{j=0}^{N_S-1} A_n \omega_{\text{rec}}^{(n)}\left(t - jT_f - c_j^{(n)} T_c\right) + \sum_{v=1}^{n-1} \sum_{j=-\infty}^{\infty} A_v \omega_{\text{rec}}^{(v)}\left(t - jT_f - c_j^{(v)} T_c - \delta d_{\lfloor j/N_s \rfloor}^{(v)} - \tau^{(v)}\right) + n(t), \quad (3.2)$$

where n is the number of users in the system, $\omega_{\text{rec}}(t)$ is the received pulse of the transmitted pulse $\omega(t)$, $\tau^{(v)}$ is the reference delay of the vth user relative to the nth user due to the asynchronous time difference, and A_v is the attenuation over the propagation path of the signal received from the vth user. $n(t)$ is the AWGN with two-sided power spectral density $N_0/2$. We assume perfect clock and sequence synchronization for the signal transmitted by the nth user and $\tau^{(n)} = 0$. The receiver is a correlation receiver with a template signal $v(t)$, where $v(t) = \omega_{\text{rec}}(t) - \omega_{\text{rec}}(t - \delta)$.

The output of the correlator, Z, is

$$Z = \sum_{j=0}^{N_S-1} \int_{\tau_i^{(n)}+jT_f}^{\tau_i^{(n)}+(j+1)T_f} r(t) v\left(t - jT_f - c_j^{(n)} T_c - \tau^{(n)}\right) dt, \quad (3.3)$$

Z is the decision statistic and $d_j^{(v)} = 0$ is decided if the decision statistic is $Z > 0$. Using the technique in [1], the output SINR can be shown to be [1]

$$\text{SINR} = \frac{S^2}{\text{MAI} + \sigma_{\text{rec}}^2}, \quad (3.4)$$

where the desired signal term is

$$S = N_s A_n m_p, \tag{3.5}$$

the multiple-access interference (MAI) term is

$$\text{MAI} = \sum_{v=1}^{n-1} N_s \sigma_a^2 A_v^2, \tag{3.6}$$

and the received noise term is

$$\sigma_{\text{rec}}^2 = N_0 N_s m_p, \tag{3.7}$$

m_p is the signal at the correlator's output during a frame interval, σ_a^2 is the power of the interference resulting from one of the active time-hopping sequences from the interfering users on one pulse and $N_0/2$ is the double-sided power spectral density of AWGN. The MAI term is due to $(n-1)$ other users in the system, excluding the user under consideration.

The SINR can be written as (similar to [1])

$$\text{SINR} = \frac{(N_s A_n m_p)^2}{\sum_{v=1}^{n-1} N_s \sigma_a^2 A_v^2 + \sigma_{\text{rec}}^2}. \tag{3.8}$$

N_s is the number of impulses dedicated to the transmission of one symbol and is given by

$$N_s = \frac{1}{R_s T_f}, \tag{3.9}$$

where R_s is the symbol data transmission rate and T_f is the frame time interval or pulse repetition time. The frame time interval is given by

$$T_f = N_c T_c, \tag{3.10}$$

where N_c is the number of time delay bins in a frame time T_f, and T_c is the basic chip time. m_p is given by [1]

$$m_p = \int_{-\infty}^{\infty} \omega_{\text{rec}}(t - \delta) v(t)\, dt, \tag{3.11}$$

where $\omega_{\text{rec}}(t)$ is a typical idealized received monocycle at the output of the antenna subsystem and δ is a delay parameter. If δ is greater than the monocycle waveform's width, the design corresponds to orthogonal signaling [2]. Using $\omega_{\text{rec}}(t)$ from [1] as

$$\omega_{\text{rec}}(t+0.35) = \left[1 - 4\pi\left(\frac{t}{\tau_m}\right)^2\right] \exp\left[-2\pi\left(\frac{t}{\tau_m}\right)^2\right], \tag{3.12}$$

it can be derived that [3]

$$m_p = \left[\frac{\pi^2\delta^4}{2\tau_m^3} - \frac{3\pi\delta^2}{2\tau_m} + \frac{3\tau_m}{8}\right] \exp\left[-\frac{\pi\delta^2}{\tau_m^2}\right] - \frac{3\tau_m}{8}. \tag{3.13}$$

The signal power of a spreading chip, σ_a^2, is given by

$$\sigma_a^2 = \frac{1}{T_f} \int_{-\infty}^{\infty} \left[\int_{-\infty}^{\infty} \omega_{\text{rec}}(t-s) v(t)\, dt\right]^2 ds. \tag{3.14}$$

Using (3.13), it can be derived that [3]

$$\begin{aligned}\sigma_a^2 = \frac{1}{T_f} \Bigg\{ \frac{105\tau_m^3}{512\sqrt{2}} + \Bigg[-\frac{\pi^4\delta^8}{512\sqrt{2}\,\tau_m^5} + \frac{7\pi^3\delta^6}{128\sqrt{2}\,\tau_m^3} - \frac{105\pi^2\delta^4}{256\sqrt{2}\,\tau_m} \\ + \frac{105\pi\tau_m\delta^2}{128\sqrt{2}} - \frac{105\tau_m^3}{512\sqrt{2}} \Bigg] \exp\left[-\frac{\pi\delta^2}{2\tau_m^2}\right] \Bigg\}. \end{aligned} \tag{3.15}$$

The last term in the denominator of (3.8), σ_{rec}^2, is due to thermal noise and is given by (3.7).

3.2.1.2 Multiclass Traffic with Variable-Bit-Rate Transmission We are concerned with the capacity of a multicode TH-PPM UWB system using multiple correlation receivers to support the services of K classes of users. The basic system considered here is similar to that in Section 3.2.1.1 except that multiple spreading codes are used by each user for varying-bit-rate traffic rather than a single spreading code per user. Figure 3.2 shows the pulse signals of two users using a multiple-TH-PPM UWB system. N_s pulses are used to represent a data bit, and there are N_c chips per frame interval T_f. Each chip time has a time interval of T_c. The solid-line pulses are for user 1; the dashed-line pulses are for user 2. When user 1 is transmitting a data bit '0', there is a shift in all its N_s pulses with respect to their chip intervals. However, there are shifts in the pulses when user 1 is transmitting a '1'. User 1 is using two time-hopping codes of {1,3,2} and {3,4,1}. In the first, second, and third frames, user 1 transmits a '0' pulse in chips 1, 3, and 2 of the respective frame using the first code. Similarly, user 1 transmits a '1' pulse in chips 3, 4, and 1 of the respective frame with shifts in the pulses using the second code as shown by the thicker solid-line pulses. On the other hand, user 2 is transmitting a data bit of '1'

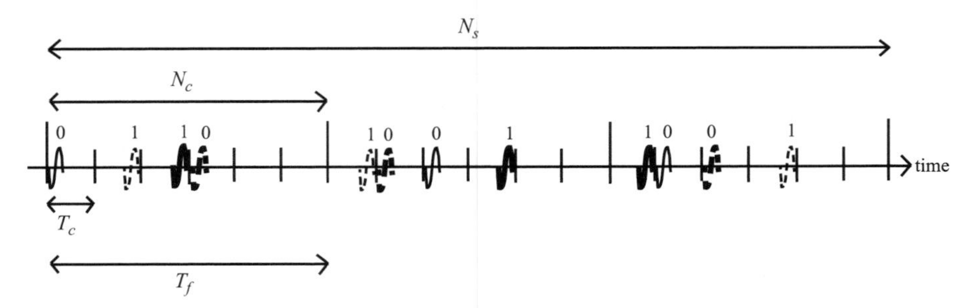

User 1 transmitting bit '0' with time-hopping code {1,3,2} and transmitting bit '1' with time-hopping code {3,4,1}
User 2 transmitting bit '1' with time-hopping code {2,1,4} and transmitting bit '0' with time-hopping code {4,2,3}

FIGURE 3.2 Multiple-time-hopping pulse position modulation UWB system.

and a data bit of '0' using time-hopping codes of {2,1,4} and {4,2,3}, as shown by the dashed and thicker dashed-line pulses, respectively.

From [16–18], a variable-bit-rate source can be modeled by a continuous-time Markov chain with finite states. Each state represents the discrete level of bit rate generated by a single source. We assume that the highest level is state M_i, and this is also matched to the maximum number of active time-hopping sequences used by a class i user. That is, we assume that each level uses one time-hopping sequence for a class i user. This means that each level has a data rate of R_s, corresponding to one class i time-hopping sequence. If $M_i = 1$, the source is an on/off source. Each level can be modeled by a two-state minisource with an increase rate of α_i and a decrease rate of β_i. The Markov chain for this minisource is shown in Figure 3.3. Thus, the continuous-time Markov chain for a single source at state m has an increase rate of $(M_i - m)\alpha_i$ and a decrease rate of $m\beta_i$. This Markov chain is shown in Figure 3.4. The steady-state probability of being in state m, denoted by P_m, is given by

$$P_m = \binom{M_i}{m} (p_i)^m (1 - p_i)^{M_i - m}, \quad m = 0, 1, 2, \ldots, M_i, \tag{3.16}$$

where

$$p_i = \frac{\alpha_i}{\alpha_i + \beta_i}, \tag{3.17}$$

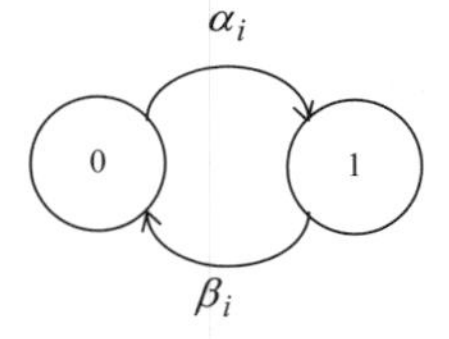

FIGURE 3.3 Two-state minisource. (From [6] © 2006, IEEE.)

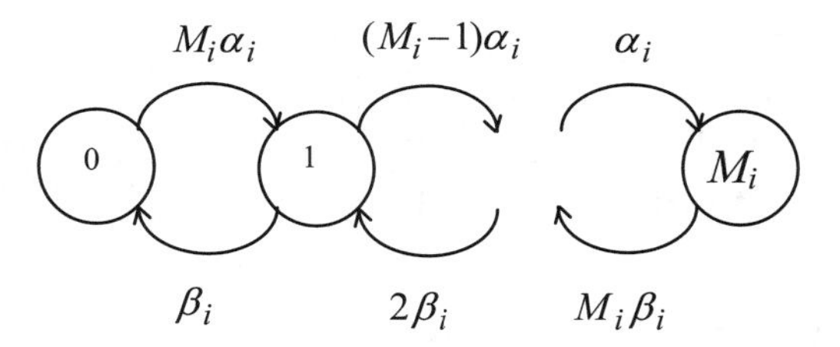

FIGURE 3.4 Variable-bit-rate source. (From [6] © 2006, IEEE.)

and its mean, second moment, and variance are $M_i p_i$, $M_i p_i[1 + (M_i - 1)p_i]$, and $M_i p_i[1 - p_i]$, respectively.

Next, let SINR_i denote the signal-to-interference-noise ratio for class i. By extending the results from Section 3.2.1.1, it is given by [3]

$$\text{SINR}_i = \frac{\left(N_s A_{in_i} m_p\right)^2}{\sum_{j=1}^{n_i-1} \psi_{ij} N_s \sigma_a^2 A_{ij}^2 + \sum_{\substack{k=1\\k\neq i}}^{K} \sum_{j=1}^{n_k} \psi_{kj} N_s \sigma_a^2 A_{kj}^2 + \sigma_{\text{rec}}^2}, \qquad i = 1, 2, \ldots, K, \tag{3.18}$$

where n_i is the number of class i users, K is the number of classes, and $\psi_{ij} \in \{0, 1, 2, \ldots, m, \ldots, M_i\}$ is a binomial random variable indicating the number of active time-hopping sequences used by the jth user of class i. The probability that m active time-hopping sequences are used by a source, denoted by $\Pr[\psi_{ij} = m]$, is given by

$$\Pr[\psi_{ij} = m] = P_m, \qquad m = 0, 1, 2, \ldots, M_i. \tag{3.19}$$

A_{ij} is the attenuation over the propagation path of the signal received from the jth user of class i. The first and second terms in the denominator of (3.18) are due to MAI. The MAI terms are due to the other users in class i and the users in the other classes.

Let BER_i^* denote the bit error rate (BER) requirement and SINR_i^* denote the SINR requirement for class i users. The system capacity is defined as the maximum $(n_1, \ldots, n_i, \ldots, n_K)$ that can be supported such that the SINR achieved is greater than or equal to the SINR_i^* required 99% of the time for all classes. That is, the outage probability is defined as

$$\begin{aligned}\Pr[\text{BER}_i \geq \text{BER}_i^*] &= \Pr[\text{SINR}_i \leq \text{SINR}_i^*] \\ &= \Pr\left[\sum_{j=1}^{n_i-1} \psi_{ij} \frac{A_{ij}^2}{A_{in_i}^2} + \sum_{\substack{k=1\\k\neq i}}^{K} \sum_{j=1}^{n_k} \psi_{kj} \frac{A_{kj}^2}{A_{in_i}^2} \geq \delta_i\right],\end{aligned} \tag{3.20}$$

where

$$\delta_i = \frac{N_s m_p^2}{\mathrm{SINR}_i^* \sigma_a^2} - \frac{\sigma_{\mathrm{rec}}^2}{N_s \sigma_a^2 A_{in_i}^2}, \qquad i = 1, 2, \ldots, K. \tag{3.21}$$

Assuming perfect power control, we have

$$A_{ij} = A_{kj} = A_{in_i}. \tag{3.22}$$

Thus, the outage probability is given by

$$\Pr[\mathrm{BER}_i \geq \mathrm{BER}_i^*] = \Pr\left[\sum_{j=1}^{n_i-1} \psi_{ij} + \sum_{\substack{k=1 \\ k \neq i}}^{K} \sum_{j=1}^{n_k} \psi_{kj} \geq \delta_i\right]. \tag{3.23}$$

The probability that l_i active time-hopping sequences are used by n_i class i sources, denoted by $\Pr[\phi_i = l_i]$, is given by

$$\Pr[\phi_i = l_i] = \binom{M_i n_i}{l_i} (p_i)^{l_i} (1 - p_i)^{M_i n_i - l_i}, \quad l_i = 0, 1, 2, \ldots, M_i n_i. \tag{3.24}$$

Using the central limit approximation and solving (3.23) by conditioning on the active time-hopping sequences used and then unconditioning the probability in (3.23) by summing up all cases for the numbers of active time-hopping sequences used of all classes, and multiplying by all the corresponding binomial probabilities of active time-hopping sequences used, we have

$$\begin{aligned}
&\Pr[\mathrm{BER}_i \geq \mathrm{BER}_i^*] \\
&\quad = \sum_{l_1=0}^{M_1 n_1} \cdots \sum_{l_i=0}^{M_i(n_i-1)} \cdots \sum_{l_K=0}^{M_K n_K} \Pr\left[l_i + \sum_{\substack{k=1 \\ k \neq i}}^{K} l_k \geq \delta_i \,\middle|\, \sum \psi_{1j} = l_1, \sum \psi_{2j} = l_2, \ldots, \sum \psi_{Kj} = l_K \right] \\
&\quad \times \prod_{k=1}^{K} \Pr\left[\sum \psi_{kj} = l_k\right] \\
&\quad = \sum_{l_1=0}^{M_1 n_1} \cdots \sum_{l_i=0}^{M_i(n_i-1)} \cdots \sum_{l_K=0}^{M_K n_K} \Pr\left[\sum_{k=1}^{K} l_k \geq \delta_i \,\middle|\, \sum \psi_{1j} = l_1, \sum \psi_{2j} = l_2, \ldots, \sum \psi_{Kj} = l_K \right] \\
&\quad \times \binom{M_i(n_i - 1)}{l_i} p_i^{l_i} (1 - p_i)^{M_i(n_i-1)-l_i} \prod_{\substack{k=1 \\ k \neq i}}^{K} \binom{M_k n_k}{l_k} p_k^{l_k} (1 - p_k)^{M_k n_k - l_k},
\end{aligned} \tag{3.25}$$

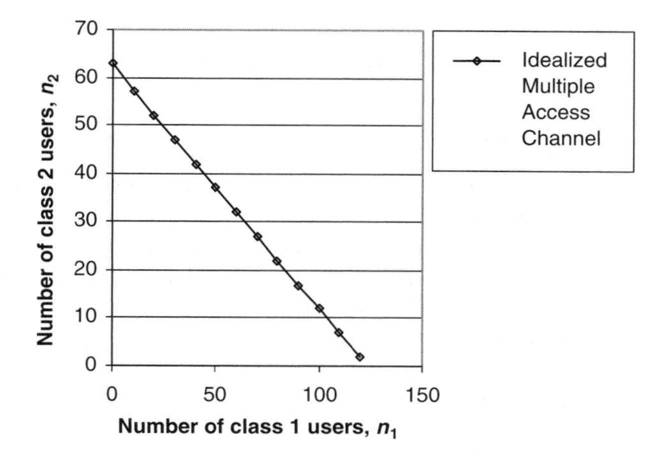

FIGURE 3.5 System capacity of an idealized multiple-access channel for MTH-PPM UWB.

where

$$\Pr\left[\sum_{k=1}^{K} l_k \geq \delta_i \mid \sum \psi_{1j} = l_1, \sum \psi_{2j} = l_2, \ldots, \sum \psi_{Kj} = l_K\right] = \begin{cases} 1 \text{ if } \sum_{k=1}^{K} l_{k1} \geq \delta_i \\ 0 \text{ if } \sum_{k=1}^{K} l_k < \delta_i. \end{cases} \tag{3.26}$$

We present results for the system capacity with two classes ($K = 2$) as an illustrative example. The parameter values used in the numerical examples are tabulated in Table 3.1. The admissible region for the system capacity is shown in Figure 3.5, where the capacities (n_1, n_2) of the system are on a "line." That is, the elements of the doublet (n_1, n_2) are bounded by this line. Thus, the combinations of the numbers of users of different classes that can be admitted to the system are possible only when these numbers are on or below this line. Numerical results can be obtained for K

TABLE 3.1 Parameter Values Used for TH-PPM UWB

Symbol	Value	Symbol	Value
M_1	1	$\Delta\tau$	1 ns
M_2	2	R_s	1.3333 Mbps
α_1	0.9	T_f	250 ns
β_1	0.1	δ	0.156 ns
α_2	0.9	τ_m	0.2877 ns
β_2	0.1	$\text{BER}_1^* = \text{BER}_2^* = \text{BER}^*$	10^{-8}
N_{tot}	100	$\text{SINR}_1^* = \text{SINR}_2^* = \text{SINR}^*$	15.0 dB
L_p	5, 20, 40, 100	σ_{rec}^2	0 (assumed negligible)

traffic classes, and the system capacity is in a K-dimensional space. $(n_1, \ldots, n_K)$ are admissible as long as they are within its system capacity.

3.2.2 Multipath Channel

In the multipath phenomenon, a signal reaches the receive antenna in two or more paths, resulting in either destructive or constructive interference. The very narrow pulses of UWB make them less sensitive to the multipath effect. One important aspect of UWB systems is the investigation of the distribution of channel parameters in a UWB channel. From this investigation, the impulse response of the channel can be obtained and the performance of different UWB systems can then be studied under this channel model. One of the channel models that has been investigated and studied is the IEEE 802.15.3a UWB indoor channel model. This channel model is described in the following subsection.

3.2.2.1 Channel Model The UWB channel under consideration is characterized by the following impulse response [5]:

$$h(t) = \sum_{l=1}^{L} \sum_{g=1}^{G} \alpha_{g,l} \delta(t - T_l - \tau_{g,l}), \tag{3.27}$$

where $\alpha_{g,l}$ is the multipath gain coefficient of the gth ray in the lth cluster, T_l is the arrival time of the first path of the lth cluster, and $\tau_{g,l}$ is the delay of the gth path within the lth cluster relative to the first path arrival time, T_l. Both the cluster and the ray arrival times are independently modeled by Poisson processes. The intercluster and interray arrival times are independent, exponentially distributed with a mean cluster arrival rate of Λ and a mean ray arrival rate of λ. That is,

$$\Pr[T_l | T_{l-1}] = \Lambda \exp\left[-\Lambda(T_l - T_{l-1})\right], \quad l > 0, \tag{3.28}$$

and

$$\Pr[\tau_{g,l} | \tau_{(g-1),l}] = \lambda \exp\left[-\lambda(\tau_{g,l} - \tau_{(g-1),l})\right], \quad g > 0. \tag{3.29}$$

$\alpha_{g,l}$ is given by $\alpha_{g,l} = p_{g,l}\beta_{g,l}$, where $p_{g,l}$ is equiprobable to the values ± 1 to account for signal polarity inversion due to reflections, and $\beta_{g,l}$ is a lognormal random variable such that $20\log_{10}(\beta_{g,l}) \propto N(\mu_{g,l}, \sigma^2)$. The exponential power decay for the amplitude of the cluster and for the amplitude of the multipath ray within the cluster is doubly exponentially decaying and is given by

$$E\left[\beta_{g,l}^2\right] = \Omega_0 \exp\left[-T_l/\Gamma\right] \exp\left[-\tau_{g,l}/\gamma\right], \tag{3.30}$$

where Ω_0 is the mean power of the first ray of the first cluster. Γ and γ are the power decay factors of the cluster and ray, respectively. The parameter settings for

TABLE 3.2 Parameter Settings for the IEEE 802.15.3a Indoor UWB Channel Model1

Scenario	CM1 LOS (0–4 m)	CM2 NLOS (0–4 m)	CM3 NLOS (4–10 m)	CM4 NLOS Extreme NLOS Multipath Channel
Λ (ns^{-1})	0.0233	0.4	0.0667	0.0667
λ (ns^{-1})	2.5	0.5	2.1	2.1
Γ	7.1	5.5	14	24
γ	4.3	6.7	7.9	12

the various scenarios in the IEEE 802.15.3a indoor UWB channel model are shown in Table 3.2.

3.2.2.2 Single-Class Traffic with Continuous Transmission

Here we consider the SINR for a TH-PPM UWB system under a UWB multipath channel. The analysis here is similar to that of Jia and Kim [4]. The RAKE receiver is used here to collect the multipath components. Assuming TH-BPPM, the typical transmitted signal of the vth user can be expressed as [4]

$$s_{\text{tr}}^{(v)}(t) = \sum_{j=-\infty}^{\infty} \sqrt{E_Z},\, w\big(t - jT_f - c_j^{(v)}T_c - \delta d_{\lfloor j/N_s \rfloor}^{(v)}\big), \tag{3.31}$$

where T_f is the frame time interval, T_c is the chip interval, N_c is the number of chips per frame time interval of T_f such that $N_c = T_f/T_c$, N_s is the number of frames per symbol, E_z is the transmitted pulse energy, $w(t)$ is the unit-energy transmitted pulse, $c_j^{(v)}$ is the time-hopping value for the jth frame of the vth user depending on the time-hopping code, δ is the modulation index or the time shift associated with the binary PPM modulation, and $\{d_{\lfloor j/N_s \rfloor}^{(v)}\}$ is the user-specific data sequence in a symbol for the vth user, and $d_{\lfloor j/N_s \rfloor}^{(v)} \in \{0, 1\}$. Assuming that bit '0' is transmitted, the received signal is given by [8]

$$\begin{aligned} r(t) = &\sum_{j=0}^{N_S-1} \sqrt{E_z} g^{(n)} \left(t - jT_f - c_j^{(n)}T_c\right) + \sum_{v=1}^{n-1} \sum_{j=-\infty}^{\infty} \sqrt{E_z} g^{(v)} \\ &\times \big(t - jT_f - c_j^{(v)}T_c - \delta d_{\lfloor j/N_s \rfloor}^{(v)} - \tau_0^{(v)}\big) + n(t), \end{aligned} \tag{3.32}$$

where n is the number of users in the system, $\tau_0^{(v)}$ is the reference delay of the vth user relative to the nth user due to the asynchronous time difference, $g^{(v)}(t) = w_{\text{rec}}(t) \otimes h^{(v)}(t)$, $v = 1, 2, \ldots, n_u$, $w_{\text{rec}}(t)$ is the received pulse, $h^{(v)}(t)$ is the channel impulse response for the vth user, and $\otimes$ denotes convolution. $n(t)$ is the AWGN with two-sided power spectral density $N_0/2$. We assume perfect clock and sequence synchronization for the signal transmitted by the nth user.

Let L_p be the number of RAKE fingers and Z_i be the output of the ith RAKE finger:

$$Z_i = \sum_{j=0}^{N_S-1} \int_{\tau_i^{(n)} jT_f}^{\tau_i^{(n)}+(j+1)T_f} r(t)v\big(t - jT_f - c_j^{(n)}T_c - \tau_i^{(n)}\big)\, dt, \tag{3.33}$$

where $v(t) = w_{\text{rec}}(t) - w_{\text{rec}}(t-\delta)$, $\tau_i^{(v)}$ is the ith time bin of the vth user. The output of the maximal ratio combiner is given by

$$Z = \sum_{i=1}^{L_p} \sqrt{E_z} A_i^{(n)} Z_i, \tag{3.34}$$

where $A_i^{(v)}$ is the sum of the channel coefficients of all multipath components that arrived in the ith time bin. Z is the decision statistic and $d_j^{(v)} = 0$ is decided if the decision statistic is $Z > 0$. Using the techniques described in [1,2,4], the *instantaneous* output signal-to-interference noise ratio (SINR) conditioned on all the channel coefficients of all users can be shown to be [4]

$$\text{SINR}(\{A_m^{(v)}\}) = \frac{S^2}{E_z^2 N_s T_f^{-1} \sum\limits_{v=1}^{n-1} G_{\text{eff}}^{(v)^2} + \sigma_{\text{rec}}^{2'}}, \tag{3.35}$$

where

$$S = N_s m_p \sum_{i=1}^{L_p} A_i^{(n)^2}, \tag{3.36}$$

$$G_{\text{eff}}^{(v)^2} = \sum_{i=1}^{L_p} \sum_{k=1}^{L_p} \sum_{m=1}^{N_{\text{tot}}} \sum_{q=1}^{N_{\text{tot}}} A_i^{(n)} A_k^{(n)} A_m^{(v)} A_q^{(v)} Q[(i-k-m+q)\Delta\tau], \tag{3.37}$$

$$\sigma_{\text{rec}}^{2'} = N_0 N_s m_p \sum_{i=1}^{L_p} A_i^{(n)^2}. \tag{3.38}$$

In the above equations, m_p is the signal at the correlator's output during a frame interval and is given by (3.13), N_{tot} is the total number of multipath bins considered for each channel realization, $\Delta\tau$ is the duration of a time bin, $Q(\cdot)$ is the autocorrelation of $R(\cdot)$, $R(\cdot)$ is the crosscorrelation function of the received signal of the transmitted pulse $w(t)$ and the template signal $v(t)$, and $N_0/2$ is the double-sided power spectral density of AWGN. Using the techniques in [4], it can be shown that the average SINR is given by

$$\mathrm{SINR} = E\{\mathrm{SINR}(\{A_m^{(v)}\})\} = \frac{(N_s m_p)^2 \Omega_0^{(n)} \overline{E}_{0,X}^{(n)} \big|_{X=L_p}}{\sum\limits_{v=1}^{n-1} N_s \sigma_a^2 + \sigma_{\mathrm{rec}}^2}, \tag{3.39}$$

where σ_a^2 is the power of the interference resulting from one of the active time-hopping sequences from the interfering users on one pulse and is given by (3.15), N_s is the number of impulses dedicated to the transmission of one symbol and is given by $N_s = 1/R_l T_f$, where R_l is the symbol data transmission rate using one time-hopping sequence, and T_f is the frame time interval or the pulse repetition time. The frame time interval is given by $T_f = N_c T_c$, where N_c is the number of time delay bins in a frame time T_f and T_c is the basic chip time. $\Omega_0^{(n)}$ is the normalization factor of user n, and $\overline{E}_{0,X}^{(n)}$ is the average channel energy of user n with X multipaths.

$$\begin{aligned}
\overline{E}_{0,X}^{(n)} &= \overline{E}_{0,X}^{(v)} \\
&= 1 + P_c \frac{\exp[-\Delta\tau/\Gamma] - \exp[-X\Delta\tau/\Gamma]}{1 - \exp[-\Delta\tau/\Gamma]} \\
&\quad + P_r \frac{\exp[-\Delta\tau/\gamma] - \exp[-X\Delta\tau/\gamma]}{1 - \exp[-\Delta\tau/\gamma]} + P_c P_r \frac{\rho^2 \exp[\Delta\tau/\Gamma]}{1-\rho} \\
&\quad \times \left[\frac{\exp[-2\Delta\tau/\gamma] - \exp[-(X+1)\Delta\tau/\gamma]}{1 - \exp[-\Delta\tau/\gamma]} \right. \\
&\quad \left. - \frac{\exp[-2\Delta\tau/\gamma] - \rho^{X-1} \exp[-(X+1)\Delta\tau/\gamma]}{1 - \exp[-\Delta\tau/\Gamma]} \right],
\end{aligned} \tag{3.40}$$

$$\Omega_0^{(n)} = \Omega_0^{(v)} = \frac{1}{\overline{E}_{0,X}^{(n)} \big|_{X=N_{\mathrm{tot}}}} = \frac{1}{\overline{E}_{0,X}^{(v)} \big|_{X=N_{\mathrm{tot}}}} \tag{3.41}$$

$\Omega_0^{(v)}$ is the normalization factor of user v. N_{tot} is the total number of multipaths, and $L_p (\leq N_{\mathrm{tot}})$ is the number of first multipaths. Note that the total average path energies for user v, $E[\sum_{l=1}^{L} \sum_{g=1}^{G} |\alpha_{g,l}^{(v)}|^2] = \Omega_0^{(v)} \overline{E}_{0,X}^{(v)} |_{X=N_{\mathrm{tot}}}$, are normalized to 1. $\alpha_{g,l}^{(v)}$ is the gain coefficient of the gth ray in the lth cluster for user v. Approximating the Poisson processes for the cluster and ray arrival rates by binomial processes, the probability that there is one cluster arrival in a time bin period of $\Delta\tau$ is given by $P_c = \Lambda \Delta\tau$, and the probability that there is one ray arrival in a time bin period of $\Delta\tau$ is given by $P_r = \lambda \Delta\tau$. ρ is defined as $\rho = \exp[\Delta\tau/\gamma - \Delta\tau/\Gamma]$. The term due to additive noise is $\sigma_{\mathrm{rec}}^2 = N_0 N_s m_p$.

3.2.2.3 Multiclass Traffic with Variable-Bit-Rate Transmission We are interested in the capacity of a multicode TH-PPM UWB system using a RAKE receiver to support the services of K classes of users. The basic system considered here is similar to that in Section 3.2.2.2 except that multiple time-hopping spreading codes are used by each user for varying-bit-rate traffic rather than a single time-hopping spreading code per user.

By extending the results from [3,4], the *channel-average* SINR for class i, SINR_i, is given by [6]

$$\text{SINR}_i = \frac{(N_s m_p)^2\,\Omega_0^{(in_i)}\overline{E}_{0,X}^{(in_i)}\big|_{X=Lp}}{\sum_{j=1}^{n_i-1}\psi_{ij}N_s\sigma_a^2 + \sum_{k=1,k\neq i}^{K}\sum_{j=1}^{n_k}\psi_{kj}N_s\sigma_a^2 + \sigma_{\text{rec}}^2}, \quad i = 1, 2, \ldots, K, \tag{3.42}$$

where $\Omega_0^{(in_i)}$ is the normalization factor of user n_i of class i and $\overline{E}_{0,X}^{(in_i)}$ is the average channel energy of user n_i of class i with X multipaths.

$$\begin{aligned}
\overline{E}_{0,X}^{(n)} &= \overline{E}_{0,X}^{(v)} = \overline{E}_{0,X}^{(in_i)} = \overline{E}_{0,X}^{(ij)} = \overline{E}_{0,X}^{(kj)} \\
&= 1 + P_c\frac{\exp[-\Delta\tau/\Gamma] - \exp[-X\Delta\tau/\Gamma]}{1-\exp[-\Delta\tau/\Gamma]} \\
&\quad + P_r\frac{\exp[-\Delta\tau/\gamma] - \exp[-X\Delta\tau/\gamma]}{1-\exp[-\Delta\tau/\gamma]} + P_cP_r\frac{\rho^2\exp[\Delta\tau/\Gamma]}{1-\rho} \\
&\quad \times\frac{\exp[-2\Delta\tau/\gamma] - \exp[-(X+1)\Delta\tau/\gamma]}{1-\exp[-\Delta\tau/\gamma]} \\
&\quad - \frac{\exp[-2\Delta\tau/\gamma] - \rho^{X-1}\exp[-(X+1)\Delta\tau/\gamma]}{1-\exp[-\Delta\tau/\Gamma]},
\end{aligned} \tag{3.43}$$

$$\Omega_0^{(n)} = \Omega_0^{(v)} = \frac{1}{\overline{E}_{0,X}^{(n)}\big|_{X=N_{\text{tot}}}} = \frac{1}{\overline{E}_{0,X}^{(v)}\big|_{X=N_{\text{tot}}}}, \tag{3.44}$$

$$\Omega_0^{(in_i)} = \Omega_0^{(ij)} = \Omega_0^{(kj)} = \frac{1}{\overline{E}_{0,X}^{(ij)}\big|_{X=N_{\text{tot}}}} = \frac{1}{\overline{E}_{0,X}^{(kj)}\big|_{X=N_{\text{tot}}}} \tag{3.45}$$

$\Omega_0^{(v)}$ is the normalization factor of user v; $\Omega_0^{(ij)}$ and $\Omega_0^{(kj)}$ are the normalization factors of user j of class i and user j of class k, respectively; N_{tot} is the total number of multipaths; and $L_p(\leq N_{\text{tot}})$ is the number of first multipaths. Note that the total average path energies for user v, $E[\sum_{l=1}^{L}\sum_{g=1}^{G}|\alpha_{g,l}^{(v)}|^2] = \Omega_0^{(v)}\overline{E}_{0,X}^{(v)}|_{X=N_{\text{tot}}}$, for user j of class i, $E[\sum_{l=1}^{L}\sum_{g=1}^{G}|\alpha_{g,l}^{(ij)}|^2] = \Omega_0^{(ij)}\overline{E}_{0,X}^{(ij)}|_{X=N_{\rm tot}}$, and for user j of class k, $E[\sum_{l=1}^{L}\sum_{g=1}^{M}|\alpha_{g,l}^{(kj)}|^2] = \Omega_0^{(kj)}\overline{E}_{0,X}^{(kj)}|_{X=N_{\rm tot}}$, are normalized to 1. $\alpha_{g,l}^{(v)}$, $\alpha_{g,l}^{(ij)}$, and $\alpha_{g,l}^{(kj)}$are the gain coefficients of the gth ray in the lth cluster for user v, user j in class i, and user j in class k, respectively. $\psi_{ij} \in \{0, 1, 2, \ldots, m, \ldots, M_i\}$ is a binomial random variable indicating the number of active spreading codes used by the jth user of class i. The probability that m active spreading codes are used by a source, denoted by $\Pr[\psi_{ij} = m]$, is given by (3.19). N_s is the number of impulses dedicated to the transmission of one symbol and is given by $N_s = 1/R_lT_f$, where R_l is the symbol data transmission rate using one time-hopping sequence, and T_f is the frame time interval or the pulse repetition time. The frame time interval is given by $T_f = N_cT_c$, where N_c is the number of time delay bins in a frame time $T_f$and T_c is the basic chip time. m_p is the signal at the correlator's output during a frame interval and is given by (3.13). σ_a^2 is the power of the interference resulting from one of the active time-hopping sequence from the interfering users on one pulse and is given by (3.15). The first term in the denominator of (3.42) is due to MAI from other users in class i, and the second term in the denominator of (3.42) is due to MAI from users in other classes. The last term in the denominator of (3.42), $\sigma_{\rm rec}^2$, is the power of the thermal noise. Note that when the average channel energy is sufficient with a large number of multipaths equal to the total number of multipaths, $N_{\rm tot}$, $\Omega_0^{(in_i)}\overline{E}_{0,X}^{(in_i)}|_{X=N_{\rm tot}} = 1$ and (3.42) reduces to (3.18) of an idealized multiple-access channel [3] with perfect power control, which is the best result that can be achieved without a multipath channel. Furthermore, note that the analytical results in [4] *agree well* with the simulation results. This justifies the reasonableness of the analysis in [4] and the analytical extension here.

Let BER_i^* denote the BER requirement for class i users and SINR_i^* denote the SINR requirement for class i users. The system capacity is defined as the maximum $(n_1, \ldots, n_i, \ldots, n_K)$ that can be supported such that the SINR achieved is greater than or equal to the SINR_i^* required 99% of the time for all classes. That is, the outage probability is defined as (3.24). However, δ_i is given by

$$\delta_i = \frac{N_s m_p^2 \Omega_0^{(in_i)}\overline{E}_{0,L_p}^{(in_i)}}{\text{SINR}_i^* \sigma_a^2} - \frac{\sigma_{\rm rec}^{2'}}{N_s\sigma_a^2}, \qquad i = 1, 2, \ldots, K. \tag{3.46}$$

Note that (3.46) is the same as (3.21) for an AWGN channel with perfect power control [3] if $\Omega_0^{(in_i)}\overline{E}_{0,X}^{(in_i)}|_{X=N_{\rm tot}} = 1$, where $A_{ij} = A_{kj} = A_{in_i} = 1$.

The probability that l_i active time-hopping sequences are used by n_i class i sources, denoted by $\Pr[\phi_i = l_i]$, is given by (3.24). Solving (3.23) by conditioning on the active time-hopping sequences used and then unconditioning the probability in (3.23) by summing up all cases for the numbers of active time-hopping sequences used by all classes, and multiplying by all the corresponding binomial probabilities of active time-hopping sequences used, we have (3.25) and (3.26).

We present results for the system capacity with two classes ($K = 2$) as an illustrative example. The parameter values used in the numerical examples are

TABLE 3.3 Average Channel Energy for CM1 to CM4

$\overline{E}_{0,X}^{(in_i)}\|_{X=N_{tot}}$.	CM1	CM2	CM3	CM4
$N_{tot} = 100$	12.1734	12.3435	31.4530	63.3426
N_{tot} sufficient	12.1734	12.3435	31.4821	64.6146

tabulated in Table 3.1. A suggestion for computing sufficient number of multipaths is $\lfloor 10(\Gamma + \gamma)/\Delta\tau \rfloor$ [3], where $\Delta\tau$ is the time bin for the multipath arrivals. This is compared with 100 multipaths in Table 3.3, where mostly sufficient channel energy is captured. There are four channel models (CMs) in IEEE 802.15.3a. The admissible region for the system capacity using (3.25) for CM1, CM2, CM3, and CM4 are shown in Figures 3.6, 3.7, 3.8, and 3.9, where the capacities (n_1, n_2) of the system are on the various curves. That is, the elements of the doublet (n_1, n_2) are bounded by these curves. Thus, the combinations of the numbers of users of different classes that can be admitted to the system are possible only when these numbers are on or below these curves. From these numerical results, it can be seen that the system capacity increases with the increase in the number of paths collected, L_p, for each channel model. The larger the number of paths, L_p, the larger the system capacity. From another viewpoint, the SINR of each class improves as the number of multipaths whose energy is collected increases. Thus, the outage probability of each class decreases and therefore the system capacity increases. In general, the system capacity of CM1 is larger than that of CM2, which in turn is larger than that of CM3, which in turn is larger than that of CM4 if the same number of multipaths, L_p, is collected. The differences in system capacity for CM1 to CM4 are due to their different multipath characteristics. At $L_p = N_{tot} = 100$ (all RAKEs), the system capacities of CM1, CM2, CM3, and CM4 are approximately the same, assuming that most of the channel energies are captured.

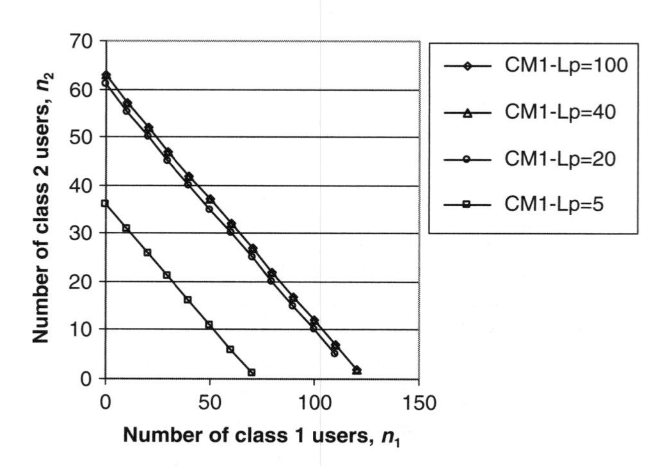

FIGURE 3.6 System capacity of CM1 for MTH-PPM UWB. (From [6] © 2006, IEEE.)

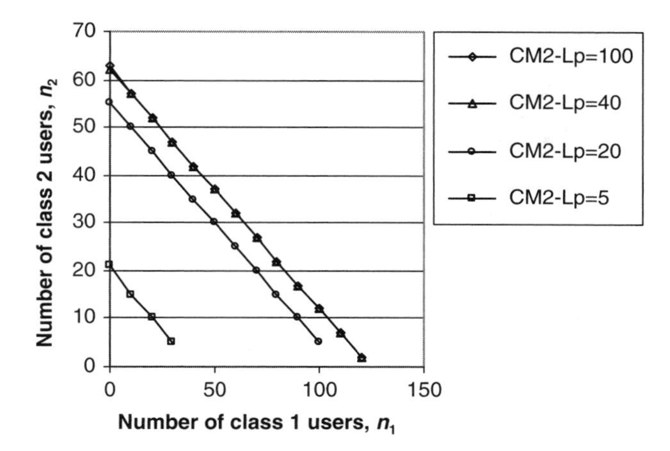

FIGURE 3.7 System capacity of CM2 for MTH-PPM UWB. (From [6] © 2006, IEEE.)

This corresponds to the case that $\Omega_0^{(in_i)}\overline{E}_{0,X}^{(in_i)}|_{X=N_{\text{tot}}}$ is maximum at a value of 1. This is also the upper bound of the system capacity for an idealized multiple-access channel. In this case, the SINR expression in (3.42) is the same as the SINR expression in an idealized multiple-access channel in (3.18) with signal propagation attenuation terms all equal to 1. Thus, the system capacity approaches that in an AWGN channel [3]. Note that the numerical results used here with 100 RAKE fingers are only to illustrate the link between the multipath channel and the AWGN channel. In practical systems, it may be difficult to have 100 RAKE fingers. Numerical results can be obtained for K traffic classes, and the system capacity is in a K-dimensional space. $(n_1, \ldots, n_K)$ are admissible as long as they are within its system capacity.

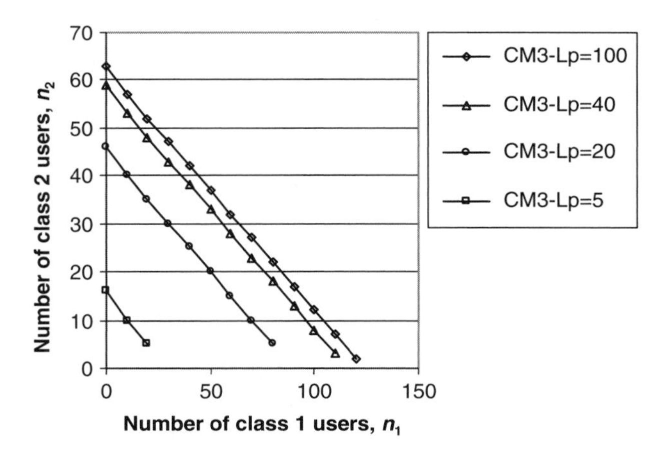

FIGURE 3.8 System capacity of CM3 for MTH-PPM UWB. (From [6] © 2006, IEEE.)

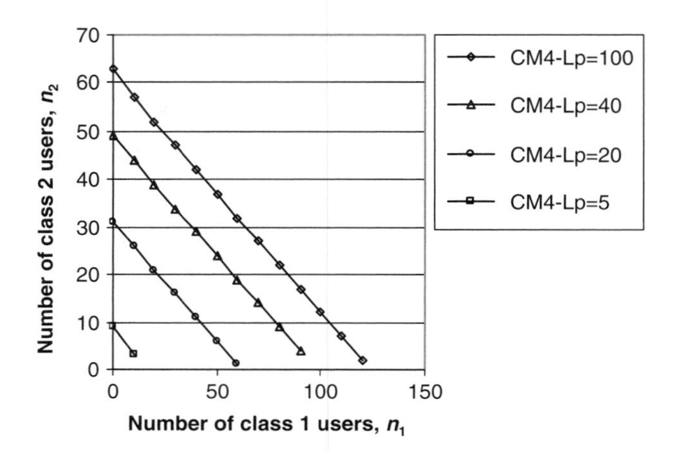

FIGURE 3.9 System capacity of CM4 for MTH-PPM UWB. (From [6] © 2006, IEEE.)

3.3 DIRECT SEQUENCE ULTRAWIDEBAND

A slightly more advanced UWB system uses direct-sequence (DS) spreading codes. The DS codes enable each user to be distinguishable from other users. The received DS-UWB signal is correlated with a template signal. The correlation operation provides a measure on the similarity of the received DS-UWB signal and the template signal. The two signals have a strong resemblance if the value of the correlation function is large. Otherwise, they have a low correlation. In this way, the transmitted signal can be recovered. More users can be supported by DS-UWB than that by TH-PPM UWB.

3.3.1 Idealized Multiple-Access Channel

3.3.1.1 Single-Class Traffic with Continuous Transmission First we consider the multipath phenomenon to determine the SINR for a DS-UWB system under an idealized multiple-access channel. The analysis here is similar to that of Boubaker and Letaief in [7].

The typical transmitted signal of the νth user for DS-UWB can be expressed as

$$s_{\text{tr}}^{(v)}(t) = \sum_{j=-\infty}^{\infty} \sum_{y=0}^{N_c-1} d_j^{(v)} p_y^{(v)} \omega(t - jT_f - yT_c), \qquad (3.47)$$

where T_f is the frame time interval or bit period, T_c is the chip interval, N_c is the number of chips per bit period of T_f such that $N_c = T_f/T_c$, $\omega(t)$ is the transmitted pulse, $\{d_j^{(v)}\}$ is the modulated data bit for the νth user, $\{p_y^{(v)}\}$ is the spreading chips of the νth user, and $p_y^{(v)} \in \{-1, 1\}$. Spreading codes are used in DS-UWB systems. For

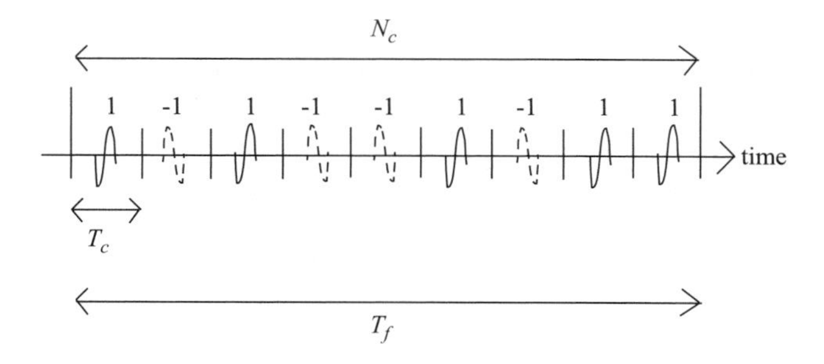

FIGURE 3.10 Direct-spread UWB system.

each spreading code in a frame time interval of T_f, there are N_c chip intervals. In each chip, the pulse has a positive or negative polarity, depending on the code sequence. Figure 3.10 shows the pulse signals of a user using a DS-UWB system with a spreading code of $\{1, -1, 1, -1, -1, 1, -1, 1, 1\}$. N_c pulses are used to represent a data bit and there are N_c chips per frame interval T_f. Each chip time has a time interval of T_c. The solid-line pulses are for the positive polarity pulses corresponding to the '1' in the spreading chip sequence, and the dashed-line pulses are for negative polarity pulses, corresponding to the '-1' in the spreading chip sequence. Thus, a user using a single spreading code uses his spreading code sequence to transmit his data sequence. A user can also use multiple codes in DS-UWB systems, as in Sections 3.3.1.2 and 3.3.2.2.

The received signal for the nth user is given by

$$r(t) = \sum_{y=0}^{N_c-1} A_n d_j^{(n)} p_y^{(n)} \omega_{\text{rec}}^{(n)}(t - jT_f - yT_c)$$
$$+ \sum_{V=1}^{n-1} \sum_{y=-\infty}^{\infty} A_V d_j^{(v)} p_y^{(v)} \omega_{\text{rec}}^{(v)} \left(t - jT_f - yT_c - \tau^{(v)}\right) + n(t), \quad (3.48)$$

where n is the number of users in the system, $\omega_{\text{rec}}(t)$ is the received pulse of the transmitted pulse $\omega(t)$, $\tau^{(v)}$ is the reference delay of the vth user relative to the nth user due to asynchronous time difference, A_V is the attenuation over the propagation path of the signal received from the vth user, and $n(t)$ is the receiver noise. We assume perfect clock and sequence synchronization for the signal transmitted by the nth user and $\tau^{(n)} = 0$.

The output Z of the correlation receiver can be expressed as

$$Z = \int_{\tau_i^{(n)}+jT_f}^{\tau_i^{(n)}+(j+1)T_f} r(t) \sum_{y=0}^{N_c-1} p_y^{(n)} \omega_{\text{rec}}\left(t - jT_f - yT_c - \tau^{(n)}\right) dt, \quad (3.49)$$

where Z is the decision statistic and $d_j^{(v)} = 0$ is decided if the decision statistic is $Z > 0$. Using the technique in [1,7], the output SINR can be shown to be [7]

$$\text{SINR} = \frac{S^2}{\text{MAI} + \sigma_{\text{rec}}^2}, \tag{3.50}$$

where the desired signal is

$$S = N_c A_n E_\omega, \tag{3.51}$$

the multiple-access interference (MAI) power is

$$\text{MAI} = \sum_{v=1}^{n-1} N_c \sigma_a^2 A_j^2, \tag{3.52}$$

and the received noise power is

$$\sigma_{\text{rec}}^2 = \sigma_n^2 N_c E_\omega. \tag{3.53}$$

In the above equations, $E_\omega = \int_{-\infty}^{\infty} \omega_{\text{rec}}^2(t)\, dt$ is the signal at the output of the correlation receiver during a chip interval, $\omega_{\text{rec}}(t)$ is a typical idealized received monocycle at the output of the antenna subsystem [1], σ_a^2 is the power of the interference resulting from one of the active spreading codes from the interfering users on one pulse, and $\sigma_n^2 = N_0/2$ is the variance of the receiver noise. The MAI is due to the $(n-1)$ other users in the system.

The SINR is given by (similar to [7])

$$\text{SINR} = \frac{(N_c A_n E_\omega)^2}{\sum_{j=1}^{n-1} N_c \sigma_a^2 A_j^2 + \sigma_{\text{rec}}^2}. \tag{3.54}$$

N_c is the number of chips per message symbol period of T_f such that $N_c T_c = T_f$ [7] and is also given by

$$N_c = \frac{1}{R_s T_c}, \tag{3.55}$$

where R_s is the system bit rate, T_f is the frame time interval, and T_c is the chip interval. N_c is also the spread-spectrum processing gain [7]. A_j is the attenuation over the propagation path of the signal received from the jth user. Using $\omega_{\text{rec}}(x)$ from

[1] as

$$\omega_{\text{rec}}(x + 0.35) = \left[1 - 4\pi\left(\frac{x}{\tau_m}\right)^2\right]\exp\left[-2\pi\left(\frac{x}{\tau_m}\right)^2\right], \tag{3.56}$$

it can be derived that [8]

$$E_\omega = \frac{3\tau_m}{8}. \tag{3.57}$$

σ_a^2 in (3.52) is given by [7]

$$\sigma_a^2 = \frac{1}{T_c}\int_{-\infty}^{\infty}\left[\int_{-\infty}^{\infty}\omega_{\text{rec}}(x - s)\omega_{\text{rec}}(x)\,dx\right]^2 ds. \tag{3.58}$$

Using (3.58), it can be derived that [8]

$$\sigma_a^2 = \frac{1}{T_c}\left(\frac{105\tau_m^3}{1024\sqrt{2}}\right). \tag{3.59}$$

The last term in the denominator of (3.54), σ_{rec}^2, is the variance of the ambient noise and is given by (3.53).

3.3.1.2 Multiclass Traffic with Variable-Bit-Rate Transmission

To determine the capacity of a multicode DS-UWB system using multiple correlation receivers to support the services of K classes of users, we consider the basic system similar to that in Section 3.3.1.1, except that multiple spreading codes are used by each user for varying-bit-rate traffic rather than a single spreading code per user.

By extending the results from Section 3.3.1.1 and [7], the SINR_i for class i is given by

$$\text{SINR}_i = \frac{\left(N_c A_{in_i} E_\omega\right)^2}{\sum_{j=1}^{n_i-1}\psi_{ij}N_c\sigma_a^2 A_{ij}^2 + \sum_{\substack{k=1\\k\neq i}}^{K}\sum_{j=1}^{n_k}\psi_{kj}N_c\sigma_a^2 A_{kj}^2 + \sigma_{\text{rec}}^2}, \qquad i = 1, 2, \ldots, K, \tag{3.60}$$

where $\psi_{ij} \in \{0, 1, 2, \ldots, m, \ldots, M_i\}$ is a binomial random variable indicating the number of active spreading codes used by the jth user of class i. The probability that m active spreading codes are used by a source, denoted by $\Pr[\psi_{ij} = m]$, is given by (3.19). A_{ij} is the attenuation over the propagation path of the signal received from the jth user of class i. The first term in the denominator of (3.60) is due to MAI from other users in class i, while the second term in the denominator of (3.60) is due to MAI from other users in the other classes.

Let BER_i^* and SINR_i^* denote, respectively, the BER requirement and the SINR requirement for class i users. The system capacity is defined as the maximum $(n_1, \ldots, n_i, \ldots, n_K)$ that can be supported such that the SINR achieved is greater than or equal to the SINR_i^* required 99% of the time for all classes. That is, the outage probability is defined as in (3.20) with

$$\delta_i = \frac{N_c E_\omega^2}{\text{SINR}_i^* \sigma_a^2} - \frac{\sigma_{\text{rec}}^2}{N_c \sigma_a^2 A_{in_i}^2}, \quad i = 1, 2, \ldots, K. \tag{3.61}$$

Assuming perfect power control, we have

$$A_{ij} = A_{kj} = A_{in_i}. \tag{3.62}$$

Thus, the outage probability is given by (3.23). The probability that l_i active spreading codes are used by n_i class i sources, denoted by $\Pr[\phi_i = l_i]$, is given by (3.24). Again, using the central limit approximation and solving (3.23) by conditioning on the active spreading codes used, then unconditioning the probability in (3.23) by summing up all cases for the numbers of active spreading codes used of all classes, and multiplying by all the corresponding binomial probabilities of active spreading codes used, we have (3.25) and (3.26).

We present results for the system capacity with two classes ($K = 2$) as an illustrative example. The parameter values used in the numerical examples are tabulated in Table 3.4. The admissible region for the system capacity is shown in Figure 3.11, where the capacity (n_1,n_2), of the system is on a "line." That is, the elements of the doublet (n_1,n_2) are bounded by this line. Thus, the combinations of the numbers of users of different classes that can be admitted to the system are possible only when these numbers are on or below this line. Similarly, numerical results can be obtained for K traffic classes, and the system capacity is in a K-dimensional space. $(n_1, \ldots, n_K)$ are admissible as long as they are within its system capacity.

TABLE 3.4 Parameter Values Used for DS-UWB

Symbol	Value	Symbol	Value
M_1	1	$\Delta\tau$	1 ns
M_2	2	T_f	350 ns
α_1	0.9	T_c	50 ns
β_1	0.1	τ_m	0.411 ns
α_2	0.9	$\text{BER}_1^* = \text{BER}_2^* = \text{BER}^*$	10^{-7}
β_2	0.1	$\text{SIR}_1^* = \text{SIR}_2^* = \text{SIR}^*$	14.3 dB
N_{tot}	50	σ_{rec}^2	0 (assumed negligible)
L_p	10, 20, 40, 50		

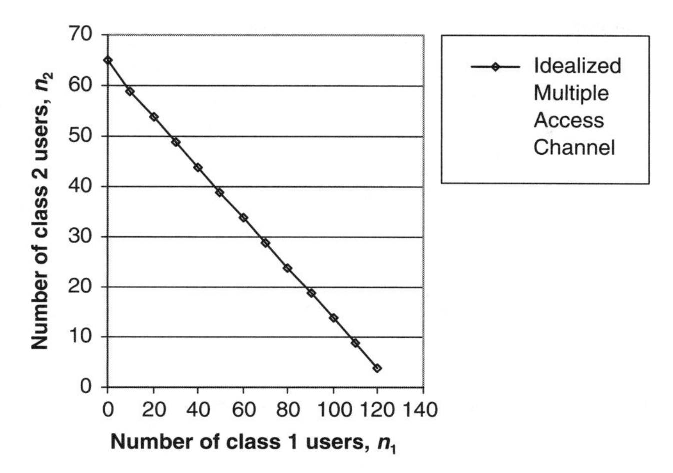

FIGURE 3.11 System capacity of an idealized multiple-access channel for multicode DS-UWB.

3.3.2 Multipath Channel

3.3.2.1 Single-Class Traffic with Continuous Transmission Using the channel model described in Section 3.2.2.1, we consider here the SINR for a DS-UWB system under a UWB multipath channel. The analysis here is similar to that of Wong et al. [9].

A RAKE receiver is used here. The typical transmitted signal of the vth user can be expressed as [9]

$$s_{\text{tr}}^{(v)}(t) = \sum_{j=-\infty}^{\infty} \sum_{y=0}^{N_c-1} \sqrt{E_z} d_j^{(v)} p_y^{(v)} w(t - jT_f - yT_c), \tag{3.63}$$

where T_f is the frame time interval or message symbol period, T_c is the chip interval, N_c is the number of chips per message symbol period of T_f such that $N_c = T_f/T_c$, E_z is the transmitted pulse energy, $w(t)$ is the unit-energy transmitted pulse, $d_j^{(v)}$ represents the modulated data symbols for the vth user, and $\{p_y^{(v)}\}$ represents the spreading chips. The received signal is given by [9]

$$\begin{aligned} r(t) = &\sum_{y=0}^{N_c-1} \sqrt{E_z} d_j^{(n)} p_y^{(n)} g^{(n)}(t - jT_f - yT_c) \\ &+ \sum_{v=1}^{n-1} \sum_{y=-\infty}^{\infty} \sqrt{E_z} d_j^{(v)} p_y^{(v)} g^{(v)} \left(t - jT_f - yT_c - \tau_0^{(v)}\right) + n(t), \end{aligned} \tag{3.64}$$

where n is the number of users in the system, $\tau_0^{(v)}$ is the reference delay of the vth user relative to the nth user due to asynchronous time difference, $g^{(v)}(t) = w_{\rm rec}(t) \otimes h^{(v)}(t), v = 1, 2, \ldots, n, w_{\rm rec}(t)$ is the received pulse, and $h^{(v)}(t)$ is the channel impulse response for the vth user. We assume perfect clock and sequence synchronization for the signal transmitted by the nth user.

Let L_p be the number of RAKE fingers and Z_i be the output of the ith RAKE finger:

$$Z_i = \int_{\tau_i^{(n)}+jT_f}^{\tau_i^{(n)}+(j+1)T_f} r(t) \sum_{y=0}^{N_c-1} p_y^{(n)} w_{\rm rec}\left(t - jT_f - yT_c - \tau_i^{(n)}\right) dt, \tag{3.65}$$

where $\tau_i^{(v)}$ is the ith time bin of the vth user. The output of the maximal ratio combiner is given by

$$Z = \sum_{i=1}^{L_p} \sqrt{E_z} A_i^{(n)} Z_i, \tag{3.66}$$

where $A_i^{(v)}$ is the sum of the channel coefficients of all multipath components that arrived in the ith time bin. Z is the decision statistic and $d_j^{(v)} = 0$ is decided if the decision statistic is $Z > 0$. Using the techniques in [1,2,4], the *instantaneous* output SINR conditioned on all the channel coefficients of all users can be shown to be [9]

$$\text{SINR}\left(\{A_m^{(v)}\}\right) = \frac{S^2}{E_z^2 N_c T_c^{-1} \sum_{v=1}^{n-1} G_{\rm eff}^{(v)^2} + \sigma_{\rm rec}^{2'}} \tag{3.67}$$

where

$$S = N_c E_\omega \sum_{i=1}^{L_p} A_i^{(n)^2}, \tag{3.68}$$

$$G_{\rm eff}^{(v)^2} = \sum_{i=1}^{L_p} \sum_{k=1}^{L_p} \sum_{m=1}^{N_{\rm tot}} \sum_{q=1}^{N_{\rm tot}} A_i^{(n)} A_k^{(n)} A_m^{(v)} A_q^{(v)} Q[(i - k - m + q)\,\Delta\tau], \tag{3.69}$$

$$\sigma_{\rm rec}^{2'} = \sigma_n^2 N_c E_\omega \sum_{i=1}^{L_p} A_i^{(n)^2}. \tag{3.70}$$

In the above equations, $E_\omega = \int_{-\infty}^{\infty} \omega_{\rm rec}^2(t)\, dt$ is the signal at the output of the correlation receiver during a chip interval, and $\omega_{\rm rec}(t)$ is a typical idealized received monocycle at the output of the antenna subsystem [2], $N_{\rm tot}$ is the total number of multipath bins considered for each channel realization, $\Delta\tau$ is the duration of a time bin, $Q(\cdot)$ is the autocorrelation of $R(\cdot)$, which is the autocorrelation function, $w_{\text{rec}}(t)$, of the received pulse of $w(t)$, and $\sigma_n^2 = N_0/2$ is the variance of the receiver noise at the correlator input of each tap. Using the techniques in [4], it can be shown that the average SINR is given by [9]

$$\text{SINR} = E\left\{\text{SINR}\left(\{A_m^{(v)}\}\right)\right\} = \frac{(N_c E_\omega)^2 \Omega_0^{(n)} \overline{E}_{0,X}^{(n)}\big|_{X=L_p}}{\sum_{v=1}^{n-1} N_c \sigma_a^2 + \sigma_{\text{rec}}^2}, \tag{3.71}$$

where $\Omega_0^{(n)}$ is the normalization factor of user n and $\overline{E}_{0,X}^{(n)}$ is the average channel energy of user n with X multipaths. Detailed expressions for these terms are as shown in (3.40) and (3.41), respectively. σ_a^2 is the power of the interference resulting from one of the active spreading codes from the interfering users on one pulse and is given by (3.59). The first term in the denominator of (3.71) is due to MAI from other users in the system. The term due to noise is $\sigma_{\text{rec}}^2 = \sigma_n^2 N_c E_\omega$.

3.3.2.2 Multiclass Traffic with Variable-Bit-Rate Transmission The basic system considered here is similar to that in Section 3.3.2.1 except that multiple spreading codes are used by each user for varying-bit-rate traffic rather than a single spreading code per user.

Let SINR_i denote the average SINR for class i. Assuming no interchip interval interference, by extending the results from Section 3.3.2.1 and from [4,9], it can be shown that

$$\text{SINR}_i = \frac{(N_c E_\omega)^2\, \Omega_0^{(in_i)} \overline{E}_{0,X}^{(in_i)}\big|_{X=L_p}}{\sum_{j=1}^{n_i-1} \psi_{ij} N_c \sigma_a^2 + \sum_{k=1,k\neq i}^{K} \sum_{j=1}^{n_k} \psi_{kj} N_c \sigma_a^2 + \sigma_{\text{rec}}^2}, \qquad i = 1, 2, \ldots, K, \tag{3.72}$$

where $\Omega_0^{(in_i)}$ is the normalization factor of user n_i of class i and $\overline{E}_{0,X}^{(in_i)}$ is the average channel energy of user j of class i with X multipaths. $\psi_{ij} \in \{0, 1, 2, \ldots, m, \ldots, M_i\}$ is a binomial random variable indicating the number of active spreading codes used by the jth user of class i. $E_\omega = \int_{-\infty}^{\infty} \omega_{\text{rec}}^2(t)\, dt$ is the signal at the output of the correlation receiver during a chip interval and is given by (3.57), and $\omega_{\text{rec}}(t)$ is a typical idealized received monocycle at the output of the antenna subsystem [1]. σ_a^2 is the power of the interference resulting from one of the active spreading codes from the interfering users on one pulse and is given by (3.59). The first term in the denominator of (3.72) is due to MAI from other users in class i, while the second term is due to MAI from other users from other classes. The term due to noise is $\sigma_{\text{rec}}^2 = \sigma_n^2 N_c E_\omega$. The probability that m active spreading codes are used by a source, denoted by $\Pr[\psi_{ij} = m]$, is given by (3.19). Note that when the average channel energy is sufficient with a large number of multipaths equal to the total number of multipaths, N_{tot}, $\Omega_0^{(1)} \overline{E}_{0,X}^{(1)}|_{X=N_{\text{tot}}} = 1$, and (3.72) reduces to (3.60) for an idealized multiple access channel [9] with perfect power control, which is the best result that can be achieved without a multipath channel. Furthermore, note that the analytical results in [4] *agree well* with the simulation results and that the BER for DS-UWB using Gaussian approximation is valid for a large number of users in the system. This justifies the reasonableness of the analysis in [4] and the analytical extension here.

Let BER_i^* and SINR_i^* denote, respectively, the BER and SINR requirements for class i users. The system capacity $\max(n_1, \ldots, n_i, \ldots, n_K)$ that can be supported such that the SINR achieved is greater than or equal to the required SINR_i^*. That is, the outage probability is defined as (3.20) with

$$\delta_i = \frac{N_c E_\omega^2 \Omega_0^{(in_i)} \overline{E}_{0,X}^{(in_i)} \,|_{X=L_p}}{\text{SINR}_i^* \sigma_a^2} - \frac{\sigma_{\text{rec}}^{2'}}{N_c \sigma_a^2}, \qquad i = 1, 2, \ldots, K. \tag{3.73}$$

Solving (3.23) by conditioning on the active spreading codes used and then unconditioning the probability in (3.23), we have (3.25) and (3.26).

We present results for the system capacity with two classes ($K = 2$) as an illustrative example. The parameter values used in the numerical examples are tabulated in Table 3.4. The parameter settings for the different scenarios in the IEEE 802.15.3a indoor UWB channel model are shown in Table 3.2. The admissible region for the system capacity using (3.26) for channel models, CM1 to CM4, with $L_p = 10$, 20, and 30 are shown in Figures 3.12, 3.13, and 3.14, respectively, where the capacities (n_1, n_2) of the system at this number of multipath fingers are shown on various curves. That is, the elements of the doublet (n_1, n_2) are bounded by these curves. Thus, the combinations of the numbers of users of different classes (n_1, n_2) that can be admitted to the system are possible only when these numbers are on or below these curves. The higher the curve, the larger the system capacity admission region. As the number

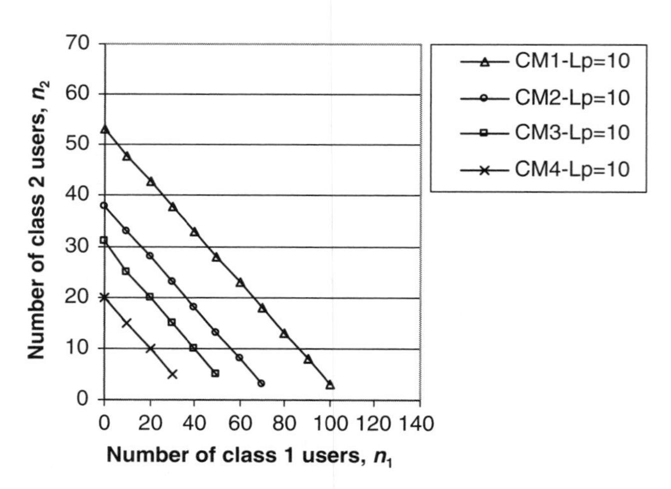

FIGURE 3.12 System capacity of CM1, CM2, CM3, and CM4 with $L_p = 10$ for multicode DS-UWB. (From [9] © 2006, IEEE.)

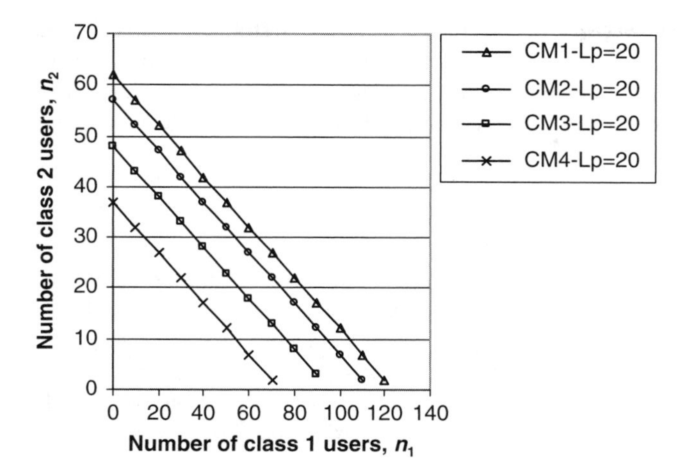

FIGURE 3.13 System capacity of CM1, CM2, CM3, and CM4 with $L_p = 20$ for multicode DS-UWB. (From [9] © 2006, IEEE.)

of multipath fingers increases, the system capacity increases. In general, CM1 has the best system capacity performance for partial RAKE receivers, followed by CM2, CM3 and CM4, respectively. The differences in system capacity for CM1 to CM4 are due to their different multipath characteristics. When $L_p = N_{\text{tot}} = 50$, the system capacities for CM1 to CM4 with full RAKE receivers are approximately the same, assuming that most of the channel energies are captured. This corresponds to the case where $\Omega_0^{(in_i)} \overline{E}_{0,X}^{(in_i)}|_{X=L_p}$ attains the maximum value of 1. This is also the upper

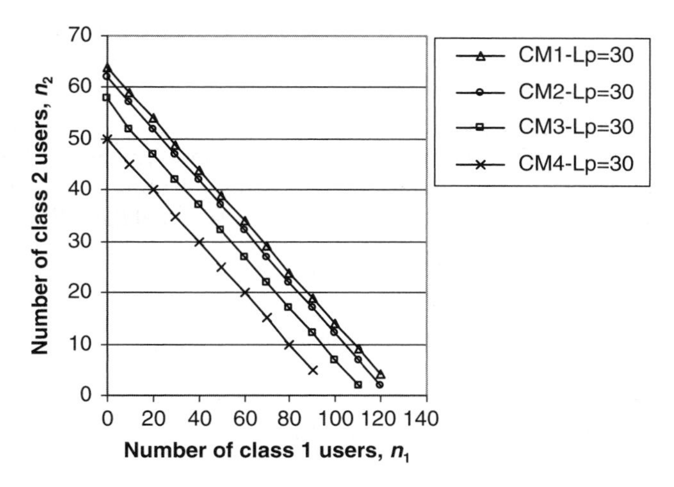

FIGURE 3.14 System capacity of CM1, CM2, CM3 and CM4 with $L_p = 30$ for multicode DS-UWB. (From [9] © 2006, IEEE.)

bound of the system capacity for an idealized multiple-access channel. In this case, the SINR expression in (3.72) is the same as the SINR expression in an idealized multiple-access channel in (3.60) with signal propagation attenuation terms all equal to 1. Thus, the system capacity approaches that for an AWGN channel.

3.4 MULTIBAND

The TH-PPM UWB and DS-UWB systems use techniques that utilize the entire available spectrum; that is, they use a single band. Another approach is to use multiple bands. This approach divides the available spectrum into several bands to comply with the FCC requirements. The main advantages of the multiband technique are lower design complexity, reduction of power consumption over smaller bandwidth, and spectral flexibility, which enables coexistence with existing services.

3.4.1 Multiband OFDM UWB

OFDM is a form of multicarrier transmission such that frequency bands of adjacent subcarriers do not overlap. Thus, the signals in the adjacent subcarriers are orthogonal. This will not result in interchannel interference (ICI) prior to transmission onto the propagation channel. The value of OFDM is the mitigation of frequency-selective fading. OFDM UWB transmits a train of short pulses by splitting them over orthogonal subcarriers. These pulses are sent over a UWB channel. At the receiver, these received UWB pulses are reassembled to get orthogonality and to recover each subcarrier separately. WiMedia uses multiband OFDM.

The transmitted signal for multiband OFDM UWB is [10]

$$x_i(t) = \frac{1}{T_S} \sum_{n=0}^{N-1} c_{n,i}\, g(t - iT_S') e^{j2\pi n(t - iT_S')/T_S}, \qquad -\infty \le i \le \infty, \tag{3.74}$$

where N is the number of subcarriers, $T_S' = T_C + T_S + T_G$, T_S is the duration of a useful OFDM symbol, T_C is the duration to eliminate intersymbol interference, T_G is the guard interval between two consecutive OFDMA symbols, $\{c_{n,i}, n = 0, \ldots, N-1\}$ is an input sequence with the OFDMA symbol index i and the subcarrier index n, and

$$g(t) = \begin{cases} 1, & T_C \le t \le T_C + T_S \\ 0, & \text{otherwise.} \end{cases} \tag{3.75}$$

Assuming perfect frequency and timing synchronization, the average signal-to-noise ratio (SNR) per QPSK symbol is given by [10]

$$\text{SNR} = \frac{E_s \sigma_H^2}{\sigma_C^2 + \sigma_S^2 + N_0}, \tag{3.76}$$

where E_s is the symbol energy; σ_H^2 is the variance of the coefficient of the fading channel given by

$$\sigma_H^2 = \Omega_0 + \frac{1}{T_S^2} A_2(T_S, 0, \lambda, \gamma) + \frac{1}{T_S^2} A_2(T_S, 0, \Lambda, \Gamma) + \frac{1}{T_S^2} A_1(T_S, 0, \Lambda, \Gamma, \lambda, \gamma); \tag{3.77}$$

σ_C^2 is the variance of the intercarrier interference (ICI) given by

$$\sigma_C^2 = E_s \sum_{\substack{n-0 \\ n \neq m}}^{N-1} \frac{1}{4\pi^2 (n-m)^2} [B_2(0, n, m, 0, \lambda, \gamma) + B_2(0, n.m, 0, \Lambda, \Gamma) + B_1(0, n, m, 0, \Lambda, \Gamma, \lambda, \gamma)]; \tag{3.78}$$

and σ_S^2 is the intersymbol interference (ISI) given by

$$\sigma_S^2 = E_s \left[\frac{1}{T_S^2} A_2(T_C + T_G, p_0, \Lambda, \Gamma) + \frac{1}{T_S^2} A_1(T_C + T_G, p_0, \Lambda, \Gamma, \lambda, \gamma) + \sum_{\substack{n=1 \\ n \neq m}}^{N-1} \frac{1}{4\pi^2 (n-m)^2} [B_2(T_C + T_G, n, m, p_0, \Lambda, \Gamma) + B_1(T_C + T_G, n, m, p_0, \Lambda, \Gamma, \lambda, \gamma) \right]. \tag{3.79}$$

Ω_0 is the mean power of the first ray of the first cluster, and $p_0 = \lfloor \Lambda(T_G + T_C) \rfloor$. Note that the first two terms in (3.77) and the first term in (3.78) do not exist for the case of non-line-of-sight (NLOS) conditions.

$A_1(\cdot)$, $A_2(\cdot)$, $B_1(\cdot)$, and $B_2(\cdot)$ are defined in the following equations:

$$\begin{aligned} A_1(T, p_0, \Lambda, \Gamma, \lambda, \gamma) = \Omega_0 [&\Gamma^2 f_3(p_0, \Lambda, \Gamma) f_1(0, \lambda, \gamma) + \gamma^2 f_1(p_0, \Lambda, \Gamma) f_3(0, \lambda, \gamma) \\ &+ T^2 f_1(p_0, \Lambda, \Gamma) f_1(0, \lambda, \gamma) + 2\Gamma\gamma f_2(p_0, \Lambda, \Gamma) f_2(0, \lambda, \gamma) \\ &- 2T\Gamma f_2(p_0, \Lambda, \Gamma) f_1(0, \lambda, \gamma) - 2T\gamma f_1(p_0, \Lambda, \Gamma) f_2(0, \lambda, \gamma)], \end{aligned} \tag{3.80}$$

$$\begin{aligned} A_2(T, p_0, \lambda_X, \gamma_X) = \Omega_0 [&\gamma_X^2 f_3(p_0, \lambda_X, \gamma_X) + T^2 f_1(p_0, \lambda_X, \gamma_X) \\ &- 2T\gamma_X f_2(p_0, \lambda_X, \gamma_X)], \end{aligned} \tag{3.81}$$

$$
\begin{aligned}
B_1(T, n, m, p_0, \Lambda, \Gamma, \lambda, \gamma) = {} & 2\Omega_0 f_1(p_0, \Lambda, \Gamma) f_1(0, \lambda, \gamma) - 2\Omega_0 \beta_T^{p_0+1} \beta_\tau \\
& \times \left[\cos\left((p_0+1)\theta_T + \theta_\tau - \frac{2\pi(n-m)T}{T_S} \right) \right. \\
& - \beta_\tau \cos\left((p_0+1)\theta_T - \frac{2\pi(n-m)T}{T_S} \right) \\
& - \beta_T \cos\left(p_0\theta_T + \theta_\tau - \frac{2\pi(n-m)T}{T_S} \right) \\
& \left. + \beta_T \beta_\tau \cos\left(p_0\theta_T - \frac{2\pi(n-m)T}{T_S} \right) \right] \Big/ \left[(1 + \beta_T^2 \right. \\
& \left. - 2\beta_T \cos\theta_T)(1 + \beta_\tau^2 - 2\beta_\tau \cos\theta_\tau) \right],
\end{aligned} \tag{3.82}
$$

$$
\begin{aligned}
B_2(T, n, m, p_0, \lambda_X, \gamma_X) = {} & 2\Omega_0 f_1(p_0, \lambda_X, \gamma_X) - 2\Omega_0 \beta_X^{p_0+1} \\
& \times \left[\cos\left((p_0+1)\theta_X - \frac{2\pi(n-m)T}{T_S} \right) \right. \\
& \left. - \beta_X \cos\left(p\theta_X - \frac{2\pi(n-m)T}{T_S} \right) \right] \Big/ \\
& \left(1 + \beta_X^2 - 2\beta_X \cos\theta_X \right),
\end{aligned} \tag{3.83}
$$

where

$$
\beta_X = \frac{\lambda_X}{\sqrt{\left(\lambda_X + \frac{1}{\gamma_X} \right)^2 + \frac{4\pi^2 (n-m)^2}{T_S^2}}} \tag{3.84}
$$

$$
\theta_X = \arctan\left(\frac{2\pi(n-m)}{T_S} \frac{\gamma_X}{\lambda_X \gamma_X + 1} \right). \tag{3.85}
$$

$f_1(\cdot)$, $f_2(\cdot)$, and $f_3(\cdot)$ are defined in the following equations:

$$
f_1(p_0, \lambda_X, \gamma_X) = \frac{(\lambda_X \gamma_X)^{p_0+1}}{(\lambda_X \gamma_X + 1)^{p_0+1}} (\lambda_X \gamma_X + 1), \tag{3.86}
$$

$$
f_2(p_0, \lambda_X, \gamma_X) = \frac{(\lambda_X \gamma_X)^{p_0+1}}{(\lambda_X \gamma_X + 1)^{p_0+1}} (\lambda_X \gamma_X + p_0 + 1), \tag{3.87}
$$

$$
\begin{aligned}
f_3(p_0, \lambda_X, \gamma_X) = {} & \frac{(\lambda_X \gamma_X)^{p_0+1}}{(\lambda_X \gamma_X + 1)^{p_0+2}} (2\lambda_X \gamma_X (\lambda_X \gamma_X + 1) \\
& + 2(p_0+1)\lambda_X \gamma_X + (p_0+1)(p_0+2)).
\end{aligned} \tag{3.88}
$$

3.5 OTHER TYPES OF UWB

Besides TH-PPM UWB, DS-UWB, and multiband OFDM UWB systems, four other types of UWB systems are considered in this section: transmitted reference (TR) UWB, chirp UWB, multicarrier UWB, and MIMO UWB systems.

3.5.1 Transmitted Reference UWB

Transmitted reference (TR) UWB transmits a first UWB pulse to serve as the template signal for the second pulse transmitted. The first pulse is unmodulated, whereas the second pulse is modulated. In this way, no template signal needs to be defined. However, bandwidth efficiency drops by half, as the first UWB pulse does not convey data information. It also has high noise vulnerability, as the reference signal is also transmitted through the channel. The advantages of the TR-UWB system are that the receiver can capture the entire signal energy for a slowly varying channel without the need for channel estimation, and it is robust to synchronization problems. Channel estimation is used to estimate the impulse response of the channel to improve the noise performance of the receiver, while synchronization is a problem of timekeeping of the received UWB pulses to allow them to be decoded.

The typical transmitted signal of the vth user can be expressed as [11]

$$s_{\text{tr}}^{(v)}(t) = \sum_{j=-\infty}^{\infty} \sqrt{E_z} b_j^{(v)} \Big[w\big(t - jT_f - c_j^{(v)} T_c - \tau_0^{(v)}\big) + d_{\lfloor j/N_s \rfloor}^{(v)} w\Big(t - jT_f - c_j^{(v)} T_c - T_d^{(v)} - \tau_0^{(v)}\Big)\Big], \tag{3.89}$$

where T_f is the frame time interval, T_c is the chip interval, N_c is the number of chips per frame time interval of T_f such that $N_c = T_f/T_c$, N_s is the number of frames per symbol, E_z is the transmitted pulse energy, $w(t)$ is the unit-energy transmitted pulse, $\{b_j^{(v)}\}$ is the pseudorandom sequence which randomizes the polarities of the reference and data-modulated pulses in one frame for the vth user, $\{c_j^{(v)}\}$ is the time-hopping value for the jth frame for the vth user depending on the time-hopping code, $T_d^{(v)}$ is the delay between the reference and the data pulses for the vth user, $\{d_{\lfloor j/N_s \rfloor}^{(v)}\}$ is the information bit for the vth user, and $d_{\lfloor j/N_s \rfloor}^{(v)} \in \{-1, 1\}$. The first pulse in equation (3.89) is the reference pulse; the second pulse in the equation is the data pulse. The information is encoded in the phase difference between the reference pulse and the data pulses. Figure 3.15 shows the pulse signals of two users using a TR-UWB system. $2N_s$ pulses are used to represent a data bit and there are N_c chips per frame interval T_f. Each chip time has a time interval of T_c. The solid-line pulses are for user 1; the dashed-line pulses are for user 2. In the following discussion we assume that the pseudorandom sequences that randomize both the reference and data-modulated pulses in one frame for users 1 and 2 are both positive as an illustrative example. Thus, their N_s reference pulses have positive polarities. User 1 is transmitting a data bit '-1' and thus the polarity of all of its N_s data pulses is negative with respect to their chip intervals. User 1 is also using a time-hopping code of $\{1,3,2\}$. In the first

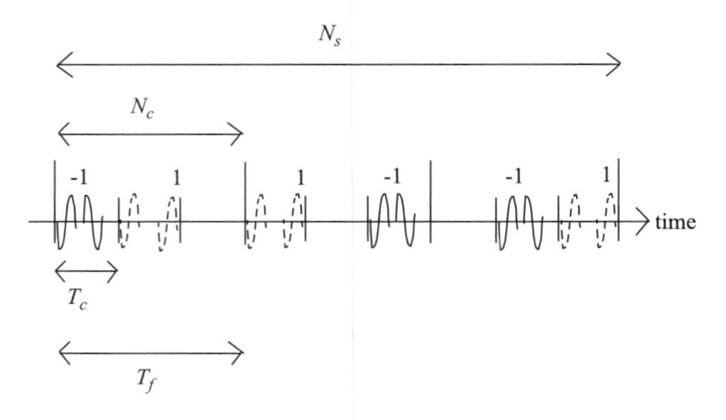

User 1 transmitting bit '-1' with time-hopping code {1,3,2}
User 2 transmitting bit '1' with time-hopping code {2,1,3}

FIGURE 3.15 Transmitted reference UWB system.

frame, user 1 transmits a positive reference pulse and a '-1' data pulse in chip 1 of the frame; in the second frame, user 1 transmits a positive reference pulse and a '-1' pulse in chip 3 of the frame; and in the third frame, user 1 transmits a reference pulse and a '-1' pulse in chip 2 of the frame. The delay between the reference pulse and the data pulse in a chip time is $T_d^{(1)}$. This delay is used when the received data signal correlates with the delayed version of the reference signal of user 1. On the other hand, user 2 is transmitting a data bit of '1', and the polarity of all of its N_s data pulses is positive with respect to their chip intervals. User 2 is using a time-hopping code of {2,1,3}. Similarly, in the first frame, user 2 transmits a reference pulse and a '1' pulse in chip 2 of the frame; in the second frame, user 2 transmits a reference pulse and a '1' pulse in chip 1 of the frame; and in the third frame, user 2 transmits a reference pulse and a '1' pulse in chip 3 of the frame. The delay between the reference pulse and the data pulse in a chip time is $T_d^{(2)}$. Similarly, this delay is used when the received data signal correlates with the delayed version of the reference signal of user 2.

The signal received for the nth user is given by [11]

$$\begin{aligned}
r(t) = &\sum_{j=0}^{N_S-1} \sqrt{E_z} b_j^{(n)} \left[g^{(n)}\left(t - jT_f - c_j^{(n)} T_c\right) + d_{\lfloor j/N_s \rfloor}^{(n)} g^{(n)}\left(t - jT_f - c_j^{(n)} T_c - T_d^{(n)}\right) \right] \\
&+ n(t) + \sum_{v=2}^{N_u} \sum_{j=-\infty}^{\infty} \sqrt{E_z} b_j^{(v)} \left[g^{(v)}\left(t - jT_f - c_j^{(v)} T_c - \tau_0^{(v)}\right) \right. \\
&\left. + d_{\lfloor j/N_s \rfloor}^{(v)} g^{(v)}\left(t - jT_f - c_j^{(v)} T_c - T_d^{(v)} - \tau_0^{(v)}\right) \right],
\end{aligned} \tag{3.90}$$

where n is the number of users in the system, $\tau_0^{(v)}$ is the reference delay of the vth user relative to the nth user due to asynchronous time difference, $g^{(v)}(t) = w(t) \otimes h^{(v)}(t)$, $v = 1, 2, \ldots, N_u$, and $h^{(v)}(t)$ is the channel impulse response for the vth user. $n(t)$ is the AWGN with two-sided power spectral density $N_0/2$. We assume perfect clock and sequence synchronization for the signal transmitted by the nth user.

The channel-averaged SINR is given by [11]

$$\begin{aligned}\mathrm{SINR}(L_p) = N_s E_z^2 \big[G^{(n)}(L_p)\big]^2 \Big\{ & N_0 E_Z G^{(n)}(L_p) + N_0^2 W L_p T_c/2 \\ & + \frac{4(n-1)}{T_f} E_z^2 Q(0) G^{(n)}(L_p) + \frac{2(n-1)}{T_f} E_z N_0 L_p T_c \\ & + \frac{4(n-1)(n-2)}{T_f^2} \Big[\int_0^{T_c} \int_y^{T_c} R^2(x)\,dx\,dy \\ & + \int_0^{T_c} \int_{-T_c}^{T_c - y} R^2(x)\,dx\,dy + Q(0)(L_p - 2)T_c \Big] \Big\}, \end{aligned} \tag{3.91}$$

where

$$G^{(n)}(L_p) = \sum_{i=1}^{L_p} E^{(n)}\{A_i^{(n)^2}\} = \Omega_0^{(1)}\, \overline{E_{0,X}^{(n)}}\Big|_{X=L_p}, \tag{3.92}$$

L_p is the number of multipaths, W is the one-sided receiver bandwidth, and $Q(\cdot)$ is the autocorrelation of $R(\cdot)$, which is the autocorrelation function of the received pulse of $w(t)$.

3.5.2 Chirp UWB

Another type of UWB uses chirp waveforms. A chirp is a signal that increases or decreases with frequency in time. The instantaneous frequency in a linear chirp varies linearly with time. Its advantages include easy generation of a UWB signal with a simple circuit, flexibility for meeting a spectrum mask, easy implementation of subband systems, and extreme robustness against multipath delay spread [12].

Assuming a single user, the signal transmitted can be expressed as [13]

$$s(t) = \sum_{i=-\infty}^{\infty} s_i(t) = \sum_{i=-\infty}^{\infty} \sum_{n=0}^{N-1} \sqrt{E_b}\, b_n[i] c_n(t - iT), \tag{3.93}$$

where $s_i(t)$ is the signal transmitted in the ith data frame, N is the number of subbands or codes, E_b is the energy per bit, $b_n[i]$ is the nth bit of the ith frame, and $b_n[i] \in \{-1, 1\}$, T is the code duration which is equal to $N_c T_c$, N_c is the number of chips per frame, T_c is the chip duration, $c_n(t)$ is the code, which is given by

$$c_n(t) = \sum_{m=0}^{N_c-1} a_{n,m} p_n \left(t - mT_c - \frac{T_c}{2} \right), \qquad n = 0, \ldots, N-1, \tag{3.94}$$

where the energy of all codes is normalized to 1 ($\int_{-\infty}^{\infty} c_n^2(t)\,dt = 1$), and the coefficient $a_{n,m}$ controls the polarity of the mth chip of $c_n(t)$ and $a_{n,m} \in \{-1, 1\}$. The chip coefficients of $a_{n,m}$, for $m = 0, \ldots, N_c$, form the time-domain spreading sequence for the nth code $c_n(t)$. Each chip of $c_n(t)$ is a linear frequency-modulated, or chirped, waveform with a chip duration of T_c and an instantaneous bandwidth of B_c. The chirp waveform in each chip of $c_n(t)$ is given by

$$p_n(t) = \begin{cases} g(t)\cos\left(2\pi(f_0 + nB_c)t + \pi\mu t^2\right), & |t| < \dfrac{T_c}{2} \\ 0, & \text{otherwise,} \end{cases} \tag{3.95}$$

where $g(t)$ is the chirp envelope having a continuous waveform, $f_0 + nB_c$ is the center frequency of the nth chirp waveform ($f_0 \gg B_c$), and μ is the chirp rate. A typical choice of $g(t)$ is a rectangular window with amplitude 1 and duration T_c centered at $t = 0$; that is, $g(t) = \Pi(t/T_c)$. The autocorrelation of $p_n(t)$ is given by [13]

$$\phi_{n,n}^{(p)}(t) = \begin{cases} \sqrt{B_s T_c}\dfrac{\sin\{\pi B_s t\,(1 - |t|/T_c)\}}{\pi B_s t}\cos\left(2\pi(f_0 + nB_c)t\right), & |t| < T_c \\ 0, & \text{otherwise,} \end{cases} \tag{3.96}$$

where B_s is the frequency-sweep range.

The channel model used is the same as that in Section 3.2.2.1. Assuming a linear time-invariant system, the received signal for the ith frame is given by [13]

$$r_i(t) = \sum_{n=0}^{N-1}\sum_{l=0}^{L-1} \sqrt{E_b}b_n[i]\alpha_{n,l}(n)c_n(t - iT - \tau_{n,l}) + n(t), \tag{3.97}$$

where L is the total number of multipath components; $\alpha_{n,l}$ is the channel gain of the nth code of the lth multipath, $\tau_{n,l} = T_h + \tau_{g,h}$; T_h is the arrival time of the first path of the hth cluster; $\tau_{g,h}$ is the delay of the gth path within the hth cluster relative to the first path arrival time, T_h; and $n(t)$ is the AWGN with a two-sided power spectral density $N_0/2$.

Using the technique in [13], the instantaneous signal-to-noise ratio (SNR) per bit is given by [13]

$$\text{SNR} = \frac{E_b \sum\limits_{l=0}^{L-1} \alpha_{n,l}^2}{N_0/2}, \tag{3.98}$$

3.5.3 Multicarrier UWB

A type of MC-UWB is based on the characteristics of frequency-coded pulse trains [14,15]. MC-UWB is also another type of UWB system in the literature. The signals are placed at multiple carriers in the frequency domain. The frequency-coded pulse train is given by

$$p(t) = \sum_{n=0}^{N-1} s(t - nT)e^{-j[2\pi c(n)t/T_c]}, \tag{3.99}$$

where $s(t)$ is an elementary pulse having unit energy with duration $T_s < T$. $p(t)$ has a duration $T_p = NT$, and $c(n)$ is a permutation of integers $\{0, 1, \ldots, N-1\}$. Each pulse is modulated with a frequency of $c(n)/T_c$.

3.5.4 MIMO UWB

MIMO UWB uses multiple transmit and receive antennas to improve the SINR. With MIMO UWB, the gain in SINR is increased by a factor of the product of the number of transmit antennas and the number of receive antennas. A MIMO UWB system is shown in Figure 3.16. There are n transmit users. Each user has N_T transmit antennas and N_R receive antennas.

3.5.4.1 Channel Model The channel model is the IEEE 802.15.3a indoor UWB channel model [5].

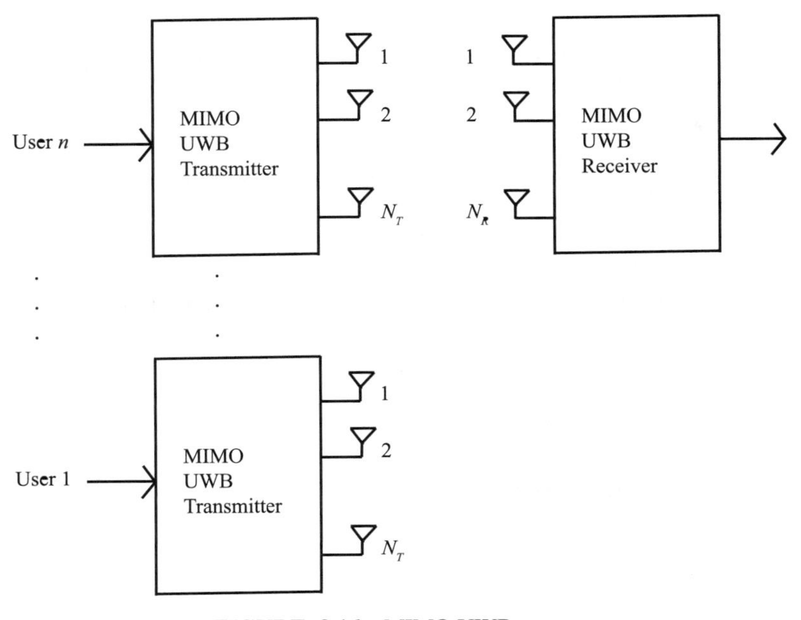

FIGURE 3.16 MIMO UWB system.

3.5.4.2 TH-PPM UWB with MIMO We consider a TH-PPM UWB with MIMO technology. The signal is transmitted over multiple transmit antennas and received by multiple receive antennas, giving a better SINR than that of a single transmit antenna and a single receive antenna.

Assuming TH-BPPM, the typical signal transmitted by transmit antenna i of the vth user can be expressed as (similar to [19])

$$s_{\mathrm{tr},i}^{(v)}(t) = \sum_{k=-\infty}^{\infty} \sqrt{\frac{E_z}{N_T}}\, w\left(t - kT_f - c_k^{(v)}T_c - \delta d_{i,\lfloor k/N_s \rfloor}^{(v)}\right), \tag{3.100}$$

where N_T is the number of transmit antennas, T_f is the frame time interval, T_c is the chip interval, N_c is the number of chips per frame time interval of T_f such that $N_c = T_f/T_c$, N_s is the number of frames per symbol, E_z/N_T is the pulse energy transmitted, $w(t)$ is the unit-energy transmitted pulse, $\{c_k^{(v)}\}$ is the time-hopping value for the kth frame for the vth user depending on the time-hopping code, δ is the modulation index or the time shift associated with the binary PPM modulation, $\{d_{i,\lfloor k/N_s \rfloor}^{(v)}\}$ is the user-specific data sequence in a symbol for the vth user of the ith transmit antenna, and $\{d_{i,\lfloor k/N_s \rfloor}^{(v)}\} \in \{0, 1\}$. Figure 3.17 shows the pulse signals transmitted on N_T transmit antennas of two users using a MIMO TH-PPM UWB system. Assuming that bit '0' is transmitted, the signal received at the jth receive antenna is given by (similar to [19])

$$r_j(t) = \sum_{i=1}^{N_T} \sum_{k=0}^{N_S-1} \sqrt{\frac{E_z}{N_T}}\, g_{i,j}^{(n)}\left(t - kT_f - c_k^{(n)}T_c\right) + \sum_{v=1}^{n-1} \sum_{i=1}^{N_T} \sum_{k=-\infty}^{\infty} \sqrt{\frac{E_z}{N_T}}\, g_{i,j}^{(v)}\big(t - kT_f$$

$$- c_k^{(v)}T_c - \delta d_{i,\lfloor k/N_s \rfloor}^{(v)} - \tau_0^{(v)}\big) + n_j(t), \tag{3.101}$$

where n is the number of users in the system, $\tau_0^{(v)}$ is the reference delay of the vth user relative to the nth user due to the asynchronous time difference ($\tau_0^{(n)} = 0$), $g^{(v)}(t) = w_{\mathrm{rec}}(t) \otimes h^{(v)}(t)$, $v = 1, 2, \ldots, n$, $w_{\mathrm{rec}}(t)$ is the pulse received, and $h^{(v)}(t)$ is the channel impulse response for the vth user. $n_j(t)$ is the AWGN with two-sided power spectral density $N_0/2$. We assume perfect clock and sequence synchronization for the signal transmitted by the nth user.

Let L_p be the number of RAKE fingers, and let $Z_{l,j}$ be the output of the lth RAKE finger of the jth receive antenna:

$$Z_{l,j} = \sum_{j=0}^{N_S-1} \int_{\tau_l^{(n)}+kT_f}^{\tau_l^{(n)}+(k+1)T_f} r_j(t)v\big(t - kT_f - c_k^{(n)}T_c - \tau_l^{(n)}\big)\, dt, \tag{3.102}$$

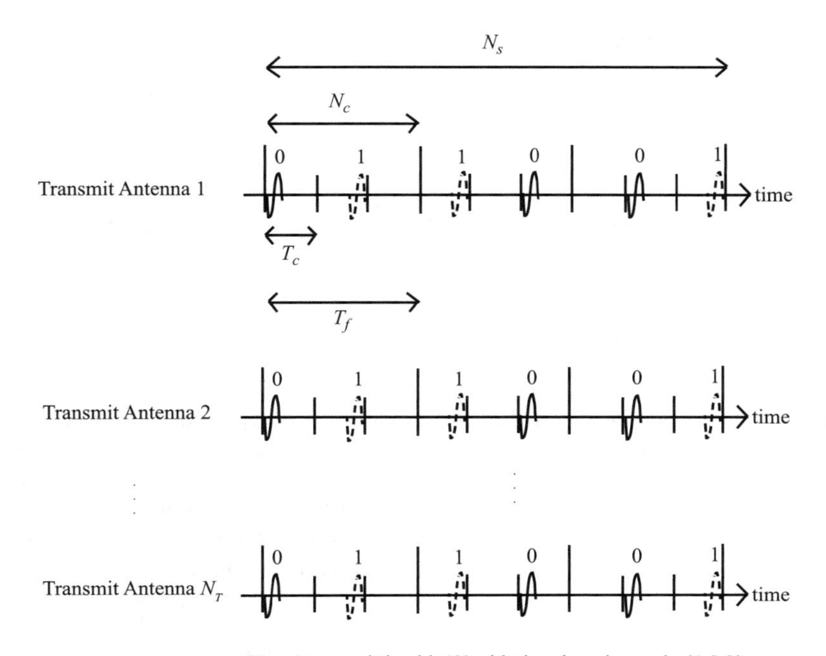

FIGURE 3.17 MIMO TH-PPM UWB system.

where $v(t) = w_{\text{rec}}(t) - w_{\text{rec}}(t - \delta)$ and $\tau_l^{(v)}$ is the lth time bin of the vth user. The output of the maximal ratio combiner is given by

$$Z = \sum_{i=1}^{N_R} \sum_{l=1}^{L_p} \sqrt{\frac{E_z}{N_T}} A_{i,j,l}^{(n)} Z_{l,j}, \tag{3.103}$$

where N_R is the number of receiving antennas and $A_{i,j,l}^{(v)}$ is the sum of the channel coefficients of all multipath components that arrived in the lth time bin from transmit antenna i to receive antenna j. Z is the decision statistic, and $d_k^{(v)} = 0$ is decided if the decision statistic is $Z > 0$. Using the techniques described in [1,2,4], the *instantaneous* output SINR conditioned on all the channel coefficients of all users can be shown to be

$$\text{SINR}\left(\left\{A_{i,j,l}^{(v)}\right\}\right) = \frac{S^2}{\dfrac{N_R}{N_T} E_z^2 N_s T_f^{-1} \displaystyle\sum_{v=1}^{n-1} G_{\text{eff}}^{(v)^2} + \sigma_{\text{rec}}^2}, \tag{3.104}$$

where

$$S = N_R N_s m_p \sum_{l=1}^{L_p} A_{i,j,l}^{(n)^2}, \tag{3.105}$$

$$G_{\text{eff}}^{(v)^2} = \sum_{l=1}^{L_p} \sum_{m=1}^{L_p} \sum_{s=1}^{N_{\text{tot}}} \sum_{q=1}^{N_{\text{tot}}} A_{i,j,l}^{(n)} A_{i,j,m}^{(n)} A_{i,j,s}^{(v)} A_{i,j,q}^{(v)} Q[(l - m - s + q)\Delta\tau], \tag{3.106}$$

$$\sigma_{\text{rec}}^{2'} = \frac{N_R}{N_T} N_0 N_s m_p \sum_{l=1}^{L_p} A_{i,j,l}^{(n)^2}. \tag{3.107}$$

In the above equations, m_p is the signal at the correlator's output during a frame interval and is given by (3.14), N_{tot} is the total number of multipath bins considered for each channel realization, $\Delta\tau$ is the duration of a time bin, $Q(\cdot)$ is the autocorrelation of $R(\cdot)$, which is the crosscorrelation function of the received pulse of $w(t)$ and the template signal $v(t)$, and $N_0/2$ is the double-sided power spectral density of AWGN. Using the techniques described in [4], it can be shown that the average SINR is given by

$$\text{SINR} = E\{\text{SINR}(\{A_{i,j,s}^{(v)}\})\} = \frac{N_T N_R (N_s m_p)^2 \Omega_0^{(n)} \overline{E}_{0,X}^{(n)} \,|_{X=L_p}}{\sum_{v=1}^{n-1} N_s \sigma_a^2 + \sigma_{\text{rec}}^2}, \tag{3.108}$$

where σ_a^2 is the power of the interference resulting from one of the active time-hopping sequences from the interfering users on one pulse and is given by (3.15), N_s is the number of impulses dedicated to the transmission of one symbol and is given by $N_s = 1/R_l T_f$, where R_l is the symbol data transmission rate using one time-hopping sequence, and T_f is the frame time interval or the pulse repetition time. The frame time interval is given by $T_f = N_c T_c$, where N_c is the number of time delay bins in a frame time T_f and T_c is the basic chip time. $\Omega_0^{(n)}$ is the normalization factor of user n, and $\overline{E}_{0,X}^{(n)}$ is the average channel energy of user n with X multipaths. They are given by (3.40) and (3.41), respectively. N_{tot} is the total number of multipaths and $L_p(\leq N_{\text{tot}})$ is the number of first multipaths. The term due to noise is $\sigma_{\text{rec}}^2 = N_0 N_s m_p$. Note that the SINR for the MIMO case is increased by a factor of $N_T N_R$ compared to the SISO case.

3.5.4.3 DS-UWB with MIMO For DS-UWB MIMO technology, the typical transmitted signal for the ith transmit antenna of the vth user can be expressed as (similar to [19])

$$s_{\text{tr},i}^{(v)}(t) = \sum_{k=-\infty}^{\infty} \sum_{y=0}^{N_c-1} \sqrt{\frac{E_z}{N_T}} d_{k,i}^{(v)} p_y^{(v)} w(t - kT_f - yT_c), \tag{3.109}$$

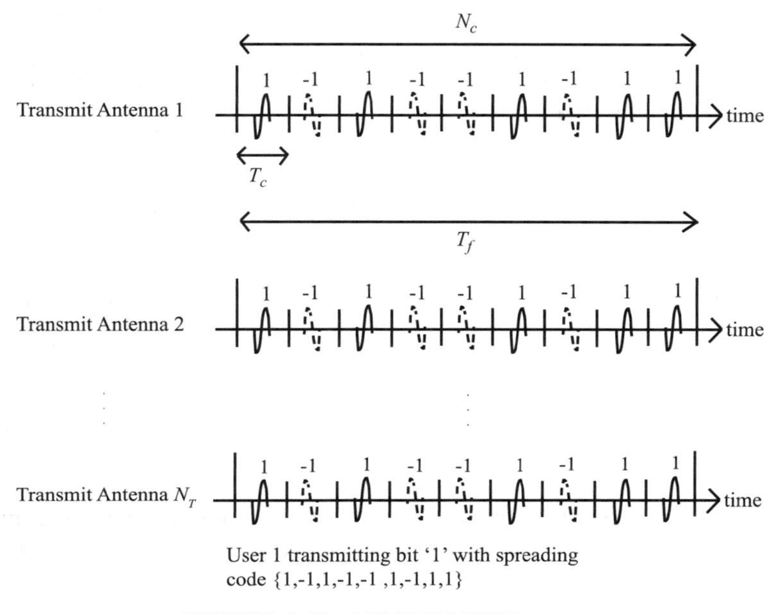

FIGURE 3.18 MIMO DS-UWB system.

where N_T is the number of transmit antennas, T_f is the frame time interval or message symbol period, T_c is the chip interval, N_c is the number of chips per message symbol period of T_f such that $N_c = T_f/T_c$, E_z/N_T is the pulse energy transmitted, $w(t)$ is the unit-energy pulse transmitted, $\{d_j^{(v)}\}$ is the modulated data symbols for the vth user, and $\{p_y^{(v)}\}$ represents the spreading chips. Figure 3.18 shows the pulse signals transmitted on N_T transmit antennas of a user using a MIMO DS-UWB system with a spreading code of $\{1, -1, 1, -1, -1, 1, -1, 1, 1\}$. The signal received is given by (similar to [19])

$$r_j(t) = \sum_{i=1}^{N_T} \sum_{y=0}^{N_c-1} \sqrt{\frac{E_z}{N_T}} d_{k,i}^{(n)} p_y^{(n)} g_{i,j}^{(n)}(t - kT_f - yT_c)$$
$$+ \sum_{v=1}^{n-1} \sum_{i=1}^{N_T} \sum_{y=-\infty}^{\infty} \sqrt{\frac{E_z}{N_T}} d_{k,i}^{(v)} p_y^{(v)} g_{i,j}^{(v)} \left(t - kT_f - yT_c - \tau_0^{(v)}\right) + n_j(t), \tag{3.110}$$

where n is the number of users in the system, $\tau_0^{(v)}$ is the reference delay of the vth user relative to the nth user due to asynchronous time difference, $g^{(v)}(t) = w_{\text{rec}}(t) \otimes h^{(v)}(t)$, $v = 1, 2, \ldots, N_u$, $w_{\text{rec}}(t)$ is the pulse received, and $h^{(v)}(t)$ is the channel impulse response for the vth user. $n_j(t)$ is the AWGN with two-sided power spectral density $N_0/2$. We assume perfect clock and sequence synchronization for the signal transmitted by the nth user.

Let L_p be the number of RAKE fingers and $Z_{l,j}$ be the output of the lth RAKE finger of the jth receiving antenna:

$$Z_{l,j} = \int_{\tau_l^{(n)}+kT_f}^{\tau_l^{(n)}+(k+1)T_f} r_j(t) \sum_{y=0}^{N_c-1} p_y^{(n)} w_{\text{rec}}\left(t - kT_f - yT_c - \tau_l^{(n)}\right) dt, \tag{3.111}$$

where $\tau_l^{(v)}$ is the lth time bin of the vth user. The output of the maximal ratio combiner is given by

$$Z = \sum_{j=1}^{N_R} \sum_{l=1}^{L_p} \sqrt{\frac{E_z}{N_T}} A_{i,j,l}^{(n)} Z_{l,i}, \tag{3.112}$$

where $A_{i,j,l}^{(v)}$ is the sum of the channel coefficients of all multipath components that arrived in the lth time bin from transmit antenna i to receive antenna j. Z is the decision statistic, and $d_k^{(v)} = 0$ is decided if the decision statistic is $Z > 0$. Using the techniques discussed in [1,2,4], the *instantaneous* output SINR conditioned on all the channel coefficients of all users can be shown to be

$$\text{SINR}\left(\{A_{i,j,s}^{(v)}\}\right) = \frac{S^2}{\frac{N_R}{N_T} E_z^2 N_c T_c^{-1} \sum_{v=1}^{n-1} G_{\text{eff}}^{(v)^2} + \sigma_{\text{rec}}^{2'}}, \tag{3.113}$$

where

$$S = N_R N_c E_\omega \sum_{l=1}^{L_p} A_{i,j,l}^{(n)^2}, \tag{3.114}$$

$$G_{\text{eff}}^{(v)^2} = \sum_{l=1}^{L_p} \sum_{m=1}^{L_p} \sum_{s=1}^{N_{\text{tot}}} \sum_{q=1}^{N_{\text{tot}}} A_{i,j,l}^{(n)} A_{i,j,m}^{(n)} A_{i,j,n}^{(v)} A_{i,j,q}^{(v)} Q[(l - m - s + q)\Delta\tau], \tag{3.115}$$

$$\sigma_{\text{rec}}^{2'} = \frac{N_R}{N_T} \sigma_n^2 N_c E_\omega \sum_{l=1}^{L_p} A_{i,j,l}^{(n)^2}. \tag{3.116}$$

In the above equations, $E_\omega = \int_{-\infty}^{\infty} \omega_{\text{rec}}^2(t)\, dt$ is the signal at the output of the correlation receiver during a chip interval, and $\omega_{\text{rec}}(t)$ is a typical idealized received monocycle at the output of the antenna subsystem [2], N_{tot} is the total number of multipath bins considered for each channel realization, $\Delta\tau$ is the duration of a time bin, $Q(\cdot)$ is the autocorrelation of $R(\cdot)$, which is the autocorrelation function, $w_{\text{rec}}(t)$, of the received pulse of $w(t)$, and $\sigma_n^2 = N_0/2$ is the variance of the receiver noise at the correlator input of each tap. Using the techniques described in [4], it can be shown that the average SINR is given by

$$\mathrm{SINR} = E\{\mathrm{SINR}(\{A_{i,j,s}^{(v)}\})\} = \frac{N_T N_R (N_c E_\omega)^2 \Omega_0^{(n)} \overline{E}_{0,X}^{(n)} \big|_{X=L_p}}{\sum_{v=1}^{n-1} N_c \sigma_a^2 + \sigma_{\mathrm{rec}}^2}, \tag{3.117}$$

where $\Omega_0^{(n)}$ is the normalization factor of user n and $\overline{E}_{0,X}^{(n)}$ is the average channel energy of user n with X multipaths. Detailed expressions for these terms are as shown in (3.40) and (3.41), respectively. σ_a^2 is the power of the interference resulting from one of the active spreading codes from the interfering users on one pulse and is given by (3.59). The term due to noise is $\sigma_{\mathrm{rec}}^2 = \sigma_n^2 N_c E_\omega$. There is also a gain in SINR of a factor of $N_T N_R$ from the SISO case to the MIMO case.

SUMMARY

The performance metrics of the TH-PPM UWB and DS-UWB systems in terms of closed-form SINR, outage probability, and system capacity have been highlighted in Sections 3.2 and 3.3. The performance metrics under AWGN channel and IEEE 802.15.3a indoor UWB channels have also been considered for both single-class traffic with continuous transmission and multiple traffic classes with variable-bit-rate transmissions. The signal-to-interference-noise ratios for TH-PPM UWB and DS UWB under AWGN and IEEE 802.15.3a UWB indoor channel models are summarized in Tables 3.5 and 3.6, respectively. The reader who is interested in the performance of TH-PPM UWB and DS-UWB systems with multiclass traffic having a two-dimensional Markov chain video source is referred to [20,21] for the AWGN results. These results can be extended to the case with an IEEE 802.15.3a indoor UWB channel in a manner similar to that described in Sections 3.2.2.3 and 3.3.2.2. In these results, a second derivative Gaussian received pulse is used. Results using a Rayleigh received pulse for TH-PPM UWB and DS-UWB systems under AWGN channels may be found in [22,23]. The results of using these pulses in TH-PPM UWB and DS-UWB are compared in Tables 3.7 and 3.8, respectively. A comparative analysis of optimum and suboptimum RAKE receivers in impulsive environment is studied in [24]. A variant of the TH-PPM UWB is the M-ary PPM UWB system. Analyses of this system may be found in [25–27].

The SINR of a multiband OFDM UWB system is explained in Section 3.4 under the assumption of perfect frequency and timing synchronization. For imperfect frequency and timing synchronization, more SNR results may be found in [10].

In a TR-UWB system, the SINR is as presented in Section 3.5.1. Other performance analyses of TR-UWB systems may be found in [28–42]. The efficiency of TR-UWB is low. Thus, differential TR-UWB systems are used to improve the efficiency [31,41,42]. In chirp UWB, the instantaneous SNR has been presented in Section

TABLE 3.5 SINR for TH-PPM UWB in AWGN and IEEE 802.15.3a Channels

Type		Signal/Interference-Noise Ratio, SINR
Idealized multiple-access channel	Single-class traffic with continuous transmission	$\mathrm{SINR} = \dfrac{(N_s A_n m_p)^2}{\sum_{j=1}^{n-1} N_s \sigma_a^2 A_j^2 + \sigma_{\mathrm{rec}}^2}$
	Multiclass traffic with variable-bit-rate transmission	$\mathrm{SINR}_i = \dfrac{(N_s A_{in_i} m_p)^2}{\sum_{j=1}^{n_i-1} \psi_{ij} N_s \sigma_a^2 A_{ij}^2 + \sum_{\substack{k=1\\k\neq i}}^{K} \sum_{j=1}^{n_k} \psi_{kj} N_s \sigma_a^2 A_{kj}^2 + \sigma_{\mathrm{rec}}^2}$
IEEE 802.15.3a channel	Single-class traffic with continuous transmission	$\mathrm{SINR} = \dfrac{(N_s m_p)^2 \Omega_0^{(n)} \overline{E}_{0,X}^{(n)} \mid_{X=L_p}.}{\sum_{v=1}^{n-1} N_s \sigma_a^2 + \sigma_{\mathrm{rec}}^2}$
	Multiclass traffic with variable-bit-rate transmission	$\mathrm{SINR}_i = \dfrac{(N_s m_p)^2 \Omega_0^{(in_i)} \overline{E}_{0,X}^{(in_i)} \mid_{X=Lp}.}{\sum_{j=1}^{n_i-1} \psi_{ij} N_s \sigma_a^2 + \sum_{k=1,k\neq i}^{K} \sum_{j=1}^{n_k} \psi_{kj} N_s \sigma_a^2 + \sigma_{\mathrm{rec}}^2}$

3.5.2. The signal of a type of MC-UWB system is presented in Section 3.5.3. Other analyses of other types of MC-UWB may be found in [43–45].

Another growing interest in the literature is MIMO UWB. Simple MIMO UWB extensions for TH-PPM UWB and DS-UWB are extended in Section 3.5.4. In general, the SINR increases by a factor that is equal to the product of the number

TABLE 3.6 SINR for DS-UWB in AWGN and IEEE 802.15.3a Channels

Type		Signal-to-Interference-Noise Ratio, SINR
Idealized multiple-acccess channel	Single-class traffic with continuous transmission	$\mathrm{SINR} = \dfrac{(N_c A_n E_\omega)^2}{\sum_{j=1}^{n-1} N_c \sigma_a^2 A_j^2 + \sigma_{\mathrm{rec}}^2}$
	Multiclass traffic with variable-bit-rate transmission	$\mathrm{SINR}_i = \dfrac{(N_c A_{in_i} E_\omega)^2}{\sum_{j=1}^{n_i-1} \psi_{ij} N_c \sigma_a^2 A_{ij}^2 + \sum_{\substack{k=1\\k\neq i}}^{K} \sum_{j=1}^{n_k} \psi_{kj} N_c \sigma_a^2 A_{kj}^2 + \sigma_{\mathrm{rec}}^2}$
IEEE 802.15.3a channel	Single-class traffic with continuous transmission	$\mathrm{SINR} = \dfrac{(N_c E_\omega)^2 \Omega_0^{(n)} \overline{E}_{0,X}^{(n)} \mid_{X=L_p}.}{\sum_{v=1}^{n-1} N_c \sigma_a^2 + \sigma_{\mathrm{rec}}^2}$
	Multiclass traffic with variable-bit-rate transmission	$\mathrm{SINR}_i = \dfrac{(N_c E_\omega)^2 \Omega_0^{(in_i)} \overline{E}_{0,X}^{(in_i)} \mid_{X=L_p}.}{\sum_{j=1}^{n_i-1} \psi_{ij} N_c \sigma_a^2 + \sum_{k=1,k\neq i}^{K} \sum_{j=1}^{n_k} \psi_{kj} N_c \sigma_a^2 + \sigma_{\mathrm{rec}}^2}$

TABLE 3.7 m_p and σ_a^2 Expressions in TH-PPM UWB with Second Derivative Gaussian and Rayleigh Pulses

Second Derivative Gaussian Pulse	
Received pulse form	$\omega_{\rm rec}(t+0.35) = \left[1 - 4\pi\left(\frac{t}{\tau_m}\right)^2\right]\exp\left[-2\pi\left(\frac{t}{\tau_m}\right)^2\right]$
m_p	$m_p = \left[\frac{\pi^2\delta^4}{2\tau_m^3} - \frac{3\pi\delta^2}{2\tau_m} + \frac{3\tau_m}{8}\right]\exp\left[-\frac{\pi\delta^2}{\tau_m^2}\right] - \frac{3\tau_m}{8}$
σ_a^2	$\sigma_a^2 = \frac{1}{T_f}\left\{\frac{105\tau_m^3}{512\sqrt{2}} + \left[-\frac{\pi^4\delta^8}{512\sqrt{2}\tau_m^5} + \frac{7\pi^3\delta^6}{128\sqrt{2}\tau_m^3} - \frac{105\pi^2\delta^4}{256\sqrt{2}\tau_m} + \frac{105\pi\tau_m\delta^2}{128\sqrt{2}} - \frac{105\tau_m^3}{512\sqrt{2}}\right]\exp\left[-\frac{\pi\delta^2}{2\tau_m^2}\right]\right\}$
Rayleigh Pulse	
Received pulse form	$\omega_{\rm rec}(x+0.35) = \left(\frac{x}{\tau_m^2}\right)\exp\left[-2\pi\left(\frac{x}{\tau_m}\right)^2\right]$
m_p	$m_p = \frac{1}{16\pi\tau_m}\left[1 - 2\pi\left(\frac{\delta}{\tau_m}\right)^2\right]\exp\left[-\frac{\pi\delta^2}{\tau_m^2}\right] - \frac{1}{16\pi\tau_m}$
σ_a^2	$\sigma_a^2 = \frac{1}{T_f}\left\{\frac{3}{512\sqrt{2}\tau_m} - \left[\frac{\delta^4}{512\sqrt{2}\tau_m^5} - \frac{3\delta^2}{256\sqrt{2}\pi\tau_m^3} + \frac{3}{512\sqrt{2}\pi^2\tau_m}\right]\exp\left[-\frac{\pi\delta^2}{2\tau_m^2}\right]\right\}$

TABLE 3.8 E_ω and σ_a^2 Expressions in DS-UWB with Second Derivative Gaussian and Rayleigh Pulses

Second Derivative Gaussian Pulse	
Received pulse form	$\omega_{\rm rec}(t+0.35) = \left[1 - 4\pi\left(\frac{t}{\tau_m}\right)^2\right]\exp\left[-2\pi\left(\frac{t}{\tau_m}\right)^2\right]$
E_ω	$E_\omega = \frac{3\tau_m}{8}$
σ_a^2	$\sigma_a^2 = \frac{1}{T_c}\left(\frac{105\tau_m^3}{1024\sqrt{2}}\right)$
Rayleigh Pulse	
Received pulse form	$\omega_{\rm rec}(x+0.35) = \left(\frac{x}{\tau_m^2}\right)\exp\left[-2\pi\left(\frac{x}{\tau_m}\right)^2\right]$
E_ω	$E_\omega = \frac{1}{16\pi\tau_m}$
σ_a^2	$\sigma_a^2 = \frac{1}{T_f}\sigma_a^2 = \frac{1}{T_c}\left\{\frac{3}{1024\sqrt{2}\pi^2\tau_m}\right\}$

of transmit antennas and the number of receive antennas. These results can be extended to variable-bit-rate transmissions using multicode as in Sections 3.2.1.2, 3.2.2.3, 3.3.1.2, and 3.3.2.2. A few articles focusing on analytical formulations may be found in [16,46–50]. A scheme that uses space–time code MIMO may be found in [16]. One other area of interest is the performance of UWB systems in the presence of narrowband systems. References [51–54] address some of these issues.

REFERENCES

[1] M. Z. Win and R. A. Scholtz, "Ultra-wide bandwidth time-hopping spread-spectrum impulse radio for wireless multi-access communications," *IEEE Trans. Commun.*, vol. 48, no. 4, pp. 679–691, Apr. 2000.

[2] R. A. Scholtz, "Multiple access with time-hopping impulse modulation," *IEEE MILCOM 1993, Conference Record*, pp. 447–450, Oct. 1993.

[3] T. C. Wong, J. W. Mark, and K. C. Chua, "Performance analysis of variable bit rate multiclass services in a multi-time-hopping pulse position modulation UWB system," *IEEE Wireless Communications and Networking Conference 2005, Conference Record*, pp. 651–656, New Orleans, LA, Mar. 13–17, 2005.

[4] T. Jia and D. I. Kim, Analysis of average signal-to-interference-noise ratio for indoor UWB rake receiving system, *IEEE Vehicular Technology Conference 2005–Spring, Conference CD-ROM*, Stockholm, Sweden, May 2005.

[5] J. Foerster, *IEEE 802.15 Channel Modeling Sub-Committee Report*, IEEE Press, Piscataway, NJ, Mar. 2003.

[6] D. T. C. Wong, J. W. Mark, and K. C. Chua, "System capacity of a multi-time-hopping PPM UWB system supporting variable bit rate multiclass services with rake receivers," *IEEE Vehicular Technology Conference 2006–Fall, Conference CD-ROM*, Montreal, Quebec, Canada, Sept. 25–28, 2006.

[7] N. Boubaker and K. B. Letaief, "Performance analysis of DS-UWB multiple access under imperfect power control," *IEEE Trans. Commun.*, vol. 52, no. 9, pp. 1459–1463, Sept. 2004.

[8] T. C. Wong, J. W. Mark, and K. C. Chua, "Capacity region of a multi-code DS-UWB system supporting variable bit rate multiclass services," *IEEE International Conference on Communications 2005, Conference Record*, pp. 2857–2861, Seoul, Korea, May 16–20, 2005.

[9] D. T. C. Wong, J. W. Mark, and K. C. Chua, "Capacity analysis of a multi-code DS-UWB system supporting variable bit rate multiclass services with rake receivers," *IEEE International Conference on Communication Systems 2006, Conference CD-ROM*, Singapore, 30 Oct.–2 Nov. 2006.

[10] H.-Q. D. Lai, "Baseband implementation and performance analysis of the multiband OFDM UWB system," M. Sc. thesis, University of Maryland, 2006.

[11] T. Jia, "Performance investigation of UWB RAKE receiving and transmitted reference systems," M. A. Sc. thesis, School of Engineering Science, Simon Fraser University, 2006.

[12] Y. Koike, S. Ishii. and R. Kohno, "Chirp UWB system with software defined receiver for industrial mobile ranging and autonomous control," *IEEE Conference on Ultra Wideband Systems, Conference Record*, pp. 381–383, May 18–25, 2004.

[13] H. Liu, "Multicode ultra-wideband scheme using chirp waveforms," *IEEE J. Sel. Areas Commun.*, vol. 24, no. 4, pp. 885–891, Apr. 2006.

[14] E. Saberinia and A. H. Tewfik, "Receiver structures for multi-carrier UWB systems," 7th *International Symposium on Signal Processing and Its Applications 2003*, vol. 1, pp. 313–316, July 2003.

[15] E. Saberinia and A. H. Tewfik, "All-digital receiver structures for MC-UWB systems," *IEEE Vehicular Technology Conference 2003–Fall*, vol. 1, pp. 289–293, Oct. 2003.

[16] B. Maglaris, D. Anastassiou, P. Sen, G. Karlsson, and J. D. Robbins, "Performance models of statistical multiplexing in packet video communications," *IEEE Trans. Commun.*, vol. 36, no. 7, pp. 834–844, July 1988.

[17] R. O. Onvural, *Asynchronous Transfer Mode Networks: Performance Issues*, Artech House, Norwood, MA, 1993.

[18] N. Ohta, *Packet Video Modeling and Signal Processing*, Artech House, Norwood, MA, 1994.

[19] W. P. Siriwongpairat, W. F. Su, M. Olfat, and K. J. R. Liu, "Space-time-frequency coded multiband UWB communication systems," *IEEE Wireless Communications and Networking 2005*, pp. 426–431, New Orleans, LA, Mar. 13–17, 2005.

[20] T. C. Wong, J. W. Mark, and K. C. Chua, "Capacity region of a multi-time-hopping PPM UWB system supporting video services," *IEEE Vehicular Technology Conference 2005–Fall, Conference CD-ROM*, Dallas, TX, Sept. 25–28, 2005.

[21] T. C. Wong, J. W. Mark, and K. C. Chua, "System capacity of a multi-code DS-UWB system supporting video services," *IEEE International Conference on Information, Communications and Signal Processing 2005, Conference CD-ROM*, Bangkok, Thailand, 6–9, Dec. 2005.

[22] T. C. Wong, J. W. Mark, and K. C. Chua, "Capacity region of a multi-time-hopping pulse position modulation UWB system with Rayleigh monocycles supporting variable bit rate multiclass services," *CDMA International Conference 2004, Conference CD-ROM*, Seoul, Korea. Oct. 25–28, 2004.

[23] T. C. Wong, J. W. Mark, and K. C. Chua, "Capacity region of a multi-code DS-UWB system with Rayleigh monocycles supporting variable bit rate multiclass services," *IEEE International Symposium on Personal, Indoor, and Mobile Radio Communications 2005, Conference CD-ROM*, Berlin, Germany, Sept. 11–14, 2005.

[24] B. S. Kim, J. Bae, I. Son, S. Y. Kim, and H. Kwon, "A comparative analysis of optimum and suboptimal rake receivers in impulsive UWB environment," *IEEE Trans. Veh. Technol.*, vol. 55, no. 6, pp. 1797–1804, Sept. 2006.

[25] F. Ramirez-Mireles, "Performance of ultrawideband SSMA using time hopping and M-ary PPM," *IEEE J. Sel. Areas Commun.*, vol. 19, no. 6, pp. 1186–1196, June 2001.

[26] R. Pasand, S. Khalesehosseini, J. Nielsen, and A. Sesay, "Exact evaluation of M-ary TH-PPM UWB systems on AWGN channels for indoor multiple-access communications," *IEE Proc.–Commun.*, vol. 153, no. 1, pp. 83–92, Feb. 2006.

[27] N. V. Kokkalis, P. T. Mathiopoulos, G. K. Karagiannidis, and C. S. Koukourlis, "Performance analysis of M-ary PPM TH-UWB systems in the presence of MUI and timing jitter," *IEEE J. Sel. Areas Commun.*, vol. 24, no. 4, pp. 822–828, Apr. 2006.

[28] J. D. Choi and W. E. Stark, "Performance of ultra-wideband communications with suboptimal receivers in multipath channels," *IEEE J. Sel. Areas Commun.*, vol. 20, no. 9, pp. 1754–1766, Dec. 2002.

[29] R. T. Hoctor, "Multiple access capacity in multipath channels of delay-hopped transmitted-reference UWB," *IEEE Conference on Ultra Wideband Systems and Technologies 2003*, pp. 315–319, Nov. 16–19, 2003.

[30] Y. L. Chao and R. A. Scholtz, "Optimal and suboptimal receivers for ultra-wideband transmitted reference systems," *IEEE GLOBECOM 2003*, vol. 2, pp. 759–763, San Francisco, Dec. 1–5, 2003.

[31] Y. L. Chao and R. A. Scholtz, "Multiple access performance of ultra-wideband transmitted reference systems in multipath environments," *IEEE Wireless Communications and Networking 2004*, vol. 3, pp. 1788–1793, Atlanta, GA, Mar. 21–25, 2004.

[32] Y. L. Chao and R. A. Scholtz, "Ultra-wideband transmitted reference systems," *IEEE Trans. Vehi. Technol.*, vol. 54, no. 5, pp. 1556–1569, Sept. 2005.

[33] S. Gezici, F. Tufvesson, and A. F. Molish, "On the performance of transmitted-reference impulse radio," *IEEE GLOBECOM 2004*, vol. 5, pp. 2874–2879, Texas, Dallas, TX, Nov. 29–Dec. 3, 2004.

[34] W. M. Gifford and M. Z. Win, "On transmitted-reference UWB communications," *38th Asilomar Conference on Signals. Systems and Computers 2004,* vol. 2, pp. 1526–1531, Pacific Grove, CA, Nov. 7–10, 2004.

[35] T. Q. S. Quek and M. Z. Win, "Performance analysis of ultrawide bandwidth transmitted-reference communications," *IEEE Vehicular Technology Conference 2004–Spring,* vol. 3, pp. 1285–1289, Milan, Italy, May 17–19, 2004.

[36] T. Q. S. Quek and M. Z. Win, "Ultrawide bandwidth transmitted-reference signaling," *IEEE International Conference on Communications,* vol. 6, pp. 3409–3413, Paris, June 20–24, 2004.

[37] T. Q. S. Quek and M. Z. Win, "Analysis of UWB transmitted-reference communication systems in dense multipath channels," *IEEE J. Sel. Areas Commun.*, vol. 23, no. 9, pp. 1863–1874, Sept. 2005.

[38] J. Romme and K. Witrisal, "Transmitted-reference UWB systems using weighted autocorrelation receivers," *IEEE Trans. Microwave Theory Tech.*, pp. 1–8, 2006.

[39] A. Rabbachin and I. Oppermann, "Comparison of UWB transmitted reference schemes," *IEE Proc.–Commun.*, vol. 153, no. 1, pp. 136–142, Feb. 2006.

[40] Z. Xu and B. M. Sadler, "Multiuser transmitted reference ultra-wideband communication systems," *IEEE J. Sel. Areas Commun.*, vol. 24, no. 4, pp. 766–772, Apr. 2006.

[41] A. A. D'Amico and U. Mengali, "GLRT receivers for UWB systems," *IEEE Commun. Lett.*, vol. 9, no. 6, June 2005.

[42] K. Witrisal, G. Leus, M. Pausini, and C. Krall, "Equivalent system model and equalization of differential impulse radio UWB systems," *IEEE J. Sel. Areas Commun.*, vol. 23, no. 9, pp. 1851–1862, Sept. 2005.

[43] L. Q. Yang and G. B. Giannakis, "Ultra-wideband multiple access: unification and narrowband interference analysis," *IEEE Conference on Ultra Wideband Systems and Technologies 2003*, pp. 320–324, Nov. 16–19, 2003.

[44] L. Q. Yang and G. B. Giannakis, "A general model and SINR analysis of low duty-cycle UWB access through multipath with narrowband interference and rake reception," *IEEE Trans. Wireless Commun.*, vol. 4, no. 4, pp. 1818–1833, July 2005.

[45] Z. Zeinalpour-Yazdi and M. Nasiri-Kenari, "Performance analysis and comparisons of different ultra-wideband multiple access modulation schemes," *IEEE Proc.–Commun.*, vol. 153, no. 5, pp. 705–718, Oct. 2006.

[46] W. P. Siriwongpairat, M. Olfat, and K. J. R. Liu, "Performance analysis of time hopping and direct sequence UWB space-time systems," *IEEE GLOBECOM 2004*, vol. 6, pp. 3526–3530, Dallas, TX, Nov. 29–Dec. 3, 2004.

[47] W. P. Siriwongpairat, M. Olfat, and K. J. R. Liu, "Performance analysis and comparison of time-hopping and direct-sequence UWB-MIMO time systems," *EURASIP J. Appl. Signal Process.*, vol. 2005, no. 3, pp. 328–345, Mar. 2005.

[48] W. P. Siriwongpairat, W. F. Su, and K. J. R. Liu, "Performance characterization of multiband UWB communication systems using Poisson cluter arriving fading paths," *IEEE J. Sel. Areas in Commun.*, vol. 24, no. 4, pp. 745–751, Apr. 2006.

[49] L.-C. Wang, W.-C. Liu, and K.-J. Shieh, "On the performance of using multiple transmit and receive antennas in pulse-based ultrawideband systems," *IEEE Trans. Wireless Commun.*, vol. 4, no. 6, pp. 2738–2750, Nov. 2005.

[50] F. Heliot, M. Ghavami, and M. R. Nakhai, "Design and performance analysis of a space-time block coding scheme for single band UWB," *IEE Proc.–Commun.*, vol. 153, no. 1, pp. 127–135, Feb. 2006.

[51] L. Zhao and A. M. Haimovich, "Performance of ultra-wideband comunications in the presence of interference," *IEEE J. Sel. Areas Commun.*, vol. 20, no. 9, pp. 1684–1691, Dec. 2002.

[52] L. Piazzo, "Performance analysis and optimization for impulse radio and direct-sequence impulse radio in multiuser interference," *IEEE Trans. Commun.*, vol. 52, no. 5, pp. 801–810, May 2004.

[53] L. Piazzo and F. Ameli, "Performance analysis for impulse radio and direct-sequence impulse radio in narrowband interference," *IEEE Trans. Commun.*, vol. 53, no. 9, pp. 1571–1580, Sept. 2005.

[54] R. Giuliano and F. Mazzenga, "Capacity analysis for UWB systems with power controlled terminals under power and coexistence constraints," *IEEE Trans. Wireless Commun.*, vol. 5, no. 11, pp. 3316–3328, Nov. 2006.

CHAPTER 4

MEDIUM ACCESS CONTROL

4.1 INTRODUCTION

Medium access control (MAC) is a sublayer of the data link layer, which is layer 2 in the ISO/OSI reference model. MAC acts as an interface for network access. There are many types of wireless networks, including wireless personal area networks (WPANs), wireless local area networks (WLANs), wireless metropolitan area networks (WMANs), and cellular networks. Each type of network uses a different type of MAC. The MAC chosen for each type of network also depends on its data rate and applications or services that it provides.

MAC enables multiple users to access a shared channel or channels. Traditional methods include time-division multiple access (TDMA), frequency-division multiple access (FDMA), and code-division multiple access (CDMA). TDMA divides time into slots which can be assigned to individual users who transmit only in their assigned slots. Users must wait their turn to access their time slots. FDMA divides the frequency band into subbands that can be assigned to individual users. CDMA allocates all resources of a channel to all simultaneous users by using orthogonal algebraic codes that enable a receiver to distinguish each user and control the transmit power levels to maintain a given SNR for the required quality of service. TDMA, FDMA, and CDMA have been used as cellular access technologies.

Wireless Broadband Networks, By David Tung Chong Wong, Peng-Yong Kong, Ying-Chang Liang, Kee Chaing Chua, and Jon W. Mark

For packet-based systems, one of the most basic MAC protocols is the ALOHA protocols, which simply allows a user to transmit a packet when it is ready to do so. If the packet collides with other packets, it will back off and then try again later after a random delay. As its maximum throughput is low due to the high number of collisions (expected), a slotted version of the ALOHA protocol was introduced that almost doubled its maximum throughput. A form of slotted ALOHA protocol is also commonly used for initial network access in cellular networks. Carrier-sense multiple access (CSMA) protocols can increase the maximum throughput further. In the wireless domain, CSMA with collision avoidance (CSMA/CA) is commonly used as the multiple-access technique. This is a widely used MAC protocol in wireless local area networks (WLANs). In IEEE 802.11 WLANs, the protocol used for medium access control is the *distributed coordination function* (DCF). Although there is also a point coordination function (PCF) as part of the IEEE 802.11 standard for real-time traffic, this version is seldom implemented by vendors. The most commonly implemented MAC in IEEE 802.11-based WLANs is DCF. One of the most important analytical papers on IEEE 802.11 is by Bianchi [1]. Another refinement of Bianchi's work is that of Xiao [2]. Polling is another form of MAC protocol, which is used in Bluetooth. Polling consists of a central controller which will decide the order for stations to get services for their pending packets. Bluetooth, which uses polling, is a wireless personal area network (WPAN) technology with a data rate of up to 1 Mbps. Reservation-type MACs are one of the oldest forms of MAC since their use in satellites. Dynamic TDMA (D-TDMA) is a form of reservation MAC [3]. Even today, reservation types of MACs continue to be used in some standards, including IEEE 802.15.4 [4] and WiMedia. The former MAC catered to energy-efficient applications as in sensor networks. ZigBee uses IEEE 802.15.4 and has a data rate of up to 250 kbps. IEEE 802.15.4 also uses a form of CSMA/CA. In reservation MAC, the frame format usually consists of a number of minislots for initial access through a contention process and a number of reserved information transmission slots for transmission of information packets. These packets can be for voice, video, and data traffic. Reservation MACs are used in satellites. Multichannel MAC is another form of MAC used to increase throughput [5]. Throughput can be increased by sending packets in multiple channels. Directional-antenna MAC protocols can also help to increase throughput [6]. A directional antenna radiates power in one or more directions, resulting in improved performance in the receiver and transmitter and reduced interference from unwanted sources. Spatial reuse is maximized as nodes limit their transmission to the smallest possible area. Thus, throughput can be increased. Multihop saturated throughput for IEEE 802.11 MAC is considered in [7,8]. The multihop saturated throughput is calculated based on the saturated throughput for a node and the average number of neighbors of a node.

This chapter is organized as follows. In Section 4.2 we give an analysis of a form of slotted ALOHA protocol. Section 4.3 covers the performance analysis of CSMA/CA for IEEE 802.11 under saturated conditions. Both saturated throughput and delay are considered. In Section 4.4 we present an analysis for polling MAC, where the average transfer delay is presented. The performance analysis of integrated voice/data traffic for D-TDMA MAC is presented in Section 4.5. The voice packet loss probability, average data throughput, and average data delay are derived. In Section 4.6 we present an analysis of the IEEE 802.15.4 MAC protocol, where the device and coordinator power consumptions are derived. This MAC protocol is also energy-efficient. A type of multichannel MAC known as dedicated channel control is presented in Section 4.7. For this multichannel MAC, the total system throughput is derived. Two directional-antenna MACs are presented and analyzed in Section 4.8. The two MAC schemes are based on directional transmission and omnidirectional reception (DTOR) and directional transmission and directional reception (DTDR). In Section 4.9 we consider the multihop saturated throughput for IEEE 802.11 MAC. Traditional forms of MAC, such as TDMA, FDMA, and CDMA, are presented in Section 4.10. Note that a few symbols may be reused in the analyses of different sections for convenience of use, and they are only applicable in each of these sections of the respective analyses. The summary includes extensive references on MAC protocols in the literature.

4.2 SLOTTED ALOHA MAC

A form of slotted ALOHA with a frame format shown in Figure 4.1 is commonly used for reservation access in cellular networks. Figure 4.1 shows a frame with period T having N minislots. The aggregate arrivals at the frame are assumed to be Poisson distributed. Each request arrival randomly picks and transmits its request in one of the N minislots. If it is the only request in the minislot, it is successful. Otherwise, there are collisions and the collided requests will each randomly pick another minislot for request transmission in the next frame. This process repeats itself until all requests are successful. This form of reservation MAC is discussed further in Section 4.5.

4.2.1 Analysis

Let C_i be the number of outstanding collisions. The sequence $\{C_i, i = 0, 1, 2, \ldots\}$ constitutes a first-order homogeneous Markov chain. The transition probabilities of

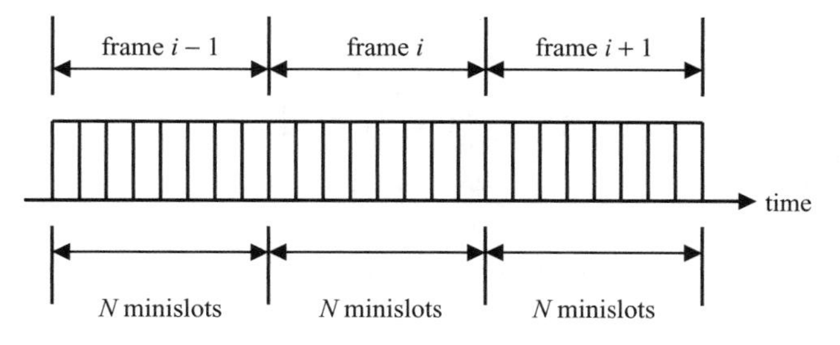

FIGURE 4.1 Frame format for a form of slotted ALOHA.

the Markov chain is given by [1]

$$P_{jk} = \Pr[C_{i+1} = k|C_i = j] = \sum_{m=0}^{\infty} \Pr[A = m]B(m + j - k, m + j; N),$$
$$i, j = 0, 1, 2, \ldots, m + j - k \geq 0, \tag{4.1}$$

where A is the number of request arrivals in a frame period,

$$\Pr[A = m] = \frac{(\lambda T)^m \, e^{-\lambda T}}{m!}, \tag{4.2}$$

λ is the mean aggregate request arrival rate, and $B(u, v; N)$ is the probability of u successful requests given that v requests were transmitted in N minislots. $B(u, v; N)$ can be computed recursively as follows. Let $\eta(u, v)$ be the number of combinations that u requests are successful given that v requests have been transmitted randomly in N minislots. Then [9]

$$B(u, v', N) = \eta(u, v)N^{-v}, \tag{4.3}$$

where

$$\eta(u, v) = \binom{N}{u}\binom{v}{u} u!\,(N - u)^{v-u} - \sum_{i=1}^{r(v)-u} \binom{u+i}{i} \eta(u + i, v), \tag{4.4}$$

$$\binom{m}{n} = \frac{m!}{n!\,(m - n)!}, \tag{4.5}$$

$$r(v) = \begin{cases} v, & v \leq N \\ N - 1, & v > N, \end{cases} \tag{4.6}$$

and $r(v)$ is the maximum number of successful reservations that v requests transmit in N minislots. Under steady-state conditions we have

$$\mathbf{\Pi} = \mathbf{\Pi\Psi}, \tag{4.7}$$

where

$$\mathbf{\Pi} = (\pi_0, \pi_1, \pi_2, \ldots), \tag{4.8}$$

$$\pi_k = \lim_{i \to \infty} \Pr[C_i = k], \tag{4.9}$$

$$\mathbf{\Psi} = [P_{jk}]. \tag{4.10}$$

The number of collisions expected is given by

$$E[C] = \sum_{k=0}^{\infty} k\pi_k. \tag{4.11}$$

The number of frames expected to resolve all collisions is given by

$$\begin{aligned} E[F] &= \sum_{f=1}^{\infty} f(1-\pi_0)^{f-1}\pi_0 \\ &= \frac{1}{\pi_0}. \end{aligned} \tag{4.12}$$

4.3 CARRIER-SENSE MULTIPLE ACCESS WITH COLLISION AVOIDANCE MAC

CSMA/CA MAC is commonly used in IEEE 802.11 wireless local area networks (WLANs). IEEE 802.11b has a data rate of up to 11 Mbps, whereas IEEE 802.11a and IEEE 802.11g have data rates of up to 54 Mbps. The upcoming IEEE 802.11n has a data rate of up to 600 Mbps. There are two access methods in CSMA/CA MAC: the basic access method and the request to send/clear to send (RTS/CTS) access method. In the basic access method, there is a two-way handshake. The source station will send its frame to the destination station in the data transmission phase. After receiving the frame correctly, the destination station will send an acknowledgment to the source station in the acknowledgment phase. Thus, this process completes the two-way handshake. In the RTS/CTS access method, there is four-way handshaking. First, the source station will send a RTS frame to the destination station. If the destination station receives the RTS frame correctly and if it is available for reception, it will reply with a CTS frame. Then the source station will send its data frame to the destination station. Upon receiving the data frame correctly, the destination station will acknowledge receipt of the data frame with an acknowledgment frame. This completes the four-way handshake. If the payload length is below a certain threshold, the basic access method is used. On the other hand, if the payload length is above the threshold, the RTS/CTS access method is used.

The CSMA/CA MAC works as follows.

- If the channel is idle for more than a distributed coordination function interframe space time (DIFS), a station can transmit immediately.
- If the channel is busy, the station will generate a random backoff period. This random backoff period is selected uniformly from zero to the current contention window size minus one. The backoff counter will decrement by one if the channel is idle for each time slot and will freeze if the channel is sensed busy. The backoff counter is reactivated to count down when the channel is sensed idle for more than a distributed coordination function interframe space time. At the initial backoff stage, the current contention window size is set at the minimum contention window size.
- If the backoff counter reaches zero, the station will attempt to transmit its frame. If it is successful, the destination station will send an acknowledgment after a short interframe space and the current contention window size is reset to the

minimum contention window size. If it is not successful, it will increase the current contention window size by doubling it only until a maximum contention window size is reached in the next backoff stage and a new random backoff period is selected as before.

- This process repeats until the frame is transmitted successfully or until the maximum retry limit is reached. If the frame is still not transmitted successfully, it is dropped.
- If a station does not receive an acknowledgment within an acknowledgment timeout period after a frame is transmitted, it will continue to attempt to retransmit the frame according to the backoff algorithm.
- In the RTS/CTS access method, if a station does not receive a CTS frame within a CTS timeout period after sending a RTS frame, it will attempt to retransmit the frame according to the RTS/CTS access method and the backoff algorithm.

4.3.1 Analysis

The analysis here follows that in [2] with some simplifications to obtain closed-form results. Let $b(t)$ be a random process representing the value of the backoff counter at time t and $a(t)$ be a random process representing the backoff stage j, $j = 0, 1, \ldots, L$ at time t, where L is the retry limit. The value of the backoff counter $b(t)$ is chosen uniformly in the range $\{0, 1, \ldots, W_j - 1\}$, with

$$W_j = \begin{cases} 2^j W, & L \leq m \\ 2^m W, & L > m \end{cases} \tag{4.13}$$

and W is the minimum contention window size. Let p denote the probability that a transmitted frame collides; p is also equal to the probability that a station in the backoff stage senses the channel busy. The two-dimensional random process $\{a(t), b(t)\}$ is a discrete-time Markov chain. Thus, the state of each station is described by $\{j, k\}$, where j stands for the backoff stage taking values from $\{0, 1, \ldots, L\}$ and k stands for the backoff delay taking values from $\{0, 1, \ldots, W_j - 1\}$ in time slots. The discrete-time Markov chain for the state transition diagram is shown in Figure 4.2. The non-null transition probabilities are listed as follows:

$$\begin{aligned}
\Pr[(0,k)|(j,0)] &= \frac{1-p}{W_0}, \quad 0 \leq k \leq W_0 - 1, \quad 0 \leq j < L, \\
\Pr[(0,k)|(j,0)] &= \frac{1}{W_0}, \quad 0 \leq k \leq W_0 - 1, \quad j = L, \\
\Pr[(j,k)|(j-1,0)] &= \frac{p}{W_j}, \quad 0 \leq k \leq W_j - 1, \quad 1 \leq j \leq L, \\
\Pr[(j,k)|(j,k)] &= p, \quad 1 \leq k \leq W_j - 1, \quad 0 \leq j \leq L, \\
\Pr[(j,k)|(j,k+1)] &= 1-p, \quad 0 \leq k \leq W_j - 2, \quad 0 \leq j \leq L.
\end{aligned}$$

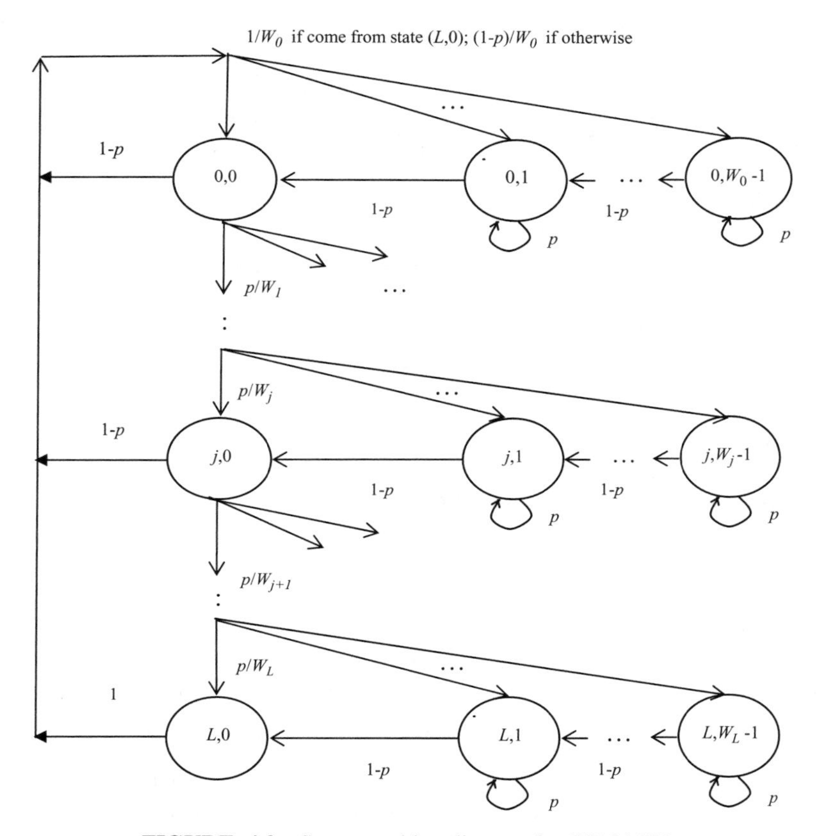

FIGURE 4.2 State transition diagram for CSMA/CA.

The first and second equations above represent the transition probability to each state in backoff stage 0, and the third equation represents the transition probability to each state in backoff stage j. The fourth equation represents the self-transition probability in a state due to the channel being busy, while the fifth equation represents the backoff counter decrementing by one as the channel is not busy. Let $b_{j,k} = \lim_{t\to\infty} \Pr[a(t) = j, b(t) = k]$ be the stationary distribution of the Markov chain. In steady state, we can derive the following relationships through chain regularities:

$$b_{j,0} = p^j b_{0,0}, \qquad k = 0, \quad 0 \le j \le L, \tag{4.14}$$

$$b_{j,k} = \frac{W_j - k}{W_j} \frac{1}{1-p} b_{j,0}, \qquad 1 \le k \le W_j - 1, \quad 0 \le j \le L. \tag{4.15}$$

Equating the sum of total probability to 1, we have

$$\sum_{j=0}^{L} \sum_{k=0}^{W_j - 1} b_{j,k} = 1. \tag{4.16}$$

Substituting (4.14) and (4.15) into (4.16), we have

$$\sum_{j=0}^{L}\left[p^j b_{0,0} + \frac{1}{1-p}\sum_{k=1}^{W_j-1}\frac{W_j-k}{W_j}p^j b_{0,0}\right] = 1. \tag{4.17}$$

Simplifying and rearranging (4.17), we have

$$\begin{aligned} b_{0,0} &= \frac{1}{\sum_{j=0}^{L}\left[1+\frac{1}{1-p}\sum_{k=1}^{W_j-1}\frac{W_j-k}{W_j}\right]p^j} \\ &= \frac{1}{\sum_{j=0}^{L}\left[1+\frac{1}{1-p}\frac{W_j-1}{2}\right]p^j}. \end{aligned} \tag{4.18}$$

Substituting (4.13) into (4.18) and simplifying, we have

$$b_{0,0} = \begin{cases} \dfrac{2(1-p)^2(1-2p)}{(1-2p)^2(1-p^{L+1})+W(1-p)[1-(2p)^{L+1}]}, & L \le m \\ \dfrac{2(1-p)^2(1-2p)}{(1-2p)^2(1-p^{L+1})+W(1-p)[1-(2p)^{m+1}]+W2^m p^{m+1}(1-2p)(1-p^{L-m})}, & L > m. \end{cases} \tag{4.19}$$

Let τ be the probability that a station transmits during a generic slot. A station transmits when its backoff counter reaches zero, i.e., the station is at any of the states $\{j, 0\}, 0 \le j \le L$.

$$\begin{aligned} \tau &= \sum_{j=0}^{L} b_{j,0} \\ &= \sum_{j=0}^{L} p^j b_{0,0} \\ &= \frac{1-p^{L+1}}{1-p} b_{0,0}. \end{aligned} \tag{4.20}$$

Substituting (4.19) into (4.20), we have

$$\tau = \begin{cases} \dfrac{2(1-p)(1-2p)(1-p^{L+1})}{(1-2p)^2(1-p^{L+1})+W(1-p)[1-(2p)^{L+1}]}, & L \le m \\ \dfrac{2(1-p)(1-2p)(1-p^{L+1})}{(1-2p)^2(1-p^{L+1})+W(1-p)[1-(2p)^{m+1}]+W2^m p^{m+1}(1-2p)(1-p^{L-m})}, & L > m. \end{cases} \tag{4.21}$$

Let n denote the number of stations. The probability p that a station in the backoff stage senses the channel to be busy is given by

$$p = 1 - (1 - \tau)^{n-1} . \tag{4.22}$$

τ and p can be solved numerically.

Let p_b denote the probability that the channel is busy. It occurs when at least one station transmits during a slot time and is given by

$$p_b = 1 - (1 - \tau)^n . \tag{4.23}$$

Let p_s denote the probability that a successful transmission occurs in a time slot and is given by

$$p_s = n\tau (1 - \tau)^{n-1} . \tag{4.24}$$

Let S denote the normalized saturated throughput. Let δ, $T_{E(L)}$, T_s, and T_c denote the duration of an empty slot, the time to transmit the average payload, the average time that the channel is sensed busy because of a successful transmission and the average time that the channel has a collision, respectively. The probability that the channel is idle for a slot time is $1 - p_b$. The probability that the channel is neither idle nor successful for a time slot is $[1 - (1 - p_b) - p_s] = p_b - p_s$. The normalized saturated throughput, S, is given by [4]

$$\begin{aligned} S &= \frac{E(\text{payload transmission time in a slot time})}{E(\text{length of a slot time})} \\ &= \frac{p_s T_{E(L)}}{(1 - p_b)\delta + p_s T_s + [p_b - p_s] T_c} . \end{aligned} \tag{4.25}$$

Let T_H, T_{ACK}, T_{SIFS}, T_{DIFS}, L^*, $T_{E(L^*)}$, and γ denote the time to transmit the header (including the MAC header, physical layer header, and the tail), the time to transmit an acknowledgment, the short interframe space, the distributed coordination function interframe space, the length of the largest frame in a collision, the time to transmit a payload with length $E(L^*)$, and the propagation time, respectively. For the basic access method,

$$T_s = T_H + T_{E(L)} + T_{\text{SIFS}} + \gamma + T_{\text{ACK}} + T_{\text{DIFS}} + \gamma \tag{4.26}$$

and

$$T_c = T_H + T_{E(L^*)} + T_{\text{DIFS}} + \gamma . \tag{4.27}$$

For the RTS/CTS access method,

$$T_s = T_{\mathrm{RTS}} + T_{\mathrm{SIFS}} + \gamma + T_{\mathrm{CTS}} + T_{\mathrm{SIFS}} + \gamma + T_H + T_{E(L)} + T_{\mathrm{SIFS}} + \gamma \\ + T_{\mathrm{ACK}} + T_{\mathrm{DIFS}} + \gamma \tag{4.28}$$

and

$$T_c = T_{\mathrm{RTS}} + T_{\mathrm{DIFS}} + \gamma. \tag{4.29}$$

Station delay is the average delay under saturation conditions and includes the medium access delay due to backoff, collisions, the transmission delay, and the interframe spaces such as T_{SIFS} [2]. The average backoff delay depends on the value of a station's backoff counter and the duration when the counter freezes due to other transmissions. Let X denote the random variable representing the total number of backoff slots that a frame encounters without the case when the counter freezes, and its mean value is given by [2]

$$E(X) = \sum_{j=0}^{L} \frac{p^j(1-p)}{1-p^{L+1}} \sum_{h=0}^{j} \frac{W_h - 1}{2}. \tag{4.30}$$

Substituting (4.13) into (4.30) and using the identities

$$\sum_{j=0}^{n} jp^j = p\left[\frac{1-p^n}{(1-p)^2} - \frac{np^n}{1-p}\right] \tag{4.31}$$

and

$$\sum_{j=m}^{n} jp^j = \frac{p^{m+1} - p^{n+2}}{(1-p)^2} - \frac{mp^m - (n+1)p^{n+1}}{1-p}, \tag{4.32}$$

we have

$$E(X) = \begin{cases} \dfrac{W(1-p)[1-(2p)^{L+1}]}{(1-2p)(1-p^{L+1})} - \dfrac{W+1}{2} - \dfrac{p(1-p^L)}{2(1-p)(1-p^{L+1})} + \dfrac{Lp^{L+1}}{2(1-p^{L+1})}, & L \le m \\ \dfrac{W(1-p)[1-(2p)^{m+1}]}{(1-2p)(1-p^{+1})} - \dfrac{W+1}{2} - \dfrac{p(1-p^m)}{2(1-p)(1-p^{L+1})} + \dfrac{mp^{m+1}}{2(1-p^{L+1})} & \\ + \dfrac{(W2^m - W2^{m-1}m)p^{m+1}(1-p^{L-m})}{1-p^{L+1}} + \dfrac{(W2^m-1)p(p^{m+1}-p^{L+1})}{2(1-p)(1-p^{L+1})} & \\ + \dfrac{(W2^m-1)p[(m+1)p^m - (L+1)p^L]}{2(1-p^{L+1})}, & L < m. \end{cases} \tag{4.33}$$

Only successful transmissions are considered in (4.33). Let B denote the random variable representing the total number of slots that a frame encounters when the counter freezes. The portion of idle slots used to decrease $E(X)$ is $(1-p)$. Thus, the mean value of B is given by

$$E(B) = \frac{E(X)}{1-p}p. \tag{4.34}$$

$E(X)$ and $E(B)$ can be considered as the total number of idle and busy slots, respectively, that the frame encounters during backoff stages. The average number of retries, $E(N)$, is

$$\begin{aligned} E(N) &= \sum_{j=0}^{L} \frac{jp^j(1-p)}{1-p^{L+1}} \\ &= \frac{1-p}{1-p^{L+1}} \sum_{j=0}^{L} jp^j \\ &= \frac{1-p}{1-p^{L+1}} p \left[\frac{1-p^L}{(1-p)^2} - \frac{Lp^L}{1-p} \right] \\ &= \frac{p(1-p^L)}{(1-p)(1-p^{L+1})} - \frac{Lp^{L+1}}{1-p^{L+1}}. \end{aligned} \tag{4.35}$$

Note that $E(N)$ is one less than the number of transmissions. Let D denote the random variable representing the frame delay. Let T_o denote the time that a station has to wait when its frame transmission collides before sensing the channel again. Let $T_{\text{ACK timeout}}$ and $T_{\text{CTS timeout}}$ denote the duration of the ACK and CTS timeouts, respectively. The average slot lengths for a station are

- δ for an idle slot at state $\{j, k\}$, $k > 0$
- $\left[\frac{p_s}{p_b} T_s + \frac{p_b - p_s}{p_b} T_c \right]$ for a busy slot at states $\{j, k\}$, $k > 0$
- $T_c + T_o$ for a failed transmission slot at state $\{j, 0\}$, $k = 0$
- T_s for a successful transmission at state $\{j, 0\}$, $k = 0$

Thus, the average frame delay is given by [2]

$$E(D) = E(X)\delta + E(B) \left[\frac{p_s}{p_b} T_s + \frac{p_b - p_s}{p_b} T_c \right] + E(N)(T_c + T_o) + T_s. \tag{4.36}$$

For the basic access method,

$$T_o = T_{\text{SIFS}} + T_{\text{ACK timeout}}. \tag{4.37}$$

For the RTS/CTS access method,

$$T_o = T_{\mathrm{SIFS}} + T_{\mathrm{CTS\ timeout}}. \tag{4.38}$$

4.4 POLLING MAC

Polling is used in Bluetooth, which is a wireless personal area network (WPAN). Polling uses one master and up to seven slaves. The maximum data rate in Bluetooth is 1 Mbps. Figure 4.3 shows a polling MAC with n stations and one central controller. The central controller sends a *polling message* to the first station in the polling sequence. After receiving the polling message, this station transmits its data to the central controller and indicates that it has completed transmitting by adding a *go-ahead message*. The central controller then polls the next station in sequence and the process continues until all stations have had an opportunity to transmit. The polling sequence is used over and over again.

4.4.1 Analysis

Let λ in packets per second be the average arrival rate at each station (the same for every station). The arrivals are assumed to be Poisson distributed. Let R be the medium bit rate in bits per second. Let $\overline{X}$ be the average packet length in bits. The service at each station is assumed to be exhaustive; that is, all stored packets are transmitted each time the station has access to the channel. Propagation and other delays between stations are assumed equal. Let τ be the maximum propagation

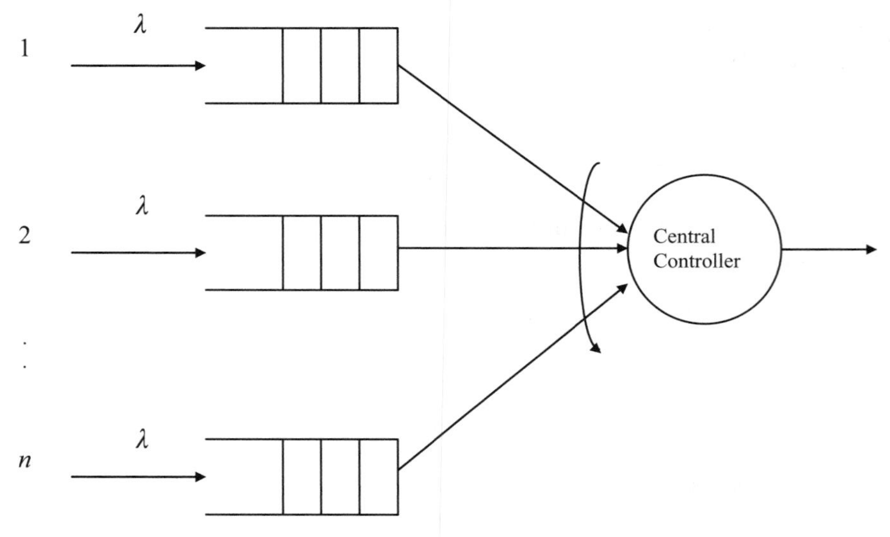

FIGURE 4.3 Polling.

delay between stations in seconds. This system can be modeled as an $M/G/1$ queue with vacation for n users. The average packet transmission time, denoted by $\overline{P}$, is given by

$$\overline{P} = \frac{\overline{X}}{R}. \tag{4.39}$$

The vacation time (station walk time) consists of the propagation time, the time to transmit control packets (*polling* and *go-ahead*), and possibly the synchronization time and turnaround time for a half duplex link. We can also consider this time as the time required to transfer the use of the channel from one station to the next. Using queueing theory and assuming that the vacation time is *constant*, the average queueing delay in the buffer, denoted by $\overline{W}$, is given by (see the Appendix)

$$\begin{aligned}\overline{W} &= \frac{\rho}{2(1-\rho)}\frac{\overline{Y^2}}{\overline{Y}} + \frac{(n-\rho)\overline{V}}{2(1-\rho)} + \frac{\mathrm{Var}[V]}{2\overline{V}} \\ &= \frac{\rho}{2(1-\rho)}\frac{\overline{Y^2}}{\overline{Y}} + \frac{(n-\rho)\overline{V}}{2(1-\rho)}.\end{aligned} \tag{4.40}$$

where $\overline{Y}$ is the mean service time, $\overline{Y^2}$ is the second moment of service time, $\rho = \lambda\overline{Y}$, $\overline{V}$ is the mean vacation period, and $\mathrm{Var}[V]$ is the variance of the vacation period. Note that $\mathrm{Var}[V]$ is zero, as the vacation period is assumed to be constant. The average transfer delay for polling, denoted by $\overline{T}$, is given by

$$\begin{aligned}\overline{T} &= \overline{P} + \overline{W} + \tau \\ &= \frac{\overline{X}}{R} + \frac{\rho}{2(1-\rho)}\frac{\overline{Y^2}}{\overline{Y}} + \frac{(n-\rho)\overline{V}}{2(1-\rho)} + \tau.\end{aligned} \tag{4.41}$$

4.5 RESERVATION MAC

We consider a form of reservation MAC known as *dynamic time-division multiple access* (D-TDMA). Figure 4.4 shows the D-TDMA frame format, where the frame is partitioned into S_r reservation slots and S payload slots, consisting of S_v voice

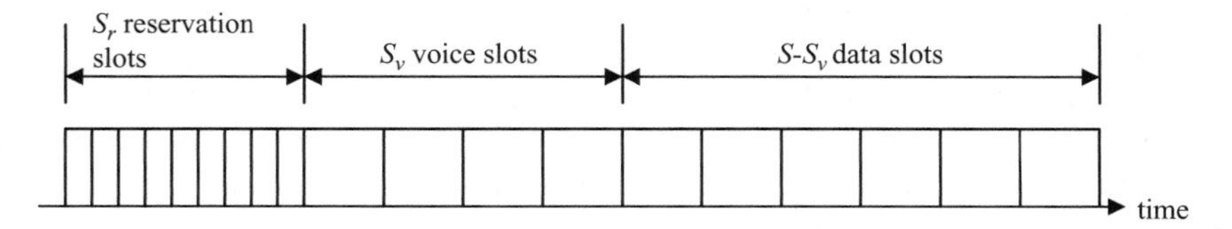

FIGURE 4.4 Frame format for dynamic time-division multiple access.

slots and $S - S_v$ data slots. When a user generates a new voice talkspurt or new data packet, an appropriate reservation packet is transmitted in a reserved slot in the next frame with probability p_t or p_r for voice or data, respectively. If there is more than one reservation packet transmitted in the same reservation slot, a collision occurs. The collided user can retry in the next reservation slot with the respective probability for voice and data. If the voice user is successful at the reservation phase, it will be assigned a voice slot to use until the end of the talkspurt. If no voice slot is available, the voice user will recontend in the next frame. If the data user is successful at the reservation phase, it can use the remaining slots, excluding the voice slots. If no data slot is available, the data user will recontend in the next frame.

4.5.1 Analysis

A voice user is modeled as having three states: the silent state (SS), the contention state (CS), and the reservation state (RS). When the voice user has no new talkspurt, it is in the *silent state*. When the voice user has a new talkspurt and it has not made a reservation, it is in the *contention state*. When the voice user contends successfully and obtains a reservation, it is in the *reservation state*. The probability that a voice user returns to the silent state before it obtains a reservation is assumed to be zero. The lengths of talkspurts and silent periods are assumed to be distributed exponentially with mean t_1 and t_2, respectively. The frame duration is assumed to be T. The probability that a talkspurt ends in a frame denoted by q_s, and the probability that a talkspurt is generated in a frame denoted by q_a, are, respectively, given by

$$q_s = 1 - \exp(-T/t_1) \tag{4.42}$$

and

$$q_a = 1 - \exp(-T/t_2). \tag{4.43}$$

The probability that a user gets a reservation in the current frame depends on permission probability, voice activity factor, channel status, and so on. [3].

The data user can be in two states: the thinking state (TH) and the backlogged state (BK). In the *thinking state*, the data user is waiting for the generation of a new data packet. In the *backlogged state*, the data user is trying to transmit the packet generated. The data packet is generated assuming a Bernoulli distribution with parameter p_0 packets/frame. The transition probability from the thinking state to the backlogged state is p_0, and the transition probability from the backlogged state to the thinking state is $p_r P_{\text{succ}}$, where P_{succ} is the transmission success probability.

With N_v voice users and N_d data users, the system can be fully described by three state variables: the number of voice users in the reservation state, N_r; the number of voice users in the silent state, N_s; and the number of data users in the backlogged state, N_b. The number of voice users in the contention state is $N_c = N_v - N_r - N_s$, and the number of data users in the thinking state is $N_t = N_d - N_b$.

Two Markov chains are considered [3]. The stationary distribution, denoted by $\mathbf{\Pi}^{(s)}$, for the state variable N_s for transition of voice users entering and leaving the silent state is given by

$$\mathbf{\Pi}^{(s)} = \{\pi(n_s)\} = \{P(N_s = n_s)\}. \tag{4.44}$$

The stationary distribution, denoted by $\mathbf{\Pi}^{(r,b|s)}$, for the state variables N_r and N_b, given N_s, for the reservation or transmission status of the system is given by

$$\mathbf{\Pi}^{(r,b|s)} = \{\pi(n_r, n_b|n_s)\} = \{P(N_r = n_r, N_b = n_b|N_s = n_s)\}. \tag{4.45}$$

Equation (4.44) can be solved using the following flow balance equations, and the sum of total probability of all states in the Markov chain has to be 1. Thus, we have

$$\mathbf{\Pi}^{(s)} = \mathbf{\Pi}^{(s)}\boldsymbol{P}^{(s)}, \tag{4.46}$$

and

$$\sum_{n_s} \pi(n_s) = 1, \tag{4.47}$$

where $\boldsymbol{P}^{(s)}$ is the one-step transition probability matrix given by [3]

$$\boldsymbol{P}^{(s)} = \{P(N_s(x+1) = j|N_s(x) = i)\}, \qquad i, j = 0, \ldots, N_v. \tag{4.48}$$

$N_s(x)$ is the number of voice users in the silent state at the beginning of the xth frame, and $P(N_s(x+1) = j|N_s(x) = i)$ is given by [3]

$$P(N_s(x+1) = j|N_s(x) = i) = \sum_{l=\max(0,i-j)}^{\min(i,N_v-j)} B(i, l, q_a)B(N_v - i, j - (i - l), q_s), \tag{4.49}$$

where l is the number of voice users leaving the silent state and becoming active in the current frame, and $B(n, m, p)$ is given by

$$B(n, m, p) = \binom{n}{m} p^m(1-p)^{n-m}. \tag{4.50}$$

Similarly, (4.45) can be solved using the following flow balance equations, and the sum of total probability of all states in the Markov chain has to be 1. Thus, we have

$$\mathbf{\Pi}^{(r,b|s)} = \mathbf{\Pi}^{(r,b|s)}\boldsymbol{P}^{(r,b|s)} \tag{4.51}$$

and

$$\sum_{n_r}\sum_{n_b}\pi(n_r, n_b|n_s) = 1, \tag{4.52}$$

where $\boldsymbol{P}^{(r,b|s)}$, the one-step transition probability matrix of the reservation process conditioned on the number of silent voice users $N_s = n_s$, is given by [3]

$$\boldsymbol{P}^{(r,b|s)} = \{P(N_r(x+1) = m, N_b(x+1) = n|N_r(x) = i, N_b(x) = j)\},$$
$$m, i = 0, \ldots, N_v, \quad n, j = 0, \ldots, N_d, \tag{4.53}$$

where

$$\begin{aligned} &P(N_r(x+1) = m, N_b(x+1) = n|N_r(x) = i, N_b(x) = j) \\ &= \sum_{l=\max(0,m-i)}^{\min(N_v-n_s-i,S_r,S_v-i)} \sum_{k=\max(0,j-n)}^{\min(j,N_d-n,S_r-l,S-(i+l))} \\ &\quad \Phi(N_v - n_s - i, j; S_r, S - i; l, k)B(i + l, m, 1 - q_s) \\ &\quad \times B(N_d - j, n - (j - k), p_0), \end{aligned} \tag{4.54}$$

l is the number of voice users who obtain reservations successfully in the current frame, and k is the number of data users transmitted successfully in the same frame. $\Phi(N_v - n_s - i, j; S_r, S - i; l, k)$ is the probability that among $N_v - n_s - i$ contending voice users and j backlogged data users, l voice users obtain reservation and k data users transmit information packets successfully in a frame with S_r reservation slots and $S - i$ available information slots. It is given by

$$\Phi(N_v - n_s - i, j; S_r, S - i; l, k) =$$
$$\times \begin{cases} \Theta(N_v - n_s - i, j; S_r; l, k), \; l < S_v - i, l + k < S - i, & \\ \displaystyle\sum_{l_1=l}^{\min(S_r-k,N_v-n_s-i)} \Theta(N_v - n_s - i, j; S_r; l_1, k), & l = S_v - i, \quad l + k < S - i, \\ \displaystyle\sum_{k_1=k}^{\min(S_r-l,j)} \Theta(N_v - n_s - i, j; S_r; l, k_1), & l < S_v - i, \quad l + k = S - i, \\ \displaystyle\sum_{l_1=l}^{\min(S_r-k,N_v-n_s-i)} \sum_{k_1=k}^{\min(S_r-l,j)} \Theta(N_v - n_s - i, j; S_r; l_1, k_1), & l = S_v - i, \quad l + k = S - i, \end{cases} \tag{4.55}$$

where $\Theta(s, t; c; l_1, k_1)$ is the probability that among s contending voice users and t backlogged data users, there are l_1 voice successes and k_1 data successes in c reservation slots. It is given by

$$\begin{aligned} \Theta(s, t; c; l_1, k_1) = {} & [1 - \xi_v(s, t) - \xi_d(s, t)]\Theta(s, t; c - 1; l_1, k_1) + \xi_v(s, t)\Theta(s - 1, t; \\ & c - 1; l_1 - 1, k_1) + \xi_d(s, t)\Theta(s, t - 1; c - 1; l_1, k_1 - 1), \end{aligned} \tag{4.56}$$

where $\xi_v(s,t)$ and $\xi_d(s,t)$ are the probabilities that with s contending voice users and t backlogged data users, there is a voice or data success, respectively, in the current reservation slot. They are given, respectively, by

$$\xi_v(s,t) = B(s,1,p_t)B(t,0,p_r) \tag{4.57}$$

and

$$\xi_d(s,t) = B(s,0,p_t)B(t,1,p_r). \tag{4.58}$$

The ending condition of the recursive evaluation in (4.56) is given by

$$\Theta(s,t;c;l_1,k_1) = \begin{cases} [1-\xi_v(s,t)-\xi_d(s,t)]^c, & l_1 = k_1 = 0 \\ 0, & c < (l_1+k_1) \text{ or } l_1 < 0 \text{ or } k_1 < 0. \end{cases} \tag{4.59}$$

The packet loss probability, denoted by P_{loss}, is the average fraction of packets in a talkspurt dropped due to delayed transmission. P_{loss} is given by [3]

$$P_{\text{loss}} = \frac{E(N_c) - E(R)}{E(N_c) + E(N_r)}, \tag{4.60}$$

where R is the number of successful voice reservations per frame,

$$E(N_c) = \sum_{n_r=0}^{\min(N_v,S_v)} \sum_{n_s=0}^{N_v-n_r} \sum_{n_b=0}^{N_d} (N_v - n_r - n_s)\pi(n_r,n_s,n_b), \tag{4.61}$$

$$E(N_r) = \sum_{n_r=0}^{\min(N_v,S_v)} \sum_{n_s=0}^{N_v-n_r} \sum_{n_b=0}^{N_d} n_r \pi(n_r,n_s,n_b), \tag{4.62}$$

$$E(R) = \sum_{n_r=0}^{\min(N_v,S_v)} \sum_{n_s=0}^{N_v-n_r} \sum_{n_b=0}^{N_d} \sum_{l=0}^{\min(N_v-n_r-n_s,S_r,S_v-n_r)} \sum_{k=0}^{\min(n_b,S_r-l,S-(n_r+l))} l$$
$$\times\, \Theta(N_v - n_r - n_s, n_b; S_c, S - n_r; l, k)\pi(n_r,n_s,n_b) \tag{4.63}$$

and

$$\pi(n_r,n_s,n_b) = \pi(n_r,n_b|n_s)\pi(n_s). \tag{4.64}$$

The average data throughput, denoted by β, is the average number of data packets transmitted successfully in a frame. It is given by

$$\beta = \sum_{n_r=0}^{\min(N_v,S_v)} \sum_{n_s=0}^{N_v-n_r} \sum_{n_b=0}^{N_d} \sum_{l=0}^{\min(N_v-n_r-n_s,S_r,S_v-n_r)} \sum_{k=0}^{\min(n_b,S_r-l,S-(n_r+l))} k \times \Theta(N_v - n_r - n_s, n_b; S_c, S - n_r; l, k)\pi(n_r, n_s, n_b). \tag{4.65}$$

The average number of backlogged data users in a frame is [3]

$$E(N_b) = \sum_{n_r=0}^{\min(N_v,S_v)} \sum_{n_s=0}^{N_v-n_r} \sum_{n_b=0}^{N_d} n_b \pi(n_r, n_s, n_b). \tag{4.66}$$

Let the *average delay*, defined as the average time that a data packet spends in the buffer until the beginning of the successful transmission, be denoted by D_d. Using Little's formula [3], we have

$$D_d = \frac{E(N_b)}{\beta} - 1. \tag{4.67}$$

From (4.66) and (4.67), it can be seen that if the average backlogged data users, $E(N_b)$, is large, the average delay, D_d, will also be long.

4.6 ENERGY-EFFICIENT MAC

In this section we consider the IEEE 802.15.4 MAC, which is energy efficient. The superframe format in a beacon-enabled mode is shown in Figure 4.5. The beacon is sent by the coordinator for synchronization and to indicate the length of the contention access period (CAP) and the length of the contention-free period (CFP), as well as to indicate if there are frames in the coordinator for the child node. The coordinator listens to the channel for the whole CAP to detect and receive any data from its child nodes. On the other hand, the child nodes only transmit a data frame and receive an optional acknowledgment when needed. This increases their energy efficiency. Uplink transmission consists of three-way handshaking: beacon, data frame by the child node, and acknowledgment from the coordinator. On the other hand, downlink

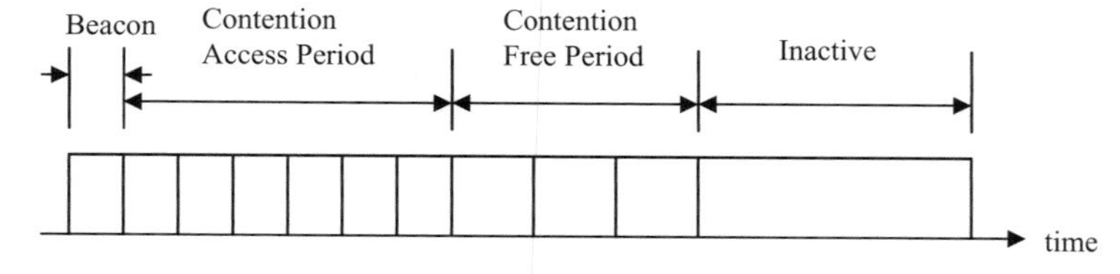

FIGURE 4.5 Superframe format for IEEE 802.15.4 in beacon-enabled mode.

transmission is indirect and consists of five-way handshaking: beacon, data request frame by the child node, acknowledgment by the coordinator, data frame by the coordinator, and acknowledgment by the child node.

IEEE 802.15.4 uses a modified slotted CSMA/CA scheme during the CAP except for acknowledgment frames, which are transmitted without carrier sensing. IEEE 802.15.4 is used by ZigBee, which has a data rate of up to 250 kbps. The scheme is modified from the IEEE 802.11 DCF protocol [1,2]. The major differences are that a channel is not sensed during a backoff time and that a new random backoff is selected if a channel is busy during carrier sensing.

- To access a channel, each node maintains three variables: NB, BE, and CW, where NB is the number of CSMA/CA backoff attempts for the current transmission, BE is the backoff exponent, and CW is the contention window length. NB is initialized to 0. BE defines the number of backoff periods a node should wait before attempting a clear channel assessment (CCA). CW defines the number of consecutive backoff periods a channel has to be silent prior to a transmission.
- Prior to a transmission, a node locates a backoff period boundary through the beacon received, waits for a random number of backoff periods in the range $0, 1, \ldots, 2^{\mathrm{BE}} - 1$, and senses the channel by CCA for CW times. CW is set at 2. Thus, two CCA analyses are performed if the first CCA is assessed to be clear.
- If the channel is idle, a transmission begins.
- Otherwise, NB and BE are increased by one and the operation returns to the random delay phase.
- If NB exceeds a threshold of macMaxCSMABackoffs (set at 4), transmission terminates with a channel access failure.
- A node may try to retransmit the frame for a maximum of aMaxFrameRetries (set at 3) times before MAC issues a frame transmission failure.

4.6.1 Analysis

The analysis in this subsection follows that in [4]. Let q_S and q_L denote the probability that a single short or long data transmission with its consecutive acknowledgment is detected by clear channel assessment (CCA) at any time in the contention access period (CAP). These are estimated by the data transmission times and the acknowledgment frames. Thus, we have [4]

$$q_S = \frac{L_S + L_{\mathrm{ACK}}}{t_{\mathrm{CAP}} R} \tag{4.68}$$

and

$$q_L = \frac{L_L + L_{\mathrm{ACK}}}{t_{\mathrm{CAP}} R}, \tag{4.69}$$

where L_S is the length of the short data frame, L_L is the length of the long data frame, L_{ACK} is the length of the acknowledgment frame, t_{CAP} is the duration of the CAP, and R is the radio data rate. Considering a cluster-tree network structure, the number of nodes hierarchically below an analyzed coordinator, denoted by n, is given by

$$n = \sum_{a=1}^{k} n_C^a (1 + n_D), \tag{4.70}$$

where k is the network depth below the coordinator being analyzed, n_C is the number of child coordinators, and n_D is the number of devices. Let d_S be the average amount of short data transmissions during CAP modeled by the number of nodes whose data is routed through the coordinator. Let u denote the number of retransmissions, I_U the uplink data transmission interval, and I_D the downlink data transmission interval, normalized to a beacon interval, denoted by I_B. To approximate the effect of indirect transmissions, I_D is divided by 2. Thus, we have

$$d_S = \left(\frac{n_D}{I_U} + \frac{n_D + n_C}{I_D/2} \right) u. \tag{4.71}$$

Let d_L be the average amount of long data transmissions during CAP, modeled by [4]

$$d_L = \frac{n L_S u}{I_U L_L}. \tag{4.72}$$

Let p_C be the probability of a clear channel by CCA. The IEEE 802.15.4 specifies two CCA analyses before attempting to transmit a frame. Thus, p_C can be modeled by

$$p_C = (1 - q_S)^{2d_S(1-h)} (1 - q_L)^{2d_L(1-h)}, \tag{4.73}$$

where h is the probability that two randomly deployed nodes in the range of a coordinator have a hidden node relationship. Let s denote the probability of successful CCA with b backoff attempts given by

$$s = \sum_{a=1}^{b} p_C (1 - p_C)^{a-1}. \tag{4.74}$$

The average number of backoffs for each frame, denoted by r, is given by

$$r = (1 - s)b + \sum_{a=1}^{b} a p_C (1 - p_C)^{a-1}. \tag{4.75}$$

Let the average backoff time as a function of a backoff component, BE, be denoted by t_{BO} (BE). It is given by

$$t_{\text{BO}}(\text{BE}) = \frac{2^{\text{BE}} - 1}{2} t_{\text{BOP}}, \tag{4.76}$$

where t_{BOP} is the length of the backoff period. Let t_{BOT} denote the total backoff time given by

$$t_{\text{BOT}} = \frac{3}{2} r (t_{\text{IR}} + t_{\text{CCA}}) + \sum_{a=0}^{r-1} t_{\text{BO}}(\min(\text{macMinBE} + a, \text{aMaxBE})), \tag{4.77}$$

where BE is initialized to macMinBE and is incremented by one until aMaxBE is reached after each unsuccessful backoff attempt. Note that CCA is performed twice only if the first CCA is assessed to be clear. Thus, $\frac{3}{2}$ CCA analyses are performed for each backoff attempt. T_{IR} is the transient time from idle mode to receive mode, and t_{CCA} is the CCA analysis time. It is assumed that a radio is in the idle mode during backoff time. Let E_{BOT} denote the total backoff energy, which can be approximated as

$$E_{\text{BOT}} = \frac{3}{2} r (t_{\text{IR}} + t_{\text{CCA}})(P_{\text{CCA}} - P_I) + t_{\text{BOT}} P_I, \tag{4.78}$$

where P_{CCA} is the power consumption during CCA and P_I is the power consumption during idle mode. Let p_h denote the probability that two nodes in a hidden node position transmit simultaneously and collide. We model p_h as

$$p_h = 2 \frac{q_L d_L + q_S d_S}{d_L + d_S}. \tag{4.79}$$

Let p_d denote the probability that two nodes select the same backoff delay and collide, modeled as

$$p_d = \frac{1}{2^{\text{BE}} - 1}. \tag{4.80}$$

Let C denote the average number of contending nodes in CAP. For approximating the amount of traffic in CAP, C is determined as

$$C = \min\left[\left(\frac{1}{I_U} + \frac{2}{I_D}\right) u, 1\right] n_D + \min\left[\left(\frac{2}{I_D} + \frac{n L_S}{I_U n_C L_L}\right) u, 1\right] n_C. \tag{4.81}$$

Let p_s denote the probability of a successful transmission, modeled as [4]

$$p_s = s(1 - p_h)^{h(d_L + d_S)}(1 - p_d)^C. \tag{4.82}$$

Let c denote the number of times a frame is transmitted before declaring a transmission failure and v denote the probability of a successful transmission after c attempts. v is given by

$$v = \sum_{a=1}^{c} p_s(1 - p_s)^{a-1}. \tag{4.83}$$

The average number of transmission attempts per frame, denoted by u, is given by

$$\overline{u} = (1 - v)c + \sum_{a=1}^{c} ap_s(1 - p_s)^{a-1}, \tag{4.84}$$

where $\overline{u} = u + 1$. A frame transmission consists of a backoff time and the actual data transmission. Let t_{TXDS} and E_{TXDS} denote the time and energy, respectively, of a short data transmission. They are given by

$$t_{\text{TXDS}} = t_{\text{SI}} + t_{\text{BOT}} + t_{\text{IT}} + \frac{L_S}{R} \tag{4.85}$$

and

$$E_{\text{TXDS}} = t_{\text{SI}} P_I + E_{\text{BOT}} + \left(t_{\text{IT}} + \frac{L_S}{R}\right) P_{\text{TX}}, \tag{4.86}$$

where t_{SI} is the transient time from sleep mode to idle mode, t_{IT} is the transient time from idle mode to transmit mode, and P_{TX} is the power consumption in the transmit mode. Similarly, let t_{TXDL} and E_{TXDL} denote the time and energy, respectively, of a long data transmission. They are given by

$$t_{\text{TXDL}} = t_{\text{SI}} + t_{\text{BOT}} + t_{\text{IT}} + \frac{L_L}{R} \tag{4.87}$$

and

$$E_{\text{TXDL}} = t_{\text{SI}} P_I + E_{\text{BOT}} + \left(t_{\text{IT}} + \frac{L_L}{R}\right) P_{\text{TX}}. \tag{4.88}$$

We assume that indirect communication utilizes only a short frame and the coordinator response time is an average of t_{BOT} and the maximum response time for a data request, t_{RES}. Let t_{RXDD} and E_{RXDD} denote the time and energy, respectively, for an indirect data transmission after a data request, which are modeled as

$$t_{\text{RXDD}} = t_I + \frac{t_{\text{BOT}} + t_{\text{RES}}}{2} + \frac{L_S}{R} + t_{\text{LIFS}} \tag{4.89}$$

and

$$E_{\text{RXDD}} = (t_{\text{RXDD}} - t_{\text{LIFS}})P_{\text{RX}} + t_{\text{LIFS}} P_I, \tag{4.90}$$

where t_I is the time due to synchronization inaccuracy, t_{LIFS} is the long interframe space, and P_{RX} is the power consumption in the receive mode. We assume that the wait duration for acknowledgment is half of the maximum allowed wait duration for acknowledgment of t_{AW}. Let t_{RXA} and E_{RXA} denote the acknowledgment reception time and energy, respectively. They are modeled by

$$t_{\text{RXA}} = t_{\text{TR}} + \frac{t_{\text{AW}}}{2} + \frac{L_{\text{ACK}}}{R} + t_{\text{SIFS}} \tag{4.91}$$

and

$$E_{\text{RXA}} = (t_{\text{RXA}} - t_{\text{SIFS}})P_{\text{RX}} + t_{\text{SIFS}} P_I, \tag{4.92}$$

where t_{TR} is the transient time from transmit mode to receive mode and t_{SIFS} is the short interframe space. Let t_{TXA} and E_{TXA} denote the acknowledgment transmission time and energy, respectively. They are modeled by

$$t_{\text{TXA}} = t_{\text{RT}} + \frac{t_{\text{AW}}}{2} + \frac{L_{\text{ACK}}}{R} \tag{4.93}$$

and

$$E_{\text{TXA}} = \left(t_{\text{RT}} + \frac{L_{\text{ACK}}}{R}\right) P_{\text{TX}} + \frac{T_{\text{AW}}}{2} P_I, \tag{4.94}$$

where t_{RT} is the transient time from receive mode to transmit mode. We assume that a node is typically in a sleep mode before the beacon reception. Let t_{RXB} and E_{RXB} denote the beacon reception time and energy, respectively. They are modeled by

$$t_{\text{RXB}} = t_{\text{SI}} + t_{\text{IR}} + (\varepsilon_{\text{RX}} + \varepsilon_{\text{TX}})I_B + t_I + \frac{L_B}{R} + t_{\text{LIFS}} \tag{4.95}$$

and

$$E_{\text{RXB}} = (t_{\text{RXB}} - (t_{\text{SI}} + T_{\text{LIFS}}))P_{\text{RX}} + (t_{\text{SI}} + T_{\text{LIFS}})P_I, \tag{4.96}$$

where t_{IR} is the transient time from idle mode to receive mode, ε_{RX} and ε_{TX} are, respectively, the crystal tolerances in receiving and transmitting nodes, and L_B is the length of a beacon frame. Let t_{TXB} and E_{TXB} denote the beacon transmission time and energy, respectively. They are modeled without CCA analysis as

$$t_{\mathrm{TXB}} = t_{\mathrm{SI}} + t_{\mathrm{IT}} + \frac{L_B}{R} \tag{4.97}$$

and

$$E_{\mathrm{TXB}} = \left(t_{\mathrm{IT}} + \frac{L_B}{R} \right) P_{\mathrm{TX}} + t_{\mathrm{SI}} P_I. \tag{4.98}$$

If a node loses contact with its associated coordinator, it may perform either an orphan device realignment procedure or reset the MAC sublayer and reassociate with the network. In the latter case, the node performs a passive scan on a single channel before reassociation in beacon-enabled networks. We assume that the energy required for the message exchange during the association is negligible compared to the scanning energy and is ignored in the following analysis. Let t_{NS} and E_{NS} denote the network scanning time and energy, respectively. They are modeled as

$$t_{\mathrm{NS}} = t_{\mathrm{IR}} + \mathrm{aBaseSuperframe}(2^{\mathrm{BO}} + 1) \tag{4.99}$$

and

$$E_{\mathrm{NS}} = t_{\mathrm{NS}} P_{\mathrm{RX}}, \tag{4.100}$$

where aBaseSuperframe = 15.36 ms and $0 \leq BO \leq 14$. Let DC_{DEV} denote the duty cycle of a device that is calculated with beacon receptions, uplink and downlink data exchanges, and network scanning. It is modeled as

$$DC_{\mathrm{DEV}} = \frac{t_{\mathrm{RXB}}}{I_B} + \frac{t_{\mathrm{TXDS}} + t_{\mathrm{RXA}}}{I_U I_B} u + \frac{t_{\mathrm{TXDS}} + t_{\mathrm{RXA}} + t_{\mathrm{RXDD}} + t_{\mathrm{TXA}}}{I_D I_B} u + \frac{t_{\mathrm{NS}}}{I_{\mathrm{NS}}}, \tag{4.101}$$

where I_{NS} is the average network scanning interval, which depends on the device speed and radio link quality. Let P_{DEV} denote the device power consumption, modeled as

$$P_{\mathrm{DEV}} = \frac{E_{\mathrm{RXB}}}{I_B} + \frac{E_{\mathrm{TXDS}} + E_{\mathrm{RXA}}}{I_U I_B} u + \frac{E_{\mathrm{TXDS}} + E_{\mathrm{RXA}} + E_{\mathrm{RXDD}} + E_{\mathrm{TXA}}}{I_D I_B} u + \frac{E_{\mathrm{NS}}}{I_{\mathrm{NS}}} + (1 - DC_{\mathrm{DEV}}) P_S, \tag{4.102}$$

where P_S is the power consumption in the sleep mode. We assume that a node has to transmit beacons and receive the CAP for communicating with nodes associated with it in order to operate as a coordinator in a beacon-enabled cluster-tree network. Furthermore, we assume that the power consumption of a coordinator during CAP is approximately equal to the reception-mode power consumption. In addition, we assume that the data flow in the uplink direction is performed by long MAC payloads containing A sensing items per frame. Let $\mathrm{DC}_{\mathrm{COOR}}$ denote the duty cycle of a coordinator, modeled as [4]

$$\begin{aligned}\mathrm{DC}_{\mathrm{COOR}} = {} & \frac{t_{\mathrm{TXB}} + t_{\mathrm{RXB}}}{I_B} + \frac{(t_{\mathrm{TXDL}} + t_{\mathrm{RXA}})(n + n_D + 1)}{I_U I_B A} u \\ & + \frac{t_{\mathrm{TXDS}} + t_{\mathrm{RXA}} + t_{\mathrm{RXDD}} + t_{\mathrm{TXA}}}{I_D I_B} u + \frac{t_{\mathrm{CAP}}}{I_B} + \frac{t_{\mathrm{NS}}}{I_{\mathrm{NS}}}, \end{aligned} \quad (4.103)$$

where t_{CAP} is the duration of CAP. Let P_{COOR} denote the coordinator power consumption, modeled as [4]

$$\begin{aligned}P_{\mathrm{COOR}} = {} & \frac{E_{\mathrm{TXB}} + E_{\mathrm{RXB}}}{I_B} + \frac{(E_{\mathrm{TXDS}} + E_{\mathrm{RXA}})(n + n_D + 1)}{I_U I_B A} u \\ & + \frac{E_{\mathrm{TXDS}} + E_{\mathrm{RXA}} + E_{\mathrm{RXDD}} + E_{\mathrm{TXA}}}{I_D I_B} u + \frac{t_{\mathrm{CAP}}}{I_B} P_{\mathrm{RX}} \\ & + \frac{E_{\mathrm{NS}}}{I_{\mathrm{NS}}} + (1 - DC_{\mathrm{COOR}}) P_S. \end{aligned} \quad (4.104)$$

For an energy-efficient MAC protocol like IEEE 802.15.4 MAC, the power consumption is very important.

4.7 MULTICHANNEL MAC

Multichannel MAC transmits in a number of channels at the same time, resulting in an increase in throughput. Figure 4.6 shows a multichannel MAC using a dedicated control channel [5]. The devices constantly monitor the control channel and keep track of idle devices and data channels. When a device has packets to send to an idle device, it sends an RTS message for that idle device on the control channel. If the idle device hears the RTS message, it replies with a CTS message. Then both the sender and the receiver tune to the agreed channel to start transmission.

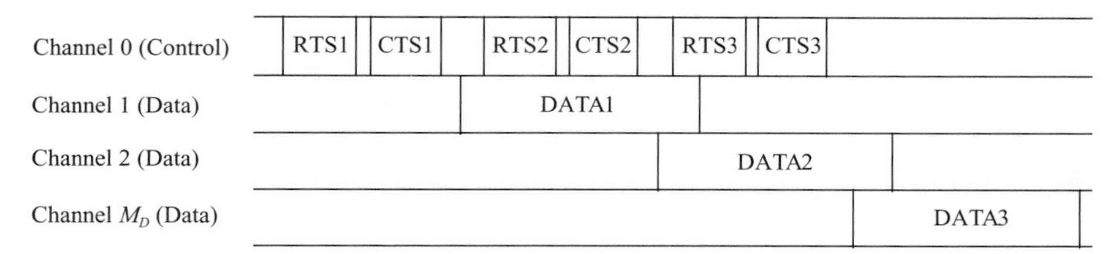

FIGURE 4.6 Dedicated control channel multichannel MAC.

4.7.1 Analysis

The following simplifications are made for all protocols:

- Time is divided into small time slots with perfect synchronization at the slot boundaries.
- For each channel agreement, the devices can transmit only one packet.
- The packet length is geometrically distributed with parameter q, and the mean packet length is $1/q$.
- Every device always has packets to send to all other devices, and an idle device, attempts to transmit with probability p in each time slot.

These simplifications allow a Markov chain to be formed with state X_t representing the number of communicating node pairs at time t. When $X_t = k$, $2k$ devices are involved in data communications while the other $N - 2k$ devices are idle, where N is the number of devices. Let M_D be the number of data channels. The state space of the Markov chain, denoted S, is bounded by the minimum of $\lfloor N/2 \rfloor$ and M_D. A state transition in the Markov chain happens when new agreements are made or when existing transfers end. Let $S_k^{(i)}$ and $T_k^{(j)}$ denote, respectively, the probability that i new agreements are made and the probability that j transfers terminate in the next slot when the state is k. An agreement is made when exactly one idle device attempts to transmit an RTS message on the control channel. Then $S_k^{(i)}$ and $T_k^{(j)}$ are, respectively, given by [5]

$$S_k^{(i)} = \begin{cases} (N-2k)p(1-p)^{(N-2k-1)}, & i = 1 \\ 1 - S_k^{(1)}, & i = 0 \\ 0, & \text{otherwise} \end{cases} \tag{4.105}$$

and

$$\begin{aligned} T_k^{(j)} &= \Pr[j \text{ transfers terminate at time } t+1 | X_t = k] \\ &= \binom{k}{j} q^j (1-q)^{k-j}. \end{aligned} \tag{4.106}$$

Let p_{kl} denote the state transition probability from state k at time t to state l at time $t+1$, given by

$$p_{kl} = \sum_{m=(k-l)^+}^{k} S_k^{(m+l-k)} T_k^{(m)}. \tag{4.107}$$

For the protocols considered in this section, $S_k^{(i)} = 0$, $\forall i > 1$. That is, at most one additional pair can meet in the next slot. Thus, we have

$$p_{kl} = T_k^{(k-l)} S_k^{(0)} + T_k^{(k-l+1)} S_k^{(1)}, \tag{4.108}$$

where $T_k^{(j)} = 0$ when $j < 0$. p_{kl} can be rewritten as

$$p_{kl} = \begin{cases} 0, & l < k+1 \\ T_k^{(0)} S_k^{(1)}, & l = k+1 \\ T_k^{(k-l)} S_k^{(0)} + T_k^{(k-l+1)} S_k^{(1)}, & 0 < l \leq k \\ T_k^{(k)} S_k^{(0)}, & l = 0. \end{cases} \tag{4.109}$$

Let M denote the number of channels. The average utilization per channel, denoted by ρ, is given by

$$\rho = \frac{\sum_{i \in S} i\pi_i}{M}, \tag{4.110}$$

where π_i the limiting probability that the system is in state i, is obtained by solving the balance equations of the Markov chain. Let C denote the channel transmission rate. The total system throughput, denoted by R, is given by

$$R = M_D C \rho, \tag{4.111}$$

where $M_D = M - 1$ is the number of dedicated channels for data transmissions, and this excludes the control channel, which is indicated by -1.

4.8 DIRECTIONAL-ANTENNA MAC

A directional antenna radiates power in one or more directions, resulting in increased performance in the receiver and transmitter and reduced interference from unwanted sources. Two directional-antenna MAC protocols are considered in this section. In both schemes, RTS, CTS, data packet, and ACK are transmitted directionally. When a node is transmitting in one direction, it is "deaf" to the other directions and cannot sense any channel activity at all. In these schemes, spatial reuse is maximized as nodes limit their transmission to the smallest possible area. The two schemes differ in the reception; one uses omnidirectional receiving, the other uses directional receiving. The former is known as the directional transmission and omnidirectional reception (DTOR) scheme; the latter is known as the directional transmission and directional reception (DTDR) scheme [6].

4.8.1 Analysis

The analysis in this section follows that in [6]. The following assumptions have been made in the analysis:

- The directional transmissions and receptions have equal beamwidth.
- In the network model, the nodes are two-dimensional Poisson distributed with density λ. The probability of finding i nodes in an area of S, denoted by $p(i, S)$,

is given by

$$p(i, S) = \frac{(\lambda S)^i}{i!} e^{-\lambda S}. \quad (4.112)$$

- The range of an omnidirectional transmission is R, and the range of a directional transmission is $R' = \gamma R$, where $\gamma \geq 1$. Let N be the average number of nodes within a circle of radius R and N' be the average number of nodes within a circle of radius R'. We have, respectively,

$$N = \lambda \pi R^2, \quad (4.113)$$

and

$$N' = \lambda \pi R'^2 = \gamma^2 N. \quad (4.114)$$

- The nodes are assumed to operate in time-slotted mode, with the length of time slot equal to a propagation delay τ, which is much smaller than the length of any packet.
- The transmission times of RTS, CTS, data, and ACK packets are normalized with respect to τ and are denoted by l_{rts}, l_{cts}, l_{data}, and l_{ack}, respectively.
- All packet lengths are assumed to be multiples of the length of a time slot.
- A node is assumed always to have a packet for transmission.
- A silent node is assumed to begin a transmission with probability p at each time slot.
- A node is assumed to become ready independently with probability p_0 at the beginning of each time slot:

$$p = p_0 \,\text{Pr[channel is sensed idle in a slot]}. \quad (4.115)$$

- A node is assumed to initiate a successful handshake with any other node with probability p_s, where $p_s < p < p_0$.
- The node model is a three-state Markov chain as shown in Figure 4.7. In the *wait state*, the node defers for other nodes or backs off. In the *succeed state*, the node can complete a successful four-way handshake with other nodes. In the *fail state*, the node initiates a handshake that is unsuccessful or cannot be completed due to collisions.
- A node is assumed to communicate directly only with other nodes that are within its omnidirectional transmission range R and communicate only indirectly with nodes outside R and inside its directional transmission range R', even though it can still be an interfering source for these nodes.

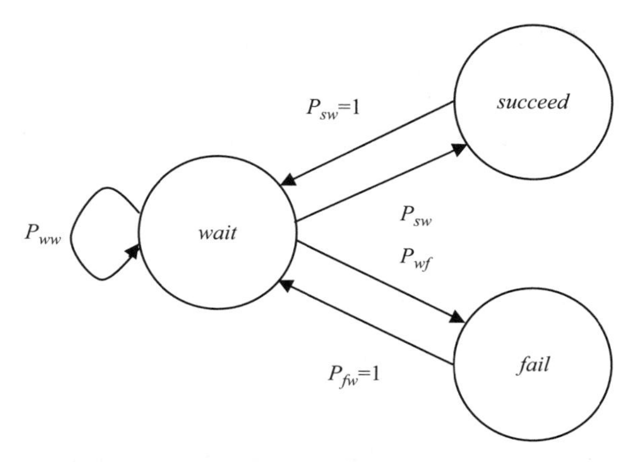

FIGURE 4.7 State transition diagram for directional antenna MAC. (From [6] © 2004, with permission from Elsevier.)

Let Th denote the throughput of each direction collision avoidance scheme, given by [6]

$$\text{Th} = \frac{\pi_s l_{\text{data}}}{\pi_w T_w + \pi_s T_s + \pi_f T_f}, \tag{4.116}$$

where π_s, π_w, and π_f are the steady-state probability in the states succeed, wait, and fail, respectively, and T_s, T_w, and T_f are the durations of the states succeed, wait, and fail, respectively. The duration in the succeed state, denoted by T_s, is given by

$$\begin{aligned} T_s &= (l_{\text{rts}} + 1) + (l_{\text{cts}} + 1) + (l_{\text{data}} + 1) + (l_{\text{ack}} + 1) \\ &= l_{\text{rts}} + l_{\text{cts}} + l_{\text{data}} + l_{\text{ack}} + 4. \end{aligned} \tag{4.117}$$

From the Markov chain, we have

$$\pi_w = \frac{1}{2 - P_{ww}}, \tag{4.118}$$

$$\pi_s = \frac{P_{ws}}{2 - P_{ww}}, \tag{4.119}$$

and

$$\pi_f = \frac{1 - P_{ww} - P_{ws}}{2 - P_{ww}}, \tag{4.120}$$

where P_{ww} is the self-transition probability of the wait state in a slot and P_{ws} is the transition state probability from the wait state to the succeed state in a time slot.

Given that a node in the wait state listens omnidirectionally, the transition probability P_{ww} that node x continues to stay in the wait state in a slot is equal to the probability that it does not initiate any transmission and there is no node around it initiating a transmission in the direction of node x. As these events are independent, from (4.112), (4.114), and (4.115) we have

$$P_{ww} = (1 - p)e^{-p'N'}, \tag{4.121}$$

where $p' = p\theta/2\pi$, and θ is the transmission and reception beamwidth. To derive P_{ws}, we need to calculate the probability $P_{ws}(r)$ that node x successfully initiates a four-way handshake with node y at a given slot when the two nodes are distance r apart.

4.8.1.1 *Directional Transmission and Omnidirectional Reception*

In the directional transmission and omnidirectional reception (DTOR) scheme, the transmit antenna is directional and the receive antenna is omnidirectional. In directional transmit antenna, the data transmission is directed in a particular direction, whereas in the omnidirectional receive antenna, the data transmission is received from all directions.

In the DTOR scheme, the probability $P_{ws}(r)$ is given by

$$P_{ws}(r) = p_x p_y \prod_{i=1}^{5} p_i, \tag{4.122}$$

where p_x is the probability that node x transmits in the time slot and is equal to p, p_y is the probability that node y does not transmit in the time slot and is equal to $1 - p$, and p_i is the probability that the nodes with area i interfere with the handshake between nodes x and y. Each of these p_i's is associated with an area i. Areas 1 to 5 are shown in Figure 4.8. The p_i's are obtained by considering area i to compute the probability that there is no interference from nodes in each of these areas using the equation that computes the probability P that no node transmits in a time slot within a planar area of size S in which nodes are randomly placed according to a two-dimensional Poisson distribution. The probability P is given by

$$\begin{aligned} P &= \sum_{i=0}^{\infty} (1 - p)^i \frac{(\lambda S)^i}{i!} e^{-\lambda S} \\ &= \sum_{i=0}^{\infty} \frac{[(1 - p)\lambda S]^i}{i!} e^{-(1-p)\lambda S} e^{-p\lambda S} \\ &= e^{-p\lambda S} \\ &= e^{-p\lambda S\pi R^2/(\pi R^2)} \\ &= e^{-pNS/(\pi R^2)}. \end{aligned} \tag{4.123}$$

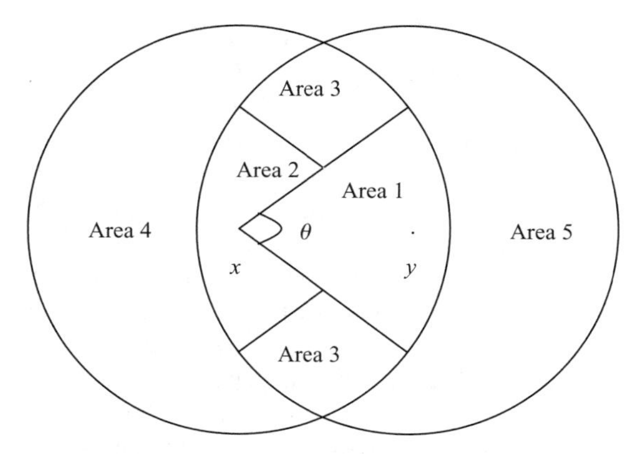

FIGURE 4.8 Areas for a DTOR scheme.

The p_i's are given by

$$p_1 = e^{-pS_1N'}, \tag{4.124}$$

$$p_2 = e^{-p'S_2N'(2l_{\text{rts}})}e^{-pS_2N'}, \tag{4.125}$$

$$\begin{aligned} p_3 &= e^{-p''S_3N'(l_{\text{rst}}+l_{\text{rst}}+1+l_{\text{cts}}+1+l_{\text{data}}+1+l_{\text{ack}}+1)} \\ &= e^{-p''S_3N'(2l_{\text{rst}}+l_{\text{cts}}+l_{\text{data}}+l_{\text{ack}}+4)}, \end{aligned} \tag{4.126}$$

$$\begin{aligned} p_4 &= e^{-p'S_4N'(l_{\text{rst}}+l_{\text{cts}}+1)}e^{-p'S_4N'(l_{\text{rst}}+l_{\text{ack}}+1)} \\ &= e^{-p'S_4N'(2l_{\text{rst}}+l_{\text{cts}}+l_{\text{ack}}+2)}, \end{aligned} \tag{4.127}$$

$$\begin{aligned} p_5 &= e^{-p'S_5N'(l_{\text{rst}}+l_{\text{rts}}+1)}e^{-p'S_5N'(l_{\text{rst}}+l_{\text{data}}+1)} \\ &= e^{-p'S_5N'(3l_{\text{rst}}+l_{\text{data}}+2)}, \end{aligned} \tag{4.128}$$

where $p'' = p\theta'/2\pi = p\theta/2\pi, \theta'$ is assumed to be θ, though the range of θ' is between θ and 2θ. The S_i's in the above equations are given by

$$S_1 = \frac{\theta}{2\pi}, \tag{4.129}$$

$$S_2 = \frac{\theta}{2\pi} - r^2\frac{\tan(\theta/2)}{2\pi}, \tag{4.130}$$

$$S_3 = \frac{2q(r/2)}{\pi} - \frac{\theta}{\pi} + r^2\frac{\tan(\theta/2)}{2\pi}, \tag{4.131}$$

$$S_4 = 1 - \frac{2q(r/2)}{\pi}, \tag{4.132}$$

$$S_5 = 1 - \frac{2q(r/2)}{\pi}, \tag{4.133}$$

where $q(t) = \arccos(t) - t\sqrt{1-t^2}$, r is normalized with respect to R by setting $R = 1$, and S_i is normalized with respect to πR^2. For p_1, the probability that no node in area 1 interferes with the handshake between nodes x and y is equal to the probability that no nodes in the area transmit in the same time slot as node x does. For p_2, no node can transmit in $2l_{\text{rts}}$ slots in the direction of node y and for no interference to exist from nodes in area 2, does not transmit in the slot when the transmission of node y arrives. For p_3, no interference exists in area 3 if no node transmits in the direction to nodes x and y during the whole handshake between the two nodes and the span angle θ' is between θ and 2θ. For p_4, no interference to nodes x and y exists from nodes in area 4 if no node in that area transmits in node x's direction when node y is transmitting. There are two such periods. One is the time when node y transmits a CTS packet to node x and the other is the time when node y transmits an ACK packet to node x. The durations of these two periods are approximately $l_{\text{rts}} + l_{\text{cts}} + 1$ and $l_{\text{rts}} + l_{\text{ack}} + 1$, respectively, assuming that nodes transmit in each time slot independently with probability p. For p_5, there is no interference to nodes in area 5 if no node in that area transmits in node y's direction when node x is transmitting. There are two such periods. One is the time when node x transmits an RTS packet to node y and the other is the time when node x transmits a DATA packet to node y. The durations of these two periods are approximately $l_{\text{rts}} + l_{\text{rts}} + 1$ and $l_{\text{rts}} + l_{\text{data}} + 1$, respectively. Therefore P_{ws} is given by

$$\begin{aligned} P_{ws} &= \int_0^1 2r P_{ws}(r)\, dr \\ &= \int_0^1 2rp(1-p)p_1p_2p_3p_4p_5\, dr. \end{aligned} \tag{4.134}$$

The duration of a node in the wait state, denoted by T_w, is given by

$$T_w = 1. \tag{4.135}$$

As the DTOR scheme cannot prevent interference from neighboring nodes, the handshake between any pair of sending and receiving nodes may be interrupted at any time. Hence, the failed period, denoted by T_f, can last from $T_1 = l_{\text{rts}} + 1$ to $T_2 = l_{\text{rts}} + l_{\text{cts}} + l_{\text{data}} + l_{\text{ack}} + 4$. Assuming that the length of the failed period follows a truncated geometric distribution with parameter p with a lower bound T_1 and an upper bound T_2, T_f can be considered as the mean value of the truncated geometric distribution and is given by

$$T_f = \frac{1-p}{1-p^{T_2-T_1+1}} \sum_{i=0}^{T_2-T_1} p^i(T_1+i). \tag{4.136}$$

4.8.1.2 Directional Transmission and Directional Reception

In the directional transmission and directional reception (DTDR) scheme, both the transmit and receive antennas are directional. The variables needed to derive the throughput for DTDR are the same as those for DTOR except for T_1 and P_{ws}. To account for more vulnerability of handshaking from interfering nodes in areas 1 and 2 in the DTDR scheme than the DTOR scheme, T_1 is given by

$$T_1 = l_{\text{rts}} + l_{\text{cts}} + 2. \tag{4.137}$$

As an approximation, P_{ws} is given by

$$P_{ws} = \int_0^1 2r\gamma(r)P_{ws}(r)\,dr, \tag{4.138}$$

where

$$\gamma(r) = \min(\gamma_1(r), \gamma_2(r)), \tag{4.139}$$

$$\gamma_1(r) = \frac{S_1 + S_2 + S_3 + S_4 + S_5}{S_1 + S_2}, \tag{4.140}$$

$$\gamma_2(r) = \frac{N(S_3 + S_4 + S_5)}{2}. \tag{4.141}$$

Note that nodes x and y are immune to the transmissions from nodes in areas 3, 4, and 5, and concurrent transmissions can continue unobstructed in these areas. A reuse factor $\gamma(t)$ is introduced in the calculation of P_{ws}. $\Gamma_1(t)$ is the ratio between the total region covered nominally by nodes x and y and the actual region covered by the handshake between nodes x and y. In theory, there can be a maximum of $\gamma_2(t)$ pairs of concurrent handshakes in areas 3, 4, and 5. Thus, $\gamma(t)$ is conservative by taking the minimum of $\gamma_1(t)$ and $\gamma_2(t)$.

4.9 MULTIHOP SATURATED THROUGHPUT OF IEEE 802.11 MAC

Data transmissions in sensor networks are usually multihop. This means that data are transmitted from node to node in the direction of the base station. In this section the multihop saturated throughput of IEEE 802.11 MAC with spatial locations of nodes is considered. First, the saturated throughput for a node is calculated using numerical methods. Then the multihop saturated throughput is calculated based on the saturated throughput for a node and the average number of neighbors of a node. Only the basic access method of the distributed coordination function (DCF) of IEEE 802.11 MAC is considered.

4.9.1 Analysis

The analysis in this section follows that in [7,8]. The following assumptions are made to facilitate the analysis:

- The nodes in a network are Poisson distributed over a plane with density λ. That is, the probability of finding i nodes in an area of A, denoted by $p(i, A)$, is given by

$$p(i, A) = \frac{(\lambda A)^i}{i!} e^{-\lambda A}. \tag{4.142}$$

- A heavy traffic load condition is assumed. That is, a node always has a packet to send and the destination is chosen randomly from one of its neighbors.
- All packet sizes are assumed to be the same.
- All nodes use the same fixed transmission range R. That is, the hearing region of any node is πR^2 and the average number of neighbors of a node, denoted as N, is

$$N = \lambda \pi R^2. \tag{4.143}$$

For notational convenience, all distances are normalized with respect to $R = 1$. Thus, we have

$$N = \lambda \pi. \tag{4.144}$$

- The ranges of communication, interference, and carrier sensing are assumed to be identical.
- The channel is assumed to be ideal and all errors are assumed to be due to collision.
- The capture effect is not considered. This means that any overlap of two transmissions arriving at a node will lead to a collision at that node.
- The collision probability of a packet is independent of the number of retransmissions (if any) and the location of the source, but is dependent on the distance between the source and the destination, denoted by r. The collision probability of a transmission with distance r is denoted as $p(r)$.
- All time-related parameters are normalized with respect to the slot time θ, which is specified by the PHY in IEEE 802.11, and θ is set to 1, for simplicity.
- The symbols l_{sifs}, l_{difs}, l_{data}, and l_{ack} are used to represent the length of SIFS, DIFS, DATA, and ACK, respectively.

The activities of a node can be modeled by a three-dimensional Markov chain $\{s(t), b(t), r(t)\}$. $s(t)$ represents the backoff stage of a node at time t and it records the number of retransmissions a packet has suffered; $b(t)$ represents the backoff counter value of a node at time t; and $r(t)$ represents the distance between a node and its destination at time t. A packet is discarded after m retries. The backoff window size in the ith backoff stage, denoted by W_i, is given by

$$W_i = \begin{cases} 2^i W, & i \leq m' \\ 2^{m'} W, & i > m', \end{cases} \tag{4.145}$$

where W is the minimum contention window size and the second condition in the equation corresponds to the maximum contention window size reached.

$R(t)$ is a continuous variable in the range [0,1]. For ease of analysis, the range is discretized into n equal intervals: $I_1 = [0, 1/n]$, $I_2 = [1/n, 2/n], \ldots, I_n = [(n-1)/n, 1]$. Correspondingly, the hearing region of a node is divided into n rings. Any value lying in the kth interval I_k is replaced with the middle point of I_k. With this conversion, we get a discrete process of $r(t)$, where $r(t) \in \{r_k | k \in [1, n]\}$, and r_k is given by

$$r_k = \frac{2k-1}{2n}. \tag{4.146}$$

Given that each source uniformly chooses one of its neighbors as a destination and that the average number of nodes with a radius r is proportional to r^2, the probability density function of the distance r between a source and its destination is given by

$$f(r) = 2r, \qquad 0 \leq r \leq 1. \tag{4.147}$$

The corresponding probability mass function is given by

$$\begin{aligned} P[r = r_k] &= 2r_k \cdot 1/n \\ &= 2r_k \, \Delta r, \end{aligned} \tag{4.148}$$

where $\Delta r = 1/n$.

Assuming that the collision probability is independent of $s(t)$ and the destination is randomly chosen, the three-dimensional discrete-time Markov chain $\{s(t), b(t), r(t)\}$ is shown in Figures 4.9 and 4.10. Figure 4.9 shows the overall Markov chain of a node. After a successful transmission or after discarding a packet, a node chooses

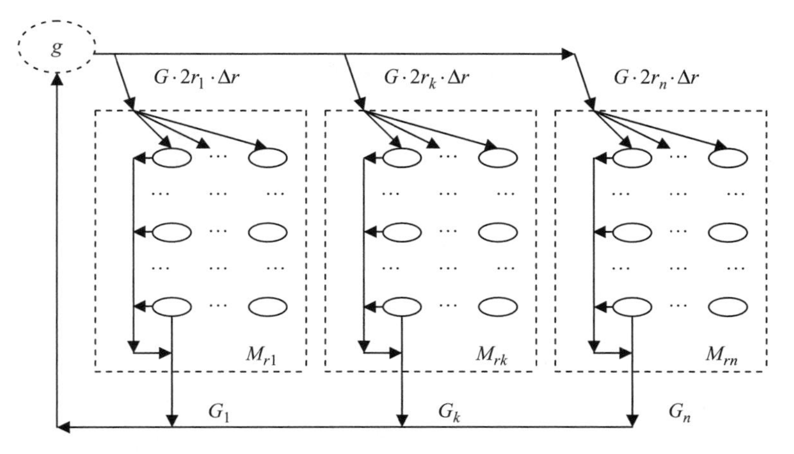

FIGURE 4.9 Overall state transition diagram for a node with spatial locations of destination nodes. (From [7] © 2006, with permission from Elsevier.)

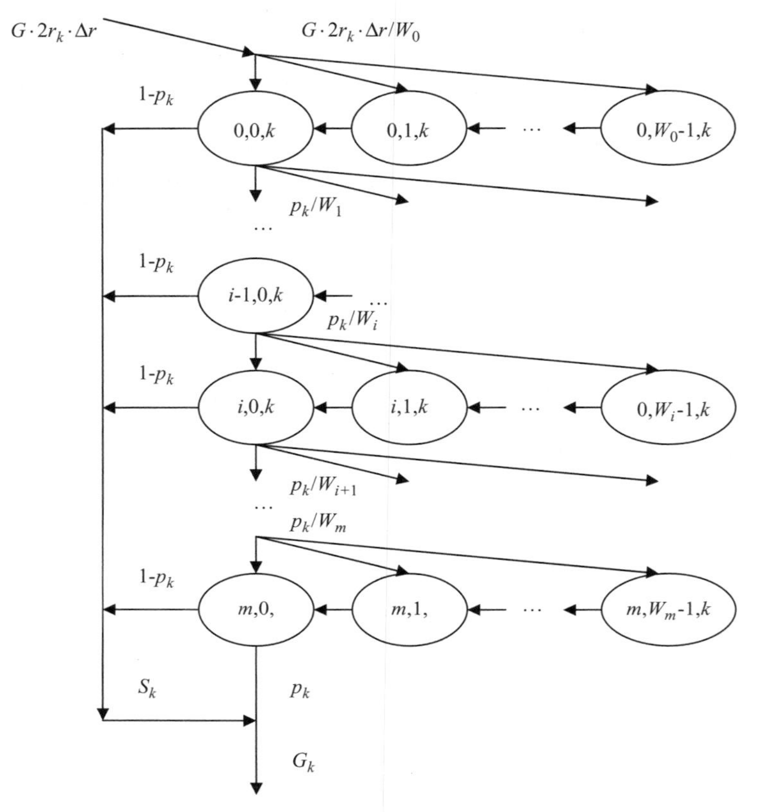

FIGURE 4.10 State transition diagram for a node with destination nodes at distance r_k, M_{rk}. (From [7] © 2006, with permission from Elsevier.)

one of its neighbors randomly as the destination of the next packet. The probability that the destination lies in the kth ring I_k is $2r_k \, \Delta r$. The state g in Figure 4.9 is a pseudostate that is introduced for the convenience of presentation. Figure 4.10 shows the detailed Markov chain for transmission with distance r_k. The states of the Markov chain can be divided into two groups: a *wait group* and a *trans group*. The two groups are as follows [7,8]:

$$\text{wait group} = \{(i, j, k) \mid i \in [0, m], \quad j \in [1, W_i - 1], \quad k \in [1, n]\}, \quad (4.149)$$

$$\text{trans group} = \{(i, j, k) \mid i \in [0, m], \quad j = 0, \quad k \in [1, n]\}. \quad (4.150)$$

When a node in the trans state transmits, the transmission outcome is either successful or collided. A node in the wait state stays in that state if it detects a busy channel and decrements its backoff counter value by one if it detects an idle slot. The time interval between two consecutive backoff time counter decrements as a macroslot is denoted as Θ. It is also the time a node dwells in the wait state. A macroslot is a variable interval consisting of a busy period and an idle slot. The length of the busy period is variable, which can be zero; the idle slot is a real slot θ.

The nonnull one-step transition probabilities in the Markov chain are [7,8]

$$\begin{aligned} P\{i, j, k|i, j+1, k\} &= 1, \quad i \in [0, m], \quad j \in [0, W_i - 2], \quad k \in [1, n], \\ P\{i, j, k|i-1, 0, k\} &= \frac{p_k}{W_i}, \quad i \in [1, m], \quad j \in [0, W_i - 1], \quad k \in [1, n], \\ P\{g|i, 0, k\} &= 1 - p_k, \quad i \in [0, m-1], \quad k \in [1, n], \\ P\{g|m, 0, k\} &= 1, \quad k \in [1, n], \\ P\{0, j, k|g\} &= \frac{2r_k \Delta r}{W_0}, \quad j \in [0, W_0 - 1], \quad k \in [1, n], \end{aligned}$$

where $p_k = p[r_k]$ is the collision probability for a transmission with distance r_k. The first equation above represents the decrement of the backoff counter value; the second equation represents the source rescheduling a collided packet; the third equation represents a successful transmission; the fourth equation represents the discarding of a packet; and the fifth equation represents the random choice of a destination for a new packet.

Let G denote the sum of all transition rates joining the pseudostate g. We have

$$G = \sum_{k=1}^{n} G_k. \tag{4.151}$$

From the local balance of the mini-Markov chain with distance r_k in Figure 4.10, we have

$$G_k = G \cdot 2r_k \Delta r. \tag{4.152}$$

Let $b_{i,j,k}$ be the stationary distribution of the Markov chain. Using the local balance equation for each stage in the mini-Markov chain, we have

$$b_{0,0,k} = G \cdot 2r_k \Delta r, \tag{4.153}$$

$$\begin{aligned} b_{i,0,k} &= p_k b_{i-1,0,k} \\ &= p_k^i b_{0,0,k}, \quad 0 < i \le m. \end{aligned} \tag{4.154}$$

From chain regularities, we have

$$b_{i,j,k} = \frac{W_i - j}{W_i} \begin{cases} G \cdot 2r_k \Delta r, & i = 0, \quad j \in [1, W_i - 1], \quad k \in [1, n] \\ p_k b_{i-1,0,k}, & 0 < i \le m, \quad j \in [1, W_i - 1], \quad k \in [1, n]. \end{cases} \tag{4.155}$$

Substituting (4.153) and (4.154) into (4.155) and simplifying, we have

$$b_{i,j,k} = \frac{W_i - j}{W_i} b_{i,0,k}, \qquad 0 \le i \le m, \quad j \in [1, W_i - 1], \quad k \in [1, n]. \tag{4.156}$$

Using the property of total probability of 1, we have

$$\sum_{i=0}^{m} \sum_{j=0}^{W_i - 1} \sum_{k=1}^{n} b_{i,j,k} = 1. \tag{4.157}$$

Simplifying (4.157), we have

$$\sum_{k=1}^{n} G \cdot 2r_k \Delta r \sum_{i=0}^{m} p_k^i \frac{W_i + 1}{2} = 1. \tag{4.158}$$

Equation (4.158) can be rearranged to yield

$$\begin{aligned} G &= \frac{1}{\sum_{k=1}^{n} 2r_k \Delta r \sum_{i=0}^{m} p_k^i \frac{W_i+1}{2}} \\ &= \frac{1}{\sum_{k=1}^{n} 2r_k \Delta r \, A(r_k)}, \end{aligned} \tag{4.159}$$

where

$$A(r_k) = \sum_{i=0}^{m} p_k^i \frac{W_i + 1}{2}. \tag{4.160}$$

Substituting (4.145) into (4.160), we have

$$A(r_k) = \begin{cases} \dfrac{2(1 - p_k)(1 - 2p_k)}{W(1 - p_k)(1 - 2p_k^{m+1}) + (1 - 2p_k)(1 - p_k^{m+1})}, & m \le m' \\[2ex] \dfrac{2(1 - p_k)(1 - 2p_k)}{W(1 - p_k)(1 - 2p_k^{m'+1}) + (1 - 2p_k)(1 - p_k^{m+1}) + W 2^{m'} p_k^{m'+1}(1 - p_k)(1 - p_k^{m-m'})}, & m > m'. \end{cases} \tag{4.161}$$

The continuous version of (4.159) is given by

$$G = \frac{1}{\int_0^1 A(r) \cdot 2r \, dr}. \tag{4.162}$$

Let $\overline{G}$ denote the probability that a node transmits in a macroslot; it is given by

$$\begin{aligned}\overline{G} &= \sum_{k=1}^{n}\sum_{i=0}^{m} b_{i,0,k} \\ &= G\sum_{k=1}^{n}\frac{1-p_k^{m+1}}{1-p_k}\cdot 2r_k\,\Delta r \\ &= G\int_0^1 \frac{1-[p(r)]^{m+1}}{1-p(r)}\cdot 2r\,dr, \end{aligned} \tag{4.163}$$

where $p(r)$ is the collision probability for a transmission with distance r; that is, it is the continuous version of p_k. Let S_k denote the successful rate of transmissions from a node to its neighbors located in the ring of r_k. S_k is given by

$$\begin{aligned} S_k &= \sum_{i=0}^{m}(1-p_k)b_{i,0,k} \\ &= G\left(1-p_k^{m+1}\right)2r_k\,\Delta r. \end{aligned} \tag{4.164}$$

Let S denote the overall successful rate of a node. We have

$$\begin{aligned} S &= \sum_{k=1}^{n} S_k \\ &= G\int_0^1 (1-[p(r)]^{m+1})\cdot 2r\,dr. \end{aligned} \tag{4.165}$$

Let T_w denote the average length of a wait state, which is equal to the average length of a macroslot Θ. As a macroslot consists of a busy period and an idle slot θ, the ratio θ/Θ is actually the probability that a channel is detected to be idle by a node when the node is not transmitting, denoted by Π_I. That is, we have

$$\Pi_I = \frac{\theta}{\Theta}. \tag{4.166}$$

Using (4.166) and $\theta = 1$, we have

$$\begin{aligned} T_w &= \Theta \\ &= \frac{1}{\Pi_I}. \end{aligned} \tag{4.167}$$

Let τ denote the transmission probability of a node in an arbitrary slot when the node is transmitting; τ is given by

$$\begin{aligned} \tau &= \frac{\overline{G}}{\overline{G}+(1-\overline{G})T_w} \\ &= \frac{\overline{G}}{\overline{G}+(1-\overline{G})/\Pi_I}. \end{aligned} \tag{4.168}$$

Let $P_{x,I}$ be the probability that a node detects that one of its neighbors x is not transmitting in an arbitrary slot. $P_{x,I}$ is given by

$$P_{x,I} = \frac{1-\tau}{T_t \tau + (1-\tau)}, \tag{4.169}$$

where T_t is the channel occupancy time of a transmission normalized with respect to a slot time θ. For the basic access method,

$$T_t = l_{\text{data}} + l_{\text{sifs}} + l_{\text{ack}} + l_{\text{difs}}. \tag{4.170}$$

For the basic access method, a node will detect an idle channel if none of its neighbors is transmitting. Assuming that transmissions of a node's neighbors are independent and the probability of having i nodes within the transmission range of a node is Poissonian, we have

$$\begin{aligned} \Pi_I &= \sum_{i=0}^{\infty} \frac{N^i}{i!} e^{-N} P_{x,I}^i \\ &= e^{-N} \sum_{i=0}^{\infty} \frac{(N P_{x,I})^i}{i!} \\ &= e^{-N(1-P_{x,I})}, \end{aligned} \tag{4.171}$$

where $N = \lambda\pi$ is the average number of neighbors of a node.

Assuming that acknowledgment is always received successfully by the source, the collision probability of a transmission is equal to the collision probability of the data frame of the transmission. Consider a transmission from node P to node Q, where they are in the common neighborhood of each other. The collision probability for a transmission with distance r is given by

$$p(r) = 1 - P_s(r), \tag{4.172}$$

where $P_s(r)$ is the probability that a transmission is successful and is given by

$$P_s(r) = P_1(r)P_2(r)P_3(r), \tag{4.173}$$

where

$P_1(r) = \Pr\{Q$ does not transmit in the same slot $\mid P$ transmits in a slot$\}$,
$P_2(r) = \Pr\{$nodes in $C(r)$ does not transmit in the same slot $\mid P$ transmits in a slot$\}$,
$P_3(r) = \Pr\{$nodes in $B(r)$ does not transmit $2l_{\text{data}} + 1$ slots $\mid P$ transmits in a slot$\}$,

where "P transmits in a slot" means that node P just detects an idle slot and will transmit at the beginning of the next slot, $C(r)$ is the common neighborhood of the sender and the receiver, and $B(r)$ is the hidden area. $C(r)$ and $B(r)$ are given by

$$C(r) = 2q\frac{r}{2}, \tag{4.174}$$

and

$$B(r) = \pi - C(r), \tag{4.175}$$

where $q(t) = \arccos(t) - t\sqrt{1-t^2}$. $P_1(r)$, $P_2(r)$, and $P_3(r)$ are given by

$$\begin{aligned}
P_1(r) &= (1-\overline{G})\,\Pr\{Q \text{ detects an idle channel} \mid P \text{ detects an idle channel}\} \\
&= (1-\overline{G})\,\Pr\{\text{none of the nodes in } B(r) \text{ transmits}\} \\
&= (1-\overline{G})e^{-\lambda B(r)(1-P_{x,I})}, && (4.176) \\
P_2(r) &= \sum_{i=0}^{\infty}(1-\tau_c)^i\frac{(\lambda C(r))^i}{i!}e^{-\lambda C(r)} \\
&= e^{-\tau_c \lambda C(r)}, && (4.177) \\
P_3(r) &= \sum_{i=0}^{\infty}((1-\tau)^{2l_{\text{data}}+1})^i\frac{(\lambda b(r))^i}{i!}e^{-\lambda B(r)} \\
&= e^{-\lambda B(r)(1-(1-\tau)^{2l_{\text{data}}+1})}, \qquad N \le 5 && (4.178) \\
&= e^{-\lambda B(r)(1-(1-\tau)^{l_{\text{data}}+1}(1-\tau_b)^{l_{\text{data}}})}, \qquad N > 5 && (4.179)
\end{aligned}$$

where

$$\tau_b = \overline{G}e^{-\lambda B(1+r/2)(1-P_{x,I})}. \tag{4.180}$$

τ_c is the average transmission probability of nodes in $C(r)$ given that node P detects an idle slot and τ_b is the average transmission probability of nodes in $B(r)$ when node P is transmitting. T_c is replaced by τ to simplify the calculation. Note that since a macroslot consists of a busy period and an idle slot, $\overline{G}$ is also the transmission probability when a node detects an idle slot. The vulnerable period for a data frame in $P_3(r)$ is $2l_{\text{data}} + 1$. Equation (4.180) is an approximation for τ_c [8].

The throughput of a node, denoted by TH, is given by

$$\text{TH} = \frac{S \cdot E[P]}{\overline{G}T_s + (1-\overline{G})T_w}, \tag{4.181}$$

where $E[P]$ is the average payload length and T_s is the average length of the successful trans state, which is given by

$$T_s = l_{\text{data}} + l_{\text{sifs}} + l_{\text{ack}}. \tag{4.182}$$

The parameters S, $\overline{G}$, and T_w are given in (4.165), (4.163), and (4.167), respectively.

To measure the throughput of a multihop channel, the authors in [7] define channel throughput as the throughput per unit of area. One natural unit of area is the hearing region of a node. Hence, the channel saturation throughput of multihop wireless networks, denoted by C, is defined as [7]

$$C = \text{TH} \cdot N, \tag{4.183}$$

where TH is the per-node saturation throughput and N is the average number of neighbors of a node.

4.10 MULTIPLE-ACCESS CONTROL

In this section we consider three traditional forms of multiple-access control technologies: TDMA, FDMA, and CDMA. *TDMA* divides time into time slots, and each user can have preassigned time slots to use. However, the users must wait their turn to access their time slots. On the other hand, *FDMA* divides a frequency band into frequency slots, and each user can have preassigned frequency slots to use. *CDMA* allocates all resources to all simultaneous users by using codes and controlling the power transmitted by each of them to the minimum required to maintain a given SNR for the quality of service required.

Let M be the number of stations and λ be the average arrival rate at each station in packets per second (the same for every station). The arrivals are assumed to be Poisson distributed. Let R be the medium bit rate in bits per second. Let $\overline{X}$ be the average packet length in bits. Service at each station is assumed to be exhaustive, that is, all stored packets are transmitted each time the station has access to the channel. Propagation and other delays between stations are assumed equal.

4.10.1 Time-Division Multiple Access

In TDMA, time is divided into slots. Figure 4.11 shows the time-slot allocation for TDMA. Each frame has a control segment, which contains a synchronization pattern to keep stations, as well as other control data, in synchronism. The guard bands ensure that the time slot assigned to each station is kept separated from the others. Each station's time slot allows for transmission of both overhead and data. The major disadvantage of TDMA is the requirement that each station have a fixed allocation of channel time, regardless of whether or not it has data to transmit.

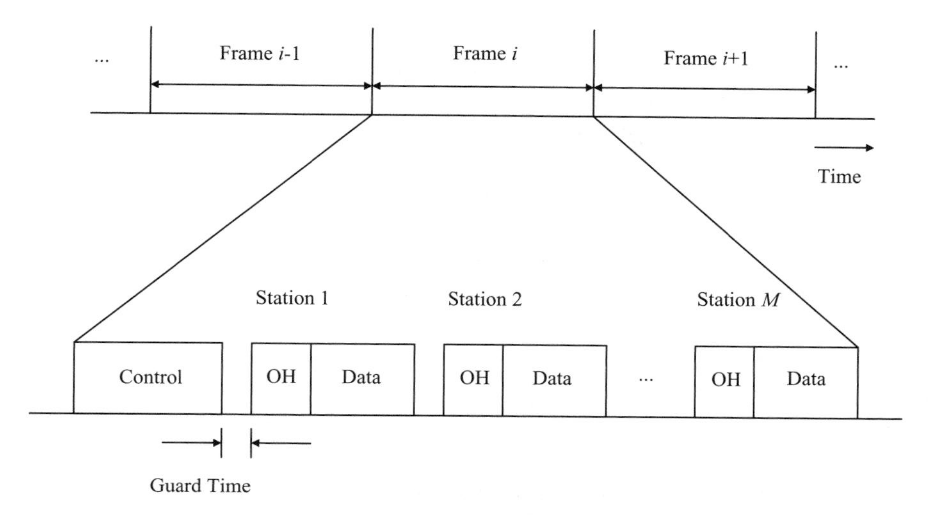

FIGURE 4.11 Frame format for time-division multiple access.

4.10.1.1 Analysis Figure 4.12 shows the analytical frame model for TDMA. The channel is always busy at its maximum rate. The average packet transmission time seen by the whole network, denoted by $\overline{P}$, is given by

$$\overline{P} = \frac{\overline{X}}{R}. \tag{4.184}$$

The duration of a frame time, denoted by T_f, is given by

$$T_f = \frac{M\overline{X}}{R}. \tag{4.185}$$

A packet arrives at random relative to a frame length and must wait for the appropriate time slot. If the packet arrival time is distributed uniformly over the length of the frame, the average wait is one-half frame time (i.e., $\frac{1}{2}T_f = M\overline{X}/2R$). Each station uses the channel $1/M$ of the time. Thus, on average, the capacity available to each

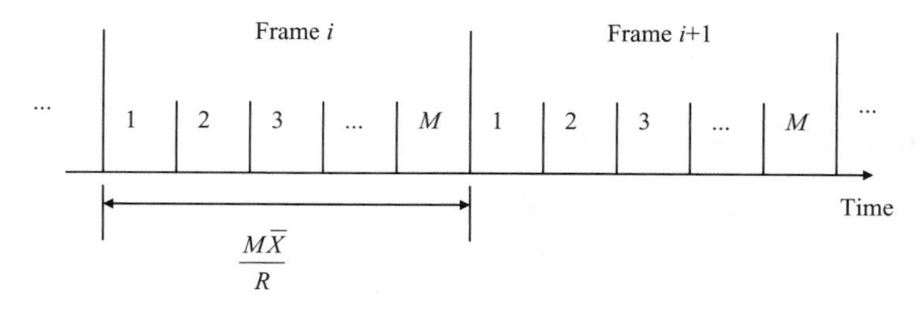

FIGURE 4.12 Simplified frame format for time-division multiple access.

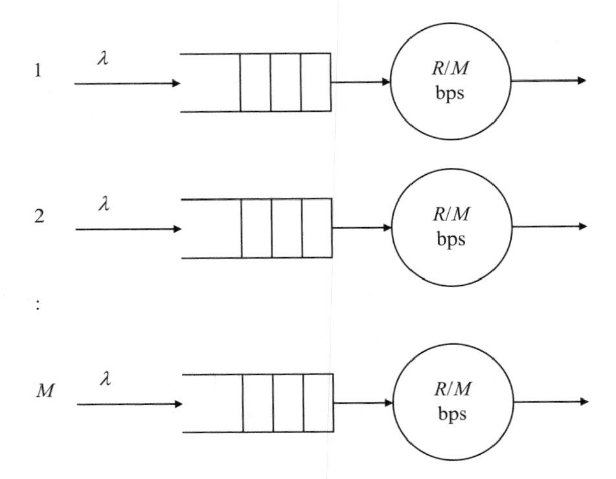

FIGURE 4.13 Queueing model for time-division multiple access.

station is R/M. To calculate the queueing delay in the buffer of each station, we can model each of them as an $M/D/1$ queue, as shown in Figure 4.13. The average service time, denoted by $\overline{Y}$, is given by.

$$\overline{Y} = \frac{\overline{X}}{R/M} = \frac{M\overline{X}}{R}. \tag{4.186}$$

From queueing theory, the average queueing delay in the buffer, denoted by $\overline{W}$, is given by

$$\begin{aligned} \overline{W} &= \frac{\rho}{2(1-\rho)}\overline{Y} \\ &= \frac{\rho}{2(1-\rho)}\frac{M\overline{X}}{R}, \qquad \rho = \lambda\overline{Y}. \end{aligned} \tag{4.187}$$

The average transfer delay for TDMA, denoted by $\overline{T}$, is given by

$$\begin{aligned} \overline{T} &= \overline{P} + \frac{1}{2}T_f + \overline{W} \\ &= \frac{\overline{X}}{R} + \frac{M\overline{X}}{2R} + \frac{\rho}{2(1-\rho)}\frac{M\overline{X}}{R}. \end{aligned} \tag{4.188}$$

The normalized average transfer delay, denoted by $\hat{T}$, is given by

$$\begin{aligned} \hat{T} &= \frac{T}{\overline{X}/R} \\ &= 1 + \frac{M}{2} + \frac{M\rho}{2(1-\rho)}. \end{aligned} \tag{4.189}$$

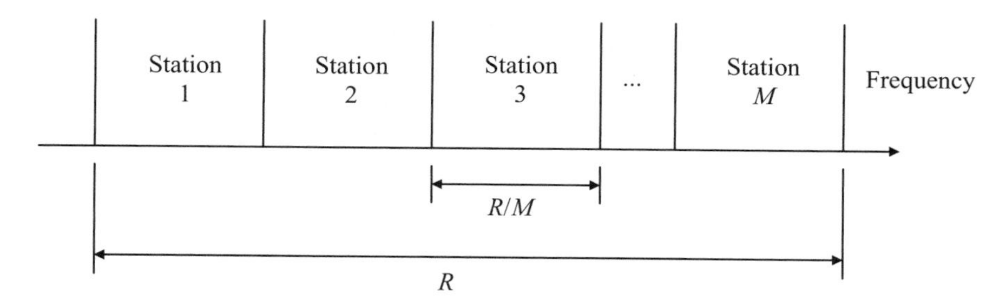

FIGURE 4.14 Frequency allocation for frequency-division multiple access.

4.10.2 Frequency-Division Multiple Access

Figure 4.14 shows the frequency band allocation for FDMA. The frequency band is divided into frequency slots. The major disadvantage of FDMA is that each station has a fixed allocation of channel bandwidth regardless of whether or not it has data to transmit. The entire channel can sustain a rate of R bps, which is equally divided among M stations (in practice, there are guard bands between frequency channels).

4.10.2.1 Analysis Since the individual bands are disjoint, there is no interference among the users' transmissions and the system can therefore be viewed as M independent $M/D/1$ queues. The average packet transmission time, denoted by $\overline{P}$, is given by

$$\overline{P} = \frac{\overline{X}}{R/M} = \frac{M\overline{X}}{R}. \tag{4.190}$$

From queueing theory, the average queueing delay in the buffer, denoted by $\overline{W}$, is given by

$$\overline{W} = \frac{\rho}{2(1-\rho)}\overline{Y} = \frac{\rho}{2(1-\rho)}\frac{M\overline{X}}{R}, \qquad \rho = \lambda\overline{Y}, \tag{4.191}$$

The average transfer delay for FDMA, denoted by $\overline{T}$, is given by

$$\begin{aligned}\overline{T} &= \overline{P} + \overline{W} \\ &= \frac{M\overline{X}}{R} + \frac{\rho}{2(1-\rho)}\frac{M\overline{X}}{R}.\end{aligned} \tag{4.192}$$

The normalized average transfer delay, denoted by $\hat{T}$, is given by

$$\hat{T} = \frac{T}{\overline{X}/R}$$

$$= M + \frac{M\rho}{2(1-\rho)}. \qquad (4.193)$$

4.10.3 Code-Division Multiple Access

CDMA does not try to allocate disjoint time or frequency resources to each user. Instead, CDMA allocates all resources to all users transmitting simultaneously by using codes and controlling the power transmitted by each of them to the minimum required to maintain a given SNR for the required quality of service. Each user uses a noiselike wideband signal occupying the entire frequency allocation during its usage. In this way, each user adds to the background noise affecting all the users, but to the minimum. Although this additional interference limits capacity, the resulting capacity is greater than that of conventional systems, as time and frequency resource allocations are unrestricted.

4.10.3.1 Analysis Assuming that each station has a mean effective packet transmission time equal to $M\overline{X}/R$ and, neglecting bit error probability, the system can be viewed as M independent $M/D/1$ queues. The effective packet transmission time, denoted by $\overline{P}$, is given by

$$\overline{P} = \frac{M\overline{X}}{R}. \qquad (4.194)$$

From queueing theory, the average queueing delay in the buffer, denoted by $\overline{W}$, is given by

$$\overline{W} = \frac{\rho}{2(1-\rho)}\overline{Y} = \frac{\rho}{2(1-\rho)}\frac{M\overline{X}}{R}, \qquad \rho = \lambda\overline{Y}. \qquad (4.195)$$

The average transfer delay for CDMA, denoted by $\overline{T}$, is given by

$$\overline{T} = \overline{P} + \overline{W}$$

$$= \frac{M\overline{X}}{R} + \frac{\rho}{2(1-\rho)}\frac{M\overline{X}}{R}. \qquad (4.196)$$

The normalized average transfer delay, denoted by $\hat{T}$, is given by

$$\begin{aligned}\hat{T} &= \frac{T}{\overline{X}/R} \\ &= M + \frac{M\rho}{2(1-\rho)}.\end{aligned} \tag{4.197}$$

SUMMARY

A form of slotted ALOHA MAC is analyzed in Section 4.2 by means of a Markov chain, and CSMA/CA MAC for IEEE 802.11 is analyzed in Section 4.3 by means of a two-dimensional Markov chain. This form of slotted ALOHA MAC is used in cellular access networks, while CSMA/CA is used in WLANs. Other analyses for IEEE 802.11 and IEEE 802.11e can be found in [2,10–13]. Multiclass CSMA/CA MAC is used in IEEE 802.11e and WiMedia. IEEE 802.11e can also be used in IEEE 802.11n WLAN. IEEE 802.11n can support a data rate of up to 600 Mbps. More details on IEEE 802.11e and IEEE 802.11n are given in Chapter 11. WiMedia is a high-rate WPAN with a data rate of up to 480 Mbps. More details on WiMedia are given in Chapter 12. The average delay for polling MAC is presented in Section 4.4. Polling is used in Bluetooth, which is a low-data-rate WPAN. A reservation MAC known as dynamic time-division multiple access is analyzed in Section 4.5 using two Markov chains. Reservation MAC is used in satellites. Other forms of reservation MACs may be found in [14–26]. In Section 4.6 an energy-efficient MAC for IEEE 802.15.4 is considered. ZigBee uses the IEEE 802.15.4 MAC and is also a low-rate WPAN. Other energy-efficient MACs may be found in [27–32]. A multichannel MAC using a dedicated control channel is considered in Section 4.7. A Markov chain is used in analytical modeling to obtain the total system throughput for this multichannel MAC. Other types of multichannel MACs are described [5,33,34]. Two directional-antenna MAC protocols are analyzed in Section 4.8. The saturated throughput for IEEE 802.11 multihop ad hoc networks is derived in Section 4.9. Other related multihop ad hoc networks are described in [35–40]. In Section 4.10 the average delays for three traditional forms of multiple access control protocols, TDMA, FDMA, and CDMA, are derived using queueing theory results. TDMA, FDMA and CDMA have been used in cellular access technologies.

REFERENCES

[1] G. Bianchi, "Performance analysis of the IEEE 802.11 distributed coordination function," *IEEE J. Sel. Areas Commun.*, vol. 18, no. 3, pp. 535–547, Mar. 2000.

[2] Y. Xiao, "Performance analysis of priority schemes for IEEE 802.11 and IEEE 802.11e wireless LANs," *IEEE Trans. Wireless Commun.*, vol. 4, no. 4, pp. 1506–1515, July 2005.

[3] X. Qiu and V. O. K. Li, "A unified performance model for reservation-type multiple-access schemes," *IEEE Trans. Veh. Technol.*, vol. 47, no. 1, Feb. 1998.

[4] M. Kohvakka, M. Kuorilehto, M. Hannikainen, and T. D. Hamalainen, "Performance analysis of IEEE 802.15.4 and ZigBee for large-scale wireless sensor network applications," *PE-WASUN 2006*, Torremolinos, Malaga, Spain, Oct. 6, 2006.

[5] J. Mo, H.-S. W. So, and J. Walrand, "Comparison of multi-channel MAC protocols," *MSWiM 2005*, Montreal, Quebec, Canada, Oct. 10–13, 2005.

[6] Y. Wang and J.J. Garcia-Luna-Aceves, "Directional collision avoidance in ad hoc networks," *Perform. Eval.*, vol. 58, pp. 215–241, 2004.

[7] J. He and H. K. Pung, "Performance modelling and evaluation of IEEE 802.11 distributed coordination function in multihop wireless networks," *Comp. Commun.*, vol. 29, pp. 1300–1308, 2006.

[8] J. He, "Fairness issues in multihop wireless ad hoc networks," Ph.D. desertation, National University of Singapore, 2005.

[9] H. W. Lee and J. W. Mark, "Combined/reservation access for packet switched transmission over a satellite with on-board processing: I. Global beam satellite," *IEEE Trans. Commun.*, vol. 31, no. 10, pp. 1161–1171, Oct. 1983.

[10] Y. C. Tay and K. C. Chua, "A capacity analysis of the IEEE 802.11 MAC protocol," *Wireless Networks*, vol. 7, pp. 159–171, 2001.

[11] E. Ziouva and T. Antonakopoulos, "CSMA/CA performance under high traffic conditions: Throughput and delay analysis," *Comp. Commun.*, vol. 25, pp. 313–321, 2002.

[12] Y. Xiao and J. Rosdahl, "Throughput and delay limits of IEEE 802.11," *IEEE Commun. Lett.*, vol. 6, no. 8, Aug. 2002.

[13] Z. N. Kong, D. H. K. Tsang, and B. Bensaou, "Performance analysis of IEEE 802.11e contention-based channel access," *J. Sel. Areas Commun.*, vol. 22, no. 10, pp. 2095–2106, Dec. 2004.

[14] J. E. Wieselthier and A. Ephremides, "Fixed- and movable-boundary channel-access schemes for integrated voice/data wireless networks," *IEEE Trans. Commun.*, vol. 43, no. 1, pp. 64–74, Jan. 1995.

[15] T.-S. P. Yum and H. Zhang, "Analysis of a dynamic reservation protocol for interactive data services on TDMA-based wireless networks," *IEEE Trans. Commun.*, vol. 47, no. 12, pp. 1796–1801, Dec. 1999.

[16] B. C. Kim and C. K. Un, "Capacity of wireless dynamic TDMA and media access control for packetized voice/data integration," *Wireless Personal Commun.*, vol. 1, pp. 313–319, 1995.

[17] K. O. Cho and J. K. Lee, "Performance analysis of the dynamic reservation multiple access protocol in the broadband wireless access system," *AsiaSim 2004*, LNAI 3398, pp. 250–259, 2005.

[18] X. Qui and V. O. K. Li, "Dynamic reservation multiple access (DRMA): a new multiple access scheme for personal communication system (PCS)," *Wireless Networks*, vol. 2, pp. 117–128, 1996.

[19] A. Iera and S. Marano, "D-RMA: a dynamic reservation multiple-access protocol for third generation cellular systems," *IEEE Trans. Veh. Technol.*, vol. 49, no. 5, Sept. 2000.

[20] H. C. B. Chan, J. Zhang, and H. Chen, "A dynamic reservation protocol for LEO mobile satellite systems," *IEEE J. Sel. Areas Commun.*, vol. 22, no. 3, pp. 559–573, Apr. 2004.

[21] S. Nanda, D. J. Goodman, and U. Timor, "Performance of PRMA: a packet voice protocol for cellular systems," *IEEE Trans. Veh. Technol.*, vol. 40, no. 3, Aug. 1991.

[22] S. Jangi and L. F. Merakos, "Performance analysis of reservation random access protocols for wireless access networks," *IEEE Trans. Commun.*, vol. 42, no. 2–3–4, pp. 1223–1234, Feb.–Mar.–Apr. 1994.

[23] G. Wu, K. Mukumoto, A. Fukuda, M. Mizuno, and K. Taira, "A dynamic TDMA wireless integrated voice/data system with data steal into voice (DSV) technique," *IEICE Trans. Commun.*, vol. E78-B, no. 8, Aug. 1995.

[24] R. Fantacci and S. Nannicini, "Performance evaluation of a reservation TDMA protocol for voice/data transmission in personal communication networks with nonindependent channel errors," *IEEE J. Sel. Areas Commun.*, vol. 18, no. 9, pp. 1636–1646, Sept. 2000.

[25] R. Fantacci and S. Nannicini, "Performance evaluation of a reservation TDMA protocol for voice/data transmission in microcellular systems," *IEEE J. Sel. Areas Commun.*, vol. 18, no. 11, pp. 2404–2416, Nov. 2000.

[26] R. Fantacci, S. Nannicini, and T. Pecorella, "Performance evaluation of a reservation TDMA protocol for voice/data transmissions in a LEO satellite communication system," *Int. J. Satellite Commun. Network.*, vol. 21, pp. 511–531, 2003.

[27] W. Ye, J. Heidemann, and D. Estrin, "Medium access control with coordinated adaptive sleeping for wireless sensor networks," *IEEE/ACM Trans. Network.*, vol. 12, no. 3, pp. 493–506, June 2004.

[28] A. El-Hoiydi and J.-D. Decotignie, "Low power downlink MAC protocols for infrastructure wireless sensor networks," *Mobile Networks Appl.*, vol. 10, pp. 675–690, 2005.

[29] C. E. Jones, K. M. Sivalingam, P. Agrawal, and J. C. Chen, "A survey of energy efficient network protocols for wireless networks," *Wireless Networks*, vol. 7, pp. 343–358, 2001.

[30] G. P. Halkes, T. V. Dam, and K. G. Langendoen, "Comparing energy-saving MAC protocols for wireless sensor networks," *Mobile Networks Appl.*, vol. 10, pp. 783–791, 2005.

[31] V. Rajendran, K. Obraczka, and J. J. Garcia-Luna-Aceves, "Energy-efficient, collision-free medium access control for wireless sensor networks," *Wireless Networks*, vol. 12, pp. 63–78, 2006.

[32] Q. Ren and Q. Liang, "Energy-efficient medium access control protocols for wireless sensor networks," *EURASIP J. Wireless Commun. Network.*, vol. 2006, ID 39814, pp. 1–17, 2006.

[33] Y. S. Han, J. Deng and Z. J. Haas, "Analyzing multi-channel medium access control schemes with ALOHA reservation," *IEEE Trans. Wireless Commun.*, vol. 5, no. 8, pp. 2143–2152, Aug. 2006.

[34] J. Chen and S.T. Sheu, "Distributed multichannel MAC protocol for IEEE 802.11 ad hoc wireless LANs," *Comput. Commun.*, vol. 28, no. 9, pp. 1000–1013, June 2005.

[35] F. Alizadeh-Shabdiz and S. Subramaniam, "Analytical models for single-hop and multi-hop ad hoc networks," *Mobile Networks Appl.*, vol. 11, pp. 75–90, 2006.

[36] Y. Chen, Q.-A. Zeng, and D. P. Agrawal, "Analytical modeling of MAC protocol in ad hoc networks," *Wireless Commun. Mobile Comput.*, in press.

[37] Y. Barowski, S. Biaz, and P. Agrawal, "Towards the performance analysis of IEEE 802.11 in multi-hop ad hoc networks," *IEEE Wireless Communications and Networking Conference 2005*, pp. 100–106, 2005.

[38] P. C. Ng and S. C. Liew, "Throughput analysis of IEEE 802.11 multi-hop ad hoc networks," *IEEE/ACM Trans. Network.*, vol. 15, no. 2, Apr. 2007.

[39] N. Gupta and P. R. Kumar, "A performance analysis of the 802.11 wireless LAN medium access control," *Commun. Inf. Syst.*, vol. 3, no. 4, pp. 279–304, Sept. 2004.

[40] R. Khalaf and I. Rubin, "Throughput and delay analysis in single hop and multihop IEEE 802.11 networks," *IEEE Broadnets 2006, Conference CD-ROM*, 2006.

CHAPTER 5

MOBILITY RESOURCE MANAGEMENT

5.1 INTRODUCTION

Mobility resource management is the management of radio resources in support of user roaming such that the grade of service (GoS) and quality of service (QoS) of a connection are maintained. Thus, mobility resource management is a combination of resource management, location management, and handoff management. The radio resources can be channels assigned to hand off calls and new calls. When a mobile terminal moves from one cell to another cell, it needs to be attached to a base station or access point. Thus, mobility resource management is needed to ensure that the connection remains intact with some GoS and QoS. When a mobile moves from one cell into another cell, it needs to hand off from the base station or access point of the current cell to the base station or access point of the next cell. Resource allocation for connection admission at the connection level typically has new call and handoff call blocking probabilities of handoff connections as GoS measures. These GoS measures, associated with connection admission, are collectively defined as *connection-level metrics*. The *blocking probability metric* is the probability that a new connection is rejected from admission into a mobile system. The *handoff call blocking probability metric* is the probability that an existing connection moving from one cell to the next cell in a mobile system is rejected from admission into the next cell due to lack of resources. The *system utilization metric* measures use of the mobile system. If the network is a homogeneous cellular network, there are

Wireless Broadband Networks, By David Tung Chong Wong, Peng-Yong Kong, Ying-Chang Liang, Kee Chaing Chua, and Jon W. Mark

three main handoff strategies, depending on who initiates the handoff or who assists in the handoff. Within homogeneous networks, handoffs are known as *horizontal handoffs*. Horizontal handoffs can be classified mainly as *hard*, *soft*, and *partial* handoffs. Handoffs between heterogeneous networks are known as *vertical handoffs*. Vertical handoffs can be divided into *upward* and *downward* handoffs, depending on the relative cell sizes of the heterogeneous networks.

Channel assignment schemes are needed to ensure certain GoS. These schemes include nonprioritized scheme, prioritized (guard channel) scheme, limited fractional guard channel scheme, fractional guard channel scheme, guard channel with buffer scheme, two-level fractional guard channel, and link-layer resource allocation scheme. The nonprioritized scheme does not differentiate between new calls and handoff calls. Therefore, the blocking probabilities for new and handoff calls are the same. The guard channel scheme gives priority to handoff calls over new calls. This results in lower handoff call blocking probability than with new call blocking probability. The reason for this is that dropping new calls is more acceptable than dropping handoff calls, which would result in call termination in the midst of a call. A limited fractional guard channel scheme gives a bit more priority to handoff calls over new calls; a fractional guard channel scheme is more general and gives a lot more priority to handoff calls than to new calls. A guard channel with buffer scheme allows queueing of calls until a channel is available or when the call hands off to another cell. This scheme is only suitable for non-real-time traffic such as data. The two-level fractional guard channel scheme is used to provide efficient priority access for handoff calls over new calls. The performance of the two-level fractional guard channel scheme is almost the same as that of a guard channel scheme under light load. However, the advantage of the two-level fractional guard channel scheme is demonstrated under heavy load, where the handoff call blocking probability is much better than that of a guard channel scheme, giving more priority and protection to handoff calls over new calls. The link-layer resource allocation scheme is based on the outage probability, which determines the maximum number of users that can be supported. These schemes can be extended for multiclass traffic, for which there are three main channel assignment schemes: complete partitioning, complete sharing, and virtual partitioning. *Complete partitioning* does not allow channels to be shared among different traffic classes, whereas *complete sharing* allows all channels to be shared among different traffic classes. *Virtual partitioning* allows for preemption of traffic classes and is like complete sharing under light load and complete partitioning under heavy load. The link-layer resource allocation scheme can also be extended for multiclass traffics. Based on the outage probabilities of multiclass traffics, the system capacity can be obtained and complete sharing or virtual partitioning can be used.

Location management is needed to ensure that packets are sent correctly to the current location of a mobile regardless of whether they are at their home network or at a foreign network. Location management consists of authentication, location update, and call delivery. Mobile IP can be used for this purpose. Cellular IP and HAWAII schemes can be used for cellular systems to handle micromobility, while mobile IP can be used to handle macromobility.

This chapter is organized as follows. In Section 5.2 we present the ramifications of horizontal and vertical handoffs. Section 5.3 provides an overview of handoff strategies. Section 5.4 covers analyses for single-traffic-class channel assignment schemes. The birth–death process or single- or two-dimensional Markov chains are used to model these channel assignment schemes, and the GoS of new and handoff blocking probabilities are derived. Channel assignment schemes for multiclass traffic are presented and analyzed in Section 5.5. Single- and multidimensional Markov chains are used for modeling these channel assignment schemes with multiclass traffic. GoS performance metrics such as new and handoff blocking probabilities and preemption probabilities are derived explicitly. In Section 5.6 we present the main components of location management. A location management scheme known as mobile IP is presented in Section 5.7. Another location management scheme of cellular networks, known as cellular IP, is presented in Section 5.8. A micromobility scheme known as HAWAII is presented in Section 5.9. The summary includes additional references to extend the topics discussed in this chapter.

5.2 TYPES OF HANDOFFS

Handoffs can be divided into horizontal and vertical handoffs. Horizontal handoffs are handoffs between base stations within the same network; vertical handoffs are handoffs between base stations and/or access points from different networks. In horizontal handoffs, two types of handoffs with single transmit and receive antennas and one type of handoff with multiple-input, multiple-output (MIMO) are considered. Handoffs of the former type are hard and soft handoffs, while the latter type of handoff is a partial handoff. In vertical handoffs, two types of handoffs are considered: upward and downward.

5.2.1 Horizontal Handoff

5.2.1.1 Hard Handoff Figure 5.1 shows a hard handoff process, with Figure 5.1(a) showing the mobile terminal (MT) connected to base station (BTS) A initially. Figure 5.1(b) shows that the MT is disconnected from BTS A. Finally, the MT is connected to BTS B in Figure 5.1(c). The MT is connected to only one BTS at any

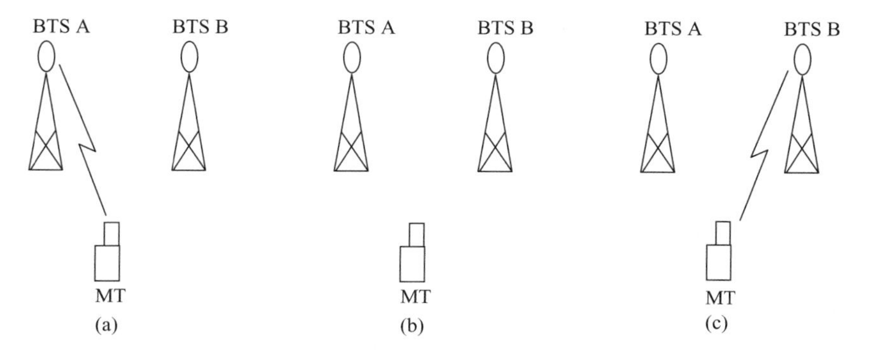

FIGURE 5.1 Hard handoff: (a) MT connected to BTS A; (b) MT disconnected from BTS A; (c) MT connected to BTS B.

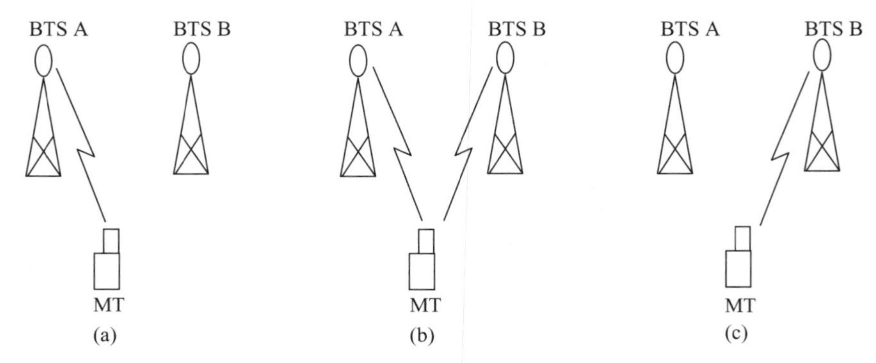

FIGURE 5.2 Soft handoff: (a) MT connected to BTS A only; (b) MT connected to both BTSs A and B; (c) MT disconnected from BTS A but still connected to BTS B.

time. The old connection is terminated before a new connection is activated. This is also known as *break before make*.

5.2.1.2 Soft Handoff Figure 5.2 shows a soft handoff process, with Figure 5.2(a) showing the MT connected to BTS A initially. Figure 5.2(b) shows that the MT is connected to both BTSs, A and B. Finally, the MT is disconnected from BTS A but still connected to BTS B in Figure 5.2(c). The MT can communicate with more than one BTS during the handoff. A new connection is made before breaking the old connection. This is also known as *make before break*.

5.2.1.3 Partial Handoff Figure 5.3 shows a partial handoff process with 2×2 MIMO links, with Figure 5.3(a) showing the MT connected initially to both transmit antennas of BTS A. Figure 5.3(b) shows that the MT has one antenna connected to one antenna of BTS A and one antenna connected to one antenna of BTS B.

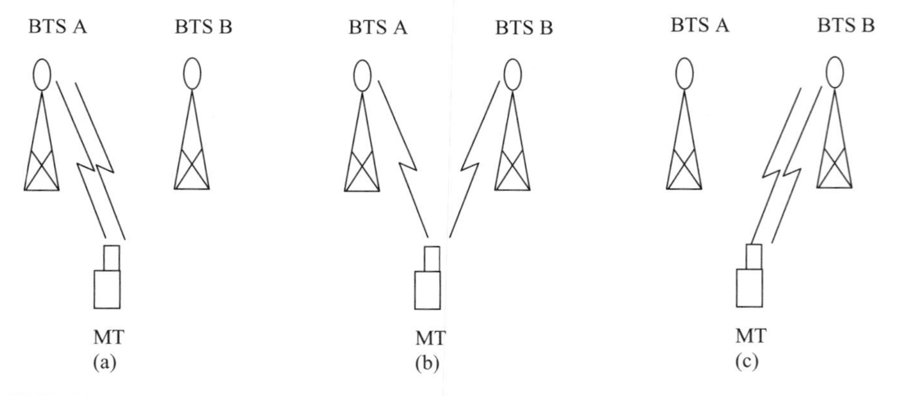

FIGURE 5.3 Partial handoff: (a) both antennas of MT connected to BTS A only; (b) one antenna of MT is connected to BTS A and one antenna of MT is connected to BTS B; (c) MT disconnected from BTS A and both antennas of MT are connected to BTS B.

Finally, the MT is disconnected from BTS A and has both antennas connected to both antennas of BTS B in Figure 5.3(c). The MT can communicate partially with more than one BTS during the handoff. A new connection is made before breaking the old connection. This is also a form of *make before break*. The advantage of partial handoff over conventional handoff is that it reduces the effect of transmit antenna correlation by separating the transmit antennas as far as possible [2]. Partial handoff can increase system capacity, link throughput, and other performance measures by reducing the effect of spatial correlation [2].

5.2.2 Vertical Handoff

5.2.2.1 Upward Handoff Upward handoff is a handoff from a smaller cell size of one network to a larger cell size of another network. Usually, the smaller cell size has a larger bandwidth and shorter coverage, while the larger cell size has a smaller bandwidth and larger coverage. An example is a handoff from a wireless local area network (WLAN) to a cellular network.

5.2.2.2 Downward Handoff Downward handoff is a handoff from a larger cell size of one network to a smaller cell size of another network. As mentioned before, the smaller cell size has a larger bandwidth and shorter coverage, whereas the larger cell size has a smaller bandwidth and larger coverage. An example is a handoff from a cellular network to a WLAN.

5.3 HANDOFF STRATEGIES

Handoff is a process of disconnecting from a source base station and connecting to a target base station. Handoff is needed when a mobile user moves from one cell to another cell or when the signal strength of the connection is weak. A *cell* is an area served by a base station. According to [1], there are three types of handoff strategies: mobile-controlled network-controlled, and mobile-assisted handoff. These handoff strategies are discussed in the following sections.

5.3.1 Mobile-Controlled Handoff

The mobile terminal (MT) monitors the signals of the surrounding base stations (BTSs) continuously and initiates the handoff process when some handoff criteria (e.g., signal measurement levels, hysteresis margin) are met. In mobile-controlled handoff (MCHO) the MT is in complete control of the handoff process [1]. The reaction time in this type of handoff strategy is short. MCHO is used in the DECT system.

5.3.2 Network-Controlled Handoff

In network-controlled handoff (NCHO), the surrounding BTSs measure the signal from the MT and the network initiates the handoff process when some handoff criteria are met. Information on the signal quality of all users is located at the mobile switching center (MSC) [1]. This information helps in resource allocation. However, the overall delay for this type of handoff strategy can be quite high. NCHO is used in CT-2 Plus and AMPS cellular access technologies.

5.3.3 Mobile-Assisted Handoff

In mobile-assisted handoff (MAHO), the network asks the MT to measure the signal (power levels) from the surrounding BTSs. The network makes the handoff decision based on reports from the MT [1]. The delay for this type of handoff strategy is between those of the MCHO and NCHO. MAHO is used in GSM and IS-95 CDMA cellular access technologies.

5.4 CHANNEL ASSIGNMENT SCHEMES

In cellular mobile systems, minimizing the handoff call droppings with good call admission schemes is very important. Maintaining an ongoing call has a higher priority than admitting a new call. Hence, handoff calls should be given a higher access priority over new calls, or a lower blocking probability than new calls. Furthermore, as new call blockings are more tolerable to users than handoff call blocking, a user is more likely to be frustrated to being dropped halfway in a call than when a user cannot get into the system at the beginning of a new call. A traditional method to provide priority access for handoff calls over new calls is the guard channels (GC) scheme [3]. By reserving a small number of guard channels for use by handoff calls only, the handoff call blocking probability is cut down significantly with the trade-off of a slight increase in the new call blocking probability. *Fractional guard channels* [4] allow new calls to be admitted with a certain probability and block new calls with a different probability. This gives *additional priority to handoff calls* as a result. Queueing can also be added to the resource allocation schemes to improve handoff call blocking probability. In this section, the terms *calls* and *connections* are used interchangeably.

Let λ_n, λ_h, $\lambda = \lambda_n + \lambda_h$, μ_c^{-1}, μ_h^{-1}, and $\mu = \mu_c + \mu_h$ denote the arrival rate of new calls, the arrival rate of handoff calls, the mean call arrival rate of new and handoff calls, the mean call holding time of a call in a cell, the mean dwell time (interhandoff time) of a call in a cell, and the mean equivalent service rate of a call in a cell, respectively.

5.4.1 Nonprioritized Scheme

In the nonprioritized scheme, there is no guard channel. Figure 5.4 shows a flowchart for a nonprioritized scheme with new and handoff connections. When a new or handoff call arrives, the system will check if there is an available channel for it. If one is not available, the call will be blocked. On the other hand, if one is available, a channel will be assigned for the call. After the call has ended, the channel will be released. Let us consider a single-class-system model for the nonprioritized scheme with new and handoff connections as shown in Figure 5.5. Both new and handoff calls can make use of any available C channels. Figure 5.6 shows a one-dimensional finite-state Markov

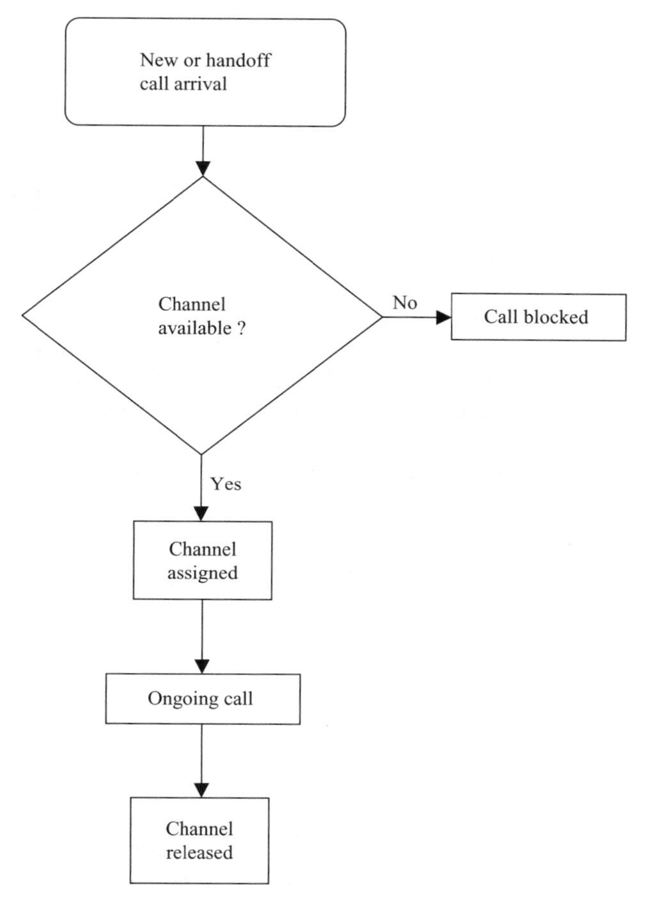

FIGURE 5.4 Flowchart for a nonprioritized scheme with new and handoff connections.

chain for the nonprioritized scheme, and there are new and handoff connections. To facilitate analytical modeling, it is necessary to make certain assumptions about the traffic parameters. It is not unreasonable to assume that the holding time has a negative exponential distribution. Although a negative exponential distribution assumption may not be as reasonable for the cell dwell time, for analytical tractability we will make the same assumption for cell dwell time (interhandoff time) and model the

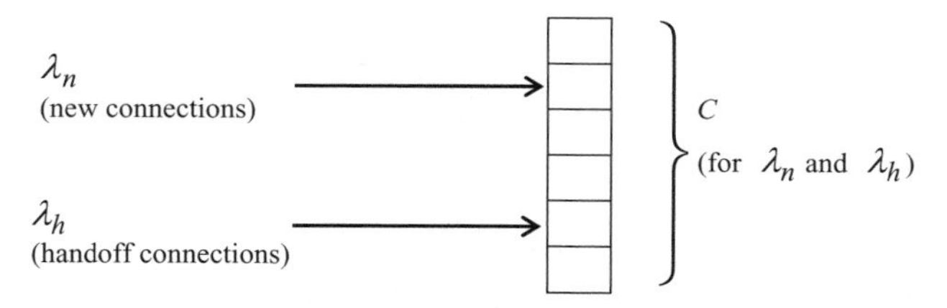

FIGURE 5.5 System model for a nonprioritized scheme with new and handoff connections.

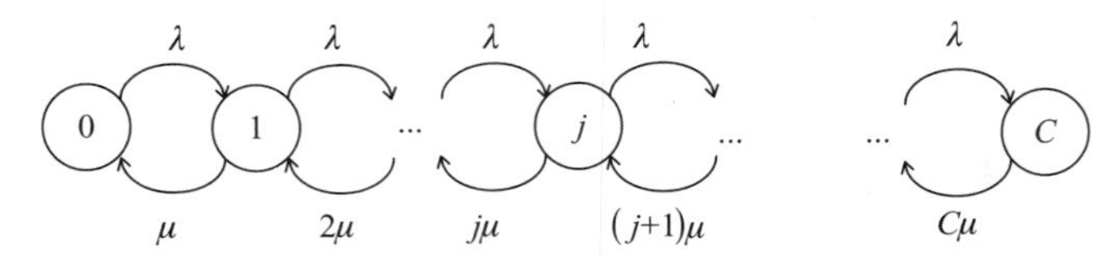

FIGURE 5.6 State transition diagram for a nonprioritized scheme with new and handoff connections.

channel as a Markov process with the connections-level parameters. It is noted in [5] that exponential and lognormal distributions give relatively close approximations, although the generalized gamma distribution provides the best approximation. A one-dimensional finite-state Markov chain is used to solve for the nonprioritized handoff scheme.

Let $P\{(j)\}$ be the steady-state probabilities of the Markov chain. Solving the Markov chain, we get

$$P(j) = \left\{ \left(\frac{\lambda}{\mu}\right)^j \frac{1}{j!} P(0), \qquad 0 < j \le C, \right. \tag{5.1}$$

where

$$P(0) = \left[1 + \sum_{j=1}^{C} \left(\frac{\lambda}{\mu}\right)^j \frac{1}{j!} \right]^{-1}. \tag{5.2}$$

The new and handoff connections blocking probabilities are given, respectively, by

$$b_n = P(j = C) \tag{5.3}$$

and

$$b_h = P(j = C). \tag{5.4}$$

The utilization is given by

$$N_u = \sum_{j=1}^{C} j P(j) = \frac{\lambda_n (1 - b_n) + \lambda_h (1 - b_h)}{\mu}. \tag{5.5}$$

Equating the handoff connections arrival rate to the product of the average handoff rate for connections and the average number of connections, we can get the handoff connections arrival rate as follows:

$$\lambda_h = \mu_h \times N_u = \mu_h \times \frac{(1-b_n)\lambda_n + (1-b_h)\lambda_h}{\mu_c + \mu_h} = \frac{\mu_h (1-b_n)\lambda_n}{\mu_c + b_h \mu_h}. \tag{5.6}$$

Thus, the class k handoff connections arrival rate can be approximated under low blocking probabilities as follows:

$$\lambda_h = \frac{\mu_h \lambda_n}{\mu_c}, \tag{5.7}$$

where $\mu_h = v/s$, v is the speed of the mobile, and s is the length of a square cell.

5.4.2 Prioritized (Guard Channel) Scheme

In the prioritized or guard channel scheme, there are a number of guard channels. These guard channels give priority to handoff calls over new calls. Figure 5.7 shows a flowchart for a guard channel scheme. When a new call arrives, the system will check if there is an available normal channel for it. There are $C - C_G$ normal channels. If it is not available, the new call will be blocked. On the other hand, if it is available, a channel will be assigned for the call. After the call has ended, the channel will be released. When a handoff call arrives, the system will check if there is an available normal channel for it. If it is available, a channel will be assigned for the call. If it is not available, the system will check if there is an available guard channel for it. There are C_G guard channels. If it is available, a channel will be assigned for the call. If it is not available, the call will be blocked. After the call has ended, the channel will be released. Let us consider a single-class-system model for the guard channel scheme with new and handoff connections as shown in Figure 5.8. Only handoff calls can make use of any available C channels, while new calls can only make use of any available $C - C_G$ normal channels. Figure 5.9 shows a one-dimensional finite-state Markov chain for the guard channel scheme, and there are new and handoff connections. The guard channels, C_G, are reserved for handoff connections only. A one-dimensional finite-state Markov chain is used to solve for the guard channel scheme.

Let $P\{(j)\}$ be the steady-state probabilities of the Markov chain. Solving the Markov chain, we get

$$P(j) = \begin{cases} \left(\dfrac{\lambda}{\mu}\right)^j \dfrac{1}{j!} P(0), & 0 < j \le C - C_G \\[2ex] \dfrac{(\lambda)^{C-C_G} (\lambda_h)^{j-(C-C_G)}}{(\mu)^j \, j!} P(0), & C - C_G < j \le C, \end{cases} \tag{5.8}$$

where

$$P(0) = \left[1 + \sum_{j=1}^{C-C_G} \left(\frac{\lambda}{\mu}\right)^j \frac{1}{j!} + \sum_{j=C-C_G+1}^{C} \frac{(\lambda)^{C-C_G} (\lambda_h)^{j-(C-C_G)}}{(\mu)^j \, j!} \right]^{-1}. \tag{5.9}$$

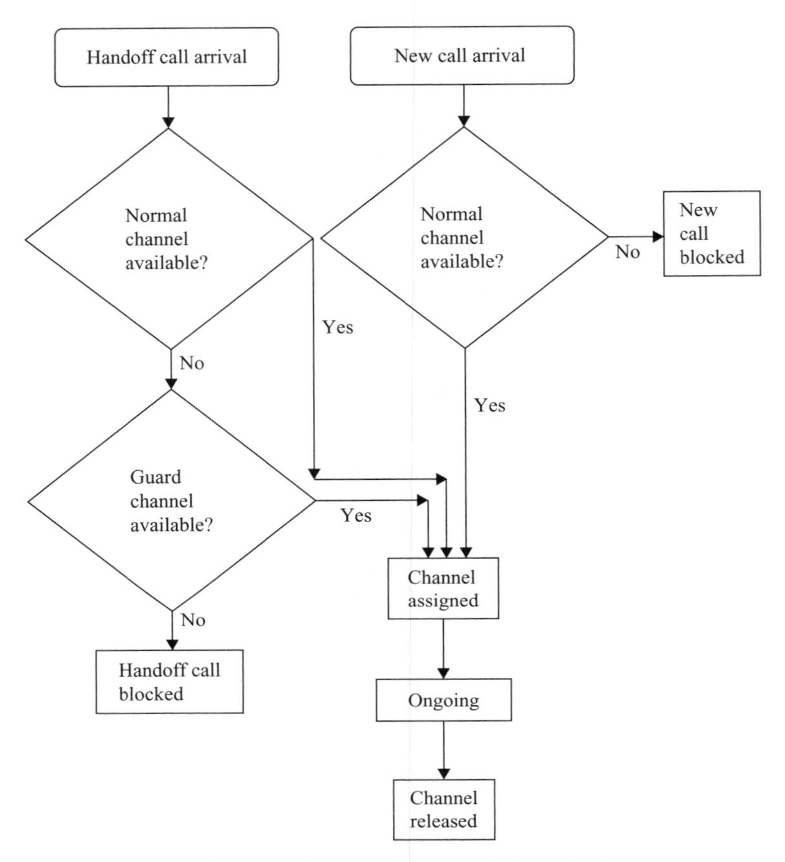

FIGURE 5.7 Flowchart for a guard channel scheme.

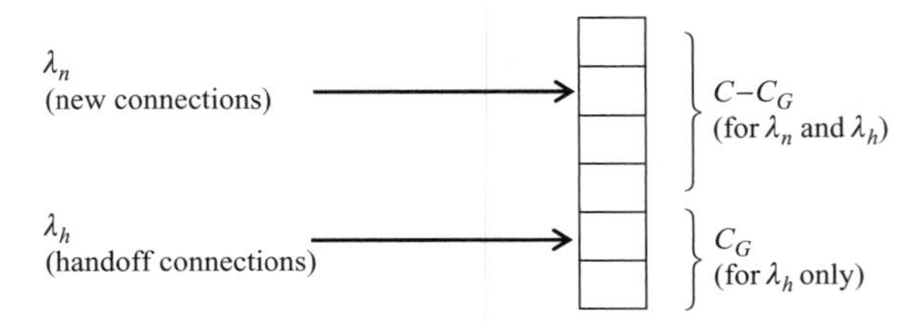

FIGURE 5.8 System model for a guard channel scheme with new and handoff connections.

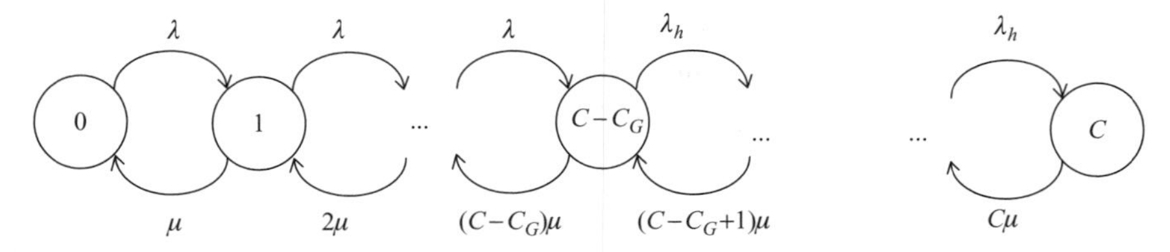

FIGURE 5.9 State transition diagram for a guard channel scheme with new and handoff connections.

The new and handoff connections blocking probabilities are given, respectively, by

$$b_n = \sum_{j=C-C_G}^{C} P(n), \tag{5.10}$$

and

$$b_h = P(n = C). \tag{5.11}$$

The utilization is given by

$$N_u = \sum_{j=1}^{C} jP(j) = \frac{\lambda_n (1 - b_n) + \lambda_h (1 - b_h)}{\mu}. \tag{5.12}$$

5.4.3 Limited Fractional Guard Channel Scheme

In the limited fractional guard channel scheme, there are a number of guard channels that gives priority to handoff calls over new calls. In addition, there is also another channel that gives priority to handoff calls by limiting this channel availability to new calls with a certain probability. Figure 5.10 shows a flowchart for the limit fractional guard channel scheme. When a new call arrives, the system will check if there is an available normal channel for it. There are $C - C_G$ normal channels. If it is not available, the new call will be blocked. On the other hand, if it is available and if it is not the last available normal channel, a channel will be assigned for the call. Otherwise, if it is the last available normal channel, a random number is generated in [0,1] and compared with β. If the random number generated is less than β, a channel will be assigned for the call. Otherwise, the call will be blocked. After the call has ended for the successful new call, the channel will be released. When a handoff call arrives, the system will check if there is an available normal channel for it. If it is available, a channel will be assigned for the call. If it is not available, the system will check if there is an available guard channel for it. There are C_G guard channels. If it is available, a channel will be assigned for the call. If it is not available, the call will be blocked. After the call has ended, the channel will be released. Let us consider a single-class-system model for the limited fractional guard channel scheme, as shown in Figure 5.11. Only handoff calls can make use of any available C channels, while new calls can only make use of any available $C - C_G - 1$ normal channels and one normal channel with a probability of β. Figure 5.12 shows a one-dimensional finite-state Markov chain for the limited fractional guard channel scheme and there are new and handoff connections. The guard channels, C_G, are reserved for handoff connections only. A one-dimensional finite-state Markov chain is used to solve for the limited fractional guard channel scheme.

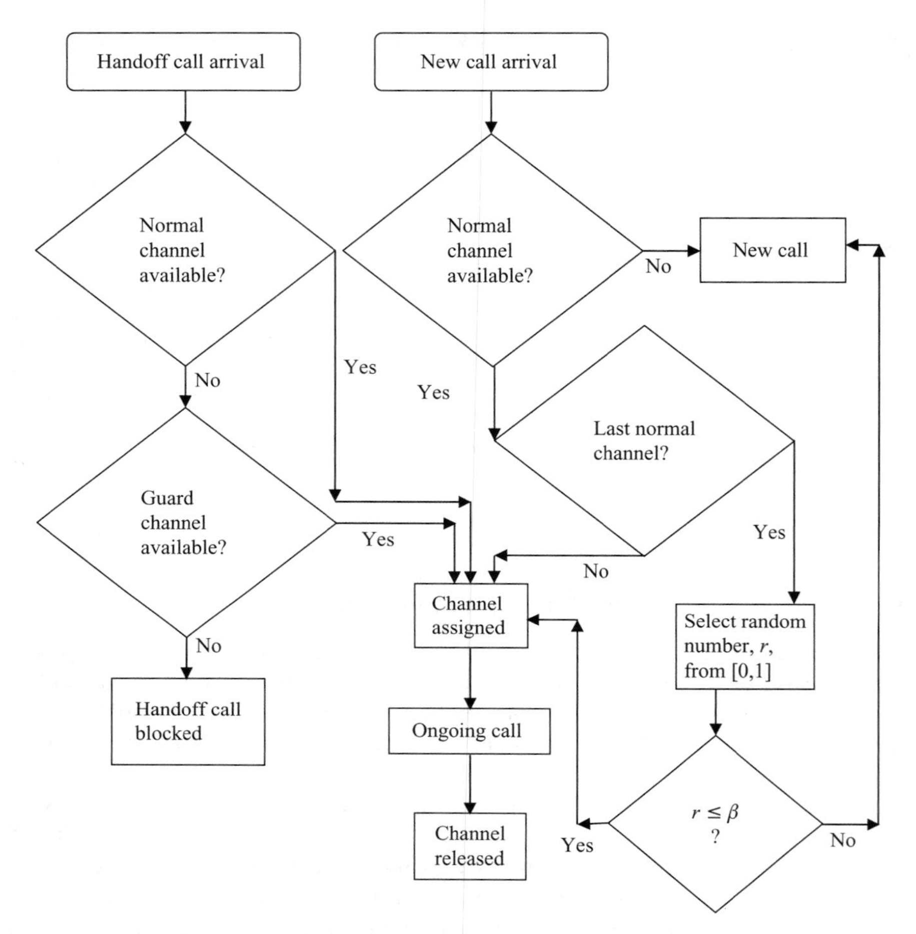

FIGURE 5.10 Flowchart for a limited fractional guard channel scheme.

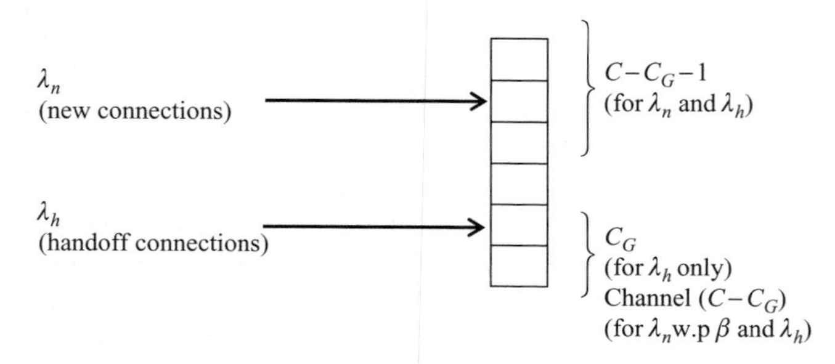

FIGURE 5.11 System model for a limited fractional guard channel scheme with new and handoff connections.

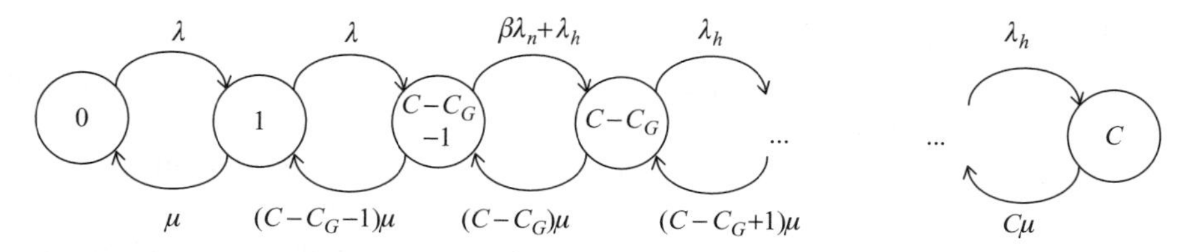

FIGURE 5.12 State transition diagram for a limited fractional guard channel scheme with new and handoff connections.

Let $\{P(j)\}$ be the steady-state probabilities of the Markov chain. Solving the Markov chain, we get

$$P(j) = \begin{cases} \left(\dfrac{\lambda}{\mu}\right)^j \dfrac{1}{j!} P(0), & 0 < j \leq C - C_G - 1 \\ \dfrac{(\lambda)^{C-C_G-1}(\beta\lambda_n + \lambda_h)}{(\mu)^j} \dfrac{1}{j!} P(0), & j = C - C_G \\ \dfrac{(\lambda)^{C-C_G-1}(\beta\lambda_n + \lambda_h)(\lambda_h)^{j-(C-C_G)}}{(\mu)^j\, j!} P(0), & C - C_G < j \leq C, \end{cases} \tag{5.13}$$

where

$$P(0) = \left[1 + \sum_{j=1}^{C-C_G-1} \left(\frac{\lambda}{\mu}\right)^j \frac{1}{j!} + \frac{(\lambda)^{C-C_G-1}(\beta\lambda_n + \lambda_h)}{(\mu)^{C-C_G}} \frac{1}{(C-C_G)!} + \sum_{j=C-C_G+1}^{C} \frac{(\lambda)^{C-C_G}(\lambda_h)^{j-(C-C_G)}}{(\mu)^j\, j!} \right]^{-1}. \tag{5.14}$$

The new and handoff connections blocking probabilities are given, respectively, by

$$b_n = (1-\beta)P(C - C_G - 1) + \sum_{j=C-C_G}^{C} P(j) \tag{5.15}$$

and

$$b_h = P(j = C). \tag{5.16}$$

The utilization is given by

$$N_u = \sum_{j=1}^{C} jP(j) = \frac{\lambda_n(1-b_n) + \lambda_h(1-b_h)}{\mu}. \tag{5.17}$$

5.4.4 Fractional Guard Channel Scheme

In the fractional guard channel scheme, there are a number of guard channels that give priority to handoff calls over new calls. In addition, the other channels give additional priority to handoff calls by limiting these channel availabilities to new calls with a certain probability associated with each channel occupancy. Figure 5.13 shows a flowchart for the fractional guard channels scheme. When a new call arrives, the system will check if a normal channel is available for it. There are $C - C_G$ normal channels. If it is not available, the new call will be blocked. On the other hand, if it is available, a random number is generated in [0,1] and compared with β_j, where j is the current channel occupancy and $0 \leq j \leq C - C_G$. If the random number generated is less than β_j, a channel will be assigned for the call. Otherwise, the call will be blocked. After the call has ended for the successful new call, the channel will be released. When a handoff call arrives, the system will check if there is an available normal channel for it. If it is available, a channel will be assigned for the

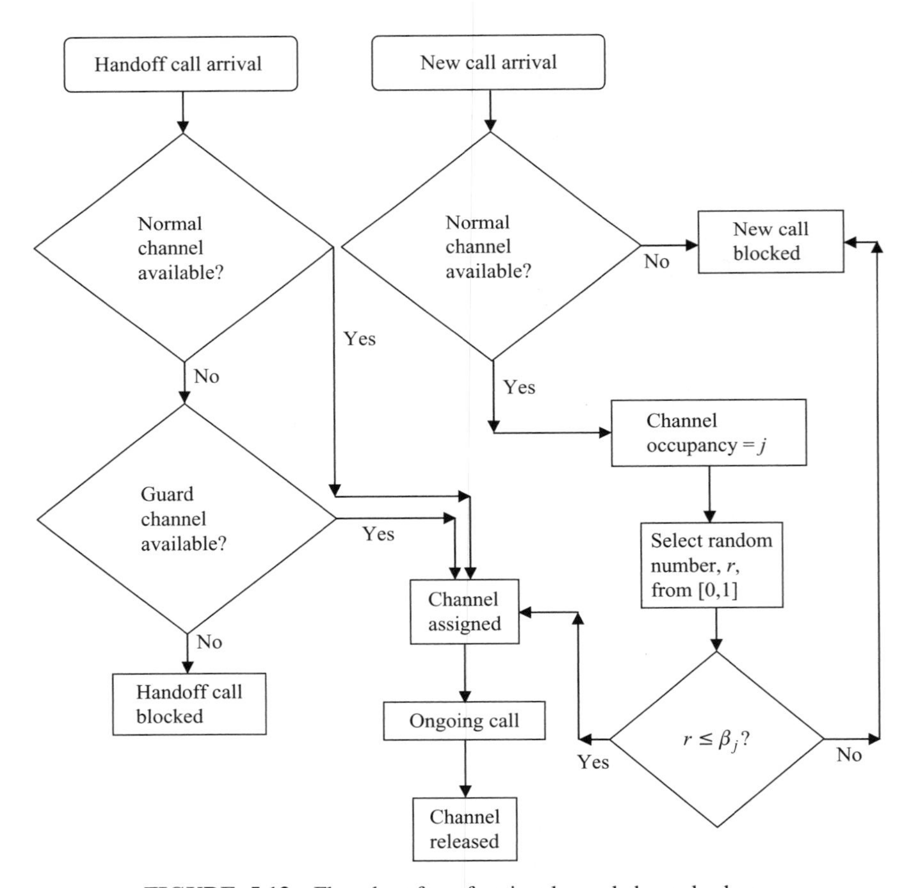

FIGURE 5.13 Flowchart for a fractional guard channel scheme.

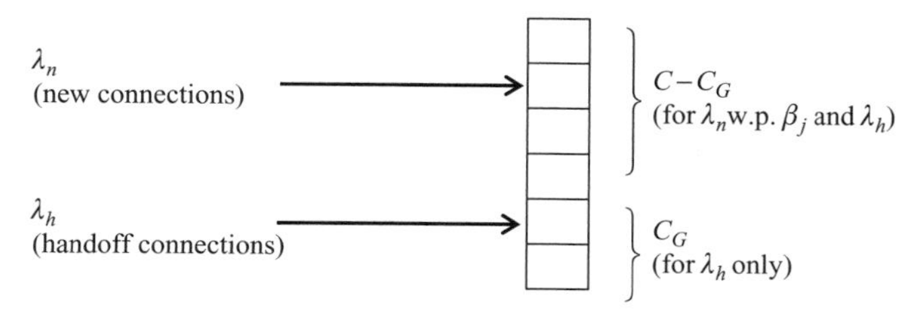

FIGURE 5.14 System model for a fractional guard channel scheme with new and handoff connections.

call. If it is not available, the system will check if there is an available guard channel for it. There are C_G guard channels. If it is available, a channel will be assigned for the call. If it is not available, the call will be blocked. After the call has ended, the channel will be released. Let us consider a single-class-system model for the fractional guard channel scheme, as shown in Figure 5.14. Only handoff calls can make use of any available C channels, while new calls can only make use of any available $C - C_G$ normal channels with probability β_j, depending on the channel occupancy. Figure 5.15 shows a one-dimensional finite-state Markov chain for the fractional guard channel scheme and there are new and handoff connections. The guard channels, C_G, are reserved for handoff connections only. A one-dimensional finite-state Markov chain is used to solve for the fractional guard channel scheme.

Let $\{P(j)\}$ be the steady-state probabilities of the Markov chain. Solving the Markov chain, we get

$$P(j) = \begin{cases} \dfrac{\prod_{i=0}^{j-1} (\beta_i \lambda_n + \lambda_h)}{(\mu)^j} \dfrac{1}{j!} P(0), & 0 < j \leq C - C_G \\[2ex] \dfrac{\left[\prod_{i=0}^{C-C_G-1} (\beta_i \lambda_n + \lambda_h)\right] (\lambda_h)^{j-(C-C_G)}}{(\mu)^j \, j!} P(0), & C - C_G < j \leq C, \end{cases} \tag{5.18}$$

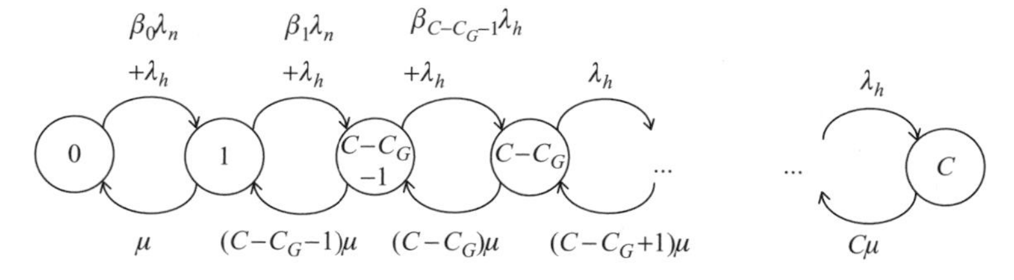

FIGURE 5.15 State transition diagram for a fractional guard channel scheme with new and handoff connections.

where

$$P(0)=\left[1+\sum_{j=1}^{C-C_G}\frac{\prod_{i=0}^{j-1}(\beta_i\lambda_n+\lambda_h)}{(\mu)^j}\frac{1}{j!}+\sum_{j=C-C_G+1}^{C}\frac{\left[\prod_{i=0}^{C-C_G-1}(\beta_i\lambda_n+\lambda_h)\right](\lambda_h)^{j-(C-C_G)}}{(\mu)^j\, j!}\right]^{-1} \tag{5.19}$$

The new and handoff connections blocking probabilities are given, respectively, by

$$b_n=\sum_{j=0}^{C-C_G-1}(1-\beta_j)P(j)+\sum_{j=C-C_G}^{C}P(j) \tag{5.20}$$

and

$$b_h=P(j=C). \tag{5.21}$$

The utilization is given by

$$N_u=\sum_{j=1}^{C}jP(j)=\frac{\lambda_n(1-b_n)+\lambda_h(1-b_h)}{\mu}. \tag{5.22}$$

5.4.5 Guard Channel Scheme with Buffer

In the guard channel scheme with buffer, there are a number of guard channels that gives priority to handoff calls over new calls. In addition, there is a queue that holds handoff calls instead of dropping them if all the channels are occupied. Figure 5.16 shows a flowchart for the guard channel scheme with queueing of handoff calls. This scheme is similar to the guard channel scheme except a buffer is used for queueing handoff calls. Let us consider a single-class-system model of guard channel scheme with queueing of handoff calls as shown in Figure 5.17. The guard channels, C_G, and the buffer, Q, are reserved for handoff calls only. We can model the channel and buffer occupancy as a Markov chain. Figure 5.18 shows a one-dimensional finite-state Markov chain for the guard channel scheme with buffer, and there is queueing of handoff calls. Note that queued handoff calls that are not serviced during its dwelling time in the cell will be handed off to the next cell. Solving the Markov chain,

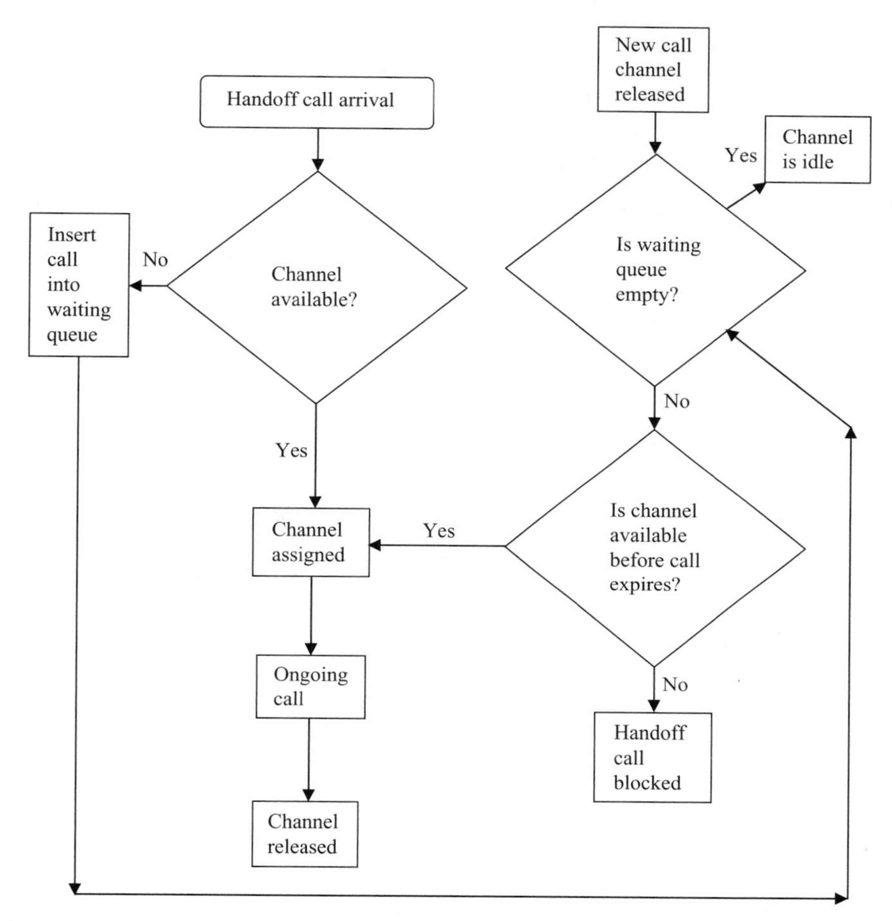

FIGURE 5.16 Flowchart for a guard channel scheme with queueing of handoff calls.

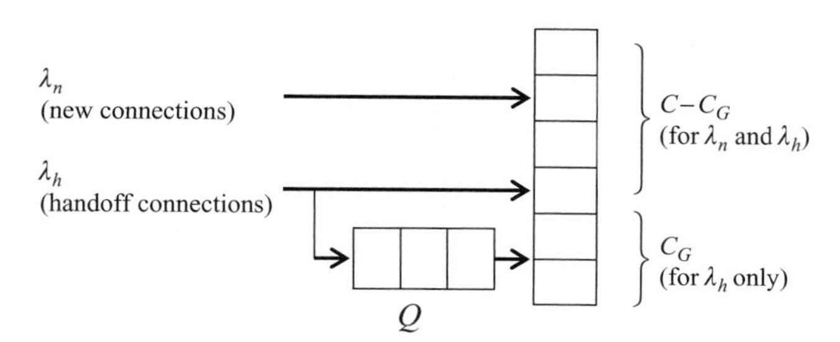

FIGURE 5.17 System model for a guard channel scheme with queueing of handoff calls.

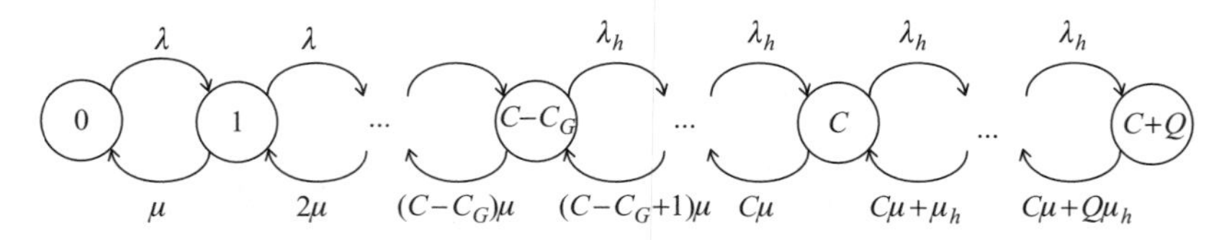

FIGURE 5.18 State transition diagram for a guard channel scheme with queueing of handoff calls.

we get

$$P(j) = \begin{cases} \left(\dfrac{\lambda}{\mu}\right)^j \dfrac{1}{j!} P(0), & 0 < j \leq C - C_G \\ \dfrac{(\lambda)^{C-C_G} (\lambda_h)^{j-(C-C_G)}}{(\mu)^j \, j!} P(0), & C - C_G < j \leq C \\ \dfrac{(\lambda)^{C-C_G} (\lambda_h)^{j-(C-C_G)}}{(\mu)^C \, C! \prod_{i=1}^{j-C} (C\mu + i\mu_h)} P(0), & C < j \leq C + Q, \end{cases} \tag{5.23}$$

where

$$P(0) = \left[1 + \sum_{j=1}^{C-C_G} \left(\frac{\lambda}{\mu}\right)^j \frac{1}{j!} + \sum_{j=C-C_G+1}^{C} \frac{(\lambda)^{C-C_G} (\lambda_h)^{j-(C-C_G)}}{(\mu)^j \, j!} \right. \\ \left. + \sum_{j=C+1}^{C+Q} \frac{(\lambda)^{C-C_G} (\lambda_h)^{j-(C-C_G)}}{(\mu)^C C! \prod_{i=1}^{j-C} (C\mu + i\mu_h)} \right]^{-1}. \tag{5.24}$$

The new call blocking probability, b_n, is given by

$$b_n = \sum_{j=C-C_G}^{C+Q} P(j). \tag{5.25}$$

The handoff call blocking probability, b_h, is given by

$$b_h = P(j = C + Q). \tag{5.26}$$

The utilization is given by

$$N_u = \sum_{j=1}^{C} \frac{jP(j)}{P_{\text{in,channels}}}, \tag{5.27}$$

where $P_{\text{in,channels}} = \sum_{j=0}^{C} P(j)$. The probability of connection queueing, P_q, is given by

$$P_q = \sum_{j=C}^{C+Q-1} P(j). \tag{5.28}$$

5.4.6 Two-Level Fractional Guard Channel Scheme

A two-level fractional guard channel (TLFGC) scheme [7] to provide efficient priority access for handoff calls over new calls in cellular systems is presented in this section. The switching between levels is controlled by a dropped call. The first level has a small number of guard channels, like the fixed guard channel (FGC) scheme, while the second level has a larger number of guard channels and a smaller number of fractional guard channels. An analytical formulation of the steady-state probabilities, new call blocking probability, handoff call blocking probability, system utilization, probabilities in each level, and the equivalent number of guard channels is presented. Numerical results show that the performance of TLFGC is almost the same as a fixed guard channel (FGC) scheme under light load. However, the advantage of the TLFGC scheme is demonstrated under heavy load, where the handoff call blocking probability is much better than that of a FGC scheme, giving greater priority and protection to handoff calls over new calls.

The TLFGC scheme works as follows. In level 1, the TLFGC scheme works similarly to a FGC scheme. A small number of guard channels, C_{G1}, in a cell is reserved for exclusive use by handoff calls, whereas the others can be used by both handoff calls and new calls. However, the TLFGC scheme differs from the FGC scheme in that the former will be triggered to move to the next level (level 2) when a handoff call is dropped because no channel is available in the cell. In level 2, more guard channels, C_{G2} ($> C_{G1}$), are allocated for exclusive use by handoff calls than in level 1; the other channels in this level can be used by both handoff calls with no restriction and new calls with a fractional probability, β_{2i}, depending on the states the system is in. β_{2i} is between 0 and 1, and the values decrease with the increase in state $(2, i), i = 1, 2, \ldots, C - C_{G2} - 1$:

$$\beta_{2i} = 1 - \frac{i}{C - C_{G2}}. \tag{5.29}$$

Thus, level 2 is like a fractional guard channel scheme [6]. When these guard channels and fractional guard channels are no longer occupied by handoff calls or new calls,

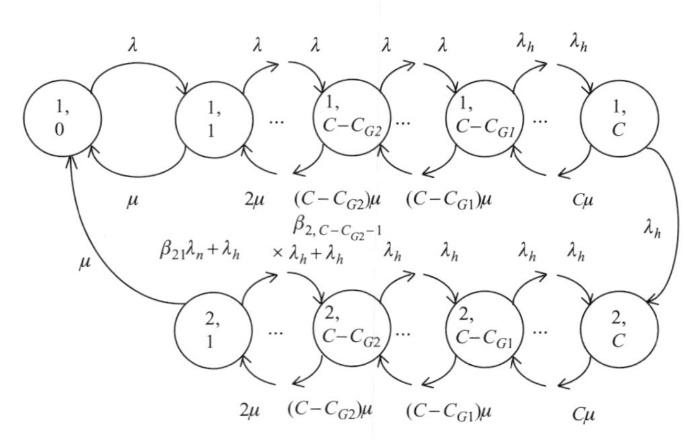

FIGURE 5.19 State transition diagram for the TLFGC scheme. (From [7] © 2006, IEEE.)

respectively, the scheme will switch back to level 1. Thus, level 2 is more favorable than level 1 for handoff calls.

Figure 5.19 shows a two-dimensional finite-state Markov chain for the TLFGC scheme. State (k, i) represents the kth level of the TLFGC scheme, and there are i channels in service in the cell. $\lambda = (\lambda_n + \lambda_h)$ is the mean call arrival rate of new and handoff calls in a cell, while λ_n and λ_h are, respectively, the arrival rate of new calls and handoff calls in the cell. $\mu = (\mu_c + \mu_h)$ is the mean equivalent service rate of a call in a cell, while $1/\mu_c$ and $1/\mu_h$ are, respectively, the mean call holding time of a call and the mean dwell time (interhandoff time) of a call in the cell. In level 1 of the TLFGC scheme, only handoff calls can be admitted in the C_{G1} channels, as indicated by the transition rates of λ_h in Figure 5.19. On the other hand, the rest of the channels can be used by both the new and handoff calls as indicated by the transition rates of λ. If the TLFGC scheme is in state $(1, C)$, it will be triggered to jump to state $(2, C)$ in level 2 by a handoff call arriving to a fully occupied cell. This transition is indicated by the state transition rate of λ_h from state $(1, C)$ to state $(2, C)$. Note that only handoff calls can enter the higher states of the Markov chain and a dropped handoff call indicates insufficient guard channels for handoff calls. Thus, a dropped call at state $(1, C)$ is an indication of a heavy load, which is dependent on both new and handoff calls, and is also an indication of the need to have more channels reserved for handoff calls. This is why more exclusive guard channels and some fractional guard channels are allocated for handoff calls in level 2. The fractional guard channels still allow new calls to be admitted, but at a lower probability than for level 1, where they can be admitted if there are channels available other than the guard channels. For each jump from level 1 to level 2, the number of guard channels is increased such that $C_{G2} > C_{G1}$. Note that in level 2, only handoff calls can use the guard channels, as indicated by the transition rates of λ_h, while the handoff calls can also use the other channels and the new calls can only use the other channels with probability β_{2i}. The latter gives additional higher-priority access to handoff calls over new calls. In state

(k, i), the channels are serviced at a mean rate of $i\mu$ and it will jump to state $(k, i-1)$ due to a call termination or a call handoff to another cell. If the TLFGC scheme is in state (2,1) in level 2, it will return to state (1,0) in level 1 with a transition rate of μ if a channel has been served due to call termination or call handoff to another cell before any new or handoff call arrival.

Consider a TLFGC model. The guard channels, C_{Gk}, are reserved for handoff connections only. To facilitate analytical modeling, it is necessary to make certain assumptions about the traffic parameters. It is not unreasonable to assume that the holding time has a negative exponential distribution [5]. As mentioned earlier, although a negative exponential distribution assumption may not be as reasonable for the cell dwell time, for analytical tractability, we will make the same assumption for cell dwell time (interhandoff time) [5] and model the channel occupancy as a two-dimensional Markov chain with the call-level transition rate parameters shown in Figure 5.19. Although this Markov chain can be modeled and solved using numerical techniques, in this section it is solved analytically instead, by equating probability flows.

Let $P_{k,i}$ be the steady-state probabilities in state (k, i) of the Markov chain, where $k = 1, 2$, and $i = 0, \ldots, C$ for $k = 1$ and $i = 1, \ldots, C$ for $k = 2$. Solving the Markov chain in Figure 5.19 by equating probability flows across surfaces of the Markov chain and using the sum of total state probabilities as 1, we get

$$= P_{1,0}\frac{1}{X}\left[\frac{\lambda_h}{\lambda} + \frac{(i+1)\mu}{\lambda}\left[\frac{\lambda_h}{\lambda} + \frac{(i+2)\mu}{\lambda}\left[\cdots\left[\frac{\lambda_h}{\lambda} + \frac{(C-C_{G1})\mu}{\lambda}\right.\right.\right.\right.$$

$$\left[1 + \frac{(C-C_{G1}+1)\mu}{\lambda_h} \times \left[\cdots\left[1 + \frac{(C-1)\mu}{\lambda_h}\right.\right.\right.$$

$$\left.\left.\left.\left.\left.\left.\left.\left.\left[1 + \frac{C\mu}{\lambda_h}\right]\right]\cdots\right]\right]\right]\cdots\right]\right]\right], \qquad 1 \le i \le C - C_{G1} - 1, \tag{5.30}$$

$$P_{1,i} = P_{1,0}\frac{1}{X}\left[1 + \frac{(i+1)\mu}{\lambda_h}\left[1 + \frac{(i+2)\mu}{\lambda_h}\left[\cdots\left[1 + \frac{(C-1)\mu}{\lambda_h}\right.\right.\right.\right.$$

$$\left.\left.\left.\left.\left[1 + \frac{C\mu}{\lambda_h}\right]\right]\cdots\right]\right]\right], \qquad C - C_{G1} \le i \le C - 1, \tag{5.31}$$

$$P_{1,i} = P_{1,0}\frac{1}{X}, \qquad i = C, \tag{5.32}$$

$$P_{2,i} = P_{1,0}\frac{1}{X}\left(\frac{\lambda_h}{\mu}\right), \qquad i = 1, \tag{5.33}$$

$$P_{2,i} = P_{1,0}\frac{1}{X}\frac{\lambda_h}{\mu}\left[\frac{1}{i} + \frac{\beta_{2,i-1}\lambda_n + \lambda_h}{i\mu}\left[\frac{1}{i-1} + \frac{\beta_{2,i-2}\lambda_n + \lambda_h}{(i-1)\mu}\left[\cdots\right.\right.\right.$$

$$\left.\left.\left.\left.\left.\left[\frac{1}{3} + \frac{\beta_{22}\lambda_n + \lambda_h}{3\mu}\left[\frac{1}{2} + \frac{\beta_{21}\lambda_n + \lambda_h}{2\mu}\right]\right]\cdots\right]\right]\right],$$

$$2 \le i \le C - C_{G2}, \tag{5.34}$$

$$
\begin{aligned}
P_{2,i} &= P_{1,0}\frac{1}{X}\frac{\lambda_h}{\mu}\left[\frac{1}{i}+\frac{\lambda_h}{i\mu}\left[\frac{1}{i-1}+\frac{\lambda_h}{(i-1)\mu}\left[\cdots\right.\right.\right.\\
&\left[\frac{1}{C-C_{G2}+1}+\frac{\lambda_h}{(C-C_{G2}+1)\mu}\left[\frac{1}{C-C_{G2}}\right.\right.\\
&\left.\left.\left.\left.\left.+\frac{\beta_{2,C-C_{G2}-1}\lambda_n+\lambda_h}{(C-C_{G2})\mu}\left[\cdots\left[\frac{1}{2}+\frac{\beta_{21}\lambda_n+\lambda_h}{2\mu}\right]\cdots\right]\right]\right]\cdots\right]\right]\right],\\
&C-C_{G2}+1\le i\le C,
\end{aligned}
\tag{5.35}
$$

where

$$
\begin{aligned}
X &= \left[\frac{\lambda_h}{\lambda}+\frac{\mu}{\lambda}\left[\frac{\lambda_h}{\lambda}+\frac{2\mu}{\lambda}\left[\cdots\left[\frac{\lambda_h}{\lambda}+\frac{(C-C_{G1})\mu}{\lambda}\left[1+\frac{(C-C_{G1}+1)\mu}{\lambda_h}\right.\right.\right.\right.\right.\\
&\left[1+\frac{(C-C_{G1}+2)\mu}{\lambda_h}\left[\cdots\right.\right.\\
&\left.\left.\left.\left.\left.\left.\left.\times\left[1+\frac{(C-1)\mu}{\lambda_h}\left[1+\frac{C\mu}{\lambda_h}\right]\right]\cdots\right]\right]\right]\right]\cdots\right]\right]\right],
\end{aligned}
\tag{5.36}
$$

and $P_{1,0}$ is given by

$$
\begin{aligned}
P_{1,0} = &\left\{1+\sum_{i=1}^{C-C_{G1}-1}\frac{1}{X}\left[\frac{\lambda_h}{\lambda}+\frac{(i+1)\mu}{\lambda}\left[\frac{\lambda_h}{\lambda}+\frac{(i+2)\mu}{\lambda}\left[\cdots\left[\frac{\lambda_h}{\lambda}+\frac{(C-C_{G1})\mu}{\lambda}\right.\right.\right.\right.\right.\\
&\left.\left.\left.\left.\times\left[1+\frac{(C-C_{G1}+1)\mu}{\lambda_h}\left[\cdots\left[1+\frac{(C-1)\mu}{\lambda_h}\left[1+\frac{C\mu}{\lambda_h}\right]\right]\cdots\right]\right]\right]\cdots\right]\right]\right]\\
&+\sum_{i=C-C_{G1}+1}^{C-1}\frac{1}{X}\left[1+\frac{(i+1)\mu}{\lambda_h}\left[1+\frac{(i+2)\mu}{\lambda_h}\left[\cdots\left[1+\frac{(C-1)\mu}{\lambda_h}\left[1+\frac{C\mu}{\lambda_h}\right]\right]\cdots\right]\right]\right]\\
&+\frac{1}{X}+\frac{1}{X}\frac{\lambda_h}{\mu}+\sum_{i=2}^{C-C_{G2}}\frac{1}{X}\left(\frac{\lambda_h}{\mu}\right)\left[\frac{1}{i}+\frac{\beta_{2,i-1}\lambda_n+\lambda_h}{i\mu}\left[\frac{1}{i-1}+\frac{\beta_{2,i-2}\lambda_n+\lambda_h}{(i-1)\mu}\left[\cdots\right.\right.\right.\\
&\left.\left.\left.\times\left[\frac{1}{3}+\frac{\beta_{22}\lambda_n+\lambda_h}{3\mu}\left[\frac{1}{2}+\frac{\beta_{21}\lambda_n+\lambda_h}{2\mu}\right]\right]\cdots\right]\right]\right]+\sum_{i=C-C_{G2}+1}^{C}\frac{1}{X}\frac{\lambda_h}{\mu}\left[\frac{1}{i}+\frac{\lambda_h}{i\mu}\right.\\
&\times\left[\frac{1}{i-1}+\frac{\lambda_h}{(i-1)\mu}\left[\cdots\left[\frac{1}{C-C_{G2}+1}+\frac{\lambda_h}{(C-C_{G2}+1)\mu}\left[\frac{1}{C-C_{G2}}\right.\right.\right.\right.\\
&\left.\left.\left.\left.\left.\left.+\frac{\beta_{2,C-C_{G2}-1}\lambda_n+\lambda_h}{(C-C_{G2})\mu}\left[\cdots\left[\frac{1}{2}+\frac{\beta_{21}\lambda_n+\lambda_h}{2\mu}\right]\cdots\right]\right]\right]\cdots\right]\right]\right]\right\}^{-1}.
\end{aligned}
\tag{5.37}
$$

The new call blocking probability, B_n, is given by

$$
B_n = \sum_{i=C-C_{G1}}^{C}P_{1,i}+\sum_{i=1}^{C-C_{G2}-1}(1-\beta_{2i})P_{2,i}+\sum_{i=C-C_{G2}}^{C}P_{2,i}. \tag{5.38}
$$

The handoff call blocking probability, B_h, is given by

$$B_h = P_{1,C} + P_{2,C}. \tag{5.39}$$

The system utilization, N_u, is given by

$$N_u = \sum_{i=1}^{C} i P_{1,i} + \sum_{i=1}^{C} i P_{2,i} = \frac{\lambda_n(1 - B_n) + \lambda_h(1 - B_h)}{\mu}. \tag{5.40}$$

The probability of being in level 1, Q_1, is given by

$$Q_1 = \sum_{i=0}^{C} P_{1,i}. \tag{5.41}$$

The probability of being in level 2, Q_2, is given by

$$Q_2 = \sum_{i=1}^{C} P_{2,i}. \tag{5.42}$$

The equivalent number of guard channels (EGC) used exclusively for handoff calls is given by

$$\text{EGC} = Q_1 C_{G1} + Q_2 C_{G2}. \tag{5.43}$$

Equating the handoff call arrival rate to the product of the average handoff rate for a call and the average number of calls, we have

$$\lambda_h = \mu_h N_u = \mu_h \times \frac{(1 - B_n)\lambda_n + (1 - B_h)\lambda_h}{\mu_c + \mu_h} = \frac{\mu_h (1 - B_n)\lambda_n}{\mu_c + B_h \mu_h}. \tag{5.44}$$

Note that the handoff call arrival rate is dependent on the new call arrival rate. Thus, the handoff call arrival rate can be approximated under low blocking probabilities as follows:

$$\lambda_h = \frac{\mu_h \lambda_n}{\mu_c}, \tag{5.45}$$

where $\mu_h = v/s$, v is the speed of the mobile, and s is the cell length of a square cell.

Note that the steady-state probabilities, $P_{k,i}$'s, are functions of the handoff call arrival rate λ_h, and λ_h is a function of the system utilization, N_u, which is a function of the $P_{k,i}$'s. We use an iterative method to determine λ_h and the $P_{k,i}$'s as follows:

1. Initialize the handoff call arrival rate, $\lambda_h(0) = \mu_h \lambda_n / \mu_c$, where the zero in parentheses means step 0.
2. Iterate between $P_{k,i}$'s as in (5.30)–(5.37) for all k and i, N_u as in (5.40), and λ_h as in (5.44) until

$$\lambda_h(n) - \lambda_h(n-1) < \varepsilon_{\lambda_h}, \tag{5.46}$$

where the variable n in parentheses denotes the nth iteration step and ε_{λ_h} is the error threshold for λ_h. Using this iterative method, the handoff call arrival rate, λ_h, is obtained iteratively. Note that the initial handoff call arrival rate, $\lambda_h(0)$, is set to the approximate handoff call arrival rate under low blocking probabilities, as in (5.45).

The multilevel dynamic guard channel (MLDGC) scheme introduced in [8] has certain salient attributes. For performance comparison with the TLFGC scheme, we use a two-level dynamic guard channel (TLDGC) scheme and the FGC scheme. We present results to examine the performance of new and handoff call blocking probabilities for the TLFGC scheme as well as the other performance measures derived above. Analytical results for the TLFGC scheme are obtained by analysis, while those for the TLDGC scheme and the FGC scheme are obtained from the analyses in [8]. Simulation results for these three schemes are obtained using C programs and the SMPL simulation kernel [9]. The parameter values used to demonstrate the performance of the proposed scheme are tabulated in Table 5.1.

In the TLFGC scheme, C_{G1} has a small value and C_{G2} is larger than C_{G1}. This allows the equivalent guard channels to adapt to the traffic load; the value will range from C_{G1} under light load to C_{G2} under heavy load. The new and handoff call blocking probabilities for the FGC scheme, the two-level MLDGC scheme, and the proposed TLFGC scheme as a function of the new call arrival rate with $C = \{10,20\}$ are shown in Figures 5.20 and 5.21, respectively. For illustration purposes, guard channels $C_{G1} = \{2,4\}$ are chosen for the FGC scheme and level 1 of the two-level MLDGC and TLFGC schemes with $C = \{10,20\}$, respectively. For the same purpose, $C_{G2} = \{4,8\}$ are chosen for the TLDGC and TLFGC schemes with $C = \{10,20\}$,

TABLE 5.1 Parameter Values Used

	$C = 10$	$C = 20$
$1/\mu_c$	1 min	1 min
v	36 km/h	36 km/h
s	200 m	200 m
$1/\mu_h$	$\frac{1}{3}$ min	$\frac{1}{3}$ min
C_{G1}	2	4
C_{G2}	4	8

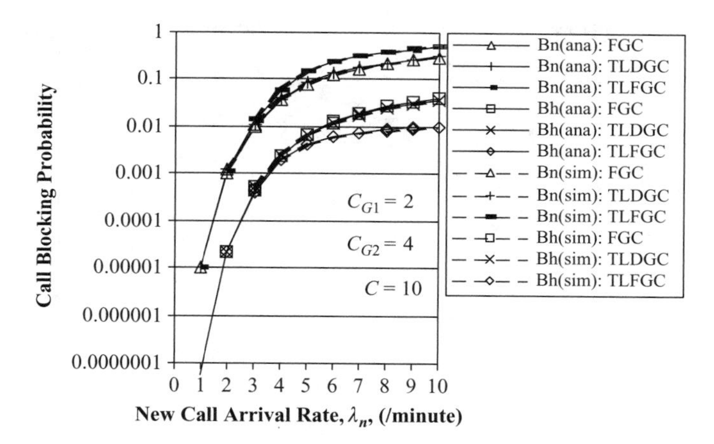

FIGURE 5.20 Call blocking probabilities with $C = 10$. (From [7] © 2006, IEEE.)

respectively. It is observed that the performances of the new and handoff blocking probabilities in the TLFGC scheme are almost the same as those in the FGC and the TLDGC schemes under light load conditions. However, the advantage of the TLFGC scheme is shown under the heavy load conditions, where the performances of the handoff call blocking probabilities of the TLFGC scheme are much lower than those of the TLDGC scheme, which are slightly better than those in the FGC scheme. These gains in handoff call blocking are obtained at the expense of a slight increase in new call blocking probabilities. Thus, the objective of giving a higher priority to handoff calls over new calls is further achieved by the TLFGC scheme in the heavy-load region. Hence, extra protection is given to handoff calls during heavy loads. Note that

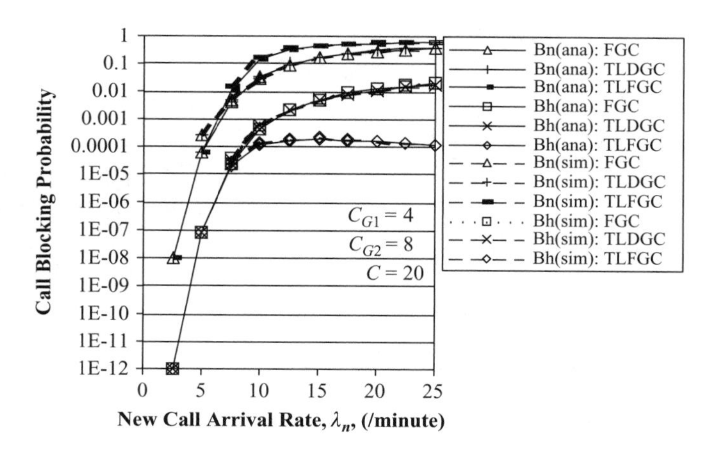

FIGURE 5.21 Call blocking probabilities with $C = 20$ (From [7] © 2006, IEEE.)

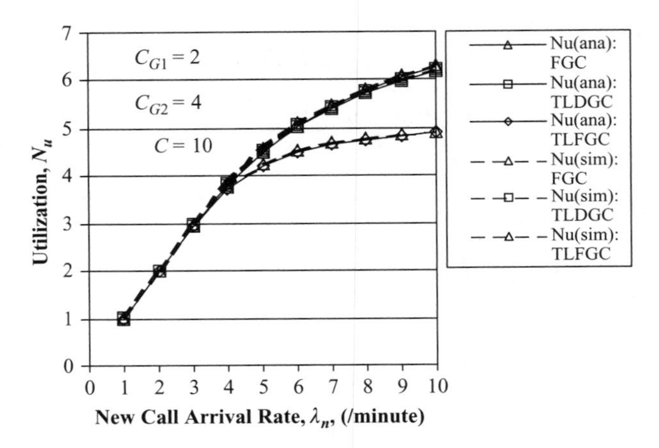

FIGURE 5.22 System utilization with $C = 10$. (From [7] © 2006, IEEE.)

the handoff blocking probability, B_h, in Figure 5.21 decreases slightly as the traffic load increases in the heavy load. This is due to the decrease in mean handoff call arrival rate, λ_h, obtained from (5.46), in this region.

Figures 5.22 and 5.23 show the system utilizations for the FGC scheme: the TLDGC scheme and the proposed TLFGC scheme as a function of the new call arrival rate for $C = \{10, 20\}$, respectively. From these figures it can be seen that the system utilizations for the TLFGC scheme are almost the same as those for the TLDGC scheme and the FGC scheme under light load. However, system utilizations for the TLFGC scheme are lower than those of the TLDGC scheme, which is slightly lower than those in the FGC scheme under heavy load. The reason for this is the increased number of guard channels in level 2 as well as the use of fractional guard

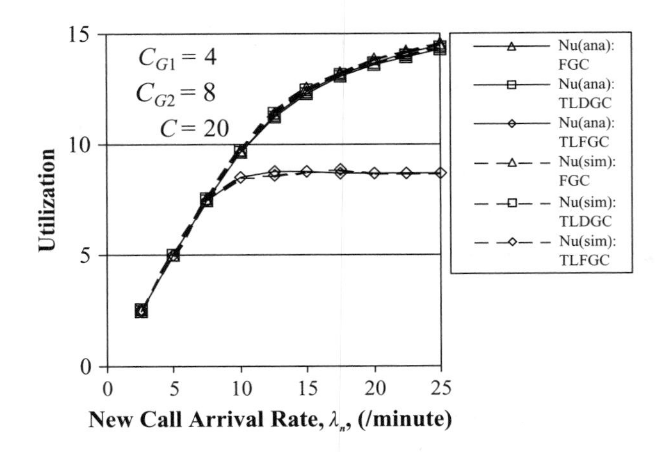

FIGURE 5.23 System utilization with $C = 20$. (From [7] © 2006, IEEE.)

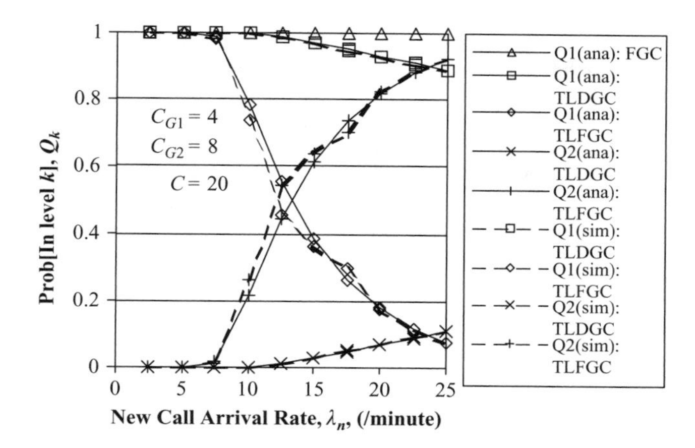

FIGURE 5.24 Probabilities of being in levels 1 and 2 with $C = 20$. (From [7] © 2006, IEEE.)

channels. Furthermore, a higher new call blocking probability means that fewer new calls are admitted and therefore the system utilization for the TLFGC scheme is lower than that for the FGC scheme. In this case, the handoff call blocking probability is lowered due to fewer new calls in the system. Thus, this is the trade-off for using the TLFGC scheme over the TLDGC and the FGC scheme. However, as seen earlier in this section, the TLFGC scheme fulfills the objective of giving higher priority and protection to handoff calls over new calls under heavy load. Thus, such a trade-off is acceptable.

Figure 5.24 shows the probabilities of being in levels 1 and 2 for TLDGC and TLFGC with $C = 20$. As the load increases, the probability of being in level 1, Q_1, decreases, while the probability of being in level 2, Q_2, increases. Thus, both TLDGC and TLFGC adapt to the traffic load, given better treatment to handoff calls in level 2 under a heavy-load scenario. The rates of change in Q_1 and Q_2 for TLFGC are faster than those of TLDGC as the load increases. Figure 5.25 shows that the equivalent number of guard channels, EGC, adaptively increases as the traffic load increases for both TLDGC and TLFGC, favoring handoff calls under heavy load, especially for the latter.

5.4.7 Link-Layer Resource Allocation Scheme with On/Off Traffic

Earlier, we considered connection-level resource allocation schemes. In this section we consider a link-layer resource allocation scheme for on/off sources. The outage probability is derived for a DS-CDMA system. From this outage probability, the system capacity can be obtained, similar to that in Chapter 3.

We consider single-class on/off traffic in the uplink of a wireless cellular DS-CDMA network. The cell site (base station) supports a single class of services that

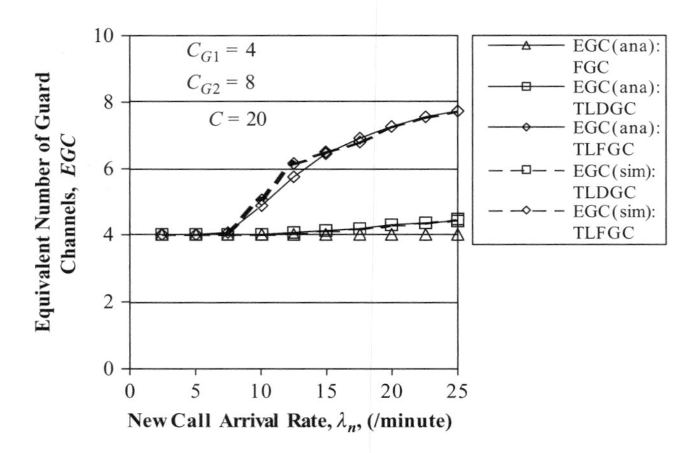

FIGURE 5.25 Equivalent number of guard channels with $C = 20$. (From [7] © 2006, IEEE.)

can originate from the mobile users in the cell. The following system parameters are used throughout this section:

W	spread-spectrum bandwidth
R	transmission rate
G	spreading gain
n	number of users per sector
BER	bit error rate
SIR	signal-to-interference ratio
BER^*	bit error rate requirement
SIR^*	signal-to-interference ratio requirement
γ	E_b/I_0 (bit energy-to-interference spectral density ratio) requirement
S	received power of a user
I	intercell interference power
η	thermal noise
ρ	density of users per unit area
α	activity factor of traffic
ψ_j	random variable for the activity factor of a user j

The following assumptions are made to facilitate the analytical formulation:

- The transmission rates of all users are the same with a basic rate.
- The processing gain, G, for users is given by W/R.
- The system is made up of hexagonal cells.
- Mobile users have omnidirectional antennas.
- The base station has three sectors in each cell.

- The sectorization in the cells is perfect.
- There is a uniform distribution of users in each cell.
- There are equal numbers of users in every cell.
- There is perfect power control in each cell.
- The spreading gain is the same for all users.
- The channel is modeled as a combination of path loss and lognormal shadowing, represented by $r^{-4} \cdot 10^{\varepsilon/10}$, where r is the distance between the mobile and the serving base station and ε is a Gaussian random variable with zero mean and variance σ^2.

We are concerned with the uplink capacity of the DS-CDMA system in terms of the number of users, n, that can be supported. The capacity region for the system is derived by considering the outage probability in terms of the signal-to-interference ratio (SIR) specification. These probabilities are expressed in terms of the number of users, their activity factor, the intracell received power for all users in the cell, the intercell interference for all users in the surrounding cells, the spreading gain, the E_b/I_0 requirement, and the background noise. The aim here is to determine the maximum number of users (i.e., the system capacity) that can be supported in the system while maintaining the required QoS. From this system capacity, the schemes in Section 5.4 can be used. First, let γ denote the E_b/I_0 (bit energy-to-interference power spectral density ratio requirement) for each user. It is given by

$$\gamma = \frac{G}{\sum_{j=1}^{n-1} \psi_j + \frac{I}{S} + \frac{\eta}{S}}, \tag{5.47}$$

where $\psi_j \in \{0,1\}$ is a binomial random variable indicating the activity factor of the jth user. The activity factor is defined by $\alpha = \Pr[\psi_j = 1]$. The numerator on the right-hand side of equation (5.47) is the processing gain. In the denominator the first term is due to intracell interference from other users in the cell, the second term is due to the intercell interference from users from other surrounding cells, and the last term is due to background noise. From [9], the intercell interference for a class i user is given by

$$\frac{I}{S} = \left(\frac{r_m}{r_d}\right)^4 \cdot 10^{(\varepsilon_d - \varepsilon_m)/10}, \tag{5.48}$$

where r_d is the distance between the intercell mobile that is causing interference and the intracell base station, r_m is the distance between the intercell mobile and its own base station, and ε_d and ε_m are Gaussian random variables with zero mean and standard deviation σ. Since ε_d and ε_m are independent, $\varepsilon_d - \varepsilon_m$ is a Gaussian random variable with zero mean and variance $2\sigma^2$. The mean and variance of I/S

are upper-bounded by [10]

$$E\left[\frac{I}{S}\right] = \alpha\rho \int\int f\left(\frac{r_m}{r_d}\right) dA \leq \mu_{11} \tag{5.49}$$

and

$$\mathrm{Var}\left[\frac{I}{S}\right] \leq \rho \int\int \left[\alpha g\left(\frac{r_m}{r_d}\right) - \alpha^2 f^2\left(\frac{r_m}{r_d}\right)\right] dA = \sigma_{11}^2, \tag{5.50}$$

where μ is the upper bound on the mean of I/S, σ^2 is the upper bound on the variance of I/S, and ρ is the density of the users per unit area, which is given by

$$\rho = \frac{2n}{\sqrt{3}}, \tag{5.51}$$

$$f\left(\frac{r_m}{r_d}\right) = \left(\frac{r_m}{r_d}\right)^4 e^{(\sigma \ln 10/10)^2}\left[1 - Q\left(\frac{40\log(r_d/r_m)}{\sqrt{2\sigma^2}} - \frac{\sqrt{2\sigma^2}\ln 10}{10}\right)\right], \tag{5.52}$$

$$g\left(\frac{r_m}{r_d}\right) = \left(\frac{r_m}{r_d}\right)^8 e^{(\sigma \ln 10/5)^2}\left[1 - Q\left(\frac{40\log(r_d/r_m)}{\sqrt{2\sigma^2}} - \frac{\sqrt{2\sigma^2}\ln 10}{5}\right)\right], \tag{5.53}$$

and

$$Q(y) = \frac{1}{\sqrt{2\pi}} \int_y^{\infty} e^{-x^2/2}\, dx. \tag{5.54}$$

Let BER* and SIR* denote, respectively, the BER and SIR requirement for the users. The system capacity is defined as the maximum n that can be supported such that the SIR achieved is greater than or equal to the SIR* required 99% of the time. That is, the outage probability is defined as

$$\begin{aligned}\Pr[\mathrm{BER} \geq \mathrm{BER}^*] &= \Pr[\mathrm{SIR} \leq \mathrm{SIR}^*] \\ &= \Pr\left[\sum_{j=1}^{n-1} \psi_j + \frac{I}{S} \geq \delta\right],\end{aligned} \tag{5.55}$$

where

$$\delta = \frac{G}{\gamma} - \frac{\eta}{S}. \tag{5.56}$$

Assuming central limit approximation and solving (5.55) by conditioning on the activity factor and then unconditioning the probability in (5.55) by summing up all cases for the numbers of active users and multiplying by the corresponding binomial probabilities of activity factor, we have

$$\begin{aligned}\Pr[\text{BER} \geq \text{BER}^*] &= \sum_{l=0}^{n-1} \Pr\left[\frac{I}{S} \geq \delta - l \,\middle|\, \sum \psi_j = l\right] \Pr\left[\sum \psi_j = l\right] \\ &= \sum_{l=0}^{n-1} Q\left(\frac{\delta - \mu}{\sigma}\right) \times \binom{n-1}{l} \alpha^l (1-\alpha)^{n-1-l}, \qquad (5.57)\end{aligned}$$

where

$$\mu = l + \mu_{11}, \qquad (5.58)$$

$$\sigma = \sqrt{\sigma_{11}^2} = \sigma_{11}. \qquad (5.59)$$

Equation (5.57) leads to the maximum number of users that can be supported in the system, which in turn can make use of the link-layer channel assignment strategies.

5.5 MULTICLASS CHANNEL ASSIGNMENT SCHEMES

In previous sections, we have considered resource allocation schemes at the connection level and link layer for single-class traffic. In this section, three multiclass connection-level channel assignment schemes are considered: the complete partitioning (CP) scheme, the complete sharing (CS) scheme, and the virtual partitioning (VP) scheme. In addition, we present a link-layer resource allocation scheme for multiclass traffic.

In the CP scheme, the channels are divided among the different traffic classes, while all channels are shared among all traffic classes in the CS scheme. VP is a call admission control (CAC) scheme that manages to combine the advantages of CS and CP and strikes a balance between unrestricted sharing in CS and unrestricted isolation in CP [11]. VP was originally proposed by Wu and Mark [12]. The concept of VP is that each individual traffic class is allocated a nominal amount of resources with the provision that underutilized resources can be used by the excess traffic of an overloaded class, subject to preemption. The underutilized resources come from the traffic classes whose arrival rates are below the thresholds set based on past statistics. In this situation, the nominal allocation for underloaded classes can be utilized by other traffic classes. However, VP performs preemption for the underloaded classes when their arrival rates revert to their thresholds. For traffic whose arrival rates are higher than the thresholds, if the overall traffic is light, the overloaded classes can use the nominal allocation of other traffic classes just as in CS. If the overall traffic becomes heavy, the overloaded classes are preempted by other traffic classes and can only use the nominal allocation for themselves just like in CP. VP behaves like unrestricted sharing when the overall traffic is light, and complete isolation when the overall traffic is heavy [5]. Thus, VP combines the best characteristics of CS and CP under different loadings. In this section we divide the *multiclass* traffic into two groups that represent RT and NRT connections. There are two proposed VP schemes: VP with preemption for group 2, which gives a higher priority to group 1 over group 2, and VP with preemption for groups 1 and 2, which treats these two groups with equal priority. VP with preemption for groups 1 and 2 is a more general scheme, which is the basis of VP with preemption for group 2.

We consider a typical generic radio cell with physical capacity C in a cellular arrangement. For easy reference, we define the basic unit of capacity as a *channel*. A user of some traffic class may transmit at a rate equal to one *channel*, while other transmission rates may require multiple number of channels. The cell site (base station) supports K classes of services that can originate from at most N mobile users. The generic cell is characterized by the following system-level parameters used throughout this section:

C	total physical capacity in a cell
K	total number of traffic classes
r_k	number of basic channels (units) required by each class k connections
C_k	nominal capacity for class k
$N = \sum_{k=1}^{K} C_k > C$	total nominal capacity in a cell (capitalizing on statistical multiplexing gain)
C_k^*	physical capacity for class k ($\sum_{k=1}^{K} C_k^* = C$ for CP)

The dynamics of a radio cell is driven by new connection requests, connection terminations, and handoffs induced by user mobility. Maintaining an ongoing connection is more important than admitting a new connection. Hence, handoff connections should be given a higher access priority, or a lower blocking probability, than new connections. Let B_n and B_h denote, respectively, the blocking probabilities of new and handoff connections. One way to facilitate this is to reserve capacity for admitting handoff connections, which is not accessible by new requests. The reserved capacity is sometimes referred to as *guard capacity*. Let C_T, C_G, and C_i denote, respectively, the total capacity, the guard capacity, and the instantaneous capacity occupancy. We have the following admission rules:

1. Admit both new and handoff connections if $C_T - C_i > C_G$.
2. Admit only handoff connections if $0 < C_T - C_i \leq C_G$, where $(C_T - C_i)$ is the free capacity for admitting new and handoff connections.

The fraction of reserved (guard) capacity for handoff connections is then

$$\alpha = \frac{C_G}{C_T} \tag{5.60}$$

with α conditioned such that $C_G = \alpha C_T$ is an integral multiple of the basic capacity unit. Deployment of the guard capacity policy has the following ramifications:

- A handoff connection is accepted as long as enough channels are available.
- A new connection is accepted as long as the number of channels available (if it is admitted) is greater than C_G.

Consider a class k user. Its QoS is specified by the new connections blocking probability, B_{nk}, handoff connections blocking probability, B_{hk}, and system utilization, N_{uk}, at the connection level. The connection- and packet-level parameters used throughout this section are as follows:

B_{nk}	new connections blocking probability for class k
B_{hk}	handoff connections blocking probability for class k
$B_n = \sum_{k=1}^{K} B_{nk}$	total new connections blocking probability
$B_h = \sum_{k=1}^{K} B_{hk}$	total handoff connections blocking probability
$B = B_n + B_h$	total connections blocking probability
n_k	number of class k sources (users) in progress
λ_{nk}	arrival rate of class k new connections
λ_{hk}	arrival rate of class k handoff connections
$\lambda_n = \sum_{k=1}^{K} \lambda_{nk}$	arrival rate of new connections
$\lambda_h = \sum_{k=1}^{K} \lambda_{hk}$	arrival rate of handoff connections
$\lambda_k = \lambda_{nk} + \lambda_{hk}$	mean connections arrival rate of class k connections
$\lambda = \sum_{k=1}^{K} \lambda_k$	total connections arrival rate
μ_{ck}^{-1}	mean connection holding time of a class k connection
μ_{hk}^{-1}	mean dwell time (interhandoff time) of a class k connection
$\mu_k = \mu_{ck} + \mu_{hk}$	mean equivalent rate of a class k connection

5.5.1 Complete Partitioning Scheme

In the complete partitioning scheme, each group of channels is only used by each traffic class. There is no sharing of channels amongst the different traffic classes. Let us consider a single class k model for the CP scheme as shown in Figure 5.26. Figure 5.27 shows a one-dimensional finite-state Markov chain for the complete partitioning scheme with class k, and there are new and handoff connections. The guard channels, C_G, are reserved for handoff connections only. A one-dimensional finite-state Markov chain is used to solve for the complete partitioning scheme with class k.

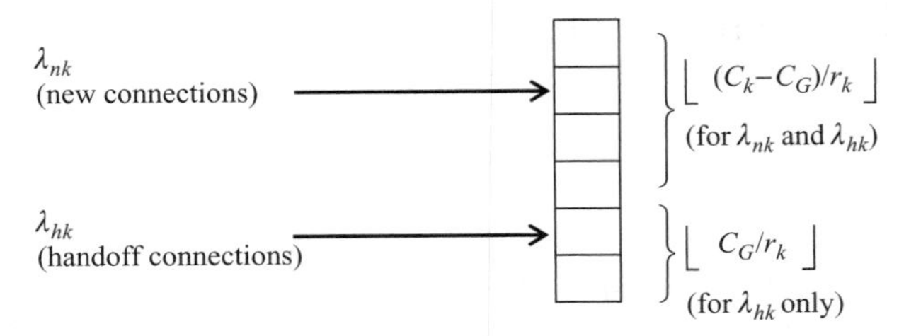

FIGURE 5.26 System model for a complete partitioning scheme considering class k with new and handoff connections.

Let $\{P(n_k)\}$ be the steady-state probabilities of the Markov chain for class k. Solving the Markov chain, we get

$$P(n_k) = \begin{cases} \left(\dfrac{\lambda_k}{\mu_k}\right)^{n_k} \dfrac{1}{n_k!} P(0), & 0 < n_k \le \left\lfloor \dfrac{C_k - C_G}{r_k} \right\rfloor \\ \dfrac{(\lambda_k)^{\left\lfloor \frac{C_k - C_G}{r_k} \right\rfloor} (\lambda_{hk})^{n_k - \left\lfloor \frac{C_k - C_G}{r_k} \right\rfloor}}{(\mu_k)^{n_k} n_k!} P(0), & \left\lfloor \dfrac{C_k - C_G}{r_k} \right\rfloor < n_k \le \left\lfloor \dfrac{C_k}{r_k} \right\rfloor, \end{cases} \tag{5.61}$$

where

$$P(0) = \left[1 + \sum_{n_k=1}^{\left\lfloor \frac{C_k - C_G}{r_k} \right\rfloor} \left(\frac{\lambda_k}{\mu_k}\right)^{n_k} \frac{1}{n_k!} + \sum_{n_k = \left\lfloor \frac{C_k - C_G}{r_k} \right\rfloor + 1}^{\left\lfloor \frac{C_k}{r_k} \right\rfloor} \frac{(\lambda_k)^{\left\lfloor \frac{C_k - C_G}{r_k} \right\rfloor} (\lambda_{hk})^{n_k - \left\lfloor \frac{C_k - C_G}{r_k} \right\rfloor}}{(\mu_k)^{n_k} n_k!} \right]^{-1}, \tag{5.62}$$

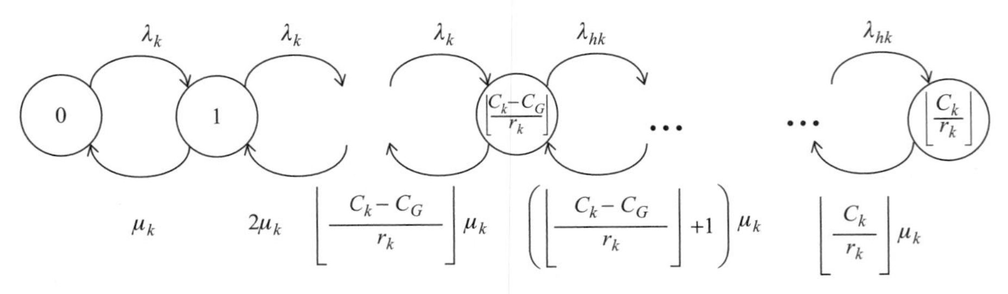

FIGURE 5.27 State transition diagram for a complete partitioning scheme considering class k with new and handoff connections.

with $\lfloor x \rfloor$ denoting the greatest integer smaller than or equal to x. The class k new and handoff connection blocking probabilities are given, respectively, by

$$b_{nk} = \sum_{n_k=\lfloor (C_k - C_G)/r_k \rfloor}^{\lfloor C_k/r_k \rfloor} P(n_k) \tag{5.63}$$

and

$$b_{hk} = P(n_k = \lfloor C_k/r_k \rfloor). \tag{5.64}$$

Let θ_{nk} denote the probability that the next arrival is a new connection from class k. Then

$$\theta_{nk} = \frac{\lambda_{nk}}{\sum_{k=1}^{K} \lambda_k}. \tag{5.65}$$

Let θ_{hk} denote the probability that the next arrival is a handoff connection from class k. We have

$$\theta_{hk} = \frac{\lambda_{hk}}{\sum_{k=1}^{K} \lambda_k}. \tag{5.66}$$

Let θ_k denote the probability that the next arrival is from class k. Then

$$\theta_k = \frac{\lambda_k}{\sum_{k=1}^{K} \lambda_k}. \tag{5.67}$$

The blocking probability for class k new connections considering all classes is given by

$$B_{nk} = b_{nk}\theta_{nk}. \tag{5.68}$$

The blocking probability for class k handoff connections considering all classes is given by

$$B_{hk} = b_{hk}\theta_{hk}. \tag{5.69}$$

The blocking probabilities of new and handoff connections, denoted by B_n and B_h, respectively, are given by

$$B_n = \sum_{k=1}^{K} B_{nk} \tag{5.70}$$

and

$$B_h = \sum_{k=1}^{K} B_{hk}. \tag{5.71}$$

Summing B_n and B_h yields the total blocking probability:

$$B = B_n + B_h. \tag{5.72}$$

The utilization for class k is given by

$$N_{uk} = \sum_{n_k=1}^{\lfloor C_k/r_k \rfloor} n_k P(n_k) r_k = \frac{\lambda_{nk}(1 - b_{nk}) + \lambda_{hk}(1 - b_{hk})}{\mu_k} r_k. \tag{5.73}$$

where b_k is the blocking probability considering class k only. The total system utilization is

$$N_u = \sum_{k=1}^{K} N_{uk}. \tag{5.74}$$

Equating the class k handoff connection arrival rate to the product of the average handoff rate for a class k connection and the average number of class k connections, we can get an approximate class k handoff connection arrival rate as follows:

$$\begin{aligned} \lambda_{hk} &= \mu_{hk} \frac{N_{uk}}{r_k} = \mu_{hk} \frac{(1 - b_{nk})\lambda_{nk} + (1 - b_{hk})\lambda_{hk}}{\mu_{ck} + \mu_{hk}} \\ &= \frac{\mu_{hk}(1 - b_{nk})\lambda_{nk}}{\mu_{ck} + b_{hk}\mu_{hk}} = \frac{\mu_{hk}\left(1 - B_{nk}/\theta_{nk}\right)\lambda_{nk}}{\mu_{ck} + (B_{hk}/\theta_{hk})\mu_{hk}}. \end{aligned} \tag{5.75}$$

Thus, the class k handoff connection arrival rate can be approximated under low blocking probabilities as follows:

$$\lambda_{hk} = \frac{\mu_{hk}\lambda_{nk}}{\mu_{ck}}, \tag{5.76}$$

where $\mu_{hk} = v_k/s$; v_k is the speed of the class k mobile, and s is the cell length of a square cell.

5.5.2 Complete Sharing Scheme

In the complete sharing scheme, all channels are shared by all traffic classes. Let us first consider a two-class transition rate diagram for the CS scheme as shown in

Figure 5.28. This extends a one-dimensional Markov chain to a two-dimensional Markov for two traffic classes. Next, let us consider a K-class CS model. Assuming that the holding time for each connection has a negative exponential distribution and the same assumption for cell dwell time (interhandoff time), we can model the channel occupancy as a K-dimensional Markov chain with connection-level parameters. This Markov chain can be modeled and solved using the techniques in [13]. Let $\boldsymbol{n} = (n_1, n_2, \ldots, n_K)$ denote the state of the system with the number of users (n_k) in each of the K classes, and let $\boldsymbol{r} = (r_1, r_2, \ldots, r_K)$ denote the number of basic channels (r_k) required for each class k connections with K classes. Let $\lambda_k(\boldsymbol{n})$ denote the arrival rate and $\mu_k(\boldsymbol{n})$ the departure rate in the system. With N denoting the total number of nominal channels, the state space of the system, denoted by S, is given by $S := \boldsymbol{n} : \boldsymbol{r} \cdot \boldsymbol{n} \leq N$. When the system is in state $\boldsymbol{n}$ and a class k connection (new or handoff) arrives, an admission policy determines whether or not the connection is admitted into the system. Here, the admission policy is a complete sharing scheme with guard channels.

We can specify the admission policy by mapping $\boldsymbol{f} := (f_1, \ldots, f_K)$ for new and handoff connections, $\boldsymbol{f}_G := (f_{G1}, \ldots, f_{GK})$ for handoff connections, where f_k and $f_{Gk} : S \to \{0, 1\}$, and $f_k(\boldsymbol{n})$ and $f_{Gk}(\boldsymbol{n})$ each takes on the value 0 or 1 if a class k connections is rejected or admitted, respectively, when the system state is $\boldsymbol{n}$. They are defined by the following equations:

$$f_k(\boldsymbol{n}) = \begin{cases} 1, & \boldsymbol{r} \cdot \boldsymbol{n} + r_k \leq N - C_G \\ 0, & \text{otherwise,} \end{cases} \tag{5.77}$$

$$f_{Gk}(\boldsymbol{n}) = \begin{cases} 1, & N - C_G < \boldsymbol{r} \cdot \boldsymbol{n} + r_k \leq N \\ 0, & \text{otherwise,} \end{cases} \tag{5.78}$$

for which $S(\boldsymbol{f}) + S(\boldsymbol{f}_G) = S$.

Let $P(\boldsymbol{n})$ denote the equilibrium probability that the system is in state $\boldsymbol{n}$. The global balance equations for the Markov process under the policies $\boldsymbol{f}$ and $\boldsymbol{f}_G$ are

$$\begin{aligned} &\sum_{k=1}^{K} [\lambda_k(\boldsymbol{n}) \{f_k(\boldsymbol{n}) + f_{Gk}(\boldsymbol{n})\} + \mu_k(\boldsymbol{n})] P(\boldsymbol{n}) \\ &\quad = \sum_{k=1}^{K} P(\boldsymbol{n} - \boldsymbol{e}_k) \lambda_k(\boldsymbol{n} - \boldsymbol{e}_k) \{f_k(\boldsymbol{n} - \boldsymbol{e}_k) + f_{Gk}(\boldsymbol{n} - \boldsymbol{e}_k)\} \\ &\quad + \sum_{k=1}^{K} P(\boldsymbol{n} + \boldsymbol{e}_k) \mu_k(\boldsymbol{n} + \boldsymbol{e}_k), \qquad \boldsymbol{n} \in S, \end{aligned} \tag{5.79}$$

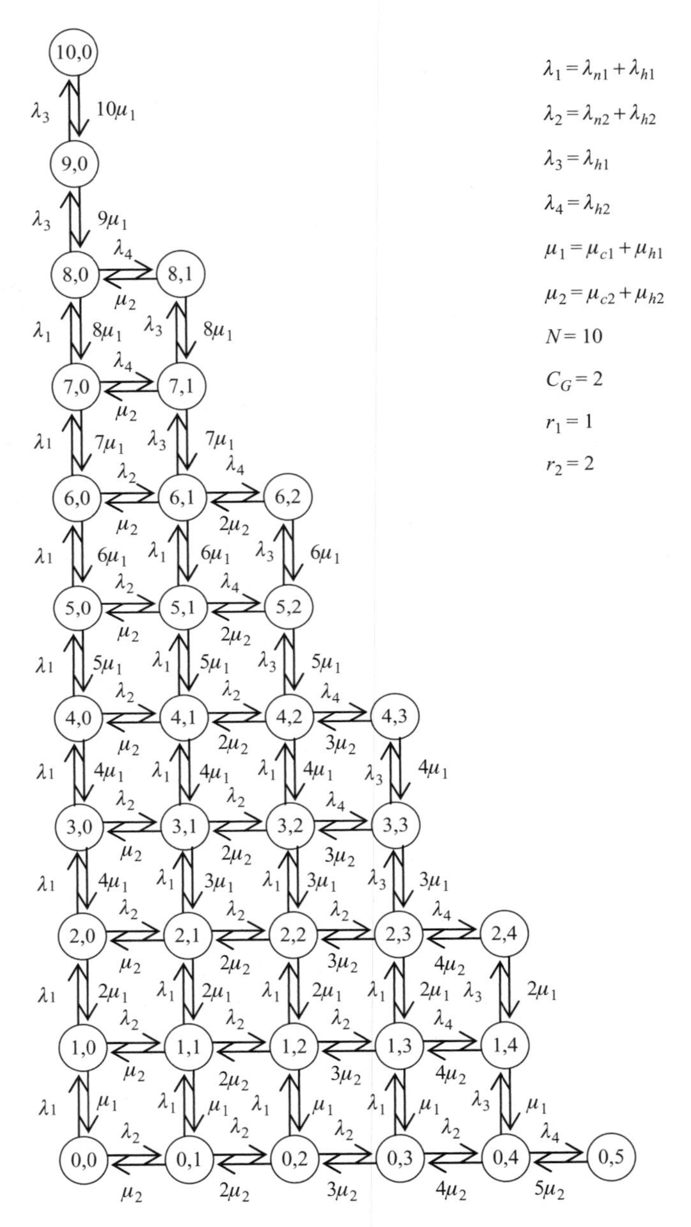

FIGURE 5.28 Two-class transition rate diagram for a complete sharing scheme with new and handoff connections.

where $\boldsymbol{e}_k$ is a K-dimensional vector of all zeros except for a 1 in the kth place,

$$\lambda_k(\boldsymbol{n}) = \begin{cases} \lambda_{nk} + \lambda_{hk} & \text{if } f_k(\boldsymbol{n}) = 1 \\ \lambda_{hk} & \text{if } f_{Gk}(\boldsymbol{n}) = 1 \end{cases} \tag{5.80}$$

and

$$\mu_k(\boldsymbol{n}) = n_k\,(\mu_{ck} + \mu_{hk}), \qquad 0 < \boldsymbol{r}\cdot\boldsymbol{n} \leq N. \tag{5.81}$$

The first condition in (5.80) allows both new and handoff connections to be admitted to $N - C_G$ channels, while the second condition allows only handoff connections to be admitted to C_G channels. Equation (5.81) allows both new and handoff connections to be serviced, with the total channel occupancy to be less than or equal to N channels. Equation (5.79) can be solved using LU decomposition together with the condition for the total probability of all states to obtain $P(\boldsymbol{n})$. LU decomposition is a common numerical technique for solving linear algebraic equations. This gives the exact solution. However, an efficient approximate computational algorithm [5] can be used for a large system state space. Results can be precomputed and stored in a lookup table for real-time applications in this situation.

A new class k connection is blocked from entering the system (and is assumed lost) if upon arrival it finds that it cannot be accommodated due to insufficient channels (excluding the guard channels) for an additional r_k channels. Therefore, the blocking probability for a new class k connection considering all classes is given by

$$B_{nk} = \sum_{n_1=0}^{\lfloor N/r_1 \rfloor} \sum_{n_2=0}^{\lfloor (N-n_1r_1)/r_2 \rfloor} \sum_{n_3=0}^{\lfloor (N-n_1r_1-n_2r_2)/r_3 \rfloor} \cdots \sum_{n_K=0}^{\left\lfloor \left(N-\sum_{k=1}^{K-1} n_kr_k\right)/r_K \right\rfloor} \times P(n_1, n_2, n_3, \ldots, n_K)\theta_{nk}, \qquad \text{if } \left(N - C_G - \sum_{k=1}^{K} n_kr_k\right) < r_k. \tag{5.82}$$

A class k handoff connection is blocked from entering the system (and is assumed lost) if upon arrival it finds that it cannot be accommodated because the number of available channels is less than r_k (including the guard channels). Therefore, the blocking probability for a class k handoff connection considering all classes is given by

$$B_{hk} = \sum_{n_1=0}^{\lfloor N/r_1 \rfloor} \sum_{n_2=0}^{\lfloor (N-n_1r_1)/r_2 \rfloor} \sum_{n_3=0}^{\lfloor (N-n_1r_1-n_2r_2)/r_3 \rfloor} \cdots \sum_{n_K=0}^{\left\lfloor \left(N-\sum_{k=1}^{K-1} n_kr_k\right)/r_K \right\rfloor} \times P(n_1, n_2, n_3, \ldots, n_K)\theta_{hk}, \qquad \text{if } \left(N - \sum_{k=1}^{K} n_kr_k\right) < r_k. \tag{5.83}$$

The blocking probabilities of new and handoff connections are given by (5.70) and (5.71), respectively. The total blocking probability is given by (5.72). The system utilization for class k is given by

$$N_{uk} = \sum_{n_1=0}^{\lfloor N/r \rfloor} \sum_{n_2=0}^{\lfloor (N-n_1r_1)/r_2 \rfloor} \sum_{n_3=0}^{\lfloor (N-n_1r_1-n_2r_2)/r_3 \rfloor} \cdots \sum_{n_K=0}^{\left\lfloor (N-\sum_{i=1}^{K-1} n_i r_i)/r_K \right\rfloor} \times n_k P(n_1, n_2, n_3, \ldots, n_K) r_k. \quad (5.84)$$

The total system utilization is given by (5.74).

5.5.3 Virtual Partitioning Scheme

In a virtual partitioning scheme, all channels are shared under light load. However, each traffic class has its own nominal channels. Under heavy-load conditions, each traffic class can preempt other traffic classes to claim back its own nominal channels.

The generic cell is characterized by the following system-level parameters used throughout this section:

- N a design parameter representing the upper bound for admitting calls (call-level capacity bound)
- C total physical capacity in a cell
- K total number of traffic classes
- N_i instantaneous channel occupancy
- n_k instantaneous channel occupancy for a class k call
- r_k number of basic channels (units) required by each class k call
- $C_i, i =$ 1, 2
  nominal capacity for each group; group 1 includes the traffic classes from 1 to p, and group 2 includes the traffic classes from $p+1$ to K and $C_1 + C_2 = N$

The dynamics of a radio cell is driven by new call requests, call terminations, and handoffs induced by user mobility. Maintaining an ongoing call is more important than admitting a new call. Hence, handoff calls should be given a higher access priority, or a lower dropping probability, than the probability of blocking new calls. One way to facilitate this is to reserve capacity for admitting handoff calls, which is not accessible by new requests. The reserved capacity is referred to as *guard capacity*.

The number of channels available for admitting new and handoff calls is $N - N_i$. Let C_G denote the number of guard channels in a cell. We have the following admission rules:

1. Admit both new and handoff calls if $N - N_i > C_G$.
2. Reject new calls and admit only handoff calls if $0 < N - N_i \leq C_G$.

As stated in Section 5.1, call-level performance measures such as blocking and dropping probabilities are known as GoS. For convenience of representation, we use QoS to represent call-level performance measures in the remaining part of this subsection.

The call-level parameters used throughout in section are as follows:

$B_{nk} = \frac{\text{number of new class } k \text{ calls blocked}}{\text{total number of call arrivals}}$	new call blocking probability for class k
$B_n = \sum_{k=1}^{K} B_{nk}$	total new call blocking probability
$B_{hk} = \frac{\text{number of handoff class } k \text{ calls dropped}}{\text{total number of call arrivals}}$	handoff call dropping probability for class k
$B_h = \sum_{k=1}^{K} B_{hk}$	total handoff call dropping probability
$B = B_n + B_h$	sum of new call blocking and handoff call dropping probabilities
$N_{uk} = \frac{\int_T n_k r_k \, dt}{T}$	system utilization for class k calls
$N_u = \sum_{k=1}^{K} N_{uk}$	total system utilization (average channel occupation, $N_u < C$)
λ_{nk}	arrival rate of new class k calls
λ_{hk}	arrival rate of class k handoff calls
$\lambda_k = \lambda_{nk} + \lambda_{hk}$	total arrival rate of class k calls
μ_{ck}^{-1}	mean class k call holding time (lifetime)
μ_{hk}^{-1}	mean class k call dwell time (inter-handoff time)

The resource allocation scheme determines the way that the users share a common resource. The traffic classes are divided into two groups. The resource allocation schemes considered in this section are VP with preemption of group 2 (case 1) and VP with preemption of groups 1 and 2 (case 2). For case 1, group 1 is offered guaranteed access, while group 2 is offered best effort service. With the nominal capacity partitioned into C_1 and C_2, the allocation is $\hat{C}_1 = \sum_{i=1}^{p} n_i r_i = C_1$ (guarantee access) and $\hat{C}_2 = \sum_{j=p+1}^{K} n_j \eta_j r_j \leq C_2$ (best effort), where η_j is the ratio of the number of class j calls admitted into the cell without being preempted during the lifetimes to the number of admitted class j calls. Group 1 is given hard capacity to guarantee access. This group is given higher QoS protection and can preempt group 2 calls. Unused capacity during any epoch is available to group 2. Within group 1, we have complete sharing (CS).

For case 2, the total channels are to be shared by groups 1 and 2 using virtual partition. If $\sum_{k=1}^{p} n_k r_k < C_1$, group 1 is in the underload state; if $\sum_{k=p+1}^{K} n_k r_k < C_2$, group 2 is in the underload state; if $\sum_{k=1}^{p} n_k r_k > C_1$, group 1 is in the overload state;

if $\sum_{k=p+1}^{K} n_k r_k > C_2$, group 2 is in the overload state. The way to implement the VP with preemption for groups 1 and 2 is as follows. When group 1 (2) is in the underload state while group 2 (1) is in the overload state, group 1 (2) can preempt group 2 (1) users, up to the channel occupancy tending to the call-level capacity bound N.

Let θ_{nk} denote the probability that the next arrival is a new call from class k:

$$\theta_{nk} = \frac{\lambda_{nk}}{\sum_{k=1}^{K} \lambda_k}. \tag{5.85}$$

Let θ_{hk} denote the probability that the next arrival is a handoff call from class k:

$$\theta_{hk} = \frac{\lambda_{hk}}{\sum\limits_{k=1}^{K} \lambda_k}. \tag{5.86}$$

Let θ_k denote the probability that the next arrival is from class k:

$$\theta_k = \frac{\lambda_k}{\sum\limits_{k=1}^{K} \lambda_k}. \tag{5.87}$$

We assume that all cells are statistically identical. Thus, the rate of handoff calls departing from a cell equals the rate at which handoff calls enter the cell. From the complete partitioning policy, equating the class k handoff call arrival rate to the product of the average handoff rate, μ_{hk}, for a call and the average number of class k calls, we can get the class k handoff call arrival rate as follows [9,10]:

$$\lambda_{hk} = \mu_{hk} \frac{N_{uk}}{r_k} = \mu_{hk} \frac{(1 - b_{nk})\lambda_{nk} + (1 - b_{hk})\lambda_{hk}}{\mu_{ck} + \mu_{hk}}. \tag{5.88a}$$

Solving for λ_{hk} in (5.88a), we get

$$\lambda_{hk} = \frac{\mu_{hk}(1 - b_{nk})\lambda_{nk}}{\mu_{ck} + b_{hk}\mu_{ck}} = \frac{\mu_{hk}\left(1 - B_{nk}/\theta_{nk}\right)\lambda_{nk}}{\mu_{ck} + \left(B_{hk}/\theta_{hk}\right)\mu_{ck}}, \tag{5.88b}$$

where $b_{nk} = B_{nk}/\theta_{nk}$ is the new class k call blocking probability considering only class k traffic; $b_{hk} = B_{hk}/\theta_{hk}$ is the handoff class k call dropping probability considering only class k traffic, and $\mu_{hk} = v_k/s$; v_k is the speed of the class k mobile, and s is the size of a square cell.

Note that the new call blocking and handoff call dropping probabilities are functions of the new call arrival rate and the handoff call arrival rate, respectively. The handoff call dropping probability can only be calculated after the handoff rate has been determined. On the other hand, the handoff call arrival rate can only be calculated using (5.88b) when B_{nk} and B_{hk} are known. To resolve this paradox, (5.88b) can be written as a set of recurrence equations. At the initial time instant $l = 1$, the handoff rate, $\lambda_{hk}(1)$, is set equal to $\mu_{hk}\lambda_{nk}/\mu_{ck}$ on the assumption that $b_{hk}(1)$ and $b_{nk}(1)$ are negligibly small. Then $\lambda_{hk}(1)$ is used to compute $b_{hk}(2)$ and $b_{nk}(2)$ using the analytical results presented in Sections 5.4.3.1 and 5.4.3.2. For example, in case 1, (5.92) is used to solve for the equilibrium system state probability $P(\vec{n})$. $P(\vec{n})$ is then used in (5.96)–(5.99) to evaluate the new call blocking and handoff call dropping probabilities, which in turn are substituted into (5.88b). Iterating in this manner, the recurrence equations can be written as

$$\lambda_{hk}(l) = \frac{\mu_{hk}(1 - b_{nk}(l))\lambda_{nk}}{\mu_{ck} + b_{hk}(l)\mu_{ck}}, \tag{5.89a}$$

$$b_{hk}(l+1) = F(\lambda_{hk}(l)), \tag{5.89b}$$

$$b_{nk}(l+1) = G(\lambda_{hk}(l)), \tag{5.89c}$$

where $F(\cdot)$ and $G(\cdot)$ are notations used to denote the analytical procedure as described above. Note that this iterative procedure can be used in the analyses of the other channel assignment schemes with handoff calls mentioned in this chapter.

5.5.3.1 *Case 1: VP with Preemption for Group 2*

Assuming Poisson-distributed arrivals for new and handoff calls and exponentially distributed call holding time, we can model the channel occupancy as a K-dimensional Markov chain and solve it using the techniques in [13–17]. The key of the techniques is to formulate the global balance equations based on a K-dimensional Markov chain. The global balance equations for CS have been presented in [10]. The difference between CS and VP lies on the preemptions in VP.

Let $\boldsymbol{n} = (n_1, n_2, \ldots, n_K)$ denote the state of the system with the number of class k users, n_k, in the kth class and $\boldsymbol{r} = (r_1, r_2, \ldots, r_K)$ denote the vector of basic channels, where r_k is the number of basic channels required for each class k call. Let $\lambda_k(\boldsymbol{n})$ denote the arrival rate and $\mu_k(\boldsymbol{n})$ the departure rate in the system. Thus, the state space of the system, denoted by S, is given by $S := \{\boldsymbol{n} | \boldsymbol{r} \cdot \boldsymbol{n} \leq N\}$. Group 1 ($G_1$,$\vec{n}_1$) contains classes from 1 to p with the nominal capacity of C_1; group 2 (G_2,$\vec{n}_2$) contains classes from $p + 1$ to K with the nominal capacity of C_2.

Focusing on the characteristic of case 1, we define five preemption rules. The five preemption rules will describe all the characteristics of preemptions that are possible in case 1. They will be used as the criteria to determine whether or not the preemption can happen in a particular state of the system. The preemption rules for case 1 are as follows:

1. Preemption could happen only under the condition that $\sum_{i=p+1}^{K} n_i r_i > C_2$ (i.e., group 2 users occupying capacity nominally allocated to group 1).

2. Preemption happens when the new class $j \in G_1$ call arrives under the conditions that $C_G + \sum_{i=1}^{K} n_i r_i = N$ or $\{C_G + \sum_{i=1}^{K} n_i r_i < N$ and $C_G + \sum_{i=1}^{K} n_i r_i > N - r_j\}$. Note that no matter which type of preemption happens, new calls can be admitted into the system only when the guard channels are not occupied.
3. Preemption happens when handoff class $j \in G_1$ call arrives under the condition that $\sum_{i=1}^{K} n_i r_i > N - r_j$.
4. When there is preemption for class $j \in G_1$ over group 2 [class j adds one user and one/some user(s) is/are deleted from group 2], the number of terminated users from each class within group 2 must be calculated individually. Here we simplify the problem to assume that the termination happens to only one class (e.g., class k). So the number of class k users terminated from the system will be less than or equal to $\lceil r_j / r_k \rceil$, where $\lceil x \rceil$ is the ceiling function, which gives the smallest integer larger than or equal to x.
5. If preemption happens over class k, it means that class k is the most overload class in group 2 and that class k will be determined by the criterion $k = \arg\min\{(C_2 - n_{p+1} r_{p+1}), \ldots, (C_2 - n_K r_K)\}$[†]. In this section we use the most overload criterion.

The admission policy for CS will be defined here. When the system is in state $\vec{n}$ and a class k call (new or handoff) arrives, an admission policy determines whether or not the call is admitted into the system. We specify the admission policy by mapping $\boldsymbol{f}(\boldsymbol{n}) = (f_1(\boldsymbol{n}), f_2(\boldsymbol{n}), \ldots, f_K(\boldsymbol{n}))$ for both new and handoff calls, excluding the guard channels, and $\boldsymbol{f}_G(\boldsymbol{n}) = (f_{G1}(\boldsymbol{n}), f_{G2}(\boldsymbol{n}), \ldots, f_{GK}(\boldsymbol{n}))$ for only handoff calls within the guard channels, where the union of $S(\boldsymbol{f}(\boldsymbol{n}))$ and $S(\boldsymbol{f}_G(\boldsymbol{n}))$ gives the system state space S, and $f_k(\boldsymbol{n})$ and $f_{Gk}(\boldsymbol{n})$ each takes on the value 0 or 1 if a class k call is rejected or admitted, respectively, when the system state is $\boldsymbol{n}$. They are defined by the following equations:

$$f_k(\boldsymbol{n}) = \begin{cases} 1, & \text{condition 1} \\ 0, & \text{otherwise,} \end{cases} \tag{5.90}$$

where condition 1 refers to: for $k \in G_2$, $\sum_{i=p+1}^{K} r_i n_i + r_k \leq N - C_G$; or for $k \in G_1$, $\sum_{i=1}^{p} r_i n_i + r_k \leq C_1 - C_G$.

$$f_{Gk}(\boldsymbol{n}) = \begin{cases} 1, & \text{condition 2} \\ 0, & \text{otherwise,} \end{cases} \tag{5.91}$$

where condition 2 refers to: for $k \in G_2$, $N - C_G < \sum_{i=p+1}^{K} r_i n_i + r_k \leq N$; or for $k \in G_1$, $C_1 - C_G < \sum_{i=1}^{p} r_i n_i + r_k \leq C_1$.

[†]Another criterion is to preempt the lowest-priority class in that group.

Due to the fact that the only difference between CS and VP is the preemption in VP, when we formulate the global balance equations for VP, the global balance equations for CS will be formulated first and then each state will be filtered by the five preemption rules. If preemption happens in that state, the corresponding state transitions will be added to complete the global balance equations. Let $P(\boldsymbol{n})$ denote the equilibrium probability that the system is in state $\boldsymbol{n}$. The global balance equations for the Markov process under the preemption rules and the admission policy for case 1 are given by [18]

$$\sum_{k=1}^{K} [\lambda_k(\boldsymbol{n}) \{f_k(\boldsymbol{n}) + f_{Gk}(\boldsymbol{n})\} + \lambda_{Pk}(\boldsymbol{n}) + \mu_k(\boldsymbol{n})]\, P(\boldsymbol{n})$$

$$= \sum_{k=1}^{K} P\,(\boldsymbol{n} - \boldsymbol{e}_k)\, \lambda_k\,(\boldsymbol{n} - \boldsymbol{e}_k)\, \{f_k\,(\boldsymbol{n} - \boldsymbol{e}_k) + f_{Gk}\,(\boldsymbol{n} - \boldsymbol{e}_k)\}$$

$$+ \sum_{k=1}^{K} P\,(\boldsymbol{n} + \boldsymbol{e}_k)\, \mu_k\,(\boldsymbol{n} + \boldsymbol{e}_k) + \sum_{k=p+1}^{K} \sum_{j=1}^{p}$$

$$\times \left[\sum_{i=1}^{\lceil r_j / r_k \rceil} P\left(\boldsymbol{n} + i\boldsymbol{e}_k - \boldsymbol{e}_j\right) \lambda_{Pk}\left(\boldsymbol{n} + i\boldsymbol{e}_k - \boldsymbol{e}_j\right) \right], \tag{5.92}$$

where $\boldsymbol{e}_k$ is a K-dimensional vector of all zeros except for a 1 in the kth place; that is, $\boldsymbol{n} + \boldsymbol{e}_k$ means to admit a class k call and $\boldsymbol{n} - \boldsymbol{e}_k$ means to terminate a class k call, because it is preempted by another call or just completes its communication,

$$\lambda_k(\boldsymbol{n}) = \begin{cases} \lambda_{nk} + \lambda_{hk} & \text{if } f_k(\boldsymbol{n}) = 1 \\ \lambda_{hk} & \text{if } f_{Gk}(\boldsymbol{n}) = 1, \end{cases} \tag{5.93}$$

$$\mu_k(\boldsymbol{n}) = n_k(\mu_{ck} + \mu_{hk}), \qquad 0 < \boldsymbol{r} \cdot \boldsymbol{n} \leq N, \tag{5.94}$$

and $\lambda_{Pk}(\boldsymbol{n})$ is defined as follows:

$$\lambda_{Pk}(\boldsymbol{n}) = \begin{cases} \lambda_{nk} & \text{by rule 2} \\ \lambda_{hk} & \text{by rule 3.} \end{cases} \tag{5.95}$$

Without the $\lambda_{Pk}(\boldsymbol{n})$ terms, (5.92) is equivalent to the global balance equations for CS. Equation (5.92) can be solved using lower triangular/upper triangular (LU) decomposition together with the condition that the sum of all the state probabilities is 1 to obtain $P(\boldsymbol{n})$. LU decomposition is a common numerical technique to solve linear algebraic equations [19].

A new class $k \in G_1$ call is blocked from entering the system (and is assumed lost) if, upon arrival, it finds that it cannot be accommodated because the number of available channels (excluding the guard channels) is less than r_k. Therefore, the blocking probability for a new class $k \in G_1$ call, considering all classes, is given by

$$B_{nk} = \sum_{n_1=\lfloor (C_1-C_G)/r_1 \rfloor}^{\lfloor C_1/r_1 \rfloor} \cdots \sum_{n_p=\lfloor (C_p-C_G)/r_p \rfloor}^{\left\lfloor \left(C_1-\sum_{i=1}^{p-1} n_i r_i\right)/r_p \right\rfloor} \sum_{n_{p+1}=0}^{\left\lfloor \left(N-\sum_{i=1}^{p} n_i r_i\right)/r_{p+1} \right\rfloor} \cdots \sum_{n_K=0}^{\left\lfloor \left(N-\sum_{i=1}^{K-1} n_i r_i\right)/r_K \right\rfloor} P(\boldsymbol{n})\theta_{nk}$$

$$+ \sum_{n_1=0}^{\lfloor (C_1-C_G)/r_1 \rfloor - 1} \cdots \sum_{n_p=0}^{\left\lfloor \left(C_1-C_G-\sum_{i=1}^{p-1} n_i r_i\right)/r_p \right\rfloor - 1} \sum_{n_{p+1}=0}^{\left\lfloor \left(N-\sum_{i=1}^{p} n_i r_i\right)/r_{p+1} \right\rfloor} \cdots$$

$$\times \sum_{n_K=0}^{\left\lfloor \left(N-\sum_{i=1}^{K-1} n_i r_i\right)/r_K \right\rfloor} P(\boldsymbol{n})\theta_{nk} - \sum_{i=p+1}^{K} P_{ni} \tag{5.96}$$

for all $\boldsymbol{n}$ satisfying $N - C_G - \sum_{i=1}^{K} n_i r_i < r_k$, where P_{ni} is the preemption probability for a class i call when a new class k call arrives, which is defined in (5.100). $\lfloor x \rfloor$ is the floor function that gives the largest integer less than or equal to x.

Similarly, the blocking probability for a class $k \in G_2$ new call, considering all classes, is given

$$B_{nk} = \sum_{n_1=0}^{\lfloor C_1/r_1 \rfloor} \cdots \sum_{n_p=0}^{\left\lfloor \left(C_1-\sum_{i=1}^{p-1} n_i r_i\right)/r_p \right\rfloor} \sum_{n_{p+1}=0}^{\left\lfloor \left(N-\sum_{i=1}^{p} n_i r_i\right)/r_{p+1} \right\rfloor} \cdots \sum_{n_K=0}^{\left\lfloor \left(N-\sum_{i=1}^{K-1} n_i r_i\right)/r_K \right\rfloor} P(\boldsymbol{n})\theta_{nk} \tag{5.97}$$

for all $\boldsymbol{n}$ satisfying $N - C_G - \sum_{i=1}^{K} n_i r_i < r_k$.

A class $k \in G_1$ handoff call is blocked from entering the system (and is assumed lost) if upon arrival it finds that it cannot be accommodated because the number of available channels (including the guard channels) is less than r_k. Therefore, the blocking probability for a class $k \in G_1$ handoff call, considering all classes, is given by

$$B_{hk} = \sum_{n_1=0}^{\lfloor C_1/r_1 \rfloor} \cdots \sum_{n_p=0}^{\left\lfloor \left(C_1-\sum_{i=1}^{p-1} n_i r_i\right)/r_p \right\rfloor} \sum_{n_{p+1}=0}^{\left\lfloor \left(N-\sum_{i=1}^{p} n_i r_i\right)/r_{p+1} \right\rfloor} \cdots \sum_{n_K=0}^{\left\lfloor \left(N-\sum_{i=1}^{K-1} n_i r_i\right)/r_K \right\rfloor} P(\boldsymbol{n})\theta_{hk} \tag{5.98}$$

for all $\boldsymbol{n}$ satisfying $C_1 - \sum_{i=1}^{p} n_i r_i < r_k$.

Similarly, the blocking probability for a class $k \in G_2$ handoff call, considering all classes, is given by

$$B_{hk} = \sum_{n_1=0}^{\lfloor C_1/r_1 \rfloor} \cdots \sum_{n_p=0}^{\left\lfloor \left(C_1 - \sum_{i=1}^{p-1} n_i r_i\right) / r_p \right\rfloor} \sum_{n_{p+1}=0}^{\left\lfloor \left(N - \sum_{i=1}^{p} n_i r_i\right) / r_{p+1} \right\rfloor} \cdots \sum_{n_K=0}^{\left\lfloor \left(N - \sum_{i=1}^{K-1} n_i r_i\right) / r_K \right\rfloor} P(\boldsymbol{n})\theta_{hk} \tag{5.99}$$

for all $\boldsymbol{n}$ satisfying $N - \sum_{i=1}^{K} n_i r_i < r_k$.

The preemption probability for a class $k \in G_2$ call when a class $j \in G_1$ new call arrives is given by

$$P_{nk} = \sum_{n_1=0}^{\lfloor C_1/r_1 \rfloor} \cdots \sum_{n_p=0}^{\left\lfloor \left(C_1 - \sum_{i=1}^{p-1} n_i r_i\right) / r_p \right\rfloor} \sum_{n_{p+1}=0}^{\left\lfloor \left(N - \sum_{i=1}^{p} n_i r_i\right) / r_{p+1} \right\rfloor} \cdots \sum_{n_K=0}^{\left\lfloor \left(N - \sum_{i=1}^{K-1} n_i r_i\right) / r_K \right\rfloor} P(\boldsymbol{n})\theta_{nj} \tag{5.100}$$

for all $\boldsymbol{n}$ satisfying rules 2 and 5.

The preemption probability for a class $k \in G_2$ call when a class $j \in G_1$ handoff call arrives is given by

$$P_{hk} = \sum_{n_1=0}^{\lfloor C_1/r_1 \rfloor} \cdots \sum_{n_p=0}^{\left\lfloor \left(C_1 - \sum_{i=1}^{p-1} n_i r_i\right) / r_p \right\rfloor} \sum_{n_{p+1}=0}^{\left\lfloor \left(N - \sum_{i=1}^{p} n_i r_i\right) / r_{p+1} \right\rfloor} \cdots \sum_{n_K=0}^{\left\lfloor \left(N - \sum_{i=1}^{K-1} n_i r_i\right) / r_K \right\rfloor} P(\boldsymbol{n})\theta_{hj} \tag{5.101}$$

for all $\boldsymbol{n}$ satisfying rules 3 and 5.

The utilization for class k is given by

$$N_{uk} = \sum_{n_1=0}^{\lfloor N/r_1 \rfloor} \sum_{n_2=0}^{\lfloor (N - n_1 r_1)/r_2 \rfloor} \cdots \sum_{n_K=0}^{\left\lfloor \left(N - \sum_{i=1}^{K-1} n_i r_i\right) / r_K \right\rfloor} n_k P(\boldsymbol{n}) r_k. \tag{5.102}$$

Thus, the total utilization is

$$N_u = \sum_{k=1}^{K} N_{uk}. \tag{5.103}$$

5.5.3.2 *Case 2: VP with Preemption for Groups 1 and 2* In mathematical terms, the admission policy for case 2 is defined as

$$f_k(\mathbf{n}) = \begin{cases} 1, & \mathbf{r} \cdot \mathbf{n} + r_k \leq N - C_G \\ 0, & \text{otherwise} \end{cases} \tag{5.104}$$

$$f_{Gk}(\mathbf{n}) = \begin{cases} 1, & N - C_G < \mathbf{r} \cdot \mathbf{n} + r_k \leq N \\ 0, & \text{otherwise.} \end{cases} \tag{5.105}$$

Preemption in case 2 is governed by the following rules:

1. Preemption could happen only under the condition that $\sum_{i=p+1}^{K} n_i r_i > C_2$ or $\sum_{i=1}^{p} n_i r_i > C_1$ (i.e., group 2 users occupying capacity nominally allocated to group 1; group 1 users occupying capacity nominally allocated to group 2).
2. Preemption happens when a new class j call arrives under the conditions that $C_G + \sum_{i=1}^{K} n_i r_i = N$ or $\{C_G + \sum_{i=1}^{K} n_i r_i < N$ and $C_G + \sum_{i=1}^{K} n_i r_i > N - r_j\}$.
3. Preemption happens when a handoff class j call arrives under the condition that $\sum_{i=1}^{K} n_i r_i > N - r_j$.
4. If preemption happens over class k, it means that class k is the most overloaded class in its own group and that class k will be determined by the following criteria:

$$k = \arg\min\left\{(C_1 - n_1 r_1), \ldots, \left(C_1 - n_p r_p\right)\right\} \qquad \text{for } k \in G_1,$$
$$k = \arg\min\left\{\left(C_2 - n_{p+1} r_{p+1}\right), \ldots, (C_2 - n_K r_K)\right\} \qquad \text{for } k \in G_2.$$

Let $P(\mathbf{n})$ denote the equilibrium probability that the system is in state $\mathbf{n}$. The global balance equations for the Markov process under the preemption rules and the admission policy for case 2 are given by [14]

$$\sum_{k=1}^{K} [\lambda_k(\mathbf{n}) \{f_k(\mathbf{n}) + f_{Gk}(\mathbf{n})\} + \lambda_{Pk}(\mathbf{n}) + \mu_k(\mathbf{n})] P(\mathbf{n})$$
$$= \sum_{k=1}^{K} P(\mathbf{n} - \mathbf{e}_k) \lambda_k (\mathbf{n} - \mathbf{e}_k) \{f_k (\mathbf{n} - \mathbf{e}_k) + f_{Gk} (\mathbf{n} - \mathbf{e}_k)\}$$
$$+ \sum_{k=1}^{K} P(\mathbf{n} + \mathbf{e}_k) \mu_k (\mathbf{n} + \mathbf{e}_k) + \sum_{k=1}^{K} \sum_{\substack{j=1 \\ j \neq k}}^{K}$$
$$\left[\sum_{i=1}^{\lceil r_j/r_k \rceil} P\left(\mathbf{n} + i\mathbf{e}_k - \mathbf{e}_j\right) \lambda_{Pk} \left(\mathbf{n} + i\mathbf{e}_k - \mathbf{e}_j\right)\right]. \tag{5.106}$$

A new class k call is blocked from entering the system (and is assumed lost) if upon arrival it finds that it cannot be accommodated because the number of available channels (excluding the guard channels) is less than r_k. Therefore, the blocking probability for a new class k call, considering all classes, is given by

$$B_{nk} = \sum_{n_1=0}^{\lfloor N/r_1 \rfloor} \sum_{n_2=0}^{\lfloor (N-n_1r_1)/r_2 \rfloor} \cdots \sum_{n_K=0}^{\left\lfloor \left(N-\sum_{i=1}^{K-1} n_ir_i\right) / r_K \right\rfloor} P(\boldsymbol{n})\theta_{nk} - \sum_{j \text{ in the other group}} P_{nj} \tag{5.107}$$

for all $\boldsymbol{n}$ satisfying $N - C_G - \sum_{i=1}^{K} n_ir_i < r_k$, where P_{nj} is the preemption probability for class j call when a new class k call arrives, as defined in (5.109).

A class k handoff call is blocked from entering the system (and is assumed lost) if upon arrival it finds that it cannot be accommodated because the number of channels available (including the guard channels) is less than r_k. Therefore, the blocking probability for a class k handoff call, considering all classes, is given by

$$B_{hk} = \sum_{n_1=0}^{\lfloor N/r_1 \rfloor} \sum_{n_2=0}^{\lfloor (N-n_1r_1)/r_2 \rfloor} \cdots \sum_{n_K=0}^{\left\lfloor \left(N-\sum_{i=1}^{K-1} n_ir_i\right) / r_K \right\rfloor} P(\boldsymbol{n})\theta_{hk} - \sum_{j \text{ in the other group}} P_{hj} \tag{5.108}$$

for all $\boldsymbol{n}$ satisfying $N - \sum_{i=1}^{K} n_ir_i < r_k$, where P_{hj} is the preemption probability for a class j call when a handoff class k call arrives, as defined in (5.110).

The preemption probability for a class k call when a class j new call arrives is given by

$$P_{nk} = \sum_{n_1=0}^{\lfloor N/r_1 \rfloor} \sum_{n_2=0}^{\lfloor (N-n_1r_1)/r_2 \rfloor} \cdots \sum_{n_K=0}^{\left\lfloor \left(N-\sum_{i=1}^{K-1} n_ir_i\right) / r_K \right\rfloor} P(\boldsymbol{n})\theta_{nj} \tag{5.109}$$

for all $\boldsymbol{n}$ satisfying rules 2 and 4.

The preemption probability for a class k call when a class j handoff call arrives is given by

$$P_{hk} = \sum_{n_1=0}^{\lfloor N/r_1 \rfloor} \sum_{n_2=0}^{\lfloor (N-n_1r_1)/r_2 \rfloor} \cdots \sum_{n_K=0}^{\left\lfloor \left(N-\sum_{i=1}^{K-1} n_ir_i\right) / r_K \right\rfloor} P(\boldsymbol{n})\theta_{hj} \tag{5.110}$$

for all $\boldsymbol{n}$ satisfying rules 3 and 4.

The utilization for class k is given by (5.102). The total utilization is given by (5.103).

5.5.4 Link-Layer Resource Allocation Scheme with Multiclass On/Off Traffic

Earlier, we considered three connection-level resource allocation schemes for multiclass traffic. In this section we consider a link-layer resource allocation scheme for multiclass traffic. The outage probability is derived and the system capacity can be obtained as in Chapter 3.

We consider multiclass on/off traffic in the uplink of a wireless cellular DS-CDMA network with *variable spreading gain*. The cell site (base station) supports K classes of services that can originate from the mobile users in the cell. The following system-level parameters are used throughout this section:

K	number of traffic classes
W	spread-spectrum bandwidth
R_i	transmission rate of class i traffic, $i = 1, 2, \ldots, K$
G_i	spreading gain of class i traffic
n_i	number of class i users per sector
BER_i	bit error rate of class i
SIR_i	signal-to-interference ratio of class i
BER_i^*	bit error rate requirement of class i
SIR_i^*	signal-to-interference ratio requirement of class i
γ_i	E_b/I_0 requirement of class i traffic
S_i	received power of class i traffic
I_i	intercell interference power of class i traffic
η	background noise
ρ_i	density of class i users per unit area
α_i	activity factor of class i traffic
ψ_{ij}	random variable for the activity factor of user j belonging to class i traffic

The following assumptions are made to facilitate the analytical formulation:

- The transmission rates of other classes are integer multiples of that for the class with the basic rate.
- The processing gain, G_i, for class i users is given by W/R_i.
- The system is made up of hexagonal cells.
- The mobile users have omnidirectional antennas.
- The base station has three sectors in each cell.
- The sectorization in the cells is perfect.
- Users are located uniformly in each cell.
- There are equal numbers of users from each class in every cell.
- There is perfect power control in each cell.
- The spreading gain can be varied for different traffic classes.
- The spreading gain is the same within the same class.

- The channel is modeled as a combination of path loss and lognormal shadowing, represented by $r^{-4} \cdot 10^{\varepsilon/10}$, where r is the distance between the mobile and the serving base station and ε is a Gaussian random variable with zero mean and variance σ^2.

We are concerned with the uplink capacity of the DS-CDMA system in terms of the number of users, n_i, that can be supported for the ith class. The capacity region for K classes is derived by considering the outage probability in terms of the signal-to-interference ratio (SIR) specification. These probabilities are expressed in terms of the number of class i users, their activity factors, the intracell received powers for K classes, the intercell interference for K classes, the spreading gains for K classes, the E_b/I_0 requirements for K classes and the background noise. The capacity of the K classes system is defined by $(n_1, \ldots, n_i, \ldots, n_K)$, where $i = 1, 2, \ldots, K$. The aim here is to determine the maximum number of users for the K classes that are allowable in the system while maintaining the required QoS. From this system capacity, schemes in Sections 5.5.2 and 5.5.3 can be used. First, let γ_i denote the E_b/I_0 for class i. It is given by

$$\gamma_i = \frac{G_i}{\sum_{j=1}^{n_i-1} \psi_{ij} + \sum_{\substack{k=1\\k\neq i}}^{K} \sum_{j=1}^{n_k} \psi_{kj} \frac{S_k}{S_i} + \sum_{k=1}^{K} \frac{I_k}{S_i} + \frac{\eta}{S_i}}, \qquad i = 1, 2, \ldots, K, \tag{5.111}$$

where $\psi_{ij} \in 0, 1$ is a Bernoulli random variable indicating the activity factor of the jth user of class i. The activity factor is defined by $\alpha_i = Pr[\psi_{ij} = 1]$. The numerator on the right-hand side of (5.111) is the class i processing gain. In the denominator, the first term is due to the intracell interference from other users in class i, the second term is due to the intracell interference from users from other classes, the third term is due to the intercell interference from all classes, and the last term is due to background noise. Rearranging (5.111), we have

$$\alpha_i(n_i - 1)S_i + \sum_{\substack{k=1\\k\neq i}}^{K} \alpha_k n_k S_k - \frac{S_i G_i}{\gamma_i} = -\sum_{k=1}^{K} I_k - \eta, \qquad i = 1, 2, \ldots, K. \tag{5.112}$$

For $i \neq k$ the power ratio can be expressed as

$$\frac{S_k}{S_i} = \left(\frac{G_i}{\gamma_i} + \alpha_i\right) \Big/ \left(\frac{G_k}{\gamma_k} + \alpha_k\right), \qquad i, k \in \{1, 2, \ldots, K\}, \quad i \neq k. \tag{5.113}$$

From [10], the intercell interference for a jth user in class i is given by

$$\frac{I_{ij}}{S_i} = \left(\frac{r_m}{r_d}\right)^4 10^{(\varepsilon_d - \varepsilon_m)/10}, \tag{5.114}$$

where r_d is the distance between the intercell mobile that is causing interference and the intracell base station, r_m is the distance between the intercell mobile and its own base station, and ε_d and ε_m are Gaussian random variables with zero mean and standard deviation σ. Since ε_d and ε_m are independent, $\varepsilon_d - \varepsilon_m$ is a Gaussian random variable with zero mean and variance $2\sigma^2$. The mean and variance of I_i/S_i are upper bounded by [10]

$$E\left[\frac{I_i}{S_i}\right] \le \alpha_i \rho_i = \mu_{ii} \tag{5.115}$$

and

$$\mathrm{Var}\left[\frac{I_i}{S_i}\right] \le \rho_i \int\!\!\int \left[\alpha_i g\left(\frac{r_m}{r_d}\right) - \alpha_i^2 f^2\left(\frac{r_m}{r_d}\right)\right] dA = \sigma_{ii}^2, \tag{5.116}$$

where μ_{ii} is the upper bound on the mean of I_i/S_i, σ_{ii}^2 is the upper bound on the variance of I_i/S_i, and ρ_i is the density of class i users per unit area and is given by

$$\rho_i = \frac{2n_i}{\sqrt{3}}, \tag{5.117}$$

$$f\left(\frac{r_m}{r_d}\right) = \left(\frac{r_m}{r_d}\right)^4 e^{(\sigma \ln 10/10)^2}\left[1 - Q\left(\frac{40\log(r_d/r_m)}{\sqrt{2\sigma^2}} - \frac{\sqrt{2\sigma^2}\ln 10}{10}\right)\right], \tag{5.118}$$

$$g\left(\frac{r_m}{r_d}\right) = \left(\frac{r_m}{r_d}\right)^8 e^{(\sigma \ln 10/5)^2}\left[1 - Q\left(\frac{40\log(r_d/r_m)}{\sqrt{2\sigma^2}} - \frac{\sqrt{2\sigma^2}\ln 10}{5}\right)\right], \tag{5.119}$$

$$Q(y) = \frac{1}{\sqrt{2\pi}}\int_y^{\infty} e^{-x^2/2}\,dx. \tag{5.120}$$

The random variable, I_k/S_i, can be expressed as

$$\frac{I_k}{S_i} = \frac{I_1}{S_1}\frac{S_k}{S_i}\left(\frac{I_k}{S_k}\bigg/\frac{I_1}{S_1}\right), \qquad i, k = 1, 2, \ldots, K. \tag{5.121}$$

Thus, the mean and variance of I_k/S_i satisfy the following inequalities:

$$E\left[\frac{I_k}{S_i}\right] \le \frac{S_k \alpha_k \rho_k}{S_i \alpha_1 \rho_1}\mu_{11}, \tag{5.122}$$

$$\mathrm{Var}\left[\frac{I_k}{S_i}\right] \le \left(\frac{S_k}{S_i}\right)^2 \sigma_{kk}^2. \tag{5.123}$$

Let BER_i^* denote the BER requirement for class i users and SIR_i^* denote the SIR requirement for class i users. The system capacity is defined as the maximum $(n_1, \ldots, n_i, \ldots, n_K)$, where $i = 1, 2, \ldots, K$, that can be supported such that the SIR achieved is greater than or equal to the SIR_i^* required 99% of the time for all classes. That is, the outage probability is defined as

$$\begin{aligned} \Pr[\text{BER}_i \geq \text{BER}_i^*] &= \Pr[\text{SIR}_i \leq \text{SIR}_i^*] \\ &= \Pr\left[\sum_{j=1}^{n_i-1} \psi_{ij} + \sum_{\substack{k=1\\k\neq i}}^{K} \sum_{j=1}^{n_k} \psi_{kj} \tfrac{S_k}{S_i} + \sum_{k=1}^{K} \tfrac{I_k}{S_i} \geq \delta_i \right], \end{aligned} \qquad (5.124)$$

where

$$\delta_i = \frac{G_i}{\gamma_i} - \frac{\eta}{S_i}, \qquad i = 1, 2, \ldots, K. \qquad (5.125)$$

Assuming central limit approximation and solving (5.124) by conditioning on the activity factors and then unconditioning the probability in (5.124) by summing up all cases for the numbers of active users of all classes and multiplying by all the corresponding binomial probabilities of activity factors, we have

$$\begin{aligned} \Pr[\text{BER}_i \geq \text{BER}_i^*] &= \sum_{l_1=0}^{n_1} \cdots \sum_{l_i=0}^{n_i-1} \cdots \sum_{l_K=0}^{n_K} \Pr\left[\sum_{k=1}^{K} \frac{I_k}{S_i} \geq \delta_i - l_i \right. \\ &\quad \left. - \sum_{\substack{k=1\\k\neq i}}^{K} l_k \frac{S_k}{S_i} \,\middle|\, \sum \psi_{1j} = l_1, \sum \psi_{2j} = l_2, \ldots, \sum \psi_{Kj} = l_K \right] \\ &\quad \times \prod_{k=1}^{K} \Pr\left[\sum \psi_{kj} = l_k\right] \\ &= \sum_{l_1=0}^{n_1} \cdots \sum_{l_i=0}^{n_i-1} \cdots \sum_{l_K=0}^{n_K} Q\left(\frac{\delta_i - \mu_i}{\sigma_i}\right) \binom{n_i - 1}{l_i} \alpha_i^{l_i} (1 - \alpha_i)^{n_i - 1 - l_i} \\ &\quad \prod_{\substack{k=1\\k\neq i}}^{K} \binom{n_k}{l_k} \alpha_k^{l_k} (1 - \alpha_k)^{n_k - l_k}, \end{aligned} \qquad (5.126)$$

where

$$\mu_i = l_i + \sum_{\substack{k=1\\k\neq i}}^{K} \left(\frac{S_k}{S_i}\right) l_k + \frac{\alpha_i \rho_i}{\alpha_1 \rho_1} \mu_{11} + \sum_{\substack{k=1\\k\neq i}}^{K} \left(\frac{S_k \alpha_k \rho_k}{S_i \alpha_1 \rho_1}\right) \mu_{11}, \qquad (5.127)$$

and

$$\sigma_i = \sqrt{\sigma_{ii}^2 + \sum_{\substack{k=1 \\ k \neq i}}^{K} \left(\frac{S_k}{S_i} \right)^2 \sigma_{kk}^2}, \qquad i = 1, 2, \ldots, K. \tag{5.128}$$

Equation (5.126) leads to an admission region which in turn can make use of the multiclass link layer channel assignment strategies.

5.6 LOCATION MANAGEMENT

Figure 5.29 shows a simplified cellular network architecture that can be used for point-to-point, point-to-multipoint, or multipoint-to-multipoint links. The user can be a fixed or mobile user. For message transfer, a mobile is associated with a home network or one of the subnetworks. The home network needs to have full knowledge

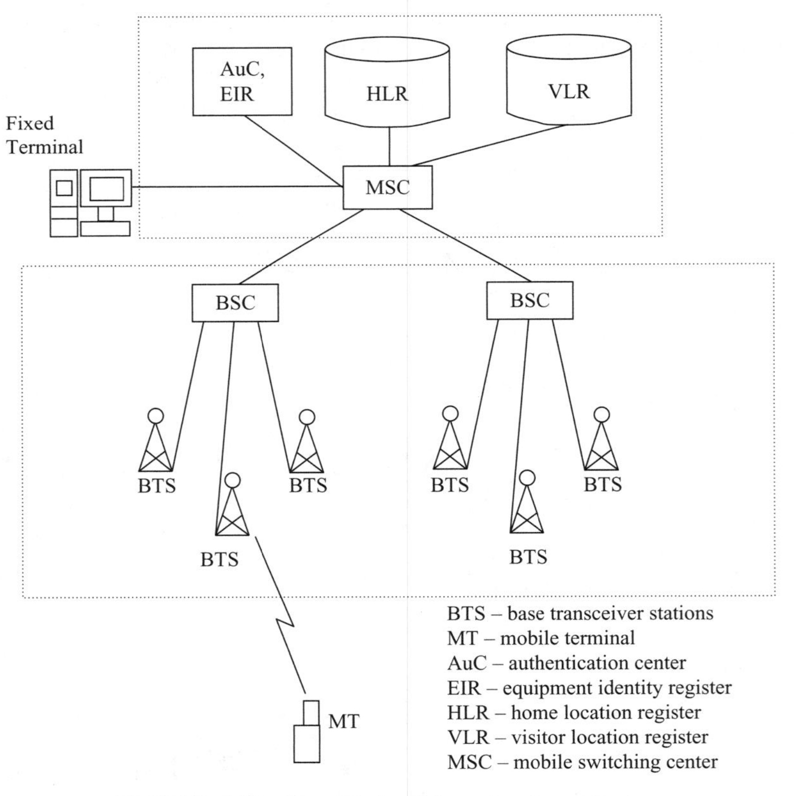

FIGURE 5.29 Simplified cellular network architecture.

of the mobile location for message transfer. A registration process is used for the association between the mobile and the home network. The mobile's identity is kept in a location register database. This register, which identifies the mobile's home network, is called the *home location register* (HLR). When the mobile moves to another subnetwork, it maintains its association with its home network through the registration process. When the mobile is away from its home network, it updates its registration with its HLR for message transfers with continuity. The subnetwork of the current mobile location is called the *foreign* or *visitor network*. The mobile also needs to register this location with a visitor location register (VLR) in the visitor network. To protect the network from fraudulent attacks, the HLR must know the true owner of the identity (ID) of the mobile, which is being kept in the database of the HLR, in order to proceed with registration update. This process is the authentication process. Location management consists of [20] (1) authentication, (2) location update, and (3) call delivery, which are described below.

5.6.1 Authentication

Secured communication in cellular networks depends on reliable authentication of network components and communicating parties. Thus, authentication is necessary. It is a process to ensure that any party's claim to be the truthful owner of a particular address is verified. Authentication depends on exchanges of cryptographic messages between the communicating parties. The authentication is one-way in GSM networks. In GSM networks, only the network can authenticate the mobile terminal, On the other hand, the authentication is two-way in UMTS networks, in which the network can authenticate the mobile terminal, and vice versa. In lay terms, encryption of a message is analogous to putting the message in an indestructible box and locking the box. A key is needed to open the box to read the message. The mobile's identity is its address. Only the mobile and its HLR know the mobile's address. In short, the identity of the address is locked in a box, and only the mobile and the HLR each have a key to open the box. GSM uses a preshared key.

5.6.2 Location Update

A registration area (RA) is covered by all the cells in a subnet of the cellular network. When the mobile moves within a registration area, it does not need to update its registration with its HLR. On the other hand, when it moves into a new registration area, it needs to do the registration update. Location update is a process in which the mobile sends periodic update messages to inform the HLR of its up-to-date location whenever it moves into a new registration area.

5.6.3 Call Delivery

The fixed user is assumed to generate a message and sends it to the mobile's permanent address for delivery to the mobile in its latest location. This can be done if the mobile has updated its location with the HLR when the mobile moves into a new registration area as mentioned in Section 5.6.2.

5.7 MOBILE IP

IP addressing is the main problem in introducing mobility to the Internet. The IP address is a special and distinct address for each network access point (e.g., in a router, in a terminal). The IP address is also used for routing packets in the routers between the source and destination nodes. Hence, the main issue for mobility in the Internet is the handling of the mobile terminal's (MT's) IP address and routing information when the mobile host hands off between two wireless access points, such as base stations, or when it roams between two network domains, such as between two network operators.

Mobile IP protocol [21] solves this issue. This protocol provides mobility support and is transparent simultaneously to the transport and higher protocol layers. Thus, mobile IP does not need changes in the existing nodes and hosts on the Internet. The mobile IP protocol lets the MT retain its IP address regardless of the point of attachment to the network. IP addresses are used primarily to identify the end system. The transmission control protocol (TCP) keeps track of a session by using the end IP addresses of the two endpoints with certain port numbers.

IP addresses are also used by routers to route the traffic from the source node to the destination node. The routes can be different for both directions for bidirectional communication. Routing in the Internet is based on the destination address and some congestion information in the intermediate network nodes. A mobile terminal must have a stable IP address to be identifiable to the other Internet hosts and nodes. Thus, mobile IP provides two IP addresses for the MT: the home address and the care-of address (CoA).

The home address is a static IP address that is used to identify higher layer connections such as TCP. The CoA is use for routing. When the mobile roams to a different network, the CoA changes. The CoA represents the IP address of the mobile terminal attachment to the network.

In mobile IPv4 management of CoA, CoA is performed by the foreign agent (FA) in the visiting network for the MT. However, the CoA is registered by the home agent (HA). Internet hosts do not need to know an MT's current location when communicating with the MT. By using the MT's home address, the MT can receive data on its home network via the HA. When the MT roams into a new network, it obtains a new CoA via the FA in that network. The new CoA is registered in the HA. Therefore, a packet addressed to the MT reaches the HA first. Then the HA tunnels the packets to the FA by using the CoA as the destination address of the packets. At the end of the tunnel, the FA decapsulates the packets, such that packets will seem to have the MT's home address as the destination IP address. After this process, the packets are sent to the MT. With such packet arrivals at the MT, mobile IP is transparent to higher layer protocols. Standard IP routing mechanisms are used to route

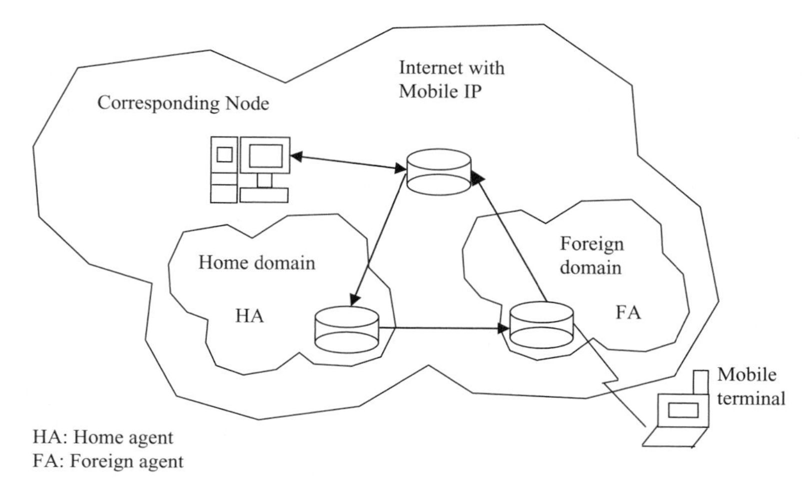

FIGURE 5.30 Mobile IP protocol.

packets sent by the MT, which uses its home address as a source address in the IP header.

CoA is a temporary address, which is used for tunneling from HA to FA when the mobile is roaming in a foreign network. The routing of packets using the mobile IP protocol forms a relay routing among the HA, FA, and the correspondent node (CN) as shown in Figure 5.30. Note that there is no FA needed to support the MTs in Mobile IPv6. IPv6 functionality is sufficient to let MTs acquire their temporary CoAs through conventional IPv6 auto-configuration mechanisms. Note that IPv6 has 128-bit addresses, whereas IPv4 has 32-bit addresses.

5.8 CELLULAR IP

Figure 5.31 shows a cellular IP network architecture. Cellular IP [22] is an extension to the mobile IP protocol. It is intended for application in the cellular access network. Cellular IP can interwork with mobile IP to support wide-area mobility such as mobility between cellular IP networks. Cellular IP is optimized for fast handoffs in the cellular network. This protocol allows integrated mobility control and location management functions at the wireless access points. Cellular IP networks are connected to the Internet through gateway routers. Mobile terminals (MTs) identified themselves to the network using the IP address of the base station (access router) as a CoA. The HA tunnels the IP packets to the gateway router of the cellular IP network as cellular IP allows mobile IP to manage macro-mobility. Within the network, packets are routed to the home address of the mobile terminal. Packets from the mobile are routed to the gateway router hop by hop in the opposite direction. Upon reaching the gateway router, packets are routed through the Internet by mobile IP. Hard or semisoft

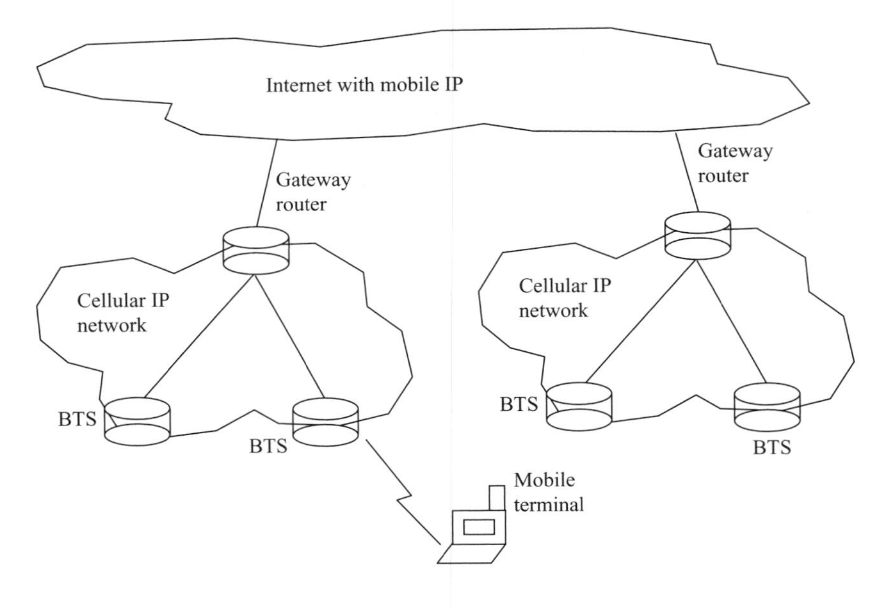

FIGURE 5.31 Cellular IP architecture.

handoff is used in cellular IP. In semisoft handoff, the MT forwards packets to both the old and new BTSs for a short time before the handoff takes place.

5.9 HAWAII

An overview of the handover-aware wireless access Internet infrastructure (HAWAII) scheme [23] is shown in Figure 5.32. The first layer of macro-mobility is handled by mobile IP, while the second layer of micro-mobility is handled by HAWAII. The HAWAII access network is divided into domains where the gateway of a HAWAII domain is the first router of the domain that uses the HAWAII routing scheme to route packets to the BTS connecting the MT. The MT does not change its IP address while moving within a HAWAII domain. The MT uses its home IP address within its home network. On the other hand, while roaming in a foreign network, it uses a CoA from the foreign network. The path to the first domain router remains unchanged, but the local path to the BTS is done by HAWAII forwarding mechanisms. Thus, there is no need to update the home address for every local movement of the MT. HAWAII supports forwarding and nonforwarding path setup schemes. In the former scheme, the MT connects to one BTS at a time. On the other hand, the MT can connect to several BTSs at the same time. Thus, the former scheme can be used for TDMA wireless systems, while the latter scheme can be used for CDMA wireless systems.

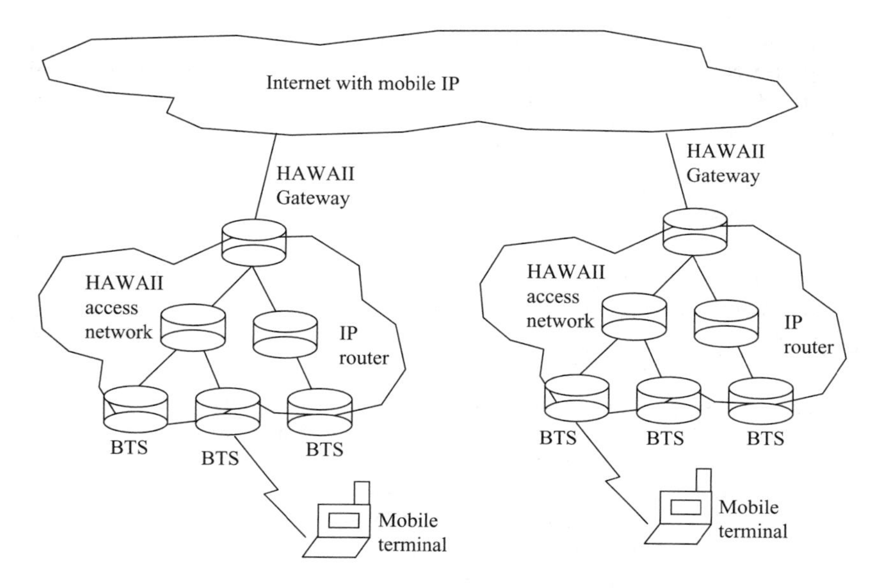

FIGURE 5.32 Handover-aware wireless access internet infrastructure (HAWAII).

SUMMARY

In Section 5.2 we presented horizontal and vertical types of handoffs. The former is for handoff within homogeneous networks, and the latter is for handoff between heterogeneous networks. The horizontal type of handoff includes mobile-controlled handoff, network-controlled handoff, and mobile-assisted handoff. On the other hand, the vertical type of handoff includes upward and downward handoff. An overview of handoff strategies is presented in Section 5.3. References to mobility management across heterogeneous networks may be found in [24–26]. Both single- and multi-class channel assignment schemes are analyzed in Sections 5.4 and 5.5.The single-class channel assignment schemes presented in Section 5.4 includes nonprioritized scheme, guard channel scheme, limited fractional guard channel scheme, fractional guard channel scheme, guard channel scheme with buffer, two-level fractional guard channel scheme, and single-class link-layer resource allocation. Further resource allocation schemes may be found in [27–32]. The multiclass channel assignment schemes presented in Section 5.5 includes complete partitioning, complete sharing, virtual partitioning, and multiclass link-layer resource allocation. An overview of location management is presented in Section 5.6. Vertical handoff analyses may be found in [33,34]. Elements of location management are authentication, location up-date, and call delivery. Finally, three location management schemes are discussed in Sections 5.7, 5.8, and 5.9, respectively. The three schemes are mobile IP for macro-mobility management and cellular IP and HAWAII for micromobility management.

Performance analysis of mobile IP may be found in [35]. Mobility management has been proposed for hierarchical state routing protocol in multilevel clustering ad hoc network [36]. In the proposed scheme, mobility management is achieved using the concept of logical subnets.

REFERENCES

[1] N. D. Tripathi, J. H. Reed and H. F. Vanlandingham, "Handoff in cellular systems," *IEEE Personal Commun.*, pp. 26–37, Dec. 1998.

[2] R. Tafazolli, *Technologies for the Wireless Future: Wireless World Research Forum (WWRF)*, vol. 2. Wiley, Hoboken, NJ, 2006.

[3] D. Hong and S. S. Rappaport, "Traffic model and performance analysis for cellular mobile radio telephone systems with prioritized and nonprioritized handoff procedures, Version 2b," *CEAS Tech. Rep. 773*, College of Engineering and Applied Sciences, State University of New York, Stony Brook, NY, June 1999.

[4] R. Ramjee, D. Towsley, and R. Nagarajan, "On optimal call admission control in cellular networks," *Wireless Networks*, vol. 3, pp. 29–41, 1997.

[5] S. C. Borst and D. Mitra, "Virtual partitioning for robust resource sharing: computational techniques for heterogeneous traffic," *IEEE J. Sel. Areas Commun.*, vol. 16, pp. 668–678, 1998.

[6] R. Ramjee, D. Towsley, and R. Nagarajan, "On optimal call admission control in cellular networks," *Wireless Networks*, vol. 3, pp. 29–41, 1997.

[7] D. T. C. Wong, J. W. Mark, and K. C. Chua, "Two-level fractional guard channels for priority access in cellular systems," *IEEE Vehicular Technology Conference 2006 – Spring, Conference Record*, pp. 383–387, Melbourne, Victoria, Australia, May 8–10, 2006.

[8] T. C. Wong, J. W. Mark, and K. C. Chua, "Multi-level dynamic guard channels for priority access in cellular networks," *IFIP Networking Conference 2005, Conference Record*, pp. 598–609, Waterloo, Ontario, Canada, May 2–6, 2005.

[9] M. H. Macdougall, *Simulating Computer Systems: Techniques and Tools*, MIT Press, Cambridge, MA, 1987.

[10] K. S. Gilhousen, I. M. Jacobs, R. Padovani, A. J. Viterbi, L. A. Weaver, Jr., and C. E. Wheatley III, "On the capacity of a cellular CDMA system," *IEEE Trans. Veh. Technol.*, vol. 40, no. 2, pp. 303–312, May 1991.

[11] D. Mitra, M. I. Reiman, and J. Wang, "Robust dynamic admission control for unified cell and call QoS in statistical multiplexers," *IEEE J. Sel. Areas Commun.*, vol. 16, no. 5, pp. 692–707, June 1998.

[12] G. L. Wu and J. W. Mark, "A buffer allocation scheme for ATM networks: complete sharing based on virtual partition," *IEEE/ACM Trans. Network.*, vol. 3, no. 6, pp. 660–670, Dec. 1995.

[13] K. W. Ross, *Multiservice Loss Models for Broadband Telecommunication Networks*, Springer, New York, 1995.

[14] T. C. Wong, J. W. Mark, and K. C. Chua, "Resource allocation in mobile cellular networks with QoS constraints," *Wireless Communications and Networking Conference 2002*, pp. 717–722, Mar. 2002.

[15] G. Haring, R. Marie, R. Puigjaner, and K. Trivedi, "Loss formulas and their application to optimization for cellular networks," *IEEE Trans. Veh. Technol.*, vol. 50, no. 3, pp. 664–673, May 2001.

[16] C. N. Wu, Y. R. Tsai, and J. F. Chang, "A quality-based birth-and-death queueing model for evaluating the performance of an integrated voice/data CDMA cellular system," *IEEE Trans. Veh. Technol.*, vol. 48, no. 1, Jan. 1999.

[17] R. P. Narrainen and F. Takawira, "A traffic model for CDMA cellular systems with soft capacity taken into account," *IEEE 6th International Symposium on Spread Spectrum Techniques and Applications*, vol. 1, pp. 325–329, 2000.

[18] J. Yao, J. W. Mark, T. C. Wong, Y. H. Chew, K. M. Lye, and K. C. Chua, "Virtual partitioning resource allocation for multiclass traffic in cellular systems with QoS constraints," *IEEE Trans. Veh. Technol.*, vol. 53, no. 3, pp. 847–864, May 2004.

[19] W. H. Press, S. A. Teukolsky, W. T. Vettering, and B. P. Flannery, *Numerical Recipes in C: The Art of Scientific Computing*, (2nd edn.), Cambridge University Press, New York, 2002.

[20] J. W. Mark and W. Zhuang, *Wireless communications and Networking*, Prentice Hall, Upper Saddle River, NJ, 2003.

[21] C. E. Perkins, "Mobile IP," *IEEE Commun. Mag.*, pp. 84–99, May 1997.

[22] A. T. Campbell, J. Gomez, S. Kim, A. G. Valko, C.-Y. Wan, and Z. R. Turanyi, "Design, implementation, and evaluation of cellular IP," *IEEE Personal Commun.*, pp. 42–49, Aug. 2000.

[23] R. Ramjee, K. Varadhan, L. Salgarelli, S. R. Thuel, S.-Y. Wang, and T. L. Porta, "HAWAII: a domain-based approach for supporting mobility in wide-area wireless networks," *IEEE/ACM Trans. Network.*, vol. 10, no. 3, pp. 396–410, June 2002.

[24] F. Siddiqui and S. Zeadally, "Mobility management across hybrid wireless networks: trends and challenges," *Comput. Commun.*, vol. 29, pp. 1363–1385, 2006.

[25] A. H. Zahran and B. Liang, "Mobility modeling for two-tier integrated wireless multimedia networks," *IEEE International Symposium on Multimedia 2005, Conference CD-ROM*, 2005.

[26] D. Cavalcanti, D. Agrawal, C. Cordeiro, B. Xie, and A. Kumar, "Issues in integrating cellular networks, WLANS, and MANETs: a futuristics heterogeneous wireless network," *IEEE Wireless Commun.*, pp. 30–41, June 2005.

[27] M. H. Ahmed, "Call admission control in wireless networks: a comprehensive survey," *IEEE Surveys*, vol. 7, no. 1, 2005

[28] M. Ghaderi and R. Boutaba, "Call admission control in mobile cellular networks: a comprehensive survey," *Wireless Commun. Mobile Comput.*, vol. 6, no. 1, Feb. 2006.

[29] C. W. Leong and W. Zhuang, "Call admission control for wireless personal communications," *Comput. Commun.*, vol. 26, pp. 522–541, 2003.

[30] L. Ortigoza-Guerrero and A. H. Aghvami, *Resource Allocation in Hierarchical Cellular Systems*, Artech House, Norwood, MA, 1999.

[31] J. Zander, S.-L. Kim, M. Almgren, and O. Queseth, *Radio Resource Management for Wireless Networks*, Artech House, Norwood, MA, 2001.

[32] H. Chen, L. Huang, S. Kumar, and C. C. J. Kuo, *Radio Resource Management for Multimedia QoS Support in Wireless Networks*, Kluwer Academic, Norwell, MA, 2006.

[33] W. Song, H. Jiang, and W. Zhuang, "Performance analysis of the WLAN-first scheme in cellular/WLAN interworking," *IEEE Trans. Wireless Commun.*, vol. 6, no. 5, pp. 1932–1943, May 2007.

[34] T. Tugcu, H. B. Yilmaz, and F. Vainstein, "Analytical modeling of CAC in next generation wireless systems," *Computer Networks*, vol. 50, pp. 3466–3484, 2006.

[35] S. Mohanty and I. F. Akyildiz, "Performance analysis of handoff techniques based on mobile IP, TCP-migrate and SIP," *IEEE Trans. Mobile Comput.*, vol. 6, no. 7, pp. 731–747, July 2007.

[36] G. Pei and M. Gerla, "Mobility management for hierarchical wireless networks," *Mobile Networks Appl.*, pp. 331–337, 2001.

CHAPTER 6

ROUTING PROTOCOLS FOR MULTIHOP WIRELESS BROADBAND NETWORKS

6.1 INTRODUCTION

In this chapter we introduce multihop wireless mesh networks as a feasible solution to provide ubiquitous wireless broadband access before focusing on their routing protocols. Routing protocols are important in finding a path between the source node and the destination node across multiple hops. In this chapter we introduce robustness and scalability as two important characteristics of routing protocols. Various routing metrics for path selection are presented. Existing routing protocols are classified based on their unique behaviors before presenting their operational details.

6.2 MULTIHOP WIRELESS BROADBAND NETWORKS: MESH NETWORKS

In the past few years we have witnessed explosive growth in the number of wireless accesses to the Internet. At the same time, advances in wireless communication technologies have enabled transmission rates of tens of Mbps at transmission ranges of 100 m, or even higher rates at shorter transmission ranges. The convergence of

Wireless Broadband Networks, By David Tung Chong Wong, Peng-Yong Kong, Ying-Chang Liang, Kee Chaing Chua, and Jon W. Mark

these two parallel developments will lead ubiquitous high-speed wireless broadband access to the Internet.

Currently, wireless broadband access is achieved primarily through localized Wi-Fi [1] hotspots installed in offices or homes with a data rate of 54 Mbps shared among all users within a hotspot. These hotspots are considered localized because of the following three characteristics: (1) each hotspot has an IEEE 802.11 access point connected directly to the wired Internet; (2) there is no direct communication between access points from different hotspots; and (3) there is a large area without coverage between these hotspots.

To provide ubiquitous wireless broadband access beyond the localized Wi-Fi hotspots, two methods can be considered. The first method is deployment of a large number of hotspots so that there is no area without coverage between any two hotspots. Assuming that each hotspot covers an area of $100 \times 100\,\text{m}^2$, we will need a few tens of thousands hotspots to cover a metropolitan area. Each hotspot has an access point with a high-speed connection to the wired Internet using T-1, DSL, cable, or fiber. These Internet connections have recurring monthly charges. Therefore, this hotspot architecture is costly and not scalable for a large area.

The second method is to bring high-speed Internet access to existing mobile telephone users. Given the fact that the cellular wireless networks already have near-ubiquitous coverage and allow high user mobility speeds, there is no problem in ensuring wireless access anytime anywhere. Examples of this method are HSPA (high-speed packet access) [2] in GSM systems and 1xEV-DO (evolution data optimized) [3] in CDMA2000 systems. However, due to limited bandwidth and large transmission range, the data rate in such cellular wireless networks is normally much lower than that in a Wi-Fi hotspot. For example, 1xEV-DO can offer a maximum downlink data rate of 2.4 Mbps shared among all users within a cell. The more challenging uplink access will have a maximum data rate of 153 kbps. These relatively moderate data rates do not come cheap—with the multibillion-dollar spectral license costs and high infrastructure costs incurred by the network operators. Also, the limited shared data rate will upper-bound the number of users supported in a cell. Hence, despite the promise to provide a genuine ubiquitous wireless access, the cellular network architecture has a significant limitation in terms of performance/cost ratio.

In view of the limitations of the two existing methods in providing ubiquitous wireless broadband access, we may look into reducing the cost of the hotspot architecture by eliminating the need for a high-speed wired connection at each access point. On the other hand, we may explore the possibility of improving the performance of the cellular network architecture by increasing its wireless link capacity without compromising the coverage area of a base station. Following the two potential developments, wireless multihop networking appears as an alternative solution to provide ubiquitous wireless broadband access. This is because it can extend radio coverage without adding new high-speed wired connections to access points, and can improve link capacity without reducing the coverage area of an expensive base station. There are two ways of using multihop networking: across multiple cells or within a cell.

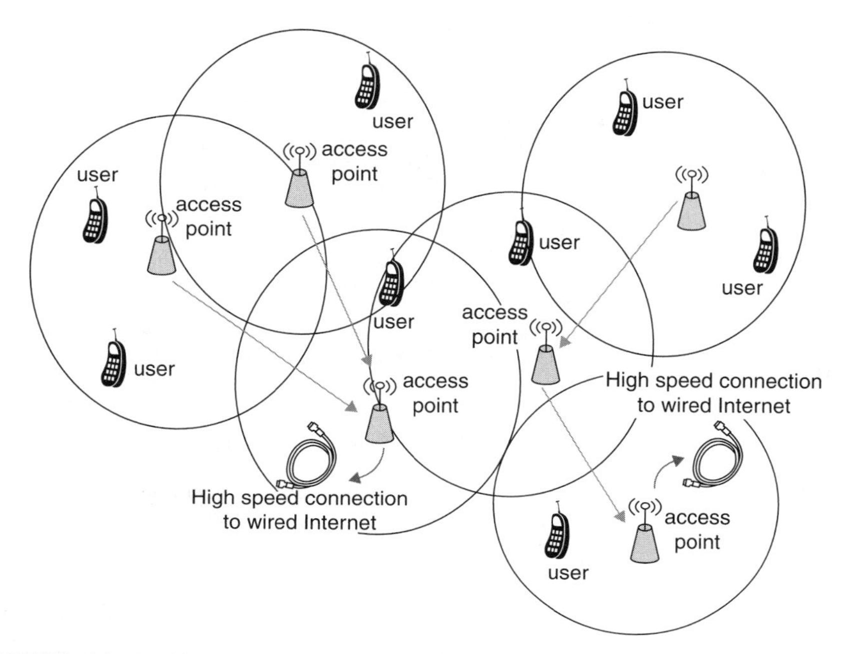

FIGURE 6.1 Multihop network across multiple cells where each access point forms a cell. Not all access points have a direct high-speed connection to the wired Internet. An access point without a high-speed connection may relay its traffic to a neighboring access point that has such a connection. A wireless link between access points may reach a further distance than the wireless link between an access point and its users.

6.2.1 Multihop Network Across Multiple Cells

As illustrated in Figure 6.1, an access point without a high-speed connection can possibly relay across multiple hops its data packets to one of its neighboring access points that has a high-speed connection. As such, the desired ubiquitous coverage can be ensured without exponentially increasing T-1 connection cost with respect to coverage area. The transit access point concept proposed by Karrer et al. [4] for a wireless broadband wireless network is an example of this way of deploying multihop networking.

6.2.2 Multihop Network Within a Cell

As illustrated in Figure 6.2, there can be multiple relay nodes between a base station and its users. The relay node is in practice a simplified low-cost version of a base station [5]. By having relay nodes, the link distance can be shortened, and thus a better link quality can be maintained for a higher link capacity. More relay nodes can be added without increasing the link distance so that the coverage of a base station can be extended without affecting the link capacity. The multihop cellular concept that

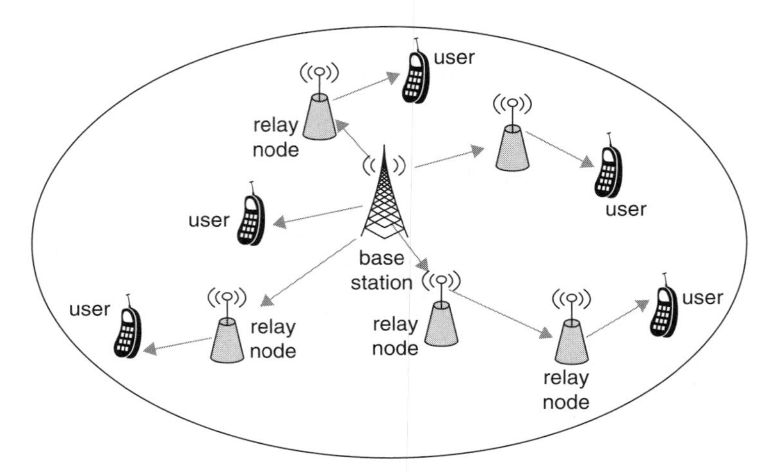

FIGURE 6.2 Multihop network within a cell. A user may be connected to base station directly or through relay nodes across multiple hops.

has been proposed and studied [6,7] is an example of this way of deploying multihop networking. In multihop cellular networks, the relays nodes are normally fixed [8,9], but they can also be mobile [10] in some less conventional usage scenarios.

A multihop relay network may also adopt a mesh topology. To be more specific, the relay network is the infrastructure wireless mesh network according to the classification in [11] because it interconnects the access points and relay nodes but not client nodes (users). Here, the client nodes connect to the access points or relay nodes via traditional wireless access technology. As such, we may collectively refer to the two types of multihop network deployments: within a cell and across cells described above as wireless mesh networks.

6.3 IMPORTANCE OF ROUTING PROTOCOLS

In a wireless mesh network, the routing protocol is very important because it helps in selecting the necessary path between the source and destination nodes across multiple wireless hops. For example, the source node can be an access point without a high-speed wired connection, and the destination node can be an access point connected directly to the Internet. Also, the source node can be a relay node and the destination is the base station. Hence, a wireless mesh network is similar to a mobile ad hoc network (MANET), which also allows communications over multiple wireless hops. As presented by Mahmud et al. [12], the major difference between the two types of networks is that nodes of a wireless mesh network are mainly static and have no power constraint, whereas nodes of a MANET are mobile and battery-powered.

Because of their similarity, routing protocols developed for MANETs may be applied to wireless mesh networks. However, there are some differences;

routing protocols for the latter should focus on improving network performance and protocol efficiency, instead of coping with node mobility or minimizing power consumption. In addition, routing protocols for a wireless mesh network should also help in enabling its self-configuration capability. Specifically, a wireless mesh network routing protocol should discover links between nodes automatically and dynamically. The information about the discovered links should then be disseminated to other nodes. Given the link information, the routing protocol at a node can identify, set up, and maintain paths autonomously from one access point to another and from a base station to a relay node across multiple hops. By being self-configurable, tedious predeployment network planning becomes unnecessary. Hence, a self-configurable wireless broadband network can be deployed rapidly to the benefit of the network operator. To achieve good performance and high efficiency, a routing protocol should have the following two characteristics: be adaptive and be scalable.

6.3.1 Adaptive Routing Protocol

During initial deployment, the path selected between a source node and its destination node should provide the best performance: for example, the highest throughput or the shortest end-to-end packet delay. However, performance of the path selected can change due to changes in the network after initial deployment. These network changes can be short term, such as temporary link failures due to the highly variable radio propagation impairments, or long term, such as the removal of a high-speed wired connection from an access point. Being adaptive, the routing protocol should detect such network changes dynamically and adjust accordingly by selecting the new best performance path when the current path no longer delivers the best performance. By striving always to select the best performance path, the routing protocol can implicitly help in providing robustness against link breakage and node failures. This is because it will quickly select another path to avoid the broken link or failed node. As such, there will be no service disruption, or even worst, single point of failure. An adaptive routing protocol can also help in providing flexibility in installation where new high-speed wired connections can be added or removed, depending on the need. The routing will automatically detect newly added (removed) high-speed wired connections and create new paths to utilize (avoid) them. Hence, adaptation capability ensures that the multihop wireless broadband network always achieves its best performance with minimum intervention from the operator.

We may measure the usefulness of an adaptive routing protocol by quantifying its ability to maintain performance robustness [13].[†] Consider link breakage and node failures as disturbances to a network. Different from conventional network robustness, which measures the ability of a network to continue operation under disturbance,

[†]Portions reprinted, with permission, from Qinghe Yin, Peng-Yong Kong, and Haiguang Wang, "Quantitative robustness metric for QoS performances of communication networks," *Proceedings of the IEEE International Symposium on Personal, Indoor and Mobile Radio Communications*, September 2006, pp. 1–5, © 2006 IEEE.

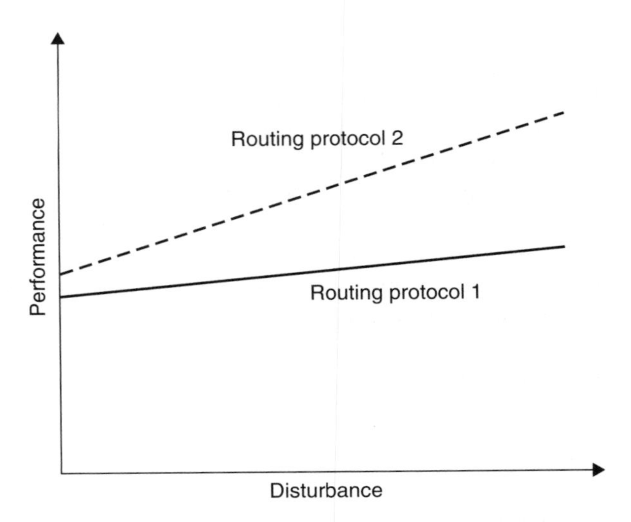

FIGURE 6.3 Example of performance robustness. Routing protocol 1 is more robust than routing protocol 2 because its performance metric changes less with respect to an increase in disturbance.

performance robustness quantifies how rapidly a desired network performance metric changes with respect to changes in a disturbance. As presented by Yin et al. [13], the idea is to define robustness based on how network performance differs from its reference point in the presence of a disturbance, as depicted in Figure 6.3. In practice, the network performance reference point can be in the form of a performance target for which the system is being designed, and usually but necessarily, the one when there is no disturbance. Let the performance reference be a random variable with a cumulative distribution function F. Let the network performance observed in the presence of disturbance be a random variable with a cumulative distribution function G. Then we need to quantify the difference between G and F. Logically, the smaller the difference, the better the performance robustness. Thus, for the purpose of defining a quantitative metric of robustness, we need to measure the distance or difference between the two distribution functions properly.

How can we define the distance between two distribution functions so that it is suitable to measure the performance robustness? A simple way is to adopt one of the existing distributional testing statistics, such as the Kolmogorov–Smirnov test statistic, χ^2 test statistic, Student-t test statistic, or even the Kullback–Leibler distance. However, we find that none of these statistics is suitable to describe robustness for communication network performance.

We do not choose χ^2 and Student-t tests because both are based on Gaussian normal distributions. It is not necessary to assume that the distribution functions involved are normal. In [14], a robustness metric has been proposed based on the Kolmogorov–Smirnov statistic. Similar to the Anderson–Darling goodness-of-fit test, the metric defined by England et al. [14] gives more weight to the tails than does the Kolmogorov–Smirnov statistic, by multiplying a specific weighting function. The application of that metric is limited. First, the Kolmogorov–Smirnov statistic is for continuous distributions. It is difficult to generalize to include discrete distributions. For example, if two distributions, F_1 and F_2, concentrate on two different constants, no matter how far or how near these two constants are, the Kolmogorov–Smirnov statistic always gives the following:

$$\sup_{-\infty < x < \infty} \{|F_1(x) - F_2(x)|\} = 1. \tag{6.1}$$

Second, even for continuous distributions, the Kolmogorov–Smirnov statistic is not sharp. For example, let F_1, F_2, and F_3 be uniform distributions on [0,1], [1,2], and [2,3], respectively. For measuring the robustness, we believe that F_2 is closer to F_1 than F_3 is to F_1. But the Kolmogorov–Smirnov statistic cannot tell the difference. The weighting functions introduced in [14] do not help in these cases. As for the Kullback–Leibler distance, which is used more commonly in information theory, it is not a real distance between two distributions, and it may not exist, which is equal to infinity even when two distributions are very close to each other. Consequently, the Kullback–Leibler distance is also not suitable for our purposes.

Another way to quantify the distance between two distributions is to take the total variation of the difference between the two distribution functions. Unfortunately, although it is much more complicated in calculation, there is no substantial difference in such a defined distance from that of the Kolmogorov–Smirnov statistic, with all the shortcomings given earlier. A better way of quantifying the distance between the distribution functions is proposed in [13] as follows.

Given a probability space $(\Omega,\boldsymbol{F},\boldsymbol{P})$, the q-distance for two random variables, X and Y, is given as follows:

$$d_p = \left[\int_\Omega |X(\omega) - Y(\omega)|^q \, d\boldsymbol{P}\right]^{1/q}, \tag{6.2}$$

where $q \geq 1$. In (6.2) it is assumed that the q-moments of X and Y exist. We take $q = 2$, since the 2-distance is a natural extension of the Euclidean distance, and it is convenient for analytical calculation. For ease of notation, we use $d(X,Y)$ to represent $d_2(X,Y)$ hereafter.

Given a distribution function F, we define a function ψ_F on [0,1] as $\psi_F = \sup\{u|F(u) \leq t\}$. If F is the uniform distribution on the interval $[a,b]$, then $\psi_F(t) = a + (b - a) \times t$. If F stands for a probability that concentrates on a constant c, then $\psi_F(t) = c$. Then ψ_F is a random variable on ([0,1], $\boldsymbol{F}$,m), where $\boldsymbol{F}$ is the Borel σ-field of [0,1] and m is the Lebesgue measure (uniform distribution) restricted on [0,1]. We call ψ_F the naturally induced random variable of F. If two distribution functions F_1 and F_2 are different, the naturally induced random variables ψ_{F1} and ψ_{F2} are different. For a naturally induced random variable we have the following property.

Lemma 6.1 Assume that F is the distribution function of a random variable X. Then we have

$$E\,|X|^q = \int_0^1 |\psi_F(u)|\,du = E\,|\psi_F|^q \tag{6.3}$$

for any q that the expectation $E|X|^q$ exists.

For simplicity, we verify Lemma 6.1 for the case that F has a density function f. In fact, the lemma should hold for all kinds of distribution functions whenever $E|X|^q$ exists.

Proof: Assume that $f(x)$ is the density function of F. By definition,

$$\psi_F(u) = \left\{ x \,\middle|\, \left| \int_{-\infty}^{x} f(v)\,dv = u \right| \right\}. \tag{6.4}$$

We have

$$E\,|\psi_F|^q = \int_0^1 |\psi_F(u)|\,du. \tag{6.5}$$

Using the variable transformation

$$u = \int_{-\infty}^{x} f(v)\,dv \tag{6.6}$$

in the right-hand-side integration, we get

$$\begin{aligned} E\,|\psi_F|^q &= \int_0^1 |\psi_F(u)|\,du \\ &= \int_{-\infty}^{\infty} |x|^q f(x)\,dx \\ &= E|x|^q. \end{aligned} \tag{6.7}$$

As such, Lemma 6.1 is verified.

Now we can define the distance between distribution functions.

Definition 6.1 Given two distribution functions F_1 and F_2, if the second moment of both ψ_{F1} and ψ_{F2} exist, we define the distance between F_1 and F_2 as follows:

$$d(F_1, F_2) = \left[\int_0^1 |\psi_{F1}(t) - \psi_{F2}(t)|\,dt \right]^{1/2}. \tag{6.8}$$

From Definition 6.1 we see that if F_1 and F_2 are, respectively, the distribution functions for random variables X and $X+c$, where c is a constant, then $d(F_1,F_2) = |c|$. In fact, by the definition of naturally induced random variables, we have $\psi_{F2} = \psi_{F1}(t) + c$ in this case. If F_1 is a distribution of a random variable X and F_2 concentrates on a constant c, then by Lemma 6.1,

$$
\begin{aligned}
[d(F_1,F_2)]^2 &= \int_{-\infty}^{\infty} |x-c|^2 \, dF_1 \\
&= E\,|X-c|^2 . \qquad (6.9)
\end{aligned}
$$

In particular, if $c = E|X|$, $[d(F_1,F_2)]^2 = E|X - E|X||^2$, which is the variance of X.

A network performance metric such as packet delay, throughput, or packet loss ratio is usually a positive (or nonnegative) random variable. With the distance between two distributions of random variable defined above, one can now define the metric of performance robustness. For the definition, we suggest the following three rules:

1. Robustness is a decreasing function of the distance between the network performance observed and its performance reference.
2. Robustness is dimensionless.
3. The value of robustness is within [0,1], and we normalize robustness such that its value is 1 when no disturbance exists.

Rule 1 is intuitive because the communication network is considered more robust with a smaller distance between the performance observed and the performance reference. The importance of rule 2 is highlighted through the following illustration. By rules 1 and 3 we may, for example, define the robustness metric as follows: $(1 + d(F^*,F))^{-1}$, where F^* is the distribution of the performance reference. Also, F is the distribution of the actual observed performance when there is disturbance. If $F = F^*$, we get 1.0 as the value robustness metric, which stands for the best robustness. From $(1+d(F^*,F))^{-1}$, while the distance $d(F^*,F)$ is increasing, robustness is dropping. Note that the value of robustness defined in this way will depend on the dimension of the random variables under consideration. For example, if we take packet delay as the performance, the value of such a robustness metric would depend on the time unit. Assume that we use milliseconds as the time unit and get the distance $d(F^*,F) = 1$; this leads to a robustness of 0.5. If we use seconds as the time unit, we would have $d(F^*,F) = 0.001$ and the robustness would be 0.999. This is not quite reasonable because the value of robustness changes abruptly despite the absence of any change in the real performance. Therefore, rule 2 is important.

Let F^0 denote the distribution of a random variable that takes the value 0 with probability 1. Then F^0, F^*, and F form the endpoints, which are actually distributions that are three vertices of a triangle. We need to measure how close or how far is the separation between the two sides of the triangle that have a common endpoint at F^0. To do this, we use the following quantity: $d(F^*,F)/(d(F^0,F^*) + d(F^0,F))$, which is the ratio of the third side of the triangle to the sum of the other two sides under consideration. The quantity $d(F^*,F)/(d(F^0,F^*)+d(F^0,F))$ is increasing with respect to $d(F^*,F)$, dimensionless, and within the interval [0,1]. This suggests that we may eventually choose $1-d(F^*,F)/(d(F^0,F^*)+d(F^0,F))$, as the robustness metric. Next, we use a few simple examples to illustrate that it is reasonable.

First, we consider network throughput as the performance. We assume that the reference network throughput which is obtained when there is no disturbance to the network at all, is a constant a. When there is disturbance, the throughput is another constant b ($b < a$). In this case, we have $d(F^0,F^*) = a$, $d(F^0,F) = b$, and $d(F^*,F) = a - b$. Then $1 - d(F^*,F)/(d(F^0,F^*)+d(F^0,F)) = 1-(a-b)/(a+b) = 2b/(a+b)$. In this situation the ratio b/a, which measures the proportion of throughput the network can really achieve compared to the ideal case, is the best metric to evaluate robustness. But it is difficult to extend it to a general case. Actually, the quantity $2b/(a+b)$ can reflect the ratio b/a correctly. In fact, if we choose b/a in the example above, $b/a = 1-(a-b)/a$ would suggest a quantity $1-d(F^*, F)/d(F^0,F^*)$. This quantity is not suitable to, for example, the case that packet delay is chosen as a performance metric.

Assume that the packet delay is a constant a when there is no disturbance to the network. The packet delay is another constant b ($b > a$) when there is disturbance to the network. The quantity $1-d(F^*,F)/d(F^0,F^*)$ in this case is $1-(b-a)/a = (2a-b)/a$, which results in a negative value when $b > 2a$. As such, rule 3 is not satisfied by $1-d(F^*,F)/d(F^0,F^*)$.

One may alternatively choose a/b as a robustness metric in this case. This time the ratio $(b-a)/b$ reflects the proportion of increase in packet delay. But with a similar reason given above for the case of network throughput against packet delay, the quantity $1-d(F^*,F)/d(F^0,F^*)$, which is the general form of $1-(b-a)/b$, cannot apply to the case of throughput. The quantity $1-d(F^*,F)/d(F^0,F^*)$ takes value $2a/(a+b)$ this time. Once again, it can reflect the ratio a/b correctly.

From the discussion above we see that the quantity $1-d(F^*,F)/d(F^0,F^*)$ satisfies rules 1 to 3, and it can reflect robustness well in the cases of network throughput and packet delay. Now, we give the following definition of robustness metric.

Definition 6.2 Assume that the performance reference is given by a distribution function F^*. The performance observed when there is disturbance $\boldsymbol{D}$ appears as a random variable $\boldsymbol{Y}(\boldsymbol{D})$ with distribution function $F^{\boldsymbol{D}}$. We define robustness as follows: $\boldsymbol{R}(\boldsymbol{D}) = 1-d(F^*,F^D)/(d(F^0,F^*)+d(F^0,F^D))$.

Remarks

1. To define the robustness metric, we adopt the 2-distance (i.e., we choose $q = 2$ in Definition 6.1). We need to assume that the second moment of the random variables considered exists. We believe that this assumption is natural.
2. In Definition 6.2 we consider robustness as a function of disturbance level $\boldsymbol{D}$, where $\boldsymbol{D}$ can be one- or multidimensional, according to the number of disturbance factors. Possibly, in some network systems the disturbance may not be quantified.

Let ψ_F be the naturally induced random variable of F. Then $d(F^0,F) = E[\psi_F^2]^{1/2}$. If F is the distribution of a random variable X, then by Lemma 1, $d(F^0,F) = E[X^2]^{1/2}$.

In practice, we may not know the distribution functions F^* and F^D, but two sets of data. As such, we need a method to calculate robustness from the data. First, we assume that we know that F^* and $\boldsymbol{Y}(\boldsymbol{D})$ appears as a set $y_1, y_2, \ldots, y_n$. In this case, by Lemma 6.1, we use $1/n \sum_{i=1}^{n} y_i^2$ to estimate $d(F^0,F)$. Furthermore, the empirical distribution of $\boldsymbol{Y}(\boldsymbol{D})$ is given by

$$\hat{F}^D(x) = \begin{cases} 0, & x < \hat{y}_1 \\ i/n, & \hat{y}_i \leq x < \hat{y}_{i+1}, \qquad i = 1, 2, \ldots, n-1 \\ 1, & x \geq \hat{y}_n, \end{cases} \tag{6.10}$$

where $\hat{y}_1 \leq \hat{y}_2 \leq \cdots \leq \hat{y}_n$ is obtained by reordering $\{y_1, y_2, \ldots, y_n\}$. We use the distance $d(F^*, \hat{F}^D)$ as an estimation of the distance $d(F^*,F^D)$. As such, $d(F^*, \hat{F}^D)$ is computed by

$$d(F^*, \hat{F}^D) = \left[\int_{-\infty}^{\hat{y}_1} |x - \hat{y}_1|^2 dF^* + \sum_{i=1}^{n-1} \int_{\hat{y}_i}^{\hat{y}_{i+1}} |x - \hat{y}_{i+1}|^2 dF^* + \int_{\hat{y}_n}^{\infty 1} |x - \hat{y}_n|^2 dF^*\right]^{1/2}. \tag{6.11}$$

If F^* also appears as a set of data $\{x_1, x_2, \ldots, x_m\}$, we use $1/m \sum_{i=1}^{m} x_i^2$ to replace $d(F^0,F^*)$ and use the distance between the two empirical distributions $d(\hat{F}^*,\hat{F}^D)$ to replace $d(F^*,F^D)$. When $m = n$, we have

$$d(\hat{F}^*,\hat{F}^D) = \left[\frac{1}{n}\sum_{i=1}^{n} (\hat{x}_i - \hat{y}_i)^2\right]^{1/2}. \tag{6.12}$$

In case $m \neq n$, let $l = mq = nr$ be a common multiple of m and n; we can make two new data sets by repeating each x_i q times and each y_j r times. These new data sets would give the same empirical distributions as the old ones. Then we calculate $d(\hat{F}^*,\hat{F}^D)$ by the two new data sets.

6.3.2 Scalable Routing Protocol

In an adaptive routing protocol, detecting network changes and creating a new path in response to the changes to keep the network performance always near-optimal require exchanges of control messages between nodes. These control messages are routing protocol overheads and consume the limited wireless bandwidth. Any bandwidth used to transmit the control messages reduces the available bandwidth for data. The lower the protocol overhead, the more efficient the routing protocol. Thus, the protocol overhead must be kept at a low level, but it will increase logically with the number of nodes and frequency of changes in the network. The ability to keep protocol overhead at an acceptably low level while some network parameters, such as node number, traffic load, and frequency of network change, increase to a very large value is called *scalability* [15].

In short, a routing protocol is scalable if its overhead can consistently be kept at an acceptable level. Let λ_i (where $i = 1, 2, \ldots, n$) be the ith network parameter, which can potentially affect the scalability, and there are a total of n parameters. Following [16], the minimum traffic load of a network is defined as the minimum amount of bandwidth required to forward packets over the path with the smallest hop count (see Section 6.4.1), assuming that all the nodes have instantaneous a priori full topology information. The defined minimum traffic load $T(\lambda_1, \lambda_2, \ldots, \lambda_n)$ is a function of network parameters λ_i. Based on the minimum traffic load, one may define a scalability factor $\theta(\lambda_i)$ for a network with respect to a given parameter λ_i as follows [16]:

$$\theta(\lambda_i) \equiv \lim_{\lambda_i \to \infty} \frac{\log T(\lambda_1, \lambda_2, \ldots, \lambda_n)}{\log \lambda_i}. \tag{6.13}$$

Further, the protocol overhead induced by a routing protocol is defined as the difference between the total amount of bandwidth actually consumed by the network running such a protocol and the minimum traffic load of the network. Similar to $T(\lambda_1, \lambda_2, \ldots, \lambda_n)$, the protocol overhead of a protocol j, which is $X_j(\lambda_1, \lambda_2, \ldots, \lambda_n)$, is also a function of network parameters λ_i. For protocol j, one may define a scalability factor $\rho_j(\lambda_i)$ for it with respect to a given parameter λ_i as follows [16]:

$$\rho_j(\lambda_i) \equiv \lim_{\lambda_i \to \infty} \frac{\log X_j(\lambda_1, \lambda_2, \ldots, \lambda_n)}{\log \lambda_i}. \tag{6.14}$$

As quantified in [16], a routing protocol is said to be scalable with respect to a network parameter if and only if, as the network parameter increases, the total protocol overhead induced by the protocol does not increase faster than the network's minimum traffic load. Therefore, for a network with $\theta(\lambda_i)$, a protocol j is scalable with respect to λ_i if and only if the following condition is satisfied:

$$\rho_j(\lambda_i) \leq \theta(\lambda_i). \tag{6.15}$$

From the above, a routing protocol that is scalable with respect to a given network parameter may not be scalable with respect to another network parameter. Also, the routing protocol scalability described above is not directly applicable to judge the scalability of a network. For a network that runs a routing protocol, it is scalable with respect to parameter λ_i if and only if its minimum traffic load does not increase faster than its transport capacity [17]. This is to ensure that the network throughput will not vanish when a scalable routing protocol runs in a scalable network.

As stated above, being adaptive is important for a node to make opportunistic use of the available resources and avoid failure so that the network performance can be improved by always operating at the near-optimal condition. For a routing protocol to adapt to the changes, it needs at least a routing metric to determine which route among all the possible routes is the best. Hence, the routing metric must be designed such that it can accurately capture the dynamic network characteristics and it can be used effectively as a parameter in the routing decision. In Section 6.4 we present a list of possible routing metrics for a wireless mesh network. Given a set of suitable routing metrics, the responsiveness of a routing protocol to a change in the network depends on how fast the change in values of a routing metric can be delivered to the node affected. To make it more responsive, a higher protocol overhead, is required, and hence the scalability of the protocol is affected. The actual size of protocol overhead differs among different types of routing protocols. For example, a proactive routing protocol that periodically sends a control packet to update neighboring nodes of its value of routing metric may consume more bandwidth than will a reactive routing protocol that requests a routing metric value update only when there is a need to set up a new path. Thus, it is important to understand how different types of routing protocols operate, and in Section 6.5 we investigate these different types of routing protocols. In Section 6.6, we describe a few selected routing protocols suitable for multihop wireless mesh networks. The description of routing protocols will serve as an illustration of the general concepts as well as some specific realizations. The chapter ends with a summary in Section 6.7.

6.4 ROUTING METRICS

The routing metric chosen must accurately capture the network dynamics and can be effectively used to make routing decisions to achieve the best performance in terms of the highest throughput and shortest end-to-end packet delay. Network dynamics include changes in path length, traffic load, data rate, error rate, interference, and so on. These parameters are usually calculated on a per-link basis, where a path consists of a set of concatenated links. Thus, for a given per-link parameter, the routing metric of a path is usually but not necessarily the summation of its values for each link of the path. The routing decision is then to choose the minimum weighted path, which is the path with the minimum value of its routing metric [18,19].

Routing metrics can be broadly classified as load sensitive or topology sensitive. *Load-sensitive metrics* assign a weight based on the traffic load on the path. Examples of load-sensitive routing metrics are degree of node activity and number of congested nodes. In general, load-sensitive metrics change frequently, as flows arrive and depart rapidly. On the other hand, *topology-sensitive metrics* assign a weight to a path based on the topological properties of the path, such as the path length, link data rate, and link error rate. In general, topology-sensitive metrics change less frequently in a network with limited node mobility.

6.4.1 Hop Count

The hop count of a path is in practice equal to the path length. It is the number of hops between the two ends of a path. Hop count also equals the number of concatenated links of the path. Thus, as a routing metric, hop count can be determined as follows:

$$\text{hop count}_p = \sum_{l \in p} I_l, \tag{6.16}$$

where l is the candidate link of path p and I_l is an indicator function for the existence of link l, defined as follows:

$$I_l = \begin{cases} 1, & \text{if } l \text{ exists} \\ \infty, & \text{otherwise.} \end{cases} \tag{6.17}$$

Hence, when a path does not exist, its hop count is infinity.

The advantage of using hop count as a routing metric is that it is simple. There is an efficient way to compute the minimum weighted path, which is the path with the smallest hop count; as long as the topology (existence of link between two nodes) is known. There is no need to collect the additional information that is normally needed for other routing metrics. Although it is easy to use, hop count does not give any information with respect to the link quality, load, and interference. It has the tendency to select the path with a large link distance because that will normally result in a small hop count. However, large link distance may be more prone to errors and thus may have a lower data rate.

6.4.2 Cumulative Round-Trip Time

The cumulative round-trip time (CRTT) measures the round-trip delay of a path. It can be taken as the sum of the per-hop round-trip delay of all links that make up the path. As such, the routing metric of a path p can be expressed as follows:

$$\text{CRTT}_p = \sum_{l \in p} \text{RTT}_l, \tag{6.18}$$

where RTT_l is the per-hop round-trip delay for the component link l of a path p.

Let the two ends of a link l be node A and node B, respectively. One way to measure RTT_l is to make node A transmit unicast probes to node B periodically. Each of these probes carries a time stamp by node A indicating the transmitting time. A failed probe will be retransmitted by node A for a maximum number of times before being discarded. Upon receiving the probe, node B will immediately transmit an acknowledgment probe echoing the received probe's time stamp. After receiving the acknowledgment probe from node B, node A can now calculate RTT_l as the difference between the time stamp carried by and the receive time of the acknowledgment probe.

The per-hop round-trip delay measured this way may be affected by traffic load as suggested in [20]. Specifically, a higher traffic load at the receiving node of a probe may cause a longer delay in transmitting the acknowledgment probe, and thus a larger round-trip delay. The traffic load may change rapidly from time to time, due to the frequent arrival and departure of traffic flows. To avoid excessive fluctuation in the value of a routing metric, an exponentially weighted moving average of the per-hop round-trip delay is used as RTT_l of link l. Let $\text{RTT}_{l,\text{new}}$ be the latest measured round-trip delay of link l. Then

$$\text{RTT}_l = (1 - \alpha)\text{RTT}_{l,\text{new}} + \alpha\text{RTT}_l, \tag{6.19}$$

where α $(0 \leq \alpha \leq 1)$ determines the weight by which the RTT_l calculated is affected by the newly measured round-trip delay. In practice, α is a controllable parameter that affects how fast the calculated RTT_l will converge to a new stationary value of round-trip delay if a change in the traffic load is permanent. A smaller α means faster convergence, which is an important characteristic because it is desirable for a routing protocol to adapt quickly to network changes. However, a smaller α also normally leads to a larger error, which is the difference between a given calculated RTT_l and its actual stationary value. Ideally, the error should be minimized so that a routing protocol will not adapt chaotically to transient changes in the network. Thus, adjusting α can lead to two clearly conflicting outcomes that need to be optimized.

Unfortunately, optimizing α to achieve the faster convergence and the smallest error is not trivial in the absence of a formula relating α to the round-trip delay. In practice, α is determined by trial and error for a given target of convergence time and tolerable error. When there is no single value of α that can simultaneously satisfy the convergence time and tolerable error requirements, an alternative is to find two values, α_{time} and α_{error}, one for each of the two requirements [22]. Usually, $\alpha_{\text{time}} < \alpha_{\text{error}}$ for the reasons given above. Then, two round-trip delays, $\text{RTT}_{l,\text{time}}$ and $\text{RTT}_{l,\text{error}}$, can be calculated as follows:

$$\begin{aligned} \text{RTT}_{l,\text{time}} &= (1 - \alpha_{\text{time}})\text{RTT}_{l,\text{new}} + \alpha_{\text{time}}\text{RTT}_{l,\text{time}} \\ \text{RTT}_{l,\text{error}} &= (1 - \alpha_{\text{error}})\text{RTT}_{l,\text{new}} + \alpha_{\text{error}}\text{RTT}_{l,\text{error}}. \end{aligned} \tag{6.20}$$

Subsequently, a flip-flop exponentially weighted moving average can be determined by dynamic selection to use one of the two round-trip delays. One selection method is given as follows:

$$\text{RTT}_l = \begin{cases} \text{RTT}_{l,\text{time}}, & \text{if } \left|\text{RTT}_{l,\text{time}} - \text{RTT}_{l,\text{error}}\right| \leq 0.1\text{RTT}_{l,\text{error}} \\ \text{RTT}_{l,\text{error}}, & \text{otherwise.} \end{cases} \tag{6.21}$$

This method of selection will choose for a faster convergence as long as the error does not exceed 10% of the stationary value.

In addition to traffic load, per-hop round-trip delay also depends implicitly on the link data rate. A higher data rate will lead to a lower transmission time for both the probe and the acknowledgment probe, and thus a shorter round-trip time. However, the round-trip delay calculation given above is not affected by the link error rate. This is because RTT_l is updated only when there is a new $\mathrm{RTT}_{l,\mathrm{new}}$, and a higher error rate simply results in fewer updates. To take the link error rate into account, RTT_l can be increased by a certain factor $\beta (0 \leq \beta \leq 1)$ when a probe is transmitted but no corrsponding acknowledgment probe is received. This method of calculating RTT_l is given as follows:

$$\mathrm{RTT}_l = \begin{cases} \alpha\, \mathrm{RTT}_{l,\mathrm{new}} + (1-\alpha)\mathrm{RTT}_l, & \text{if acknowledgment probe is received} \\ (1+\beta) \times \mathrm{RTT}_l, & \text{otherwise.} \end{cases} \tag{6.22}$$

When the error rate is low, RTT_l depends mainly on both the link data rate and the traffic load at a node.

6.4.3 Cumulative Expected Transmission Count and Time

The cumulative expected transmission count (CETX) measures the average number of transmissions required by the sender to deliver a packet successfully over a path across multiple links. The number of transmissions includes retransmissions of failed packets. The routing metric is calculated as the sum of the expected transmission count of all the component links of a path. Let ETX_l be the expected transmission count of link l. Then the routing metric for a path p can be determined as follows:

$$\mathrm{CETX}_p = \sum_{l \in p} \mathrm{ETX}_l. \tag{6.23}$$

Calculation of ETX_l was proposed originally in [23]. Consider each link that connects any two nodes: say, node A to node B. This consists of a forward link and a reversed link. The forward link is used by the sender, node A, to send data packets to the receiver, node B; the reversed link is used by node B to send the corresponding link-layer acknowledgments of the data packets back to node A. For a given link l, Conto et al. [23] have suggested that ETX_l depends on both the forward and reversed link because a data packet transmitted successfully on the forward link will still be retransmitted in the absence of a acknowledgment received successfully on the reversed link. Let $p_{f,l}$ and $p_{r,l}$ be the packet error probabilities of the forward and reversed links, respectively. Then

$$\mathrm{ETX}_l = \frac{1}{(1-p_{f,l})(1-p_{r,l})}. \tag{6.24}$$

In equation (6.24), $p_{f,l}$ and $p_{r,l}$ can be determined, respectively, by nodes A and B through active probing. Compared to passive probing that requires no injection of additional probe traffic into the network but depends on existing data packets acting as probes, active probing will send explicit probe packets into the network. Thus, active probing will increase the traffic load but has better control on the time to transmit probes. The timing control is important in accounting for silent failures: where the receiver of a probe cannot detect a failed transmission that appears only as noise.

One way to realize active probing is for a node to send broadcast probes which are different from unicast probes. A failed broadcast probe will not be retransmitted. Practically, a node will send probes periodically once every τ seconds and its neighboring nodes will keep track of the number of probes received from the node over a certain time interval, T ($\geq 2\ \tau$) seconds. In this case, node A (node B) will count the number of probes received from node B (node A). At the end of each time interval T, this count of probes received will be carried in the probe sent by the node so that the sender of the probe can be informed of the number. For example, the probes from node A will carry the count of probes received from node B, and vice versa. Say, for the time interval T, node A has transmitted $n = T/\tau$ probes and node B reports receiving only m ($\leq n$) probes from it. Upon receipt of the report from node B, node A determines $p_{f,l}$ using the window mean with exponentially weighted moving average as follows:

$$p_{f,l} = (1 - \alpha)\frac{mT}{\tau} + \alpha p_{f,l} \tag{6.25}$$

Note that the new value (mT/τ) is an average over the time window T. According to [24], window mean with exponentially weighted moving average is better than the simple exponentially weighted moving average in terms of faster convergence to a new stationary value. This is because the window average is, in fact, a form of loss-pass filtering that can remove high-frequency fluctuations.

With the method described above, nodes A and B can both compute their values of $p_{f,l}$. Then, for link l connecting node A to node B, $p_{r,l}$ at node A is the value of $p_{f,l}$ as determined and reported by node B. The method of determining $p_{f,l}$ and $p_{r,l}$ based on counting the number of probes received at the link layer is not perfect. This is because the packet error rate seen by the probes may be different from the packet error rate of data packet, as the probes are much smaller than data packets.

In practice, $p_{f,l}$ and $p_{r,l}$ may vary due to propagation impairments. Thus, the error rates can also be determined by measuring the signal strength or signal/noise ratio (SNR) received. Instead of merely counting the number of probes received within a time interval, a node can calculate and report the simple moving average of SNR for these probes. Let $\mathrm{SNR}_{f,l}(i)$ be the SNR for the ith probe received by node B on the forward path of link l that connects node A to node B. Consider a given time interval T within which the number of probes m received cannot exceed the number of probes n expected. Then the average SNR for the forward link for the given time interval is determined as

$$\mathrm{SNR}_{f,l} = \frac{\sum_{i=1}^{m} \mathrm{SNR}_{f,l}(i)}{n}. \tag{6.26}$$

With the $\mathrm{SNR}_{f,l}$ calculated, the bit error rate for the forward link can be determined for a given modulation scheme. For example, with binary phase-shift-keying modulation, the bit error rate is given by $\text{BER}_{f,l} = 0.5\,\text{erfc}(\sqrt{\text{SNR}_{f,l}})$, where $\text{erfc}(\cdot)$ is the complementary error function. Then $p_{f,l}$ can be determined as

$$p_{f,l} = (1 - \text{BER}_{f,l})^L, \tag{6.27}$$

where L is the fixed size of a data packet as generated at node A. Although this method takes into account the size of a data packet, a change in packet size and modulation scheme will result in a change in $p_{f,l}$.

When $p_{f,l}$ and $p_{r,l}$ cannot be computed using any of the methods described above, ETX_l can be determined by having the sender keep a counter to track the number of retransmissions before receiving an acknowledgment successfully from the receiver. In this case, the transmission count for each data packet is the counter value, which is reset to zero for each new data packet transmitted. Therefore, ETX_l is the average value of transmission counts for all data packets.

The time taken to deliver a packet successfully is proportional to ETX_l. Hence, as suggested in [21], the expected transmission time (ETT) of a packet can be determined as follows:

$$\begin{aligned} \text{ETT}_l &= \frac{L_d}{(1 - p_{f,l})(1 - p_{r,l})R_{f,l}} \\ &= \text{ETX}_l \frac{L_d}{R_{f,l}}, \end{aligned} \tag{6.28}$$

where L_d is the size of a data packet and $R_{f,l}$ is the data rate on the forward link. The link data rate can be determined practically by sending two back-to-back packets over the link. The first data packet is small and the second packet is relatively large. The time difference between the receptions of the two packets is recorded. Then the link data rate can be calculated by dividing the size of the larger packet by the recorded time difference. With ETT_l calculated above, another routing metric, called the *cumulative expected transmission time* (CETT), can be defined as follows:

$$\text{CETT}_p = \sum_{l \in p} \text{ETT}_l. \tag{6.29}$$

As a routing metric based only on transmission time, CETT does not include the different kinds of delay, such as queuing delay and access delay. The queuing delay is affected by traffic load at the sender nodes, and the access delay depends on the scheduling scheme employed in the medium access control protocol. Specifically, CETT does not take into account the time taken between two consecutive retransmission attempts at the sender node. Also, CETT does not consider the time taken at the receiver node between reception of a probe and subsequent transmission of a corresponding acknowledgment probe. As such, ETT is practically a link performance measure that depends on error rate and data rate, but not traffic load.

6.4.4 Monotonicity and Isotonicity of Routing Metrics

Many routing protocols, such as the pure path (or distance) vector protocols and the standard link-state protocols (see Section 6.5.3), essentially use a form of the Bellman–Ford algorithm or Dijkstra's algorithm to compute and select the minimum weighted path. Here the weight of a path is the value of the routing metric used by the protocol. For example, if a routing protocol uses path length as its routing metric and a path has three hops, the weight of the path is 3. It is usual for a routing protocol to use a routing metric which ensures that its minimum weighted path can deliver the best network performance. Thus, the minimum weighted path is also the optimal performance path. However, the routing metric that ensures the best network performance at the minimum weighted path may not guarantee that the minimum weighted path can be found efficiently. For Bellman–Ford's and Dijkstra's algorithms, the routing metric must possess two properties, monotonicity and isotonocity [25,26], to ensure that the minimum weighted path can be computed efficiently. If the routing metric is not monotone and isotone, only algorithm with exponential complexity can compute the minimum weighted path, and thus the path computation may become intractable even for a small network. The four routing metrics, hop count, CRTT, CETX, and CETT, described earlier in this section are monotone and isotone.

Monotonicity means that the weight of a path cannot decrease when the path is extended. *Isotonicity* means that the relationship between the weights of any two paths with the same origin is preserved when both are extended to the same node. For ease of exposition, consider a path a with its weight $f(a)$. Further define an operation $\oplus$ that denotes the concatenation of two paths such that the concatenation of path a and path b is given by $a \oplus b$. We call $a \oplus b$ an *extended path* of path a or path b. Monotonicity implies that $f(a) \leq f(a \oplus b)$ and $f(b) \leq f(a \oplus b)$. On the other hand, isotonicity implies that if $f(a) \leq f(b)$, then $f(a \oplus c) \leq f(b \oplus c)$ and $f(c \oplus a) \leq f(c \oplus b)$, where c is any third path other than a and b.

6.5 CLASSIFICATION OF ROUTING PROTOCOLS

The task of a routing protocol is to facilitate the establishment of a new path between a source node and a destination node as long as at least one path exists between the two nodes. To perform the task, a routing protocol needs to do two things: (1) select the best path based on its routing metrics, and (2) disseminate network information required in path selection. In a wireless mesh network, the routing protocols can be classified according to how they perform path selection and disseminate network information.

6.5.1 Topology-Based Protocols Versus Position-Based Protocols

Topology-based routing protocols select paths based on topological information such as link status between nodes and path length. In the literature, this class of routing protocols requires a path to be set up before a packet can be sent from a source node to its destination node. Examples of topology-based routing protocols include ad hoc distance vector (AODV) [27], optimized link state routing (OLSR) [28], and dynamic source routing (DSR) [29].

Different from topology-based routing protocols, *position-based routing protocols* [30] do not require a path to be set up between a source node and its destination node before the source node sends its packet. Instead, it makes forwarding decisions on a packet-by-packet basis, and two consecutive packets may be forwarded along two different paths. The packet-forwarding decision is based on the geographical positions of the forwarding node, of its neighbors, and of the destination node. As such, position-based routing protocols are also called *geographic routing protocols*. In position-based protocols, nodes are referred to by their physical or geographical positions, not by their network addresses. The position information of a node is provided by a location service unit, which may not be an integral part of the routing protocol.

Given the position information, a simple packet-forwarding algorithm can be used to determine the next hop for each packet, where the next hop is selected such that the packet will be closest to its destination node. Unfortunately, this simple algorithm may cause a packet to get stuck in a local minimum and never reach its destination although a path to the destination does exist. In view of the local minimum problem, there is need for a fallback strategy that can help to move a stuck packet out. One method is to segment a network into multiple faces where links in different faces do not cross into each other. Packets can proceed out of a local minimum by going around these faces toward the destination [31]. Examples of position-based routing protocols are location-aided routing (LAR) [33] and greedy perimeter stateless routing (GPSR) [31]. GPSR combines greedy forwarding with face routing as fallback.

6.5.2 Proactive Protocols Versus Reactive Protocols

Proactive routing protocols are also called *table-driven routing protocols*. This class of routing protocols maintains a path to all other nodes in the network even though the paths maintained are not needed or in use. The proactive maintenance of all routes is done by sending periodic routing control messages to update the link status and the routing tables at nodes. As such, there is no route acquisition latency when a new path is needed. However, the maintenance of the routing information of all available routes at all times can incur a larger control overhead. This is not productive, especially when many of the paths maintained are never used. In proactive routing protocols, the frequency of the periodic control messages is critical in capturing the network dynamics accurately. When the update frequency is not high enough, there is a possibility of using a stale or obsolete path, resulting in packet losses. Examples of proactive routing protocols are OLSR [28], topology broadcast based on reverse-path forwarding (TBRPF) [34], and DSDV [35].

Reactive routing protocols are also known as *on-demand routing protocols*. Reactive protocols do not send periodic control messages to update link states for all available paths. Instead, they compute a route only when it is needed and maintain only the paths that are currently in use. This reduces the control overhead when

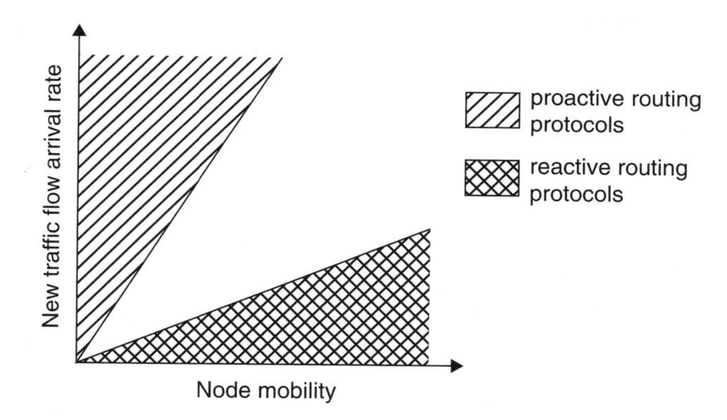

FIGURE 6.4 Design space for proactive and reactive routing protocols. (From [37].)

only a small subset of all available paths is in use at any one time. The reduction of routing control overhead remains the key motivation behind the design of on-demand protocols because a high overhead usually has a significant performance impact in low-bandwidth wireless links. However, setting up a path on demand introduces route acquisition latency because a route discovery process must be performed to set up a new path before a packet can be transmitted. The route acquisition latency may lead to a large delay for the initial packets awaiting to be sent when the route is being set up. Examples of reactive routing protocols are the temporally ordered routing algorithm (TORA) [36], AODV, and DSR.

In a multihop wireless network with low mobility, proactive routing protocols may perform better than reactive routing protocols because there will be little link breakage. Thus, routing control updates can be sent less frequently while retaining the advantage of minimum route acquisition latency. As illustrated in Figure 6.4, the benefit of proactive routing protocol is especially obvious with a high new traffic flow arrival rate, where each new flow may cause reactive routing protocols to initiate a separate on-demand path setup.

Hybrid routing protocols try to combine the advantages of both proactive and reactive routing protocols. In one concept, called *multiscoping*, information from a nearby node will be used more frequently and be more valuable compared to information from a distant node. As such, a hybrid routing protocol may proactively maintain a routing table at each node for all paths within a limited-size local neighborhood while setting up routes reactively for destination nodes outside the local neighborhood. An example of a hybrid protocol is the zone routing protocol (ZRP) [38]. In ZRP, each node proactively maintains all paths for a neighborhood of k hops. Within the neighborhood, a path is readily available when a node wants to send a packet. When a node wants to send a packet to a destination outside the k-hop neighborhood, it sends a route request to its border nodes, which are nodes at the edge of the neighborhood. These border nodes will know if the destination node is within their own k-hop neighborhood and decide either to reply to the route request or forward

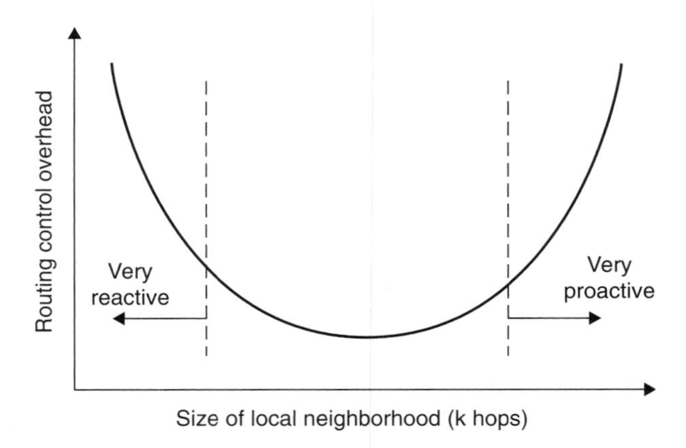

FIGURE 6.5 Optimal zone radius for ZRP, which is neither very proactive nor very reactive. (From [37].)

the route request to their respective border nodes. This forwarding of route request will continue until a path to the destination node is found. In ZRP, one challenge is to determine the size of the local neighborhood k. Note that $k = 0$ signifies a purely reactive protocol, and $k = \infty$, a purely proactive protocol. If k is too small, ZRP may not be proactive enough. On the other hand, if k is too big, excessive control overhead in maintaining too many unused routes may result. Figure 6.5 suggests that there exists an optimal k where the routing control overhead is neither predominantly reactive nor predominantly proactive. The optimal value of k may be dependent on the network traffic within the neighborhood. Besides a nonuniform network where the node density can differ significantly around different nodes, network traffic experienced by different nodes can be different, thus affecting the optimal value of k. Thus, [37] proposes that nodes in ZRP can independently determine their own optimal size of local neighborhood.

6.5.3 Distance Vector Protocols Versus Link State Protocols

Distance vector routing protocols require each node to broadcast to all its immediate neighbor nodes its distance to all other nodes in the network. Based on the distance information received from all neighbors, a node may select the shortest path to each node in the network. It is not necessary that the broadcasts to neighbor nodes contain only distance information. As a matter of fact, the broadcast may carry routing metrics such as CRTT, CETX, or CETT other than distance (path length or hop count), as described in Section 6.4. These other routing metrics can still be used as weights in computing the shortest path. Since the routing metric is not always path length, members of this class of routing protocols are more generally called *path vector routing protocols*. Examples of distance (path) vector routing protocols are AODV and DSDV.

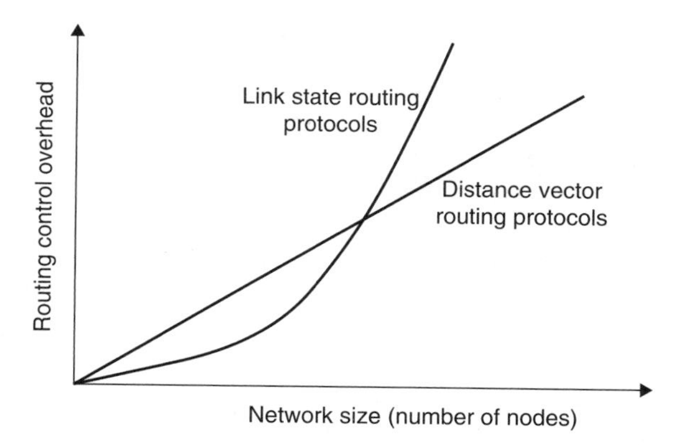

FIGURE 6.6 Routing overhead comparison between link state protocols and distance vector protocols. (From [39].)

Link state routing protocols require each node to broadcast its link status with its immediate neighbor nodes not just to its immediate neighbor nodes but to all other nodes in the network. As such, each node will make a complete picture of the network based on the most recent link information from all other nodes. With the complete network information, a node may select the shortest path to all other nodes in the network. An example of a link state routing protocol is OLSR.

Distance vector routing protocols may have lower routing control overhead than link state routing protocols because they broadcast distance information only to immediate neighbors. Figure 6.6 compares the routing overheads between distance vector and link state routing protocols.

6.5.4 Hop-by-Hop Routing Protocols Versus Source Routing Protocols

Hop-by-hop routing protocols require each node to have a routing table. The table indicates the next hop for the routes to all other nodes in the network. For a packet to reach its destination, the packet only needs to carry the destination address. Intermediate nodes forward the packet along its path based only on the destination address. Due to its simple forwarding and low message overhead, hop-by-hop routing protocols are popular in MANET and wireless mesh networks. Examples of hop-by-hop routing protocols are AODV and OLSR.

Source routing protocols do not need routing tables at all intermediate nodes. Instead, only the source node knows the complete hop-by-hop route to the destination node. These routes are calculated and stored in a route cache at the source node. Before sending a packet, the source node extracts the source route from its route cache and puts the entire path in the header. Intermediate nodes do not have routing tables but only need to forward the packet based on the path information in the packet headers. This leads to a large packet size, which is more vulnerable in a wireless channel with a high error rate. An example of a source routing protocol is DSR.

6.5.5 Flat Protocols Versus Hierarchical Protocols

Thus far, all the routing protocols we have seen in this chapter assume a flat network architecture where all nodes are peers to other nodes through one or multiple hops. In this flat network architecture, each node is obliged to be informed of the changes in network topology that have an effect on its routing table. When the network topology changes frequently, a large amount of routing control traffic will be generated regardless of whether the routing protocol is proactive or reactive. Thus, these flat routing protocols are scalable with respect to network changes in terms of number of nodes, link quality, and node mobility.

In contrast to a flat network architecture, a hierarchy of layers can be imposed on a multihop wireless broadband network where a subset of nodes can be selected to form a virtual wireless backbone. All other nodes that are not part of the virtual backbone may communicate through the virtual backbone. In this case, where there is a change in the network topology, a nonbackbone node only needs to inform its nearest backbone node, and a backbone node only needs to inform other backbone nodes. As such, routing control overhead can be reduced substantially compared to a flat routing protocol [49].

One way to construct the virtual wireless backbone and perform hierarchical routing is to adopt a cluster-based routing concept, where nodes are divided into clusters [52]. Each cluster has a cluster head, and all cluster heads form the virtual backbone. In this cluster-based hierarchical routing protocol, only cluster head nodes need to know the topology information. The cluster head knows the paths to other cluster head nodes and the nodes in its cluster by proactive routing within the cluster. When a cluster head receives a packet from a member node within its cluster, it just looks for the destination in its routing table. If the destination node is a cluster head node, it sends the packet directly along the shortest path. Otherwise, it sends the packet to one of the cluster heads, whose neighbor list includes the destination node. Thus, when a non–cluster head node has a packet to send, it simply sends the packet to its cluster head.

In this hierarchical architecture, it is possible that a cluster head can be connected to other cluster heads only through non–cluster head nodes when all the nodes have a limited transmission range. In this case, the non–cluster head node that links two adjacent clusters is called the *gateway node*. Gateway nodes ensure connectivity between all the clusters in the network.

One of the most challenging problems in a cluster-based hierarchical routing protocol is to form the clusters and select cluster heads efficiently. In the literature, nodes can be clustered and cluster heads can be selected based on node degree [50,51], node ID [52,53], node mobility, and energy.

6.5.6 Single-Path Protocols Versus Multipath Protocols

In all the routing protocols described so far, only one end-to-end path is maintained for a given pair of source and destination nodes. These single-path routing protocols are different from another class of routing protocols that may keep multiple end-to-end paths. These multipath routing protocols are useful in improving the end-to-end reliability of a multihop wireless broadband network. For example, with a multipath routing protocol, when a path is broken, other paths may come in immediately to help in end-to-end transmissions, thus avoiding a service disruption.

An example of multipath routing protocols is ad hoc on-demand multipath distance vector routing (AOMDV) [44], which is a multipath version of the single-path AODV routing protocol. Different from AODV, which maintains only a single end-to-end path, AOMDV maintains multiple link-disjoint end-to-end paths for a given source and destination pair. Although multiple paths are maintained, at any one time only one of the paths is used for packet transmission. Other maintained paths serve as backup. One of the backup paths will be selected to replace the current path when it is no longer suitable in a dynamically changing network environment. Compared to AODV, AOMDV can help to improve network performance by reducing the path rediscovery latency significantly when the current path is broken. This method of using multipath routing as backup has also been proposed [45–47].

In [47], the multipath routing algorithm will form a routing tree with one degree of redundancy, where each node x maintains two parents, default y and backup z, toward the destination D as illustrated in Figure 6.7, and switches dynamically between the default and backup links, depending on the dynamic link quality.

The default parent is selected as the node that has a link with the minimum CETX, where CETX is as described in Section 6.4.3. For a node x, this selection criterion for selecting its default path $P_{x,d}$ is written as

$$P_{x,d} = \arg \max_{i \in R} \left\{ \mathrm{CETX}_{x,i} \right\}, \tag{6.30}$$

where R is the set of all possible routes from node x to the destination D, and $\mathrm{CETX}_{x,i}$ is the CETX for a path i from node x to node D.

Compared to the default path, which is selected as the path with the smallest CETX value among all available paths, the backup path is selected as the path with

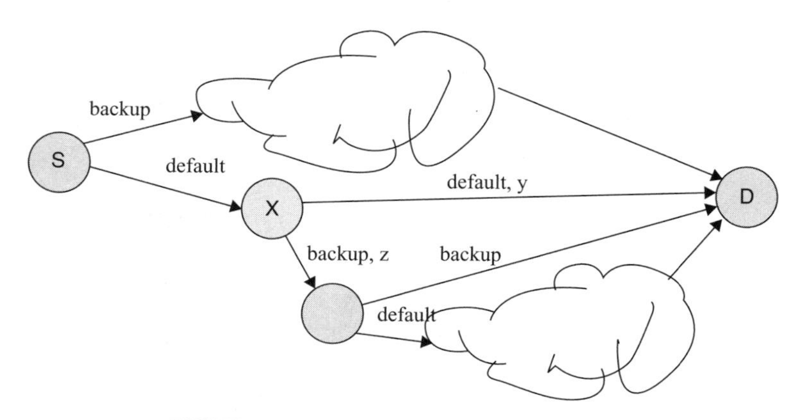

FIGURE 6.7 Multipath routing with backup.

the smallest CETX value among all available links that have a received signal strength above a threshold, S^*. For the node x, the backup path $P_{x,b}$ is selected as follows:

$$P_{x,b} = \arg\max_{i \in R} \left\{ \text{CETX}_{x,i} | (S_i \geq S^*) \text{ and } (i \neq P_{x,d}) \right\}, \tag{6.31}$$

where S_i is the average signal strength received at link i.

With the selection criterion, there is a chance that routing loops may form in the following two cases: (1) if a node selects one of its children as the parent for the backup link, a routing loop is formed when the backup link is turned on; and (2) if node x and node z happen to be the parents of backup link for each other, a routing loop is formed when both backup links are used.

Let the default path from node x be $i = \{\{x, y\}, \{y, \cdot\}, \ldots \{\cdot, D\}\}$. Further define another path, $j = \{\{y, \cdot\}, \ldots \{\cdot, D\}\}$, such that $i = j + \{x, y\}$. Then CETX on the default path i from node x (i.e., $\text{CETX}_{x,d}$) can be written as $\text{CETX}_{x,d} = \text{CETX}_{x,i} = \text{ETX}_a + \text{CETX}_{y,j}$. Note that link a is $\{x, y\}$, which connects nodes x and z.

Let $\text{CETX}_{x,d}$ and $\text{CETX}_{x,b}$ be the CETX values for node x's default and backup paths, respectively. Then a routing loop can be avoided if $\text{CETX}_{x,b} < \text{CETX}_{x,d} + 1.0$ [47]. Therefore, to avoid a routing loop, we have refined and replaced (6.31) with the following criteria for backup parent selection:

$$P_{x,b} = \arg\max_{i \in R} \left\{ \text{CETX}_{x,i} | (S_i \geq S^*) \text{ and } (i \neq P_{x,d}) \text{ and } (\text{CETX}_{x,i} < \text{CETX}_{x,d} + 1.0) \right\}. \tag{6.32}$$

Let the maximal number of retransmissions allowed per hop be k. The multipath routing algorithm in [47] depends on the finite-state machine shown in Figure 6.8 to determine to use either the default link or a backup link. Conceptually, a node will

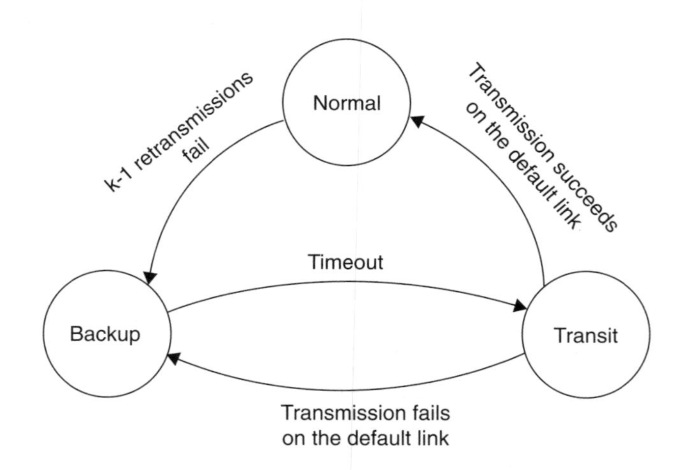

FIGURE 6.8 State transition diagram for multipath backup routing.

begin in the normal state, where the default link is used. For every new packet, the node will count the number of retransmissions attempted for the packet. If the number reaches $k - 1$, the node transitions to the backup state. In the backup state, the backup link is used for the current packet as well as all the subsequent new packets. The node will stay in the backup state for a duration given by $W \times T_0$, where W is the number of time slots and T_0 is the duration of a time slot. At the end of the duration, the node will change to the transit state, within which it will decide to move back to either the normal state or the backup state. The states and their transition conditions are described in detail next.

- *Normal state*. The node uses the default link to transmit its current packet. If the transmission fails, the node will retransmit the packet using the same default link. The node keeps track of the number of transmissions attempted for the current packet. When the transmissions via the default link fails after $k - 1$ retries for the current packet, the node assumes that propagation impairments have affected the default link and switches to the backup state only if a backup link exists.
- *Backup state*. At the entry to the backup state, the node doubles W, which is upper bounded by $W_{\max}$ as follows:

$$W = \min\{2 \times W, W_{\max}\}.$$

  The node switches to the backup link for the current and all subsequent new packets. Doubling W is to avoid sending packets over a link that is blocked for a long time. The maximum value $W_{\max}$ is used to prevent the node from staying too long in the backup state and thus missing the more efficient default link. The node keeps track of the time spent in the backup state. When the time reaches $W \times T_0$, the node switches to the transit state.
- *Transit state*. The node switches to use the default link for the current packet. If the packet transmission on the default link succeeds within $k - 1$ retransmission attempts, the node transits to the normal state after setting $W = W_{\min}$, where $W_{\min}$ is the minimum value of W. The minimum value $W_{\min}$ is useful in ensuring that a node can stay in the backup state long enough to avoid an impaired default link. If retransmission fails after $k - 1$ retries in the transit state, the node transitions back to the backup state.

In practice, the backup link may not be available due either to sparse node density at certain locations or to the fact that no node satisfies the requirements given by (6.32). In such a case, the routing protocol becomes identical to a single-path routing algorithm.

Different from backup routing, Tsirigos and Haas [48] have proposed to use the available multiple paths simultaneously. According to the method proposed in [48], each packet at the source node is coded and divided into multiple smaller packets. These smaller coded packets are sent concurrently over multiple routing paths. At the destination node, it is not necessary to receive all but a subset of the smaller coded packets to reconstruct the original packet. Specifically, as long as the number of smaller coded packets is more than a certain threshold, the original packet can be recovered.

6.6 MANET ROUTING PROTOCOLS

In this section we describe a few selected routing protocols suitable for multihop wireless mesh networks: AODV, DSR, and OLSR. These protocols are chosen based on the fact they are actively pursued in the MANET working group of the Internet Engineering Task Force (IETF) [40]. The working group has been formed to develop a routing framework for ad hoc networks. The description of these routing protocols serves as an illustration of the general concepts as well as some specific realizations.

6.6.1 AODV (Ad Hoc On-Demand Distance Vector)

AODV [27] is a reactive distance vector routing protocol that performs hop-by-hop routing. As a hop-by-hop routing protocol, AODV uses the traditional routing table to maintain routing information at each node, where there is only one entry in the table for each destination. The routing table is used to determine the next hop for each packet locally at each intermediate node. As a reactive protocol, paths are set up on demand, and only paths that are needed or in use are maintained. Specifically, a node does not have to discover and maintain a route to another node until it needs to communicate or participate in a packet transmission. Furthermore, nodes that are not parts of any route that is in use do not need to maintain any routing information or participate in any periodic routing information exchange. This helps in reducing the routing control overhead, but results in a higher route acquisition latency, which will also lead to a higher initial packet delay because all data packets are buffered before a path is acquired or set up.

To set up a new path, AODV uses a simple route discovery process that depends on two-way exchanges of request and reply. The path discovery process is initiated by a source node S when it wants to send data packets to a new destination node D, but it does not have an existing path to node D in its routing table. The source node S broadcasts a route request (RREQ) to its neighbors, which will further broadcast to their respective neighbors the RREQ received. As such, the RREQ will eventually reach the destination node if a path exists between the source and destination nodes. Upon receiving the first RREQ, the destination node will unicast a route reply (RREP) back to the source node along the path traversed by the RREQ received. The RREP sets up the path by updating the routing table at the intermediate nodes with information regarding the path it has traversed. The source node can start transmitting its first packet after receiving the RREP.

The RREQ contains several fields, including the source address, source sequence number, broadcast identity, destination address, destination sequence number, hop count, destination only flag, and gratuitous RREP flag. Source address and destination address are the network addresses for node S and node D, respectively. The source address and broadcast identity pair is used to identify a RREQ uniquely; the source node will increase the broadcast identity by one for every new RREQ issued. It is important to identify a RREQ uniquely so that an intermediate node will not process and not rebroadcast a duplicate RREQ. Thus, a broadcast storm is avoided. The hop count field indicates the number of hops the RREQ has traversed so far. Specifically, the hop count starts with zero in a new RREQ at the source node and is incremented by one at all other nodes that receive the RREQ subsequently. The source and destination sequence numbers are used to ensure freshness of a route such that a new route will not be replaced by an outdated route. The source and destination nodes are responsible for increasing the source and destination sequence numbers, respectively. In general, a new route has a higher sequence number. The source sequence number is used to maintain freshness information about the reverse path to the source node as traversed by the RREQ. The destination sequence number in a RREQ is the last-known destination sequence number at the source node, and it is used as an indicator of how fresh a forward path to the destination node must be before it can be accepted by the source node.

When a node receives a RREQ, it checks the source address and broadcast identity to decide if this is a duplicate RREQ. A duplicate RREQ is simply discarded. For a nonduplicate RREQ, the node will increase the value of its hop count field by one and create a record for the reverse path, indicating the neighbor from which it receives the RREQ. This reverse path record is needed later for the RREP to traverse back to the source node. This record must be maintained long enough for the RREQ to traverse the network and produce a RREP to the source node. The reverse route expiration time may depend on the network size, and Perkins and Royer [27] used 3 seconds for a network of 1000 nodes. Upon timeout of the reverse route expiration time, the reverse path records from those nodes that do not lie on the path from the source to the destination will be purged.

With the hop count incremented and reverse path record created, the node checks if it is the desired destination node and if it has a fresh route to the destination node. A route is considered fresh if its entry in the routing table has a destination sequence number that is not smaller than the destination sequence number carried in the RREQ received. If the node is not the destination node and has no fresh route to the destination node, it will rebroadcast the RREQ to its neighbors if the time to live (TTL) in the RREQ's IP packet header is greater than or equal to 1 after being reduced by one. If the node is the destination node, it will create and unicast a RREP to the source node along the path traversed by the received RREQ. If the node is not the destination node but has a fresh route to the destination node, it checks whether the destination only flag is set. If the flag is not set, the node will unicast a RREP to the source node. In this case, since the RREP is sent by an intermediate node that does not rebroadcast the received RREQ, the destination node will not receive the RREQ and thus does not have a reverse path to the source node. To avoid this situation, the intermediate node that sends the unicast RREP to the source will also send a gratuitous RREP to the destination node if the gratuitous RREP flag is set. This helps to create at the destination node a reverse path to the source node.

A RREP contains several fields, including the source address, destination address, destination sequence number, and hop count. When a node receives for the first time a RREP for a given source–destination pair, it will create an entry in the routing table, setting up a path to the destination node. The hop count field in the RREP, which indicates the number of hops to the destination node, is incremented by one, and the updated RREP will be forwarded to the source node following the reserve path traversed by the corresponding RREQ. As such, the source node will eventually receive a RREP if a path exists to the destination node. For all subsequent RREPs for a given source–destination pair, the node will process and forward the RREP only if it contains either a greater destination sequence number than the previous RREP or has the same destination sequence number but with a smaller hop count.

The routing table entry created at a node upon receipt of a RREP has a lifetime. It is maintained for only as long as the upstream node from which it will receive data packets to the destination node is active. A node is called *active* if it has originated or relayed at least one data packet for that destination node within the most recent active route expiration time. A route entry is considered active if it is in used by any active node. A path is considered active if it uses active route entries. The active route expiration time may depend on the traffic flow characteristic, where heavy traffic will require a shorter expiration. In [27], the expiration time is set as 3 seconds. A route entry that has expired will not be maintained, but removed from the routing table.

In AODV, local connectivity information is provided and maintained by periodic broadcasting of routing protocol messages such as RREQ. If a node has not sent a broadcast message within the last hello interval, it may broadcast a hello message. The hello interval may depend on implementation; [27] sets the interval at 1 second. A hello message is actually a RREP containing its own address as the destination address and has a TTL value of 1. Neighbor nodes that receive this hello message update their connectivity information to the node. Receiving a hello message from a new node or failing to receive a hello message from a previously connected node is an indication of network changes. If an active node fails to receive a hello message from its next-hop node for a few consecutive number of hello intervals, it considers the link between the two nodes broken. In [27], the number of this consecutive hello interval without a hello message has been suggested as 2.

When a link breakage occurs, the node upstream to the broken link checks whether any active route had used this link. If not, nothing will be done. Otherwise, the upstream node may attempt a local repair. It sends out a RREQ to establish a new second half of the path to the destination. The node performing the local repair buffers the data packets while waiting for any RREP in response to its RREQ. If local repair fails or has not been attempted, the node generates a route error (RERR). This error message is an unsolicited RREP with a fresh sequence number and hop count of infinity for all active destinations that have become unreachable because of link breakage. The RERR is sent to all neighbors that are upstream nodes of the unreachable destinations on this node. A node receiving a RERR invalidates the corresponding entries in its routing table. It removes all destinations that do not have the transmitter of the RERR as next-hop from the list of unreachable destinations. If there are upstream nodes to the destinations in this pruned list, the updated RERR message is forwarded to them.

Recent advances in AODV [41] include an optimization technique to control the flooding of RREQ in the route discovery process. This is done by limiting the maximum number of allowed rebroadcasts as indicated by the TTL value in the IP header of the RREQ packets. If no RREP is received in response to the RREQ within a timeout period, the node will retransmit the RREQ with an increased TTL value. The TTL value begins normally from 1 for a new route, and will be increased linearly for each retransmission of RREQ. To be more intelligent, if the route to a previously known destination is needed, rather than 1 the known path length is used as the initial TTL value. This technique of controlling flooding while searching for a new route is called expanding ring search. The danger of employing the expanding ring search technique is that to reach a far-away destination node, several timeouts and RREQ retransmissions with increasing TTL value may have to be performed. The process of waiting for timeout can be time consuming.

6.6.2 OLSR (Optimized Link State Routing)

OLSR [28] is a proactive link state routing protocol that performs hop-by-hop routing. As a hop-by-hop routing protocol, OLSR uses a routing table at each node to store routing information that is used to decide for each packet its next hop to be forwarded to. As a proactive routing protocol, the routing table contains entries for all reachable destinations in the network, including nodes that are currently not communicating. By maintaining all known routes, OLSR can immediately provide a path when a node wants to start a new communication, and thus reduces route acquisition latency as well as initial packet delay. However, this is at the cost of a higher control overhead needed in maintaining the unused paths. The higher control overhead is limited in OLSR through a key concept of optimized broadcast mechanism for the network-wide distribution of the necessary link state information. The optimization is achieved in two ways. First, OLSR reduces flooding of link state control packets. Instead of all nodes that receive a broadcast link state control packet from a neighbor, only selected nodes called *multipoint relays* rebroadcast the control packets. Second, OLSR reduces the size of link state control packets. Instead of declaring information of all links to all neighbors, a node will declare its links only with its neighbors who have selected it as the multipoint relay.

The multipoint relays of a node are a set of one-hop neighbors of that node through which the node can reach all its two-hop neighbors. As mentioned earlier, the neighbors that are not in the multipoint relay set will read and process, but not retransmit, a received broadcast link state control packet. To do this, each node must maintain a list of it neighbors that have selected it as their multipoint relay, and these neighbor nodes are called *multipoint relay selectors*. Then the scheme becomes simple, as a node will only rebroadcast a link state control packet received from its multipoint relay selectors, and the link state control packet contains only link information of a node with its multipoint relay selectors.

From the above it is clear that determining the multipoint relay set is a critical task in OLSR. To facilitate this, each node broadcasts hello messages periodically. The frequency of these hello messages may depend on the node mobility supported.

Each hello message from a node contains a list of addresses of the neighbors to which the node has a valid link and the respective link status. The link status can be unidirectional, bidirectional, or MPR. The link status MPR implies that the link with the neighbor is bidirectional and the neighbor has been selected by this node as its multipoint relay. Thus, based on the hello message received, each node can construct its multipoint relay selector set. Compared to the link state control packets that will be rebroadcast, hello messages received by a node are not rebroadcast but are used to update its local neighbor table. In the neighbor table, each node records its one-hop neighbors, a list of two-hop neighbors that these one-hop neighbors can reach, and the status of the links with these one-hop neighbors. When a one-hop neighbor has been selected by this node as its multipoint relay, the respective link status in the neighbor table will be set to MPR. The content of this local neighbor table is used to construct the periodic hello messages for the node.

Based on the neighbor table, each node can select its own set of multipoint relays independently. In the selection, the only requirement is that the entire two-hop neighborhood of the node will receive broadcast messages only if the multipoint relays selected forward them, and only bidirectional links are considered. The smaller the multipoint relay set, the more optimal is the routing protocol. However, it is not necessary that the multipoint relay set be minimal. In the literature, Clausen and Jacquet [42] have proposed a simple heuristic for the multipoint selection that is described below, but other selection algorithms can also be used.

1. Initialize the multipoint relay set to an empty set. Initialize $N1$ to include all one-hop neighbor nodes. Initialize $N2$ to include all two-hop neighbor nodes. Nodes that already exist in $N1$ are not included in $N2$.
2. Move from $N1$ to the multipoint relay set all the nodes that are the only nodes through which a node in $N2$ can be reached. Remove those nodes from $N2$ which are now covered by at least a node in the multipoint relay set.
3. If $N2$ is an empty set, go to step 6. Otherwise, go to step 4.
4. For each node in $N1$, calculate its reachability, which is the number of nodes in $N2$ that are not yet covered by at least one node in the multipoint relay set and that are reachable through this node in $N1$. Select the $N1$ node with the highest reachability. In case of multiple nodes having the same reachability, select the node with the highest node degree. The node degree of a node is calculated as the number of nodes in its one-hop neighbor, excluding all nodes in the initial $N1$ set. Move the selected node from $N1$ to the multipoint relay set. Remove those nodes from $N2$ which are now covered by a node in the multipoint relay set.
5. Go to step 3.
6. For each node in the multipoint relay set, check if its removal will result in any node in the initial $N2$ set becoming unreachable. If all the nodes in the initial $N2$ set will still be reachable, the node may be removed from the multipoint relay set.

Assume that each node has already determined its multipoint relay set and knows its own multipoint relay selector set. A node with a nonempty multipoint relay selector set will periodically broadcast link state control packets called *topology control messages*. The interval between transmitting two consecutive topology control messages depends on whether the multipoint selector set has changed since the last topology control message. If there is a change, the next topology control message may be broadcast immediately as long as a minimum duration has elapsed since the last topology control message. If there is no change, the next topology control message will be sent only after a predetermined interval, which may depend on the supported node mobility.

Recall that topology control messages are actually link state control messages. Each declares the multipoint relay selector set of its originator and a sequence number which indicates the freshness of the message. A topology control message with a smaller sequence number compared to that of a message received previously from the same node is considered not fresh and will not be accepted. A fresh topology message received by a node will be used to update the topology table of the node. Each entry of the topology table consists of an address of a potential destination, address of a last-hop node to that destination, and the respective sequence number. The potential destination is a multipoint selector in the topology control message received, and the last-hop node is the originator of the topology control message.

The topology table is a used to create entries in an OLSR routing table. Each entry in the routing table consists of destination address, next-hop address, and estimated distance in terms of hops to the destination. The entries are recorded in the table for each known destination for which there is a known path. The following procedure is used to construct the routing table [28]:

1. Clear all existing entries in the routing table.
2. For each entry in the neighbor table whose link status is not unidirectional, create a new entry in the routing table such that the destination and next-hop addresses in the new routing table entry are both set to the address of the neighbor as indicated in the neighbor entry. For each new routing table entry, set the estimated distance to destination to 1.
3. Initialize a parameter h to 1. The parameter represents the estimated distance to the destination. For each entry in the topology table, if its destination address does not match the destination address of any entry in the routing table and its last-hop address corresponds to the destination address of an entry in the routing table with its estimated distance equal to h, create a new entry in the routing table. For the new routing table entry, its destination is set to the destination address as indicated in the topology table entry. Also, its next-hop address is set to the next-hop address of the routing table entry whose destination is equal to the last-hop address of the topology table entry. For the new routing table entry, the estimated distance to destination is set to $h + 1$.
4. If no new routing table entry is created in step 3, the procedure stops. Otherwise, h is increased by one, and steps 3 and 4 are repeated.

All entries of the information repositories in OLSR, which include the neighbor table, topology table, and routing table, have an associated expiration time. This soft-state mechanism provides OLSR with robustness against the loss of control packets.

6.6.3 DSR (Dynamic Source Routing)

DSR [29] is a reactive distance vector routing protocol that does not perform hop-by-hop routing. DSR does not have a routing table at each intermediate node to make local forwarding decision for each packet. Instead, for every packet, the source node constructs a source route in the packet's header. The source route indicates the address of each intermediate node through which the packet must be forwarded in order to reach the destination node. Upon receiving a packet, a node will check if it is the destination node. If the node is not the destination node, it will simply forward the packet to its next hop as identified in the source route carried in the packet's header.

As a reactive routing protocol, DSR sets up and maintain a path only if one is in use or needed. All the paths that have been learned and set up by a node are stored locally in its route cache. When a source node wants to send a packet to a destination node, it will first check its route cache for an existing path. If the node cannot find such a path in its route cache, it will begin to set up a new path by initiating a route discovery process that depends on a two-way exchange of request and relay. Specifically, the source node starts by broadcasting a route request to its neighbor nodes. Each route request consists of fields, including source address, destination address, request identity, and route record. The route record is a sequence of addresses of nodes that have been traversed by the route request on its way to the destination node. The request identity is a unique sequence number maintained locally by the source node. The sequence number is increased by one only for each new route request issued by the source node. Each node maintains a list of <source address, request identity> pairs for all route requests that it has received recently so that the node can detect duplicate route request. When a node receives a route request, it will be processed using the following steps:

1. If the <source address, request identity> pair for the received route request has already appeared in the node's list of recently received route requests, it is simply discarded. Subsequently, go to step 5.
2. If the node's address has already appeared in the route record carried by the route request, the route request received is simply discarded. Subsequently, go to step 5.
3. If the node's address equals the destination address, extract the route record and use it to construct a route reply. Send the constructed route reply to the source node. Subsequently, go to step 5.
4. Append the node's own address to the route record. Subsequently, rebroadcast the route request with an updated route record.
5. End.

In step 2, checking for the existence of a node's own address is useful in detecting routing loops. Simply by ensuring that a node's address does not appear more than once, no other special mechanism is needed to avoid routing loops.

According to the process above, upon receiving the new route request, the neighbor nodes will further rebroadcast the route request. Thus, the route request will reach the destination node eventually if a path exists between the source and destination nodes. This process is similar to AODV. However, instead of setting up reverse paths in the routing tables of the intermediate nodes, these DSR route requests simply collect in a route record the addresses of the traversed nodes on its way to the destination. The destination node will respond to a route request with a route reply that contains the route record carried by the route request. The route reply may or may not traverse along the same path (in reversed order) as traversed by the route request. When the route reply reaches the source node, the route record is extracted and stored in the route cache for immediate and future use.

To send the route reply back to the source node, the destination node must have in its route cache a path to the source node. The route reply will follow the existing path from the route cache. If there is no existing path in the route cache, the destination node may reverse the route record and use it as the path to send the route reply. In the absence of an existing path in the route cache, the destination node is not allowed to buffer the constructed route reply while performing a route discovery process to the source node. This is because neither of the two end nodes has an existing path and will lead to an endless process with potential infinite number of route replies being buffered.

In DSR, route breakage can be detected when a node cannot forward a packet to its next hop after a maximum number of attempts. The node that detects route breakage will send a route error message to notify the source node of the packet. The route error message indicates the addresses of the two end nodes of the broken link. Note that the route error message is sent by the upstream node of the broken link. If the upstream node has a path to the source node in its route cache, that path will be used to send the route error message. If not, the upstream node may reverse the route carried in the packet header and use it to send the route error message. Another option is for the upstream node to buffer the route error message while performing a route discovery process to find a route to the source node. Upon notification by the route error message, the source node removes all routes that use the broken link from its route cache. A new route discovery process must be initiated by the source node if this route is still needed or in use.

DSR makes very aggressive use of source routing and route caching. A forwarding node caches the source route in a packet it forwards for possible future use. With the rich information in the route cache, several improvements on DSR have been proposed to make DSR more efficient [43]. Specifically, an intermediate node can use an alternate route from its own cache when a data packet meets a failed link on its source route. This is a form of local repair to avoid sending route error message. When a route error message cannot be avoided, a node that receives a route error message may piggyback the error message in its subsequent route request. This helps to clean up the caches of other nodes in the network that may have the failed link in one of the cached source routes.

SUMMARY

The special characteristics of multihop wireless broadband networks require and allow optimizations in order to meet the performance goals of different usage scenarios. It is likely that different usage scenarios will call for different optimizations and thus different types of routing protocols. There will not be a single routing protocol that is optimal in every useful scenario. Thus, a mechanism needs to be designed and included in the routing protocol so that it can adapt to changes in the network to achieve robust performance while being scalable and efficient. However, the routing protocol cannot be too adaptive such that the additional overhead incurred is more than the gain in performance improvement.

High network throughput and network capacity are important requirements in practical deployments. New routing metrics have to be designed to take network throughput and network capacity into account so that routes can be selected accordingly. Mobility comes into play when user nodes are mobile. In a mobile environment, heterogeneous devices with different numbers of channels and radio interfaces may exist. For such a heterogeneous multihop wireless broadband network, new routing metrics are needed to take different node capabilities into account. Cross-layer design in routing protocol is important to get better access to the lower protocol layers—the medium access control layer and physical layer—which exert a strong influence on the routing protocol performance.

REFERENCES

[1] Wi-Fi Alliance Knowledge Center, http://wi-fi.org/knowledge_center_overview.php.

[2] High speed packet access on UMTS/3GSM, http://hspa.gsmworld.com/.

[3] Mobile broadband internet access on CDMA2000, http://www.cdg.org.

[4] R. Karrer, A. Sabharwal, and E. Knightly, "Enabling large-scale wireless broadband: the case for TAPs," *ACM SIGCOMM Comput. Commun. Rev. Archive*, vol. 34 , no. 1, pp. 27–32, Jan. 2004.

[5] J. Sydir, "IEEE 802.16 Broadband Wireless Access Working Group: harmonized contribution on 802.16j (mobile multihop relay) usage models," *IEEE 802.16j Working Group Document 802.16j-06/510*, Sept. 2006.

[6] Y. Lin and Y. Hsu, "Multihop cellular: a new architecture for wireless communications," *Proceedings of IEEE INFOCOM*, pp. 1273–1282, Mar. 2000.

[7] R. Ananthapadmanabha, B. S. Manoj, and C. S. R. Murthy, "Multi-hop cellular networks: the architecture and routing protocols," *Proceedings of the IEEE International Symposium on Personal, Indoor and Mobile Radio Communications*, pp. 78–82, Sept. 2001.

[8] C. Qiao, and H. Wu, "iCAR: an integrated cellular and ad hoc relay system," *Proceedings of the IEEE International Conference on Computer Communication Network*, pp. 154–161, Oct. 2000.

[9] H. Wu, C. Qiao, S. De, and O. Tongus, "Integrated cellular and ad hoc relaying systems: iCAR," *IEEE J. Sel. Areas Commun.*, vol. 19, no. 10, pp. 2105–2115, Oct. 2001.

[10] G. Aggelou and R. Tafazolli, "On the relaying capability of next generation GSM cellular networks," *IEEE Personal Commun. Mag.*, vol. 8, no. 1, pp. 40–47, Feb. 2001.

[11] I. F. Akyildiz, X. Wang, and W. Wang, "Wireless mesh networks: a survey," *Comput. Networks*, vol. 47, no. 4, pp. 445–487, Mar. 2005.

[12] S. A. Mahmud, S. Khan, S. Khan, and H. Al-Raweshidy, "A comparison of MANETs and WMNs: commercial feasibility of community wireless networks and MANETs," *Proceedings of the ACM International Conference on Access Networks*, Sept. 2006.

[13] Q. Yin, P.-Y. Kong, and H. Wang, "Quantitative robustness metric for QoS performances of communication networks," *Proceeding of the IEEE International Symposium on Personal, Indoor and Mobile Radio Communications*, pp. 1–5, Sept. 2006.

[14] D. England, J. Weissman, and J. Sadagopan, "A new metric for robustness with application to job scheduling," *IEEE International Symposium on High Performance Distributed Computing*, pp. 135–143, July 2005.

[15] O. Arpacioglu, T. Small, and Z. J. Haas, "Notes on scalability of wireless ad hoc networks," *IETF Internet Draft Document draft-irtf-and-scalability-notes-00.txt*, Aug. 2003.

[16] C. A. Santivanez, B. McDonald, I. Stavrakakis, and R. Ramanathan, "On the scalability of ad hoc routing protocols," *Proceedings of IEEE INFOCOM*, pp. 1688–1697, Mar. 2002.

[17] P. Gupta and P. R. Kumar, "The capacity of wireless networks," *IEEE Trans. Inf. Theory*, vol. 46, no. 2, pp. 388–404, Mar. 2000.

[18] Y. Yang, J. Wang, and R. Kravets, "Designing routing metrics for mesh networks", *Proceedings of the IEEE Workshop on Wireless Mesh Networks*, 2005.

[19] R. Draves, J. Padhye, and B. Zill, "Comparison of routing metrics for static multi-hop wireless networks," *Proceedings of ACM SIGCOMM*, pp. 133–144, Aug. 2004.

[20] A. Adya, P. Bahl, J. Padhya, A. Wolman, and L. Zhou, "A multi-radio unification protocol for IEEE 802.11 wireless networks," *Proceedings of the International Conference on Broadband Networks*, pp. 344–354, 2004.

[21] R. Draves, J. Padhye, and B. Zill, "Routing in multi-radio multi-hop wireless networks," *Proceedings of ACM MOBICOM*, pp. 114–128, Sept. 2004.

[22] M. Kim and B. Noble, "Mobile network estimation," *Proceedings of ACM MOBICOM*, 2001.

[23] D. S. J. Couto, D. Aguayo, J. Bicket, and R. Morris, "A high-throughput path metric for multi-hop wireless routing," *Proceedings of ACM MOBICOM*, Sept. 2003.

[24] A. Woo and D. Culler, "Evaluation of efficient link reliability estimators for low-power wireless networks," *Tech. Rep. UCB/CSD-03-1270*, Electrical Engineering and Computer Sciences Department, University of California–Berkeley, 2003.

[25] J. L. Sobrinho, "Algebra and algorithm for QoS path computation and hop-by-hop routing in the internet," *IEEE/ACM Trans. Network.*, vol. 10, no. 4, pp. 541–550, Aug. 2002.

[26] J. L. Sobrinho, "Network routing with path vector protocols: theory and applications," *Proceedings of ACM SIGCOMM*, pp. 49–60, Aug. 2003.

[27] C. E. Perkins and E. M. Royer, "Ad-hoc on demand distance vector routing," *Proceedings of the IEEE Workshop on Mobile Computing Systems and Applications*, pp. 90–100, Febr. 1999.

[28] P. Jacquet, P. Muhlethaler, T. Clausen, A. Laouiti, A. Qayyum, and L. Viennot, "Optimized link state routing protocol for ad hoc networks," *Proceedings of the IEEE Multi Topic Conference*, pp. 62–68, 2001.

[29] D. B. Johson and D. A. Maltz, "Dynamic source routing in ad hoc wireless networks," in *Mobile Computing*, Kluwer Academic, Norwell, MA, Chap. 5, 1996.

[30] M. Mauve, J. Widmer, and H. Hartnstein, "A survey on position-based routing in mobile ad hoc networks," *IEEE Network*, vol. 15, no. 6, pp. 30–39, Nov. 2001.

[31] P. Bose, P. Morin, I. Stojmenovic, and J. Urrutia, "Routing with guaranteed delivery in ad hoc wireless networks," *ACM Wireless Networks*, vol. 7, no. 6, Nov. 2001.

[32] B. Karp and H. T. Kung, "Greedy parameter stateless routing for wireless networks," *Proceedings of ACM MOBICOM*, pp. 243–254, Aug. 2000.

[33] Y.-B. Ko and N. H. Vaidya, "Location-aided routing (LAR) in mobile ad hoc networks," *ACM/Baltzer Wireless Networks*, vol. 6, no. 4, pp. 307–321, 2000.

[34] B. Bellur, R. Ogier, and F. Templin, "Topology broadcast based on reverse-path forwarding (TBRPF)," IETF Internet Draft Document draft-ietf-manet-tbrpf-01.txt, Mar. 2001.

[35] C. E. Perkins and P. Bhagwat, "Highly dynamic destination sequenced distance vector routing (DSDV) for mobile computers," *Proceedings of ACM SIGCOMM*, pp. 234–244, Oct. 1994.

[36] V. D. Park and M. S. Corson, "A highly adaptive distributed routing algorithm for mobile wireless networks," *Proceedings of IEEE INFOCOM*, pp. 1405–1413, 1997.

[37] P. Samar, M. R. Pearlman, and Z. J. Haas, "Hybrid routing: the pursuit of an adaptable and scalable routing framework for ad hoc networks," in *The Handbook of Wireless Ad Hoc Networks*, CRC Press, Boca Raton, FL, pp. 245–262, 2003.

[38] Z. J. Haas and M. R. Pearlman, "The performance of query control schemes for the zone routing protocol," *ACM/IEEE Trans. on Newwork.*, vol. 9, no. 4, pp. 427–438, Aug. 2001.

[39] M. Audeh, "Metropolitan-scale Wi-Fi mesh networks," *IEEE Comput.*, vol. 37, no. 12, pp. 119–121, Dec. 2004.

[40] Internet Engineering Task Force, http://www.ietf.org/html.charters/manet-charter.html.

[41] C. E. Perkins, E. M. Belding-Royer, and S. R. Das, "Ad hoc on-demand distance vector (AODV) routing," *IETF RFC 3561*, July 2003.

[42] T. H. Clausen and P. Jacquet, "Optimized link state routinng protocol (OLSR)," *IETF RFC 3626*, Oct. 2003.

[43] D. Maltz, J. Broch, J. Jetcheva, and D. Johnson, "The effects of on-demand behavior in routing protocols for multi-hop wireless ad hoc networks," *IEEE J. Sel. Areas Commun.*, 1999.

[44] M. K. Marina and S. R. Das, "On-demand multipath distance vector routing for ad hoc networks," *Proceedings of the International Conference for Network Protocols (ICNP)*, Nov. 2001.

[45] S.-J. Lee and M. Gerla, "AODV-BR: backup routing in ad hoc networks," *Proceedings of the IEEE Wireless Communication and Networking Conference (WCNC)*, pp. 1311–1316, Mar. 2000.

[46] D. Tian and N. D. Georganas, "Energy efficient routing with guaranteed delivery in wireless sensor networks," *Proceedings of the IEEE Wireless Communications and Networking Conference (WCNC)*, Mar. 2003.

[47] H. Wang, P.-Y. Kong, and W. Seah, "A robust and energy efficient routing scheme for wireless sensor networks," *Proceedings of the IEEE International Conference on Distributed Computing Systems*, pp. 83–89, July 2006.

[48] A. Tsirigos and Z. J. Haas, "Analysis of multipath routing: I. The effect on the packet delivery ratio," *IEEE Trans. Wireless Commun.*, vol. 3, no. 1, pp. 138–146, Jan. 2004.

[49] J. Sucec and I. Marsic, "Hierarchical routing overhead in mobile ad hoc networks," *IEEE Trans. Mobile Comput.*, vol. 3, no. 1, Jan. 2004.

[50] L. Jia, R. Rajaraman, and T. Suel. "An efficient distributed algorithm for constructing small dominating sets," *Proceedings of the ACM Symposium on Principles of Distributed Computing (PODC)*, Aug. 2001.

[51] R. Sivakumar, P. Sinha, and V. Bharghavan, "CEDAR: a core-extraction distributed ad hoc routing algorithm," *IEEE Sel. Areas Commun.*, vol. 17, no. 8, pp. 1454–1464, Aug. 1999.

[52] A. Amis, R. Prakash, T. Vuong, and D. T. Huynh, "Max min dcluster formation in wireless ad hoc networks," *Proceedings of IEEE INFOCOM*, pp. 32–41, Mar. 2000.

[53] C. R. Lin and M. Gerla, "Adaptive clustering for mobile wireless networks," *IEEE J. Sel. Areas Commun.*, vol. 15, no. 7, pp. 1265–1275, Sept. 1997.

CHAPTER 7

RADIO RESOURCE MANAGEMENT FOR WIRELESS BROADBAND NETWORKS

7.1 INTRODUCTION

Wireless broadband communications have gained increased interest, fueled by demand for high-data-rate access for various emerging wireless and mobile applications. These applications require wireless broadband networks that support data rates in excess of 2 Mbps per radio channel, shared by multiple users or sessions. In addition, wireless broadband networks need to provide acceptable levels of quality of service (QoS) to heterogeneous traffic, ranging from voice over Internet protocol (VoIP) packets and interactive multimedia streams to encrypted e-commerce data packets, over the error-prone wireless channels. To provide the diverse QoS requirements in a robust and efficient manner, radio resource management schemes are necessary. Admission control and packet scheduling are two different types of radio resource management schemes that operate in different time scales, complementing each other. Specifically, admission control schemes are per-connection based and are normally activated once for the entire lifetime of a connection, to ensure that there are sufficient radio resources in the network for the QoS requested by the connection before it is admitted into the network. On the other hand, packet scheduling schemes work on a per-packet basis and are normally activated once for each packet or a burst of packets, ensuring that the tight QoS requirements of each packet can be met while exploiting the network dynamics and time-varying quality of the error-prone wireless channel.

Wireless Broadband Networks, By David Tung Chong Wong, Peng-Yong Kong, Ying-Chang Liang, Kee Chaing Chua, and Jon W. Mark

In this chapter we present various packet scheduling algorithms for wireless networks before focusing on admission control schemes.

7.2 PACKET SCHEDULING

Wireless bandwidth is a very scarce resource. Therefore, it must be managed efficiently through appropriate radio resource allocation mechanisms. For instance, the bandwidth should be allocated to backlogged users only where a user is considered backlogged when it has something to transmit. One of the most important and flexible mechanisms for efficient radio resource allocation is packet scheduling. It is flexible because compared to other resource allocation mechanisms that work on a longer time scale, packet scheduling can decide dynamically, on a packet-by-packet (a short time interval) basis, which backlogged user should first be allocated the radio resource.

Packet scheduling over a wireless channel must take into consideration the phenomenon that the channel qualities perceived are not the same for all users at all times, where channel quality is considered high when its bit error rate (BER) is low [1,2]. For example, a user that experiences high channel quality at one moment may experience low channel quality at a later moment, and vice versa. Also, at any moment, a user may experience high channel quality while other users experience low channel quality. This peculiar time-varying channel quality exists because in a wireless communication environment, channel errors are bursty, location dependent, and mobility dependent. These are due to radio propagation impairments such as shadowing and multipath fading, as well as interference from neighboring systems and users.

The time-varying wireless channel quality described above is a physical layer characteristic, while packet scheduling is usually implemented in the data link layer. The requirement of using information from other layers in making a packet scheduling decision is called *vertical* (cross-layer) *coupling*. This type of coupling is necessary to ensure efficient allocation of the precious wireless resource because transmissions during times of low channel quality tend to fail and hence lead to a waste of the channel resource, leading to lower network throughput.

Vertical coupling can be performed in the following ways:

1. Control the physical layer parameters, such as transmission power, modulation scheme, and coding level with respect to the channel quality at the scheduled transmission time.
2. Control the scheduled time to avoid transmission during low channel quality.

One way to control the transmission parameters at the physical layer depends on the time-varying channel quality. For example, at the time of deciding which user to transmit, a scheduling algorithm necessitate adjusting the modulation level and transmission power of the user appropriately with respect to its instantaneous channel quality [3–5]. More specifically, when a user is scheduled to transmit but its channel quality is low, its transmission power and modulation level will have to be adjusted accordingly to make the transmission more resilient to channel errors [6]. This approach of vertical coupling is useful to wireless broadband networks with readily available information on the wireless channel condition. For example, WiMAX networks will implement adaptive modulation and coding schemes together with scheduling. This adaptive modulation, coding, and scheduling synergy is more effective in enhancing network throughput than when transmissions are scheduled without considering channel quality. However, this vertical coupling approach is not portable because different wireless communication systems apply different interface technologies, but an adjustable physical-layer parameter may not exist in all these different technologies. Hence, a scheduling algorithm developed for one system by controlling a particular physical-layer transmission parameter may not be applicable in other systems.

Another method of vertical coupling is to control the transmission time with respect to the occurrence of high or low channel quality. This is done such that transmissions are avoided when channel quality is low to reduce the number of transmission failures and resource wastage. For example, when a channel is of low quality, transmissions in the channel are simply deferred until its quality improves. This vertical coupling approach, which avoids transmitting in a low-quality channel, is called *channel error avoidance* (CEA) *packet scheduling*. Since CEA packet scheduling does not control any physical layer transmission parameter, it is more portable than is the other vertical coupling approach. In this chapter we focus mainly on CEA packet scheduling algorithms.

Many CEA packet scheduling algorithms have emerged in the literature [6–9]. These CEA algorithms can be classified according to their performance objectives:

1. Improving network throughput
2. Achieving fairness in bandwidth sharing
3. Meeting deadline of time-sensitive packets
4. Providing differentiation in performance guarantees

Improving network throughput by avoiding transmission in a low-quality channel is taken a step further in some CEA algorithms by setting their performance objective in terms of fairness in bandwidth sharing. The shared bandwidth can be the raw channel bandwidth that is consumed by all transmissions, or the effective bandwidth that results in successful transmissions. When the effective bandwidth is considered in fair sharing, ensuring strict fairness potentially contradicts the basic idea of CEA because more transmissions will have to be scheduled for a low-quality channel, so that a fair number of successful transmissions can be achieved after a sufficient number of retransmissions. As such, there are also CEA algorithms that look beyond strict fairness in bandwidth sharing and trade the strict fairness requirement for better overall network throughput as long as the fairness falls within an acceptable limit. Although fairness and throughput are important, there are other CEA algorithms that set their performance objectives as providing QoS in terms of packet delay guarantees to time-sensitive traffic with deadlines. There are also CEA algorithms with the performance objectives of providing differentiation in QoS guarantees.

7.2.1 Wireless Channel Model

There are many existing packet scheduling algorithms proposed in the literature for wired networks. However, these wired network scheduling algorithms are not suitable for wireless networks because these algorithms do not consider the unique time-varying characteristics of wireless channels as described earlier in the chapter. In short, the different wireless channel characteristics is the primary impetus leading to the need to design new wireless packet scheduling algorithms instead of using those of a wired network. Then it becomes necessary to understand the wireless channel behavior before understanding the motivation in designing different CEA scheduling algorithms.

In a wireless network, there is one wireless channel (link) between each pair of spatially distributed nodes (users). The noise on each of these channels is assumed to be statistically independent of that on other channels. As such, we may focus on understanding the behavior of one wireless channel because the behavior will also be statistically representative of all other channels. For each wireless channel, instead of modeling the different wireless propagation impairments separately, a simpler way is to treat the channel as one with a time-varying bit error probability. The time-varying characteristic is a result of collective effects from the impairments. Time-varying error probability means that the probability is a function of times. Note that this is a different concept from that of another commonly used wireless channel model where error probability does not change with time. To differentiate the two concepts, the one with constant error probability is termed *random error*, and the one with time-varying error probability is termed *bursty error*. The bursty nature comes about because the error probability will stay at a value for more than a few packets' transmission time. Hence, as illustrated in Figure 7.1, bursty error becomes a sequence of random errors with different error probabilities.

In the literature, the time-varying quality of a bursty error wireless channel is commonly modeled using the Gilbert–Elliott model [10,11]. This model is a two-state Markov chain where each of the two states corresponds to high or low channel quality and is called good state or bad state, respectively. The *good state* is completely noiseless (i.e., BER equals zero) and the *bad state* is totally noisy (i.e., BER equals unity). In other words, if a packet is transmitted and the channel is in a good state, the transmission will be successful. On the other hand, transmission during a bad channel

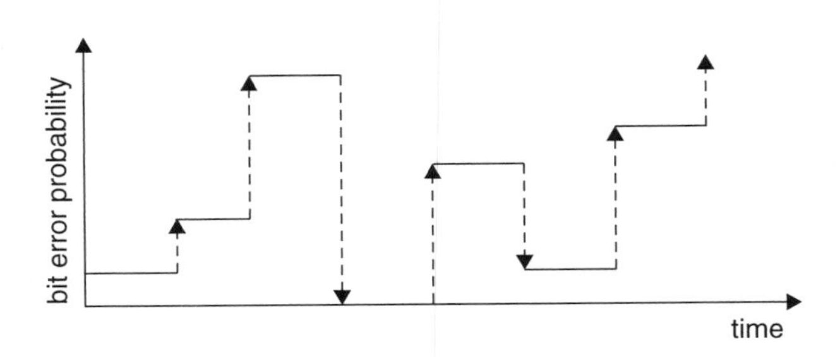

FIGURE 7.1 Bursty error wireless channel.

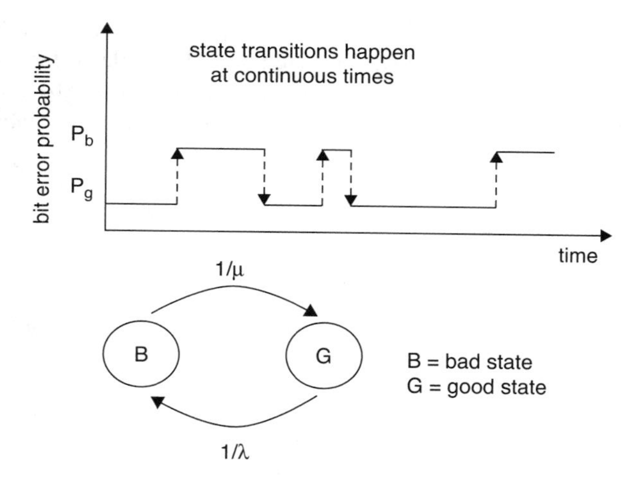

FIGURE 7.2 Fluid Gilbert–Elliot model.

state will certainly fail. The simple two-state Gilbert–Elliot model can be either a continuous- or a discrete-time Markov chain depending on whether the channel can change state in continuous or discrete time, as illustrated in Figures 7.2 and 7.3. The continuous- and discrete-time Markov chain are called the *fluid Gilbert–Elliot model* and the *discrete Gilbert–Elliot model*, respectively.

In Figures 7.2 and 7.3 the bit error probabilities in good and bad states are denoted by P_g and P_b, respectively. Recall that for Gilbert–Elliot models, $P_g = 0$ and $P_b = 1.0$. For the fluid Gilbert–Elliot model (see Figure 7.2), the duration that a wireless channel stays in a state before transiting to another state is an exponentially distributed random variable. Let the average duration that a wireless channel stays

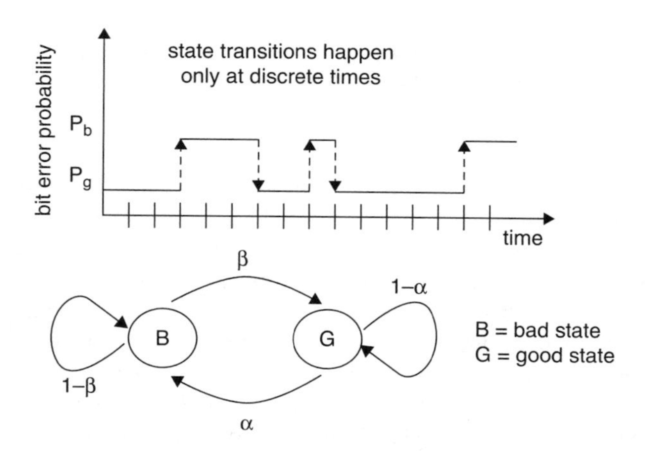

FIGURE 7.3 Discrete Gilbert–Elliot model.

in the good and bad states be denoted by λ and μ, respectively. Then the transition rate from the good state to the bad state is given by λ^{-1}. Similarly, the transition rate from the bad state to the good state is given by μ^{-1}. By solving the balance equations of the two-state Markov chain, we can determine the stationary probabilities of the good state, π_g, and bad state, π_b, as follows:

$$\begin{aligned} \pi_g + \pi_b &= 1, \\ \pi_g \lambda^{-1} &= \pi_b \mu^{-1}, \\ \pi_g &= \frac{\lambda}{\lambda + \mu}, \\ \pi_b &= \frac{\mu}{\lambda + \mu}. \end{aligned} \tag{7.1}$$

For the discrete Gilbert–Elliot model illustrated in Figure 7.3, the duration that a wireless channel stays in a state before transiting to the other state is a geometrically distributed random variable that indicates the integer number of discrete time slots. At the beginning of each time slot, let the transition probability from good state to bad state be given by α. Similarly, the transition probability from bad state to good state is given by β. Then the average duration that a wireless channel stays in the good and bad states is denoted by α^{-1} and β^{-1}, respectively. By solving the balance equations of two-state Markov chain, we can determine the stationary probabilities of the good state π_g and bad state π_b as follows:

$$\begin{aligned} \pi_g + \pi_b &= 1, \\ \pi_g \alpha &= \pi_b \beta, \\ \pi_g &= \frac{\beta}{\alpha + \beta}, \\ \pi_b &= \frac{\alpha}{\alpha + \beta}. \end{aligned} \tag{7.2}$$

With the stationary probabilities of being in the good and bad states, the average error probability of a channel, P_e, can be determined as follows:

$$P_e = \pi_b P_b + \pi_g P_g. \tag{7.3}$$

Modeling a radio communication channel using the Gilbert–Elliott model is not always adequate because transmissions during the good state may not always be error free, and transmissions during the bad state may not always be erroneous. In practice, transmissions carried out during a good state can still have errors, and those during a bad state may survive. Hence, modifications can be made to reduce BER in the bad state (P_b) and to increase BER in the good state (P_g) as long as the BER in the bad state is higher than that in the good state ($P_b > P_g$). Also, we may extend the channel model to use more than two states such that each state i has its own error probability P_i and stationary probability π_i [12]. The values of P_i and π_i can be determined by solving the set of balance equations of the Markov chain as long as the transition probability matrix is known. Then, for an N-state Markov chain model, the average error probability of the channel, P_e can be determined as follows:

$$P_e = \sum_{i=0}^{N-1} \pi_i P_i. \tag{7.4}$$

As such, the Gilbert–Elliott model becomes a special case of the generic N-state Markov chain model described above.

7.2.2 Channel Error Avoidance Scheduling for Throughput Improvement

In this section we present existing CEA algorithms that have the performance objectives of improving network throughput. The *best user algorithm* [13] has been proposed to maximize the network throughput in a system model as depicted in Figure 7.4. In the system model, the scheduling algorithm ("scheduler") resides at the base station that serves a group of M users in downlink transmissions. At the base station, each user has a separate queue to hold its packets awaiting transmission and has a separate Gilbert–Elliot wireless channel connecting it to the base station. The objective of the best user algorithm is achieved by scheduling first at each time t the transmission from the user i that has the highest probability of being in the good state:

$$i = \underset{j = 1, 2, \ldots, M}{\arg\max} \{p_j(t)\}, \tag{7.5}$$

where $p_j(t)$ is the probability of user j seeing the good channel state at time t. The good state probability $p_j(t)$ is computed according to the time of the last successful

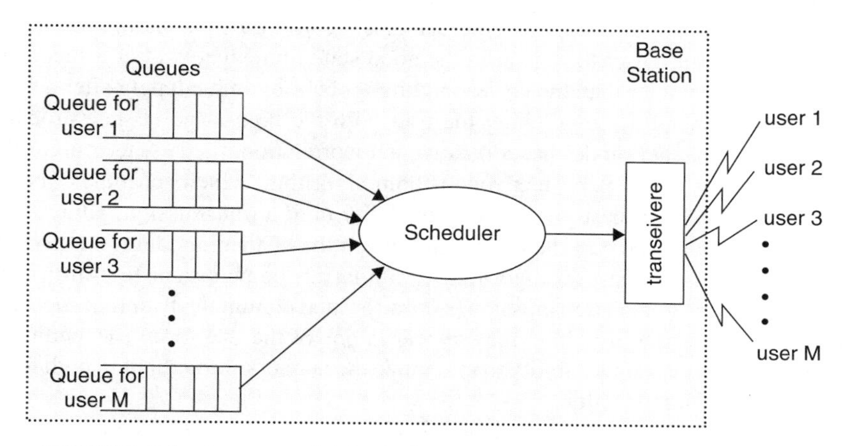

FIGURE 7.4 System model for the best user wireless scheduling algorithms.

transmission to the user j. For example, consider the fluid Gilbert–Elliot channel model: The probability for a user j to be in good state at current time t given that its previous transmission at time τ had been successful is given as follows:

$$\begin{aligned} p_j(t) &= P(\text{good state at time } t \mid \text{successful transmission at time } t-\tau) \\ &= P(\text{good state at time } t) \times P(\text{no change in state within duration } \tau) \\ &= \pi_g e^{-\alpha\tau} \\ &= \frac{\lambda e^{-\lambda\tau}}{\lambda+\mu}, \end{aligned} \tag{7.6}$$

where λ and μ are as defined for Figure 7.4. Compared to the fluid Gilbert–Elliot channel model, the discrete Gilbert–Elliot model has a different description for $p_j(t)$ as follows:

$$\begin{aligned} p_j(t) &= P(\text{good state at time } t \mid \text{successful transmission at time } t-\tau) \\ &= P(\text{good state at time } t) \times P(\text{no change in state within duration } \tau) \\ &= \pi_g (1-\alpha)^{\text{slot}(\tau)} \\ &= \frac{\beta(1-\alpha)^{\text{slot}(\tau)}}{\alpha+\mu}, \end{aligned} \tag{7.7}$$

where α and β are as defined earlier for Figure 7.3. Further, slot(τ) is a function that gives the number of discrete time slots within the duration τ.

Consider the discrete Gillbert–Elliot channel model: When $\alpha + \beta < 1$, the wireless channel is said to have a positive autocorrelation; on the other hand, if $\alpha + \beta > 1$, the channel is said to have a negative autocorrelation. A *positive autocorrelation* means that the probability for a user to remain in the good state immediately after a successful transmission is greater than the corresponding probability immediately after a failed transmission. On the other hand, a *negative autocorrelation* means that the probability for a user to remain in the good state immediately after a successful transmission is smaller than the corresponding probability immediately after a failed transmission. Therefore, by picking the user with the highest good state probability, when the channel model has a positive autocorrelation, the best user algorithm becomes similar to the persistent round-robin algorithm, which schedules first the same user with the highest good state probability until a transmission failure to the user has been observed. On the other hand, when the channel model has a negative autocorrelation, the best user algorithm is similar to the persistent first-in-first-out algorithm, where a failed transmission is retransmitted continuously until it succeeds. As such, the best user algorithm may appear in one of the two forms, depending on the channel process, and its ability to maximize the network throughput in both forms has been proved analytically.

The best user algorithm has assumed that the computed good state probability of a user is completely accurate, which means that the $p_j(t)$ calculated for user j equals

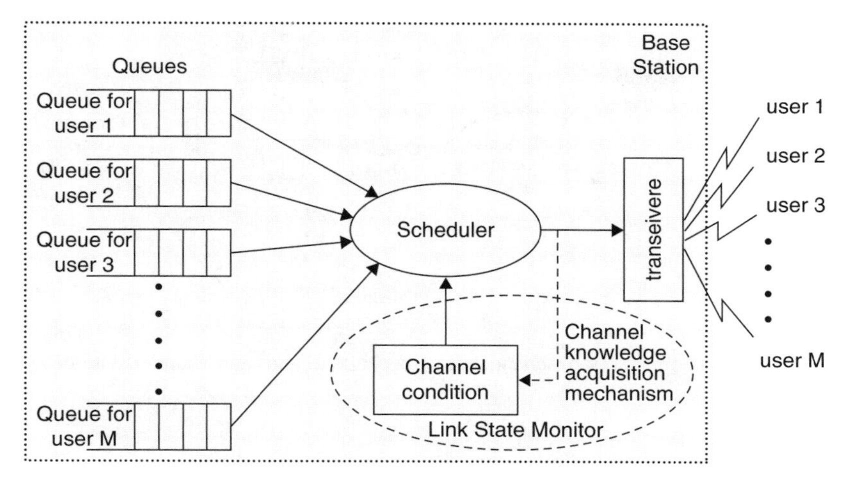

FIGURE 7.5 System model for channel-state-dependent wireless packet scheduling algorithms.

the actual probability. This assumption may not hold if the best user algorithm is evaluated through simulation where the actual channel state process is unknown. In view of this limitation, different types of wireless packet scheduling algorithms have been proposed that require the scheduler to acquire practical knowledge of the wireless channel through monitoring and to use the acquired channel knowledge in making scheduling decisions. These algorithms are called channel-state-dependent scheduling algorithms [14,15], due to the fact that the scheduling decision depends on the channel state of a user as acquired by the link state monitor, as illustrated in Figure 7.5.

Figure 7.5 is a modification of Figure 7.4, notably with the addition of the link state monitor, which determines and reports to the scheduler the channel quality of a user according to the method specified by a channel knowledge acquisition mechanism. One simple mechanism is to determine if a wireless channel is bad immediately after a transmission failure is observed over the channel. After the quality of a channel is determined to be bad, the link state monitor will reset the quality to good after a time period equal to the average bad channel duration, which is μ seconds for the fluid Gilbert–Elliot model. Then, as long as two wireless channels are reported to be in a good state, there is no difference between them in terms of probability of being in the good state. Using this method, a channel-state-dependent algorithm can dynamically mark a user as good or bad according to the user's channel state. Then, as depicted in Figure 7.6, the scheduler will select the next user to transmit only among all the users with a good channel. While selecting among the good state users, the channel-state-dependent algorithm may use one of the existing criteria and algorithms, such as round robin, earliest time stamp first, first-in-first-out, or longest queue first.

For a channel-state-dependent algorithm using the criteria round robin, earliest time stamp first, and longest queue first, simulation results described in [14] and [15] have shown that better network throughput and a lower average packet delay

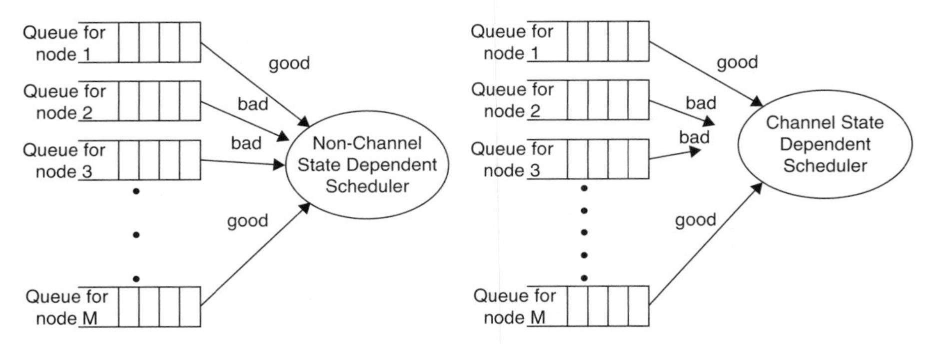

FIGURE 7.6 Comparison of channel-state-dependent and non-channel-state-dependent algorithms.

can be achieved than when using a single-queue non-channel-state-dependent first-in-first-out algorithm. This improvement is due to suppression of the head-of-queue blocking problem experienced by a single-queue first-in-first-out algorithm. The blocking problem exists when a user at the head of a single queue continuously fails to transmit. Recall from the best user algorithm [13] presented earlier that the persistent first-in-first-out algorithm is the ideal CEA algorithm when the channel process has a negative autocorrelation, as it can maximize network throughput. Hence, the simulation results on the channel-state-dependent algorithm [14,15] which say, for example, that the throughput can be improved by avoiding head-of-queue blocking may only be valid for a wireless channel with nonnegative autocorrelation.

Among the three user selection criteria simulated for the channel-state-dependent algorithm, only longest queue first has been analyzed [16,17] to show that it is not only capable of improving network throughput but also the optimal CEA algorithm that can maximize the stability region for operation. Here, the stability region is defined by the range of parameters within which the number of packets in the system will never grow to infinity. In practice, a larger stability region means a higher capability in handling more traffic and users, leading to a more efficient system.

Thus far, simulation results have indicated that channel-state-dependent algorithms [14,15] can indeed improve network throughput by alleviating the head-of-queue blocking problem. Different from these channel-state-dependent algorithms, which make scheduling decisions on a time slot-by-time slot basis, two CEA algorithms [18–20] that work on a frame-by-frame basis have been proposed. As illustrated in Figure 7.7, each frame is actually a group of contiguous time slots. Similar to channel-state-dependent algorithms, the two frame-by-frame algorithms are also for throughput improvement. They are called monopolistic allocation to the first terminal (MAF) and equal sharing (ES), respectively. MAF is identical to single-queue first-in-first-out, but MAF may look inside the first-in-first-out queue for the next user in a good state so that packets from these good state users can be scheduled to fill up all the available time slots in a frame. Hence, MAF can avoid the head-of-queue blocking problem by default. On the other hand, ES divides the number of time slots

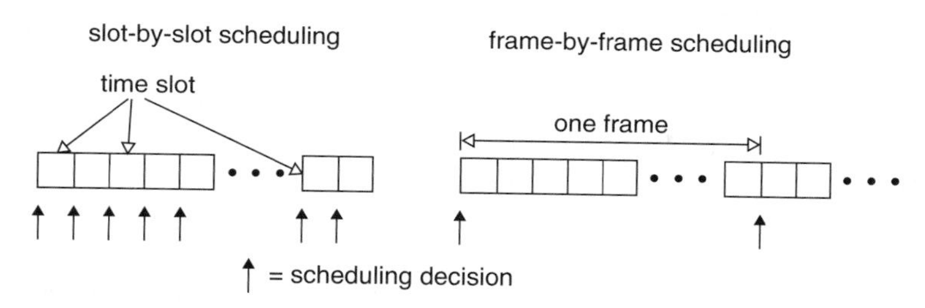

FIGURE 7.7 Comparison of slot-by-slot and frame-by-frame scheduling.

within a frame equally among all users in the good state. Simulation results have shown that both MAF and ES algorithms can improve network throughput compared to their non-CEA counterparts.

Channel adaptive open scheduling (CHAOS) [21] is another frame-by-frame CEA scheduling algorithm proposed to improve network throughput. CHAOS is different from MAF and ES [18–20] in the sense that CHAOS takes into consideration traffic arrival in addition to the time-varying channel quality. Traffic arrival at a user is assumed to be represented fully by its queue length. Also, the channel quality of a user is a real value that depends on both its current and predicted channel states in the scheduled frame. For example, when the channel is predicted to be bad in the frame, the real value is higher if the current state is good compared to the case where the current state is bad. CHAOS uses the channel quality of a user as a weighting factor in its queue length. Thus, CHAOS serves first the user with the largest weighted queue length such that weighted queue lengths of all the users in a good state can be equalized. Hence, conceptually, CHAOS is also a variation of the channel-state-dependent algorithm with longest-queue-first user selection criteria [15,16] presented earlier in the chapter. The difference between the two algorithms is that queue length in longest queue first is not weighted. CHAOS has been evaluated through simulations, and the results have confirmed that it is indeed capable of improving network throughput compared to the case where scheduling decisions are made without considering channel quality.

7.2.3 Channel Error Avoidance Scheduling for Fair Bandwidth Sharing

The CEA scheduling algorithms presented in Section 7.2.2 can improve or maximize the overall network throughput by taking into account the time-varying quality of the wireless channel. However, all these CEA algorithms do not have any mechanism to ensure that each user will be scheduled to transmit its packets within a time period, because logically a user encountering a bad channel state will avoid transmitting to achieve a better overall network throughput. Such a user may suffer bandwidth starvation if it happens to experience a bad channel state for an extended period. This is not fair to the user who may have paid the same fees as other users for use of the wireless broadband network. To be fair in sharing the wireless bandwidth, a user should be scheduled to transmit a negotiated number of packets (bits) within a time period, and the measurement of how close the actual number of packets transmitted is to the value negotiated is called *fairness*. It is perfectly fair if the number of packets transmitted is the same as the number of packets negotiated. In this section we present existing CEA algorithms that have the performance objectives to ensure fairness in bandwidth sharing.

As proposed by Fragouli et al. [22], one way to share the wireless bandwidth fairly among users in a CEA packet scheduling algorithm is to adopt class-based queuing [23], designed originally for the wired Internet. In class-based queuing, each user or group of users that forms a class is allocated a share of the channel bandwidth and is allowed to transmit only if the allocated bandwidth has not been exhausted. This original form of class-based queuing does not consider channel errors and thus requires modifications to make it capable of avoiding channel error. One important modification leading to the CEA class-based queuing is to allocate the fair share of bandwidth in terms of effective bandwidth but not the raw channel bandwidth. In CEA class-based queuing, the effective bandwidth is defined as the total number of successful transmissions (bits) per unit time and is computed dynamically over a predetermined time interval. To implement fair bandwidth sharing with class-based queuing, a user is scheduled to transmit next if its dynamically computed successful-byte sent/time ratio is lower than its allocated share of effective bandwidth. Simulation results have indicated that fairness can be assured accurately in CEA class-based queuing.

In addition to class-based queuing, deficit round robin [25] can be adopted to provide fairness in wireless packet scheduling algorithms. However, similar to the original class-based queuing, deficit round robin does not consider channel error by default, and thus modifications must be made to render it capable of avoiding channel error. We examine the original deficit round robin in detail before studying the CEA deficit round robin [26].

In the original deficit round robin, all users are examined in a round-robin fashion, but only the users that have accumulated sufficient credits are allowed to transmit their packets. Specifically, as illustrated in Figure 7.8, when a user's accumulated credits are not less than its packet size when it is examined, the packet will be transmitted. After transmitting a packet, the user's credit is reduced by as much as the size of the transmitted packet if the user is still backlogged. Here, a user is considered backlogged if its queue is not empty. If the user is not backlogged after transmitting a packet, its credit is reset to zero. Otherwise, the user may continue to transmit another packet as long as its remaining credits are sufficient. The deficit round robin will move on to examine the next user only if the current user has no packet to transmit or has insufficient credit to transmit its next packet. From the description above, deficit round robin is a credit-based scheduling algorithm because each user must accumulate enough points (credits) before transmitting its packet. deficit round robin uses a deficit counter DC_i to keep track of the credit of a user i. When user i is backlogged, the value of DC_i is increased in each service round, where a service round involves examining each user once in a round-robin manner. A service round

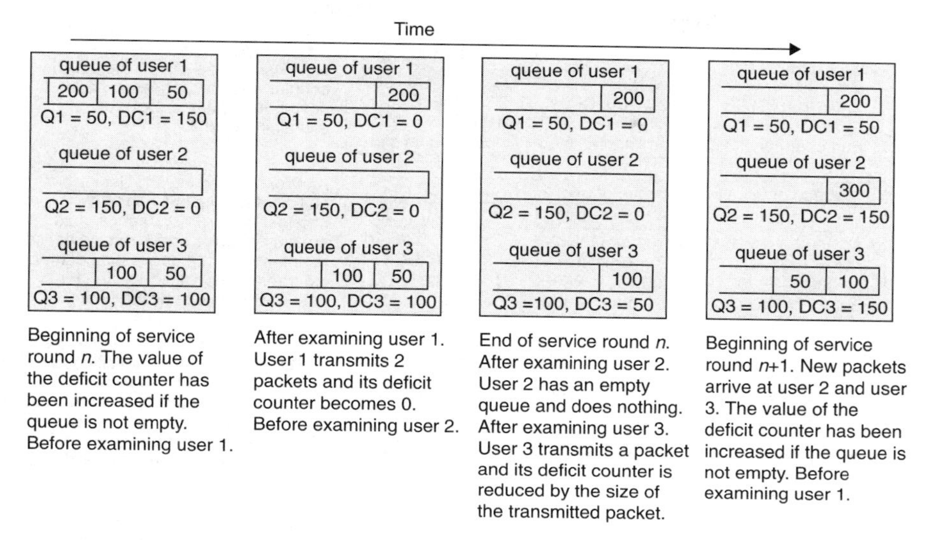

FIGURE 7.8 Operation of the deficit round robin with three users where values in the queue indicate the respective packet sizes.

begins and ends when the first and last users are examined, respectively. Practically, DC_i is increased by a quantum Q_i at the beginning of a new service round if user i is backlogged. Note that different users may has different values of Q_i, which determine the users' fair shares of bandwidth. A larger Q_i means a larger bandwidth share.

In the original form described above, deficit round robin is fair but does not consider the existence of a time-varying channel and therefore is not capable of avoiding channel errors. In view of this, CEA deficit round robin has been proposed such that a user with sufficient credits is allowed to transmit its packets if and only if it is in a good channel state. When a user with sufficient credits is examined but cannot transmit its packet due to a bad channel state, the CEA deficit round robin will proceed to examine the next user in the normal round-robin manner. In this case we say that the bad state user has lost its eligible transmission opportunity. To be fair, the user must be compensated for the lost opportunity when it returns to the good channel state. For the purpose of compensation, CEA deficit round robin uses a compensation counter CC_i to track the amount of lost transmission opportunity suffered by a user i. Similar to DC_i, CC_i must be reset to zero if user i has an empty queue. Practically, when user i with enough credit is examined but not scheduled to transmit its packet due to the bad channel state, the value of CC_i is increased by a quantum of T_i. As illustrated in Figure 7.9, the increase in CC_i is accompanied by a corresponding decrease in DC_i with the same value, which is T_i. Hence, the physical meaning of T_i is the rate at which the credit is transferred from DC_i to CC_i when user i cannot transmit its packet to avoid a channel error. The transferred credit will be moved back to DC_i at the rate $\theta \times CC_i (0 < \theta \leq 1)$ when user i returns to the good channel state.

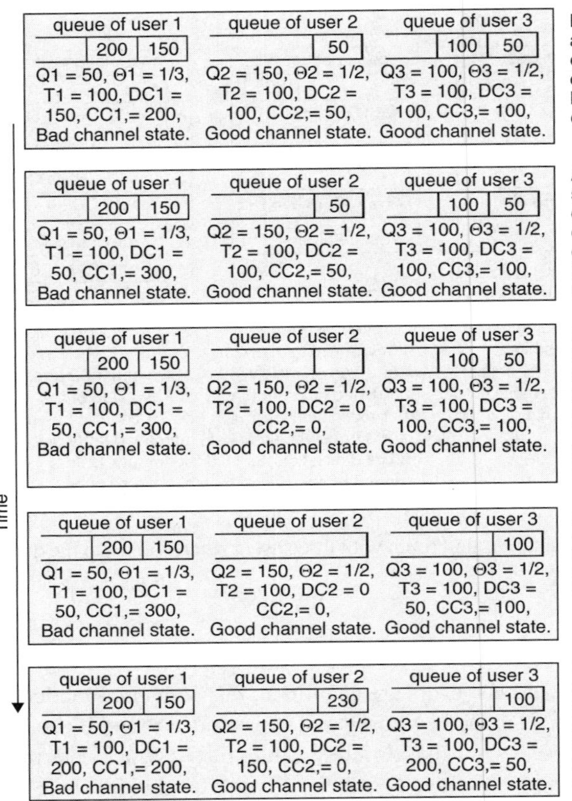

FIGURE 7.9 Operation of CEA-deficit round robin with three users where values in the queue indicate the respective packet sizes.

This procedure of returning credit to DC_i is to compensate for the lost transmission opportunity experienced by user i so that the user can still receive its allocated share of bandwidth despite a time-varying channel. The compensation is controlled through T_i and θ_i. In [26], T_i and θ_i have fixed values. As such, the compensation mechanism is rigid. For greater flexibility in the compensation mechanism, T_i and θ_i could be adjusted dynamically. A scheme has been proposed [26] to adjust T_i and θ_i dynamically according to the network load. More specifically, a fast compensation (large θ_i and small T_i) is preferred when the network is lightly loaded, and a slow compensation (small θ_i and large T_i) is favored when the network is heavily loaded.

Different from CEA deficit round robin, which is a credit-based approach, a debt-based CEA packet scheduling algorithm to ensure fair bandwidth sharing has been proposed [26,27]. In contrast to a credit-based method, where a user must have accumulated enough credit before transmitting its packet, the debt-based method allows a user to transmit a packet before earning sufficient credits for it. In the debt-based method, the short fall between the accumulated credits and the size of transmitted packets is called the *debt*. The debt of a user i is tracked by a counter DT_i. Here, as long as the queue of user i is backlogged, the value of DT_i decreases at a fixed rate which equals the user's allocated bandwidth share, say R_i. Thus, the debt counter value can be positive or negative. Unlike CEA deficit round robin, which examines all users once in each service round in a round-robin manner, the debt-based scheduling algorithm determines dynamically the user j to transmit its packet next based on the smallest counter value:

$$j = \underset{i = 1, 2, \ldots, M}{\arg\min} \{DT_i\}, \tag{7.8}$$

where M is the number of users being serviced by the scheduling algorithm. As illustrated in Figure 7.10 (compared to Figures 7.8 and 7.9), this dynamic selection of the next user to transmit is significantly different from the simple round-robin manner. In Figure 7.10, user 3 is first scheduled to transmit its packet with a size of 30 while its debt counter value DT_3 is the smallest among all the three users. After user 3 transmits its packet, the counter values for all three users are updated according to the size of the packet transmitted and their respective bandwidth share R_i. Specifically, the value of DT_i is updated as follows after each scheduled transmission from

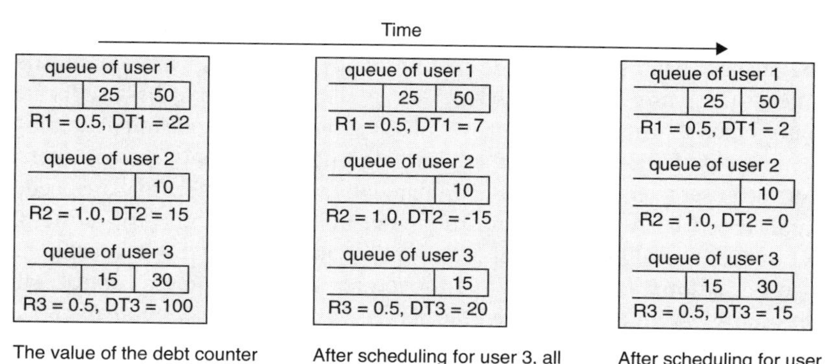

FIGURE 7.10 Operation of debit-based CEA scheduling algorithm with three users where values in the queue indicate the respective packet sizes.

user j:

$$\mathrm{DT}_i = \begin{cases} \mathrm{DT}_i^{\mathrm{old}} - R_i L_j & \text{if } i \neq j \\ \mathrm{DT}_i^{\mathrm{old}} - R_i L_j + L_j & \text{otherwise,} \end{cases} \tag{7.9}$$

where $\mathrm{DT}_i^{\mathrm{old}}$ is the current value of DT_i before the update, and L_j is the size of the packet transmitted from user j. Note that in Figure 7.10, after the scheduled transmission, user 2 has its debt counter reset to zero because its queue is no longer backlogged.

With a time-varying channel quality, a transmission from the user scheduled by the debt-based algorithm may not succeed. Ideally, only if the transmission is successful, is the debt counter increased by the transmission size. On the other hand, if the transmission fails and a retransmission is required, the debt counter is increased by $1 - \varphi$ of the transmission size as follows:

$$\mathrm{DT}_i = \begin{cases} \mathrm{DT}_i^{\mathrm{old}} - R_i L_j & \text{if } i = j \\ \mathrm{DT}_i^{\mathrm{old}} - R_i L_j + L_j & \text{if } i \neq j \text{ and transmission succeeds} \\ \mathrm{DT}_i^{\mathrm{old}} - R_i L_j + (1 - \varphi) L_j & \text{if } i \neq j \text{ and transmission fails} \end{cases} \tag{7.10}$$

where φ is a compensation constant that determines the extend of fairness toward the effective throughput. More specifically, when $\varphi = 1$, additional bandwidth must be allocated to compensate for the failure until the retransmission succeeds. This leads to fairness at the effective throughput level rather than the raw channel bandwidth level. On the other hand, when $\varphi = 0$, no additional bandwidth should be allocated (no compensation) to retransmit a failed transmission, and this represents fairness at the raw channel bandwidth level. In addition, a negative compensation constant ($\varphi < 0$) means that a user in the bad channel state will be punished by being allocated less resources than its entitlement because it has can potentially transmit and fail.

In the debt-based CEA packet scheduling algorithm [27], only one counter is needed for each user to ensure fairness, compared to the two counters that are needed in the credit-based CEA deficit round robin scheduling algorithm [26]. Different from these scheduling algorithms that use counters, another fair CEA algorithm has been proposed [28] that needs no counter for compensation in fairness control. This counterless fair CEA algorithm, called the *server-based fairness approach*, creates a dedicated virtual user to take care of the compensation matter. As illustrated in Figure 7.11, if the scheduler is supporting M users, it will have to handle $M + 1$ queues, where the additional queue is for the virtual user.

Assume that each packet is assigned a time tag which indicates the time the packet should be scheduled for transmission in order to be fair. The time tag t_i^k for packet k of user $i = 1, 2, \ldots, M, M + 1$, is calculated as follows:

$$t_i^k = t_i^{k-1} + \frac{L_i^k}{R_i}, \tag{7.11}$$

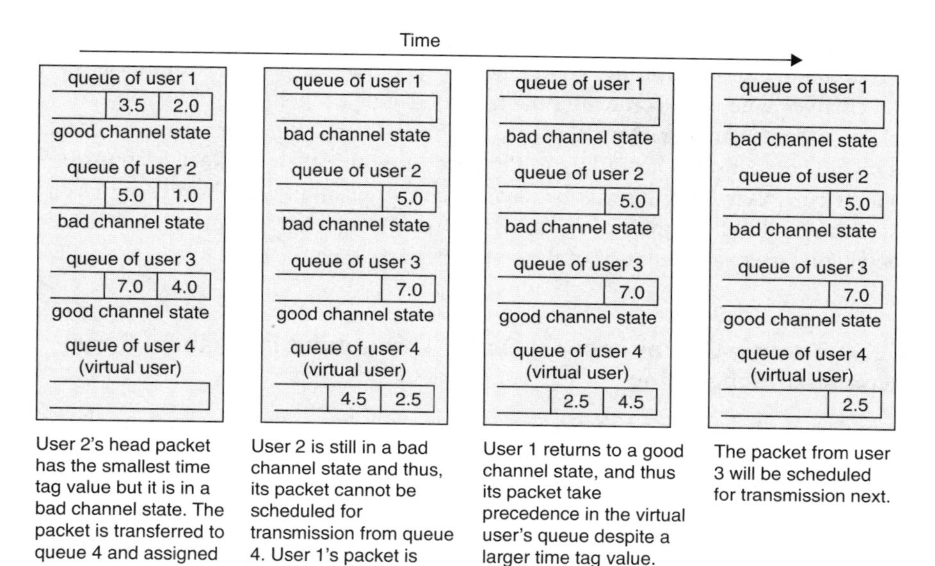

FIGURE 7.11 Operation of the counterless fair CEA scheduling algorithm with three users where values in the queue indicate the respective time tag values of the queued packets.

where L_i^k is the size of the packet k of user i, and R_i is the bandwidth share of user i such that $\sum_{i=1}^{M+1} R_i$ equals the total system bandwidth. Note that the virtual user is also allocated a fair share of bandwidth. The scheduler schedules first the packet from user j with the smallest time tag value among all users:

$$j = \underset{i=1,2,\ldots,M+1}{\arg\min} \left\{ t_i^k | k \text{ is the head of queue packet} \right\}. \tag{7.12}$$

In (7.12), instead of searching all packets in all the queues, the scheduler checks only the head of queue packets because the next packet in a queue cannot have a smaller time tag value as that computed using (7.11). Checking only the head-of-queue packets greatly reduces the complexity of the scheduling algorithm. When packet k of user i is scheduled for transmission and user i is in the good channel state, packet k is transmitted. If packet k of user i has the smallest time tag value but the user is in the bad channel state, the packet is transferred from user i's queue to the queue of the virtual user. The transferred packet will be assigned a new time tag calculated using the virtual user's bandwidth share, R_{M+1}. Since the queue of the virtual user is shared by all the actual users, it is possible that a user in the bad channel state may have a packet with a smaller time tag value in the good channel state. Thus, packets of the virtual user are sorted in an increasing order of time tag values among all users in the good channel state, so that packets from users in the bad channel state must come after the last good state packet.

Since there is only one virtual user to be shared by all actual users supported, these users are compensated according to the order in which their packets are transferred to the virtual user. This may not be ideal when there is multiclass traffic. In this case, it is beneficial to have multiple virtual users, one for each traffic class, and to assign each actual user a virtual user of the same traffic class.

7.2.4 Trading Off Fairness for Better Throughput in Channel Error Avoidance Scheduling

Thus far, we have learned that it is necessary to have a compensation mechanism regardless of whether credit based or debt based, for fair bandwidth sharing in CEA packet scheduling. This compensation mechanism is to provide users that have experienced transmission failures or low channel quality with additional bandwidth for retransmissions immediately or at a later time. To achieve fairness in effective throughput [27], a CEA scheduling algorithm may need to allocate more bandwidth to users in bad channel states, so that a fair number of successful transmissions can be achieved after several retransmissions for each failed packet. This is counterintuitive considering that the original goal of CEA packet scheduling is to avoid a low quality channel, and achieving fairness in effective throughput can lead to lower total network throughput [22]. This implies that there is a trade-off between fairness and total effective throughput.

In view of the fairness–throughput trade-off, Liu et al. [29] have proposed a CEA algorithm that not only exploits the time-varying channel quality and achieves resource allocation fairness at the same time, but also optimally maximizes the average total effective throughput. To do so, the algorithm uses a stochastic process to model the time-varying worth of a user, where the worth is defined as the throughput generated by the user. For example, a user is worth very little when it is in a bad state because of the high probability of seeing a transmission failure and producing no throughput. With the stochastic process model, the decision as to which user to serve first is formulated as an optimization problem to maximize the instantaneous total worth subject to a fairness constraint, where the fairness is defined as the fraction of time allocated over a predetermined time interval.

Results from simulation have indicated that the method proposed in [29] can outperform other CEA algorithms in terms of total effective throughput. Although analytically, it is proven optimal in effective throughput, its fairness performance is similar to that of other CEA algorithms [22, 24] in the sense that the fairness is ensured only in the long term, which is, more than a few packets' transmission time. Long-term fairness is not sufficient as a throughput (or bandwidth) guarantee measure from which a tight delay upper guarantee can be derived. This delay guarantee, which is critical to time-sensitive traffic, can be delivered only when there is short-term fairness.

7.2.5 Moving from Long- to Short-Term Fairness in Channel Error Avoidance Scheduling

Fair packet scheduling algorithms for wired networks are well known for their abilities in providing short-term fairness through strict isolation among users. Hence, a straightforward approach to providing short-term fairness in wireless networks is to adopt these fair scheduling algorithms for a wired network and make them capable of avoiding channel error. A few such adaptations exist. In general, these CEA fair scheduling algorithms use the fair scheduling algorithm adopted for wired networks as a reference model. By referring to that model, a user that has received too few services due to the time-varying channel quality can be tracked for compensation so that short-term fairness, as in wired networks, can be approximated as closely as possible.

Idealized wireless fair queuing [30,31] has been proposed as a CEA scheduling algorithm to provide short-term fairness in wireless networks. Idealized wireless fair queuing uses weighted fair queuing, which is also known as packetized generalized processor sharing for wired networks, as its reference model. In *idealized wireless fair queuing*, each arriving packet is assigned a start and a finish tag to indicate the times the packet should, respectively, start and finish its transmission according to the reference model. Among all the packets of all users in the good state, idealized wireless fair queuing schedules first the packet with the smallest finish tag. As such, service compensation is implicit in idealized wireless fair queuing. This is because a user that has not been scheduled for transmission for some time, due to its low channel quality, will have packets with small finish tags. Thus, the user's packets will be given service precedence for a short time once the user returns to the good channel state.

Although idealized wireless fair queuing is simple, as it needs no explicit mechanism for service compensation, this compensation behavior violates the isolation property that ensures short-term fairness in the reference model. Hence, to maintain short-term fairness capability, the compensation must be upper- and lower-bounded to alleviate the impact of the violation. The service compensation is upper-bounded by limiting the number of packets with a finish tag that is lower than the current time of the reference model. If the limit on the number of packets is exceeded, the packets that most recently arrived are discarded ant not inserted into the queue. On the other hand, the service compensation is lower-bounded by ensuring that all nondiscarded packets are scheduled for transmission before the difference between their finish tags and the current reference time becomes larger than a predetermined value. If a user has a continuous low channel quality and the difference between the finish tag of its packet and the current reference time becomes larger than the predetermined value, the finish tag is changed to the current reference time.

Analytical results have indicated that idealized wireless fair queuing [31] is capable of delivering short-term fairness in a channel with time-varying quality. However, the short-term fairness is a function of the stochastic channel process and the service compensation bounds. More specifically, the short-term fairness becomes looser, with a longer low-channel-quality period and a larger compensation bound. Thus, with an increasing service compensation bound, the short-term fairness may eventually become long-term fairness. A large upper bound can bring about another problem: denial of service to other users in the good channel state. For example, a user that has been in the bad channel state for an extended period will have small finish tags for its packets. Hence, the user will be given a strict priority immediately after returning to the good channel state and will for some time block all other users from receiving service.

Different from idealized wireless fair queuing, which uses weighted fair queuing as its reference model, another CEA fair scheduling algorithm called *channel-condition-independent fair queuing* [32] uses start-time fair queuing as its reference model. Channel-condition-independent fair queuing has a solution to the denial of service problem in idealized wireless fair queuing. The solution is to avoid giving strict priority while compensating users that have received too few services compared to the reference model. This is done by guaranteeing to the users that have received too many services compared to the reference model that they will still receive a minimum amount of service, which is $1 - \kappa$ times their allocated fair shares of bandwidth. Since the users will not lose all their services in one shot, this is called a *graceful service degradation property*. When a user gracefully gives up a fraction κ of its services, the excess bandwidth is distributed to all other good state users that have received too few services. The distribution of excess bandwidth is done proportionally according to these users' service shares.

Wireless fair service [33,34] is another CEA fair scheduling algorithm that is proposed to provide short-term fairness and to prevent the denial of service problem. Wireless fair service uses a modified weighted fair queuing as its reference model and achieves graceful service degradation using a method similar to channel-condition-independent fair queuing [32]. More specifically, a user that has received too much service will gracefully relinquish a fraction of its services. The services relinquished are to be shared proportionally among all other good state users that have received too few services. In addition to merely approximating the reference model for short-term fairness, wireless fair service adopts the idea of separating slots (right to transmit) from packets (data to be transmitted), where each packet has a corresponding slot. This separation is achieved by creating a virtual user to hold slots (right to transmit) for each actual user, and it helps in providing greater flexibility in choosing packet-discarding schemes on a per-user basis. Packet discarding is needed to enforce a compensation bound as explained earlier for idealized wireless fair queuing [33]. With the improved flexibility, the choice of discarding scheme may depend on the traffic requirements. For loss-sensitive users, wireless fair service may choose to drop all arrived packets and their slots as soon as their buffer is full. On the other hand, for time-sensitive users, a packet that has violated its delay bounds may be dropped from the head of the buffer space without dropping its slot, and thus its service preference is maintained.

In addition to improving flexibility in selecting packet discarding schemes as presented above by separating slots from packets, wireless fair service [34] proposes to decouple the bandwidth and delay requirements of each user so that time-sensitive users can be treated differently from loss-sensitive users. This decoupling is achieved by computing the start and finish tags of each of a user's slots (right to transmit) using the user's rate weight and delay weight, respectively, where the rate weight indicates its committed share of bandwidth and the delay weight implies its relative urgency. As such, wireless fair service serves first the slot with the smallest finish tag among all good state slots whose start tags are within a predetermined look-ahead range. By assigning two (delay and rate) weights to each user and by controlling the look-ahead range, wireless fair service can schedule transmissions based on rate or delay or a mix of both. For example, when the look-ahead range is infinity, wireless fair service is completely delay-bandwidth decoupled and is equivalent to the earliest due date [35,36] packet scheduling algorithm.

7.2.6 Moving Beyond Fairness in Channel Error Avoidance Scheduling

As described in Section 7.2.5, the level of short-term fairness provided by idealized wireless fair queuing [31], channel-condition-independent fair queuing [32], and wireless fair service [34] is related to the time-varying channel quality and the service compensation bounds. For example, a longer duration in low channel quality and a larger upper bound in service compensation will lead to a larger delay bound. This relationship between delay bound and channel characteristic is not suitable at all for a practical network because to guarantee a delay bound, a network cannot control its user's behavior that affects the perceived channel quality. For example, a user may like to walk in a certain pattern that induces a long duration in low-channel-quality. It is not reasonable to instruct the user to change his or her walking pattern to reduce the low-channel-quality period in order to deliver an acceptable delay bound. This delay bound dependency on compensation and channel quality does not exist when the look-ahead range in wireless fair service equals infinity and all packets that have exceeded their delay bounds are dropped. This implies that the earliest due date [35] algorithm can be more practical for scheduling than those algorithms that use short-term fairness approaches in providing a delay bound over a wireless channel with time-varying quality.

In the earliest due date method, also known as *shortest time to extinction* or *earliest deadline first*, traffic is modeled as a stream of packets. For user i, each packet k is assigned an expiration time d_i^k beyond which the packet is no longer useful, and whose computation is given as

$$d_i^k = \tau_i^k + D_i, \tag{7.13}$$

where τ_i^k is the time packet k from user i that arrives at the scheduler, and D_i is the target delay bound for user i. Hence, earliest due date needs to schedule to transmit each packet before the respective expiration time. When not all the packets can be scheduled for transmission before their deadline, earliest due date needs to minimize the number of packets that suffer from expiration. This is achieved by scheduling first the next packet from user j with the smallest deadline among all packets:

$$j = \underset{i=1,2,\ldots,M}{\arg\min} \left\{ d_i^k \mid k \text{ is the head of queue packet of user } i \right\}. \tag{7.14}$$

In earliest due date, since all packets that cannot meet a deadline are discarded, all packets transmitted can meet the delay-bound requirement. By discarding expired packets, earliest due date has delay bounds that are not derived from the short-term fairness guarantee given in terms of bandwidth. This means that a user can reserve a small bandwidth and still obtain a small delay bound, and therefore it is completely bandwidth-delay decoupled. In this original form, earliest due date is not capable of avoiding channel error. Thus, Shakkottai and Srikant [37] have proposed adopting the original earliest due data approach to a CEA algorithm. The algorithm proposed, called *feasible earliest due date*, schedules first the packet from user j with the smallest deadline among all users in good channel state:

$$j = \underset{i=1,2,\ldots,M}{\arg\min} \left\{ \begin{array}{l} d_i^k \mid k \text{ is the head of queue packet of} \\ \text{user } i \text{ and user } i \text{ is in good channel state} \end{array} \right\}. \tag{7.15}$$

Feasible earliest due date has been shown optimal in terms of minimizing the number of expired packets. However, this optimality is only for a static operation that is defined as the case when the scheduling algorithm works only on the snapshot of the user status, and the interval between two consecutive snapshots is the duration to finish scheduling all the packets in the first snapshot. The optimality has been proved only for this static case. By definition, feasible earliest due date is a dynamic scheduling algorithm, but it is very difficult to analyze it, considering dynamic traffic arrivals. With dynamic traffic arrivals, feasible earliest due date has been evaluated only through simulations. The simulation results indicate that feasible earliest due date outperforms CEA longest queue first [14,16,17], introduced in Section 7.2.2 in terms of lower number of expired packets.

From the above, feasible earliest due date is idealistic and reactive. It is idealistic because it assumes that the scheduler knows exactly the actual channel conditions for all users at all times. This is not a realistic assumption, because channel knowledge must be acquired through a separate mechanism and is not normally exact. Feasible earliest due date is reactive because it stops scheduling for a user only after its channel quality turns bad.

Compared to the reactive nature of feasible earliest due date, the network throughput can be improved by making earliest due date proactive. Specifically, it is beneficial to adjust a packet's deadline in anticipation of an upcoming degradation in channel quality. This new variant of the CEA earliest due date scheduling algorithm, which adjusts a packet's deadline proactively, is called *proactive earliest due date* [38,39]. Proactive earliest due date does not schedule packets from the queues in a bad channel condition. Further, proactive earliest due date adjusts a packet's deadline to a smaller value proactively, in anticipation of an upcoming degradation in its channel quality before scheduling the packet with the smallest adjusted deadline. Although proactive earliest due date makes scheduling decisions based on the adjusted deadlines, packets are considered expired only if their original deadlines are exceeded. The packet deadline adjustment is performed as illustrated in Figure 7.12. In short, if a packet has a deadline that falls within the bad channel duration, it will be assigned a

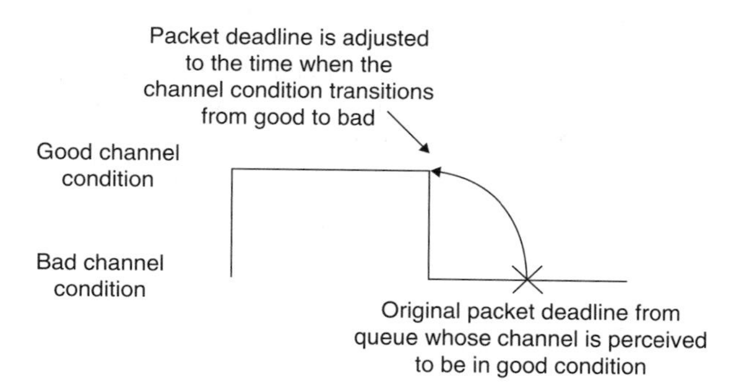

FIGURE 7.12 Packet deadline adjustment in the proactive earliest due date scheduling algorithm.

new deadline that is the time right before its channel turns bad. This packet deadline adjustment is necessary to reflect the actual urgency in the presence of time-varying channel conditions, because it is not productive to transmit a packet during a bad channel condition. To explain this further, consider a packet with an original deadline that is after a good-to-bad transition in the channel condition. According to proactive earliest due date, the deadline is adjusted to the transition time. Without the adjustment, the packet may be transmitted right before its original deadline and makes no contribution to the system throughput. With the adjustment, the packet is given a higher priority than those of packets that do not experience a bad channel condition. Through this deadline adjustment, proactive earliest due date aims at improving system throughput.

As a result of performing the deadline adjustment, packets from users with upcoming good-to-bad channel transitions may be given extra service opportunities. This may lead to unfairness to other users. However, the unfairness occurs only in the short term. In the long term, proactive earliest due date is fair to all users, because a user that receives extra service at one time may experience the opposite at another time. From the above, the adjusted deadline in proactive earliest due date should be a more accurate representation of the urgency than the deadline used in feasible earliest due date because the adjusted deadline takes into consideration not only the delay upper-bound requirement but also the time-varying channel conditions.

Similar to feasible earliest due date, proactive earliest due date is idealistic by definition. To consider a more realistic and pragmatic scenario where channel knowledge needs to be acquired explicitly, another variant of the CEA earliest due date scheduling algorithm, called *realistic-proactive earliest due date* [39], has been proposed. Realistic-proactive earliest due date differs from proactive earliest due date in the sense that it requires a separate channel knowledge acquisition mechanism. Before describing the channel knowledge acquisition mechanism, we introduce here how the acquired channel information is used in making scheduling decisions. Realistic-proactive earliest due date uses the acquired information in the following two ways: (1) to decide when to mask off which queue from being scheduled for transmission, and (2) to proactively adjust the packet deadlines for all queues that are not masked off. Contrary to proactive earliest due date with exact channel knowledge, realistic-proactive earliest due date does not have exact knowledge through acquisition. Since the channel information is not completely accurate, realistic-proactive earliest due date needs to estimate or predict the next time instant at which channel conditions will turn from good to bad so that the packet deadlines can be adjusted accordingly. Realistic-proactive earliest due date suggests predicting the next transition time T_i for user i as

$$\int_0^{T_i} f_{g,i}(x)\,dx = \gamma, \tag{7.16}$$

where $f_{g,i}(x)$ is the probability density function of good channel durations experienced by user i, and γ is the probability threshold. The probability density function is constructed dynamically for each user based on the channel knowledge acquired about the user. On the other hand, γ is a design parameter that will affect the performance of the scheduling algorithm. The immediate problem now is to decide how to construct $f_{g,i}(x)$ for each support user i so that T_i can be determined for a given γ. Realistic-proactive earliest due date proposes two strategies: (1) to construct $f_{g,i}(x)$ through channel probing and (2) to mask off all queues that are probed to be in a bad channel duration for as long as the run-time-averaged bad channel period. For a user i, the average bad period σ_i is computed as follows:

$$\sigma_i = \frac{1}{n}\left[\sigma_i^{\text{old}}(n-1) + t_{b,i}\right], \tag{7.17}$$

where σ_i^{old} is the previous average value and $t_{b,i}$ is the nth measured bad channel duration. The value of $t_{b,i}$ is measured from the outcomes of channel probing, as illustrated in Figure 7.13. In short, the measured bad channel duration is the time elapsed between the first bad channel instant detected and the next first good channel instant detected. The good channel duration is measured in the same way as the bad channel duration. All available measurements of good channel duration are used to form a histogram that represents $f_{g,i}(x)$.

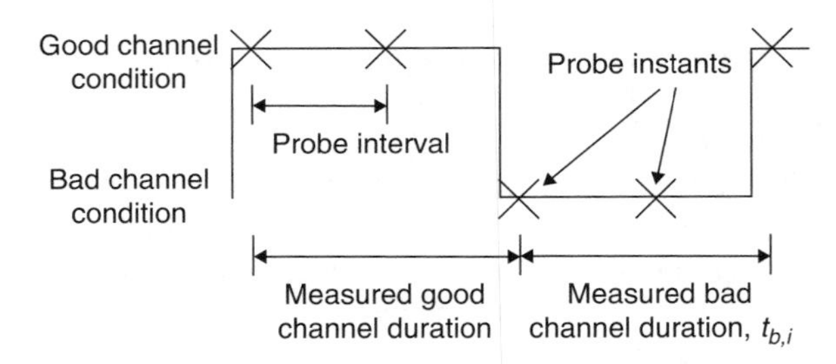

FIGURE 7.13 Measurement of good and bad channel durations through channel probing in realistic-proactive earliest due date.

Given the above, both the construction of $f_{g,i}(x)$ and the masking off of queues rely on channel probing. Other than channel probing, there are existing methods to predict channel conditions through signal-processing techniques. For example, Zhou et al. [40] suggest that radio signal strength exhibits self-similar characteristics and propose a mechanism to predict the upcoming signal strength given the past signal strengths. In general, this signal-processing-based method is not suitable for realistic-proactive earliest due date because (1) this method assumes that each user will transmit or receive continuously in all time instants (it is this continuous transmission assumption that eliminates the need for channel probing), and (2) this method needs to know the signal strengths that are available in the physical layer, but the CEA scheduling algorithm sits in a higher layer.

Conceptually, realistic-proactive earliest due date probes the channel by sending a small probe packet to each user at every probe interval. If the scheduler receives a response from a user with respect to the probe, the user is perceived as in the good channel state. Otherwise, it is in the bad channel state. With this probing method, there will be a difference between the actual and measured good or bad channel durations. This difference is a form of error in the channel information acquired. The error increases with a longer probe interval. Thus, a smaller probe interval yields more accurate acquired channel information. Unfortunately, we cannot afford to reduce the probe interval indefinitely, because each probe packet consumes bandwidth, and a smaller probe interval implies a smaller bandwidth for data packets [41].

Ideally, to conserve bandwidth for data packets, realistic-proactive earliest due date would like to make the probe interval as long as possible as long as the acquired channel knowledge is accurate enough. An optimal probe interval is determined [39] that can minimize a cost function composed of the errors in the acquired channel information and the probe traffic intensity. To reduce the probe traffic intensity further so that more bandwidth can be saved for data packet transmissions, realistic-proactive earliest due may send a probe packet to a user only if a data packet has been transmitted to the user within in the current probe interval. Specifically, if the past data packet transmission has been completed without error, the scheduling algorithm may assume that the channel condition is good at the time the user is supposed to be probed. Similarly, if the past data packet transmission to the user fails, the scheduling algorithm may assume the channel condition to be bad at the time the user is supposed to be probed.

7.2.7 Channel Error Avoidance Scheduling with Quality-of-Service Differentiation

So far, we have presented CEA scheduling algorithms with a predetermined QoS target, more specifically a committed share of throughput or a delay upper bound. Practically, given an infinitely small time scale, we can always find a case where the perceived QoS degraded by channel errors falls below the target. For this reason, guaranteeing absolute QoS is not regarded as a good choice over a wireless channel with time-varying quality. Some people believe that wireless QoS assurance should be rather loose. In this aspect, QoS differentiation appears more desirable. Differentiated

QoS may take the form of relative differentiation [42] or proportional differentiation [43,44].

Generally, relative differentiation ensures that when there are several priority classes, the traffic in a given priority class must be completely scheduled for transmission before traffic in the lower-priority classes will be scheduled. As such, the higher-priority classes will always receive better or at least no worse service than that received by lower-priority classes. Proportional differentiation is a refinement of relative differentiation. Proportional differentiation ensures that the quality spacing between those different service classes is adjustable, controllable and independent of the traffic loads [43]. For example, if q_i is the performance measure for class i, proportional differentiation means that the following constraints are met from all pairs of service classes:

$$\frac{q_i}{q_j} = \frac{\phi_i}{\phi_j} \qquad i, j = 1, 2, \ldots, N, \tag{7.18}$$

where N is the number of service classes supported, and $\phi_1 < \phi_2 < \cdots < \phi_N$ are the generic quality parameters. So even though the actual quality level of each class may vary in the presence of time-varying channel quality, the proportional commitment assures that the quality ratio between different service classes will remain fixed and controlled.

Jeong et al. [45] have proposed a CEA scheduling algorithm, called *wireless waiting-time priority* to deliver proportionally differentiated average packet delay at the scheduler among all users supported. In wireless waiting-time priority, for each user the waiting times of its packets are normalized to its differentiation weight. Then wireless waiting-time priority schedules first the good state user with the highest normalized waiting time, so that the normalized waiting time for all users can be equalized in the long run. Wireless waiting-time priority has been evaluated through simulations and the results have shown that this simple scheduling algorithm can provide accurate proportional differentiation over a wide range of channel characteristics and load levels. But similar to other CEA scheduling algorithms in the literature, when the channel quality is continuously bad and thus the low-channel-quality duration is long, wireless waiting-time priority cannot provide an accurate proportional differentiation.

Providing proportional differentiation in average packet delay is not sufficient for time-sensitive traffic such as voice and interactive video. For these time-sensitive traffics, a packet delay upper bound is crucial, but providing such an absolute QoS guarantee over a time-varying wireless channel through short-term fairness or deadline scheduling is not effective, as described above. In view of this, Kong et al. [46] have proposed providing compound QoS, where each service commitment consists of two components: a guaranteed QoS component and a differentiated QoS component. An example of a compound QoS commitment can be stated: User i and user j have packet delays never exceeding 60 and 90 ms, respectively, and user i's packet losses are two times better than the packet losses of user j. The guaranteed QoS component, such as a deterministic or statistic delay upper bound, or a hard or probabilistic minimum throughput guarantee, will fulfill the needs of all time-sensitive applications. On the other hand, the differentiated QoS component, such as a relative or proportional packet dropping ratio assurance, can be used as an approach to absorb the impact of time-varying wireless channel quality which cannot be determined at the time of admission control of the users, and to distribute the impact in a controlled manner among all admitted users. These two components must work together and can never be independent. As such, the differentiated QoS component becomes a quality measurement in meeting the guaranteed QoS component. For example, a higher relative packet loss, a higher proportional delay variance, and so on, mean a lower quality in meeting an identical delay upper bound among different users.

With the definition given above, the compound QoS commitments proposed have the following advantages:

- *It provides a practical approach to dealing with the highly variable wireless channel quality.* As a result of the time-varying wireless channel quality, the acceptable network load before congestion occurs may vary. Since the performance of differentiated QoS components are less subjective, if not completely independent of network load a compound QoS commitment allows the overall network performance to remain under control as committed while the limited resources are used to deliver the guaranteed QoS components.
- *It has two degrees of freedom and therefore increases the flexibility in committing an achievable QoS.* Let us assume that the performance metrics of either the guaranteed QoS commitments or the differentiated QoS commitments are represented by a single-dimensional matrix. Then, with compound QoS commitments, the performance metrics can actually be represented by a two-dimensional matrix, where the guaranteed QoS component and the differentiated QoS component are the dimensions. With the two-dimensional matrix, there are more possible ways to give a QoS commitment and thus the admissibility may be improved. For example, a user i that has been rejected admission when it initially requests 60 ms upper delay bound and an identical packet drop ratio as that of user j may be given admission when it requests the same upper delay bound but allows two times the packet drop ratio of user j.

Kong et al. [47] have proposed a CEA scheduling algorithm to realize compound QoS in terms of guaranteed packet delay upper bound and proportionally differentiated packet drop ratio among different supported users. The *delay* of a packet is defined as the time between its generation at the user and its subsequent arrival at the base station in uplink transmission [47]. For a user i, let D_i be its target delay upper bound and $S_i(t)$ its average packet drop ratio at time t. More specifically, $S_i(t)$ is the fraction of total traffic accumulated up to time t that is dropped by user i in order to meet D_i within a jumping time window $(\tau, \tau + T_s]$. Here T_s is the time window duration such that $\tau < t \leq \tau + T_s$.

Let I be the set of all users supported, and let $B(t)$ be a subset of users in I that are backlogged and perceive a good channel state at time t. The scheduling algorithm proposed in [47] aims to guarantee D_i for each user i in I and to ensure that $S_i(t)$ is proportionally differentiated among all the users in $B(t)$:

$$\frac{S_i(t)}{S_j(t)} = \frac{\phi_i}{\phi_j} \qquad \forall i, j \in B(t), \tag{7.19}$$

where ϕ_i is the weighting factor of user i, also called the controllable generic quality parameter, and can be negotiated among the different users. Thus, ϕ_i is a quality metric of the delay guarantee where a higher ϕ_i indicates a poorer quality in the delay guarantee D_i due to the drop of a larger proportion of packets in order to meet D_i.

Essentially, the scheduling algorithm achieves the foregoing proportional differentiation in drop ratios among users while guaranteeing a delay upper bound for each user by performing, on a frame-by-frame basis (refer to Figure 7.6), the following operations: (1) selecting which packet from which user is to be scheduled for transmission based on equalizing the normalized expired-to-eligible packet ratios (EERs) among all users, and (2) dropping all excess packets.

For operation (1) we say that a packet is *eligible* when it becomes necessary to be transmitted immediately so that it can meet its target delay upper bound. An eligible packet that is not transmitted is said to have *expired*. For a user i, its normalized EER, β_i is computed as follows:

$$\beta_i = \frac{L_i^e - L_i^s}{\phi_i L_i^e}, \tag{7.20}$$

where L_i^e and L_i^s count cumulatively the number of eligible packets and scheduled packets, respectively, for user i within the jumping time window. Since these counter values change dynamically within the jumping time window whenever packets become eligible or scheduled for transmission, the normalized EER also changes dynamically. The scheduling algorithm repeatedly selects the first packet of the user that is in a good channel state and has the highest normalized EER for transmission, until all time slots available in a frame have been fully allocated. This scheduling criterion to a user j at time t is given as follows:

$$j = \underset{i \in I}{\arg\max} \left\{ \beta_i(t) | k \text{ is the head of queue packet of user } i \text{ and } i \in B(t) \right\}. \tag{7.21}$$

Li and Kong [48] proposed a CEA scheduling algorithm to provide compound QoS in a single wireless hop architecture as illustrated in Figure 7.7, Kong et al. [48,49] further proposed a wireless scheduling algorithm to provide that compound QoS cross multiple hops in ad hoc networks in an end-to-end manner.

7.3 ADMISSION CONTROL

In a wireless network, the limited resources are shared among all users that have been admitted into the network for the purpose of meeting their individual QoS requirements. The QoS requirement can be specified at the call level, burst level, or packet level. At these different levels, the QoS requirement specifies the performance, such as packet delay experienced by a call, burst, or packet. To fulfill the QoS requirement of all users admitted, it is important to ensure that the network resources available are sufficient to support all the users admitted. This is the role of an admission control algorithm.

7.3.1 Model- and Measurement-Based Admission Control

The admission control algorithm must maintain the QoS of all existing users while admitting a new user. The simplest way to ensure this is to allocate enough resources to meet the worst-case QoS requirements of each user, including the newly admitted user. To allocate resources for the worst case, the admission control algorithm first needs to determine the worst-case traffic characteristic of all the users. Normally, the worst-case traffic characteristic of a user can be stated in terms of peak traffic rate, average traffic rate, and traffic burst size. Simply, the admission control algorithm will decide if the performance requirement of all the existing users and the new user can be satisfied when the worst-case traffic occurs simultaneously at all the users.

Unfortunately, allocating resource and performing admission control based on the worst-case scenario will result in low network utilization. This is because in practice not all the users will produce the worst-case traffic at the same time. Also, it is not efficient when the user has smooth traffic most of the time and its worst-case traffic happens very rarely. Therefore, instead of specifying the QoS requirement in the worst case, an admission control algorithm may specify a probabilistic or relative performance requirement for better network utilization.

An admission control algorithm with probabilistic QoS does not provide for the worst-case scenario. Instead, it guarantees a bound on the rate of loss or late packets based on statistical characterization of traffic. In this method, each user is allocated resources in terms of an effective bandwidth that is larger than the user's average traffic rate but less than the user's peak traffic rate. The effective bandwidth is computed based on a statistical model or on a fluid approximation of the user's traffic. To get the effective bandwidth precisely, it is critical to characterize the user's traffic precisely a priori.

Admission control algorithms with both worst-case QoS performance and probabilistic QoS are called *model-based admission control*. This is because the admission control algorithm depends on the user traffic models. However, the exact user traffic is difficult to model. For example, the traffic rate of a video stream depends on the motion of objects, where faster movement usually leads to a higher traffic rate. Similarly, the traffic rate for a voice source with activity detection depends on the silence periods. For example, in a human conversation, the silent period differs from one person to the next.

Different from model-based admission control, which depends on user traffic models, measurement-based admission control does not require such a traffic model. Instead, measurement-based admission control uses real-time measurements to characterize the traffic of existing users in the network. In this method, new users may still be admitted based on their traffic models because the network has yet to perform any measurement of their traffic. Since users are admitted based on measurement, occasional QoS violations may occur when newly admitted users suddenly inject more traffic into the network.

7.3.2 Estimating Network Capacity for Resource-Based Admission Control

In model-based admission control and measurement-based admission control, different methods are used to determine the level of network traffic. An admission control decision is made based on the criteria that the total amount of network traffic is not higher than the network capacity. However, there is no suggestion as to how the actual network capacity can be determined for admission.

In the literature for multicode-deficit round robin (multicode-DRR), Kong et al. [50] proposed a method to estimate the network capacity for the purpose of admission control in the presence of a time-varying wireless channel. In multicode-DRR, a multicode-CDMA wireless network is considered, and it is observed that the time-varying capacity of such a network has a deterministic lower bound. Hence, the idea is to separate the time-varying capacity into two components: a deterministic component and a variable component. The deterministic capacity component provides the basis for admission control for time-sensitive flows with a delay upper-bound requirement, and the remaining capacity is used to serve non-time-sensitive flows that have no specific delay requirement.

Before describing the admission control scheme, we first need to understand how to determine the deterministic lower bound in capacity and how to share the deterministic capacity among all the time-sensitive flows to meet their individual delay requirements. We first consider a homogeneous network before applying the findings to a heterogeneous network. Consider a homogeneous network where all devices (nodes, users, or flows) have an identical number of code channels, says M. Here a device with an M-code channel is called an *M-rate device*. In such a network, the transmission capacity at time t is given by

$$\varepsilon_M(t) = MN_M(t) + \delta_M, \tag{7.22}$$

where δ_M is the residual transmission rate, which is less than M ($\delta_M < M$). From (7.22), the value of $\varepsilon_M(t)$ becomes maximum when $N_M(t)$ reaches its maximum (i.e., when all the homogeneous devices always have something to transmit). Let N_M be the number of these M-rate homogeneous devices in the network. Then, for a target

BER, this maximum transmission capacity is given as follows:

$$\varepsilon_M = \max_{t\geq 0} \{\varepsilon_M(t)\}$$
$$= \begin{cases} M\left(\left\lfloor \dfrac{3G\left(1-2\gamma_o\sigma^2 B_b\right)}{2M\gamma_o} \right\rfloor + 1\right) + \delta_M & \text{for } N_M \geq \left\lfloor \dfrac{3G\left(1-2\gamma_o\sigma^2 B_b\right)}{2M\gamma_o} \right\rfloor + 1 \\ MN_M & \text{for } N_M < \left\lfloor \dfrac{3G\left(1-2\gamma_o\sigma^2 B_b\right)}{2M\gamma_o} \right\rfloor + 1, \end{cases} \tag{7.23}$$

where γ_o is the SINR required by the target BER, σ is the standard deviation of the Gaussian noise, and B_b is the baseband bandwidth for a basic rate bit stream. From the equation, the transmission capacity is limited by the number of existing devices when fewer than the maximum simultaneous devices are allowed.

Now consider a heterogeneous network in which different devices have a different number of code channels and A is the set of the numbers of code channels. Let N_A be the number of devices in the heterogeneous network. We define $M^* \in A$ as the number of code channels that may not be unique and that yield the lowest transmission capacity assuming that the network is homogeneous:

$$\varepsilon_{M^*} = \min_{M\in A} \{\varepsilon_M \,|\, \text{assume that } N_M = N_A\}. \tag{7.24}$$

Further, let $N_A(t)$ be the number of heterogeneous devices transmitting at time t. Then, provided that the devices always have something to transmit, the lower-bound transmission capacity, ε_{A^*}, in the heterogeneous network is

$$\forall t \mid N_A(t) \geq \min\left\{\left\lfloor \frac{3G\left(1-2\gamma_o\sigma^2 B_b\right)}{2M^*\gamma_o} \right\rfloor + 1, N_A\right\}$$
$$\varepsilon_A(t) \geq \varepsilon_{A^*} \geq \begin{cases} \varepsilon_{M^*} & \text{for} \left\lfloor \dfrac{3G\left(1-2\gamma_o\sigma^2 B_b\right)}{2M^*\gamma_o} \right\rfloor < 1 \\ \varepsilon_1 & \text{otherwise.} \end{cases} \tag{7.25}$$

In order to dedicate the deterministic capacity component ε_{A^*} determined above, we separate all the flows into two groups, one group for time-sensitive flows and another group for non-time-sensitive flows. Then the dedication is achieved through isolation by giving the time-sensitive group nonpreemptive strict priority. For the isolation to work well, an admission control mechanism must be in place so that the total capacity required by all the time-sensitive flows does not exceed the dedicated deterministic component. The admission control scheme depends on how the time-sensitive flows share the dedicated capacity through a scheduling algorithm. Thus, we will present the admission control mechanism after describing the scheduling algorithm.

The scheduling algorithm is a credit-based method that enables sharing of the capacity among different time-sensitive flows. The scheduling algorithm adopts an idea from deficit round robin (DRR) [25], which is an existing fair scheduling algorithm. The idea is to use only a single counter for each flow, to control its transmissions so that it greatly reduces the computational complexity to O(1) for each scheduling attempt. Although simple, the original DRR is for a generic network without simultaneous transmissions, and it provides only proportional sharing in transmission rate without any delay guarantee. The credit-based method that multicode-DRR uses in developing the admission control algorithm is different from DRR in the sense that it takes into consideration simultaneous transmissions in a multicode-CDMA network and provides delay upper-bound guarantees by allocating appropriate amounts of credits to different flows. As a credit-based scheme, the admission control is done by ensuring that the total credits allocated do not exceed those achievable by the deterministic capacity component.

For scheduling in multicode-DRR, a service list that is a sequence of indices of backlogged flows is used to determine which flow to serve next. Multicode-DRR isolates time-sensitive flows from non-time-sensitive flows by keeping two separate service lists, *H-list* and *L-list*, for time-sensitive flows and non-time-sensitive flows, respectively. Multicode-DRR serves one non-time-sensitive flow in the L-list only after serving once every time-sensitive flow in the H-list. The end of the H-list is identified by the final flow in the list, which is referred to as the *marker*. The marker is determined dynamically whenever a new backlogged flow is added to the list. Thus, multicode-DRR defines a service cycle as a periodic process that serves once every flow in the H-list until the marker, and then serves one flow in the L-list. In each service cycle, multicode-DRR gives a quantum of service credit Q_i to each backlogged flow i visited, and a state variable μ_i called the *deficit counter* is used to keep track of the number of service credits. Here the μ_i of each backlogged flow i visited can be increased by as much as Q_i in a service cycle. When a flow, say i, is visited, it is allowed to transmit its head of queue packet if the packet size is not larger than μ_i. After scheduling the transmission, μ_i is reset to zero if there is no other packet in flow i. Otherwise, μ_i is decremented by the size of the packet scheduled. Multicode-DRR stops serving a backlogged flow i and moves to the next flow only if the size of the head-of-queue packet of flow i is larger than the service credit μ_i available. When a flow i is not able to send its head-of-queue packet, due to insufficient service credits, these credits are carried forward to the next service cycle. Hence, flows that are short changed in a service cycle are compensated for in subsequent cycles.

With the multicode-DRR scheduling algorithm given above, we quantify the service cycle durations. The duration of a cycle k, T^k, can be decomposed into T_{ts}^k and T_{nts}^k, which are the time taken to serve once all time-sensitive flows in the H-list and one non-time-sensitive flow in the L-list, respectively. The service cycle duration is at its maximum when both T_{ts}^k and T_{nts}^k are at their respective maximum values. This occurs when all flows in the service lists are backlogged and consume their maximum service credits while the capacity is at it minimum value. Let $L_{\max}$ be the maximum packet size in the network. Then the maximum service credits at a flow i is given as

$$\mu_i^{\max} = Q_i + L_{\max} - 1. \tag{7.26}$$

Let I and J be sets of time-sensitive flows and non-time-sensitive flows, respectively. Then $i^* \in I$ and $j^* \in J$ are two flows such that

$$\mu_{i^*}^{\max} = \max_{i \in I} \left\{\mu_i^{\max}\right\}, \quad \text{and} \quad \mu_{j^*}^{\max} = \max_{j \in J} \left\{\mu_j^{\max}\right\}. \tag{7.27}$$

Also, let M_i be the number of code channels of flow i. Further, let R_b be the transmission rate of a code channel. Given the few definitions above, the maximum values of T_{ts}^k and T_{nts}^k are as follows:

$$\begin{aligned} T_{ts}^{\max} &= \max_{k \geq 0} \left\{T_{ts}^k\right\} \\ &= \frac{\sum_{i \in I, i \neq i^*} \mu_i^{\max}}{\varepsilon_{I^*} R_b} + \frac{\mu_{i^*}^{\max}}{\min_{i \in I} \{M_i\} R_b}. \\ T_{nts}^{\max} &= \max_{k \geq 0} \left\{T_{nts}^k\right\} \\ &= \frac{\mu_{j^*}^{\max}}{\min_{j \in J} \left\{M_j\right\} R_b}. \end{aligned} \tag{7.28}$$

where ε_{I^*} is the deterministic capacity component governed by the set I and it is the same as the ε_{A^*} derived earlier in this section using the set A.

As a result of simultaneous transmissions, $T_{ts}^{\max}$ and $T_{nts}^{\max}$ may or may not overlap. In the nonoverlap case, the longest service cycle duration, $T^{\max}$, is given by $T_{ts}^{\max} + T_{nts}^{\max}$. In the overlap case, $T^{\max}$ is given by $T_{ts,\,nol}^{\max} + \max\{T_{ts,\,ol}^{\max},\ T_{nts}^{\max}\}$, where $T_{ts,\,ol}^{\max}$ is the portion of $T_{ts}^{\max}$ shared with $T_{nts}^{\max}$, and $T_{ts,\,nol}^{\max}$ is the nonoverlapping portion of $T_{ts}^{\max}$. Consider the worst case, where overlapping occurs when only the last time-sensitive flow in the H-list transmits alone and its transmission rate permits simultaneous transmissions. According to the SINR constraint, this transmission rate is given by $\min_{i \in I}\{M_i\} \leq \lfloor 3G(1 - 2\gamma_o \sigma^2 B_b)/2\gamma_o \rfloor$. Then the maximum duration of a service cycle can be quantified as follows:

$$\begin{aligned} T^{\max} &= \max_{k \geq 0}\{T^k\} \\ &= \begin{cases} T_{ts,nol}^{\max} + \max\left\{T_{ts,ol}^{\max}, T_{nts}^{\max}\right\}, & \text{if } \min_{i \in I}\{M_i\} \leq \left\lfloor 3G(1 - 2\gamma_o \sigma^2 B_b)/2\gamma_o \right\rfloor \\ T_{ts}^{\max} + T_{nts}^{\max}, & \text{otherwise.} \end{cases} \end{aligned} \tag{7.29}$$

With the maximum cycle duration determined, we now consider a time interval (0,t]. The minimum amount of service credits that a flow may receive at the end of the interval is given by $\mu_i[0, t] = \lfloor t/T^{\max} \rfloor \mu_i^{\max}$, where $\lfloor t/T^{\max} \rfloor$ gives the minimum number of service cycles within the time interval. We know that the total packet size transmitted from flow i within the interval cannot exceed $\mu_i[0, t]$ because multicode-DRR is a credit-based algorithm where a flow must have sufficient service credits before it is allowed to send its packet. Since packet sizes vary and only the head-of-queue packet with size not larger than the deficit counter can be transmitted, the worst case occurs when the head-of-queue packet of the flow always has the maximum packet size $L_{\max}$, but the deficit counter is lacking one unit. Consequently, the minimum packet sizes that can be transmitted by flow i within the interval $(0, t]$ becomes

$$\begin{aligned} U_i[0, t] &= \max\{\mu_i[0, t] - L_{\max}, 0\} \\ &= \max\left\{\left\lfloor \frac{t}{T^{\max}} \right\rfloor \mu_i^{\max} - L_{\max}, 0\right\}, \end{aligned} \tag{7.30}$$

which is $\mu_i[0, t] - L_{\max}$ when $\mu_i[0, t] \geq L_{\max}$, and zero when $\mu_i[0, t] < L_{\max}$, because $U_i[0, t]$ can never be negative. We call the nondecreasing function $U_i[0, t]$ the guaranteed service curve of flow i at any time t provided that the flow has been allocated the service quantum Q_i.

We now study how different values of Q_i can be allocated to different flows to guarantee different service curves. Also, we examine how these service curve guarantees can be transformed into different delay upper-bound guarantees and how admission control can be done at the same time to ensure that ε_{I^*} is sufficient for all admitted time-sensitive flows.

Let flow i's arrival traffic within the interval $(0, t]$, which is a nondecreasing function $S_i[0, t]$, be constrained as follows:

$$S_i[0, t] \leq \theta_i + \rho_i t, \tag{7.31}$$

where $\theta_i \geq 0$ and $\rho_i \geq 0$ are typical token bucket constants that characterize the traffic of flow i. Then the delay $d_i(t)$ experienced by each packet of the flow is

$$\begin{aligned} d_i(t) &\leq \max_{t>0}\{\min\{\Delta | \Delta \geq 0 \text{ and } S_i[0, t] \leq U_i[0, t + \Delta]\}\} \\ &\leq \max_{t>0}\{\min\{\Delta | \Delta \geq 0 \text{ and} (\theta_i + \rho_i t) \leq U_i[0, t + \Delta]\}\}. \end{aligned} \tag{7.32}$$

That is, $d_i(t)$ is upper bounded by the maximum value of Δ and Δ is the minimum duration required after time t to send out all the traffic accumulated up to time t. Let the target delay upper bound of flow i, with D_i defined as the maximum time that elapses between traffic arrival at a flow and its successful receipt at its destination. Also, let D_i^r be the *maximum reservation delay* of flow i, which is defined as the time that elapses between the arrival of a packet to a flow and its successful resource reservation. This resource reservation is necessary because we focus on uplink scheduling in a centralized network architecture where the base station is assumed to be the destination of all traffic. Based on these few definitions, in order to meet the delay upper bound D_i of a flow i, the following must be valid:

$$d_i(t) \leq D_i - D_i^r, \qquad \forall t \geq 0 \tag{7.33}$$

Comparing (7.32) and (7.33), Δ in (7.32) can be replaced by $D_i - D_i^r$, to obtain the following:

$$(\theta_i + \rho_i t) \leq U_i[0, t + D_i - D_i^r]. \tag{7.34}$$

The variable t in (7.34) can also be replaced by $nT^{\max}$, where n is a positive integer. This is because the minimum service guaranteed by $U_i[0, t]$ is only valid at the end of discrete intervals that are multiples of $T^{\max}$. Since both $S_i[0, nT^{\max}]$ and $U_i[0, nT^{\max}]$ are nondecreasing monotonic step functions with fixed step sizes, the inequality is valid for every value of n if and only if it is valid for $n = 1$. Therefore, the delay target D_i of a flow i can be guaranteed if and only if the following condition is met:

$$(\theta_i + \rho_i T^{\max}) \leq U_i[0, T^{\max} + D_i - D_i^r]. \tag{7.35}$$

We now examine a more specific case, which is the condition required to meet the delay target for only the head-of-queue packet. Let τ_i be the time required to transmit a head-of-queue packet of flow i, and the maximum size of the head of queue packet is $L_{\max}$. Then we may write the following expression assuming that the head-of-queue packet has the maximum size:

$$\left\lfloor \frac{\tau_i}{T^{\max}} \right\rfloor \mu_i^{\max} \leq L_{\max}, \tag{7.36}$$

$$\left\lfloor \frac{\tau_i}{T^{\max}} \right\rfloor \leq \frac{L_{\max}}{Q_i + L_{\max} - 1}, \tag{7.37}$$

and because Q_i and $\tau_i / T^{\max}$ are both integer numbers larger than zero ($Q_i \geq 1$ and $\lfloor \tau_i / T^{\max} \rfloor \geq 1$), as implied by (7.36), $\tau_i \leq T^{\max}$. Here we use inequality instead of equality because the actual head-of-queue packet size may be smaller than $L_{\max}$. So the maximum time needed to transmit a head-of-queue packet from any flow i will never exceed $T^{\max}$, independent of the service quantum Q_i allocated to the flow. We also know that the amount of time needed to transmit a head-of-queue packet must not exceed the maximum delay $D_i - D_i^r$ of any flow i so that the packet can meet its delay upper-bound target. Thus, for I, the following conditions must be satisfied so that each flow in the set can meet its delay upper-bound target:

$$T^{\max} \leq D^{\min} - D^{r,\max}, \tag{7.38}$$

where

$$D^{\min} = \min_{i \in I}\{D_i\}, \tag{7.39}$$

$$D^{r,\max} = \max_{i \in I}\left\{D_i^r\right\}. \tag{7.40}$$

So far we have seen that different service quanta have to be allocated while explicitly taking into account the delay requirements of different flows so that they receive only the right amount of service. As non-time-sensitive flows do not have any specific delay requirement, we assume that multicode-DRR allocates the smallest possible service quantum to them arbitrarily as follows: $Q_j = 1; \forall j \in J$. As such, all non-time-sensitive flows may share equally the remaining capacity not used by time-sensitive flows.

For a set of time-sensitive flows $\{Q_j | i \in I\}$, a set of service quanta is feasible if and only if (7.35) and (7.38) are satisfied by all flows in the set, so that a delay guarantee can be delivered using ε_{I^*}. Thus, a new time-sensitive flow is admissible by multicode-DRR if and only if a feasible set of service quanta exists for the new set of flows. A service quantum allocation algorithm is an algorithm that can produce feasible sets of service quanta. The following is a heuristic algorithm in multicode-DRR which not only allocates service quantum but also evaluates the admission criteria and the existence of a new set of feasible quanta at the same time.

1. *Initialize*. The service quantum for each traffic flow will be initialized to unity, which is the smallest possible value.
2. *Test, then suggest*. The condition given by (7.35) will be used to verify the feasibility of each service quantum set. To do this, each flow $i \in I$ is first tested as follows:

$$\left\lfloor \frac{T^{\max} + D_i - D_i^r}{T^{\max}} \right\rfloor \mu_i^{\max} - L_{\max} \geq \theta_i + \rho_i T^{\max}. \tag{7.41}$$

   When the test above fails for any flow $i \in I$, multicode-DRR concludes that the current set of service quantum is not feasible. Then, a new service quantum will be suggested for a flow by increasing the current quantum by as much as the difference between the current value and an expected value. This expected value is suggested by (7.35) as follows:

$$Q_i = \frac{\theta_i + \rho_i T^{\max} + L_{\max}}{\left\lfloor \dfrac{T^{\max} + D_i - D_i^r}{T^{\max}} \right\rfloor} + L_{\max}. \tag{7.42}$$

3. *Terminate*. The test-then-suggest routine given in step 2 will terminate when a feasible set of service quanta is found or (7.38) is violated. When the condition specified by (7.38) is not met, no feasible set of service quanta exists and an empty set of service quanta will be generated.

The admission control algorithm described above for multicode-DRR will stop after a finite number of iterations. While multicode-DRR requires knowledge of the available capacity to perform its admission control, there are other existing admission control schemes that adopt an admit-then-test strategy without knowing the capacity available. In general, these schemes will admit a new flow into the network and give it a lower service priority than that of all existing flows already admitted into the network. Then the admission control scheme will assess the impact of the newly admitted flow on the existing flows. If the impact is acceptable, assuming that the new flow is given the same service priority as the existing flows, the new flow will be formally admitted. Otherwise, the new flow's admission request will be rejected.

SUMMARY

In this chapter we have presented CEA packet scheduling and admission control as two types of radio resource management schemes that operate in different time scales. CEA packet scheduling takes into account the time-varying nature of a wireless channel quality as it avoids transmitting in a low-quality channel. Analytical and simulation results have indicated that CEA scheduling algorithms can improve network throughput. However, avoiding channel errors alone does not ensure fairness in bandwidth sharing such that each user may not get its committed share of bandwidth. To deliver fairness, CEA scheduling algorithms need to have a compensation mechanism that provides users that have lost their services due to low channel quality with additional bandwidth once they return to the good channel state. Various compensation mechanisms, ranging from counter based to non-counter based and from credit based to debt based, exist. The majority of these CEA scheduling algorithms with compensation can provide long-term fairness but not short-term fairness. Short-term fairness is desirable because it provides bandwidth guarantee from which delay guarantee can be derived. Short-term fairness can be achieved by approximating in a wireless communication environment, a fair scheduling algorithm originally designed for wired networks. In these approximation schemes, short-term fairness is a function of the channel characteristics and the compensation bounds. When short-term fairness is used to derive a delay upper bound, the value derived is also a function of the channel characteristics. This means that longer low channel quality duration will lead to a larger delay upper bound. The relationship between a delay bound and channel characteristics is not acceptable because the channel characteristics depend on user behavior over which a network operator has no control. Deadline-based scheduling algorithms such as feasible earliest due date and proactive earliest due date are better solutions than the short-term fairness approach in providing delay bound over a wireless channel with time-varying quality. This is because the deadline-based method does not depend on bandwidth guarantee to determine delay guarantee. Although bandwidth and delay are important performance objectives for CEA scheduling algorithms, there is always a case where the perceived throughput and delay can be degraded by channel errors and fall below the level guaranteed. For this reason, QoS differentiation is more manageable. There are CEA scheduling algorithms for simple proportional differentiation in average packet delay. Also, there are CEA scheduling algorithms that go beyond simple QoS differentiation to compound QoS that provide QoS differentiation together with guaranteed QoS commitment.

Compared to CEA packet scheduling, admission control has been covered briefly in this chapter. Multicode DRR has been presented as a scheme that performs admission control by first estimating the deterministic value of a time-varying capacity. By dedicating the deterministic capacity component to time-sensitive traffic through admission control, multicode DRR has been shown to provide packet delay guarantees.

REFERENCES

[1] D. Eckhardt and P. Steenkiste, "Measurement and analysis of the error characteristics of an in-building wireless network," *Proceedings of ACM SIGCOMM*, pp. 243–254, Aug. 1996.

[2] K.-W. Lee, M. Cheng, and L. F. Chang, "Wireless QoS analysis for a rayleigh fading channel," *Proceedings of the IEEE International Conference on Communications*, pp. 1089–1093, June 1998.

[3] M. A. Arad and A. Leon-Garcia, "A generalized processor sharing approach to time scheduling in hybrid CDMA/TDMA," *Proceedings of IEEE INFOCOM*, pp. 1164–1171, Mar. 1998.

[4] A. Stamoulis and G. B. Giannakis, "Packet fair queueing scheduling based on multirate multipath-transparent cdma for wireless networks," *Proceedings of IEEE INFOCOM*, pp. 1067–1076, 2000.

[5] I. Koutsopoulos and L. Tassiulas, "Channel state-adaptive techniques for throughput enhancement in wireless broadband networks," *Proceedings of IEEE INFOCOM*, Apr. 2001.

[6] Q. Liu, S. Zhou, and G. B. Giannakis, "Cross-layer combining of adaptive modulation and coding with truncated ARQ over wireless links," *IEEE Trans. Wireless Commun.*, vol. 3, no. 5, pp. 1746–1755, Sept. 2004.

[7] V. Bharghavan, S. Lu, and T. Nandagopal, "Fair queueing in wireless networks: issues and approaches," *IEEE Personal Communi.*, pp. 44–53, Feb. 1999.

[8] Y. Cao and V. O. K. Li, "Scheduling algorithms in broadband wireless networks," *Proc. IEEE*, vol. 89, no. 1, pp. 76–87, Jan. 2001.

[9] L. Wischhof and J. W. Lockwood, "Packet scheduling for link-sharing and quality of service support in wireless local area networks," *Tech. Rep. WUCS-01-35*, Department of Computer Science, Washington University, St. Louis, MO, Nov. 2001.

[10] E. N. Gilbert, "Capacity of a burst-noise channel," *Bell Syst. Tech. J.*, vol. 39, pp. 1253–1265, Sept. 1960.

[11] E. O. Elliott, "Estimates of error rates for codes on burst-noise channels," *Bell Syst. Tech. J.*, vol. 42, pp. 1977–1997, Sept. 1963.

[12] H. S. Wang and N. Moayeri, "Finite-state Markov channel: a useful model for radio communication channels," *IEEE Trans. Veh. Technol.*, vol. 44, no. 1, pp. 163–171, Feb. 1995.

[13] G. Koole, Z. Liu, and R. Righter, "Optimal transmission algorithms for noisy channels," *Oper. Res.*, pp. 892–899, 2001.

[14] P. Bhagwat, P. Bhattacharya, A. Krishna, and S. K. Tripathi, "Enhancing throughput over wireless LANs using channel state dependent packet scheduling," *Proceedings of IEEE INFOCOM*, pp. 1133–1140, Mar. 1996.

[15] P. Bhagwat, P. Bhattacharya, A. Krishna, and S. K. Tripathi, "Using channel state dependent packet scheduling to improve TCP throughput over wireless LANs," *ACM/Baltzer Wireless Network*, vol. 3, no. 1, pp. 91–102, Mar. 1997.

[16] L. Tassiulas and A. Ephremides, "Dynamic server allocation to parallel queues with randomly varying connectivity," *IEEE Trans. Inf. Technol.*, vol. 39, no. 2, Mar. 1993.

[17] N. Bambos and G. Michailidis, "On the stationary dynamics of parallel queues with random server connectivities," *Proceedings of the IEEE Conference on Decision and Control*, Dec. 1995.

[18] M. Inoue, G. Wu, and Y. Hase, "Channel state dependent resource scheduling for wireless message transport," *Proceedings of the IEEE Vehicular Technology Conference*, pp. 1264–1268, 1998.

[19] M. Inoue, G. Wu, and Y. Hase, "Link adaptive resource scheduling for wireless message transport," *Proceedings of the IEEE Global Telecommunications Conference*, pp. 2223–2228, Nov. 1998.

[20] M. Inoue, G. Wu, and Y. Hase, "Resource scheduling with channel state information for wireless message transport," *Proceedings of IEEE ICUPC*, pp. 249–253, Oct. 1998.

[21] Baiocchi, F. Cuomo, and C. Martello, "Joint channel and traffic adaptive packet scheduling over multiaccess radio interfaces," *Proceedings of the IEEE International Conference on Communications*, June 2001.

[22] C. Fragouli, V. Sivaraman, and M. B. Srivastava, "Controlled multimedia wireless link sharing via enhanced class-based queuing with channel-state-dependent packet scheduling," *Proceedings of IEEE INFOCOM*, pp. 572–580, Mar. 1998.

[23] S. Floyd and V. Jacobson, "Link-sharing and resource management models for packet networks," *IEEE/ACM Trans. Network.*, vol. 3, no. 4, pp. 365–386, Aug. 1995.

[24] J. Gomez, A. T. Campbell, and H. Morikawa, "The havana framework for supporting application and channel dependent QoS in wireless networks," *Proceedings of IEEE ICNP*, pp. 235–244, Nov. 1999.

[25] M. Shreedhar and G. Varghese, "Efficient fair queuing using deficit round-robin," *IEEE/ACM Trans. Network.*, vol. 4, no. 3, pp. 375–385, June 1996.

[26] J. Gomez, A. T. Campbell, and H. Morikawa, "A system approach to prediction, compensation and adaptation in wireless networks," *Proceedings of the ACM Workshop on Wireless Mobile Multimedia*, pp. 91–100, Oct. 1998.

[27] G. Miklos and S. Molnar, "Fair allocation of elastic traffic for a wireless base station," *Proceedings of the IEEE Global Telecommunications Conference*, pp. 1673–1678, Dec. 1999.

[28] P. Ramanathan and P. Agrawal, "Adapting packet fair queueing algorithms to wireless networks," *Proceedings of ACM/IEEE MOBICOM*, pp. 1–9, Oct. 1998.

[29] X. Liu, E. K. P. Chong, and N. B. Shroff, "Transmission scheduling for efficient wireless utilization," *Proceeding of IEEE INFOCOM*, Apr. 2001.

[30] S. Lu, V. Bharghavan, and R. Srikant, "Fair scheduling in wireless packet networks," *Proceedings of ACM SIGCOMM*, pp. 63–74, Aug. 1997.

[31] S. Lu, V. Bharghavan, and R. Srikant, "Fair scheduling in wireless packet networks," *IEEE/ACM Trans. Network.*, vol. 7, no. 4, pp. 473–489, Aug. 1999.

[32] T. S. E. Ng, I. Stoica, and H. Zhang, "Packet fair queueing algorithms for wireless networks with location-dependent errors," *Proceedings of IEEE INFOCOM*, pp. 1103–1111, Mar. 1998.

[33] S. Lu, T. Nandagopal, and V. Bharghavan, "A wireless fair service algorithm for packet cellular networks," *Proceedings of ACM/IEEE MOBICOM*, pp. 10–20, Oct. 1998.

[34] S. Lu, T. Nandagopal, and V. Bharghavan, "Design and analysis of an algorithm for fair service in error-prone wireless channels," *ACM/Baltzer Wireless Networks*, Feb. 1999.

[35] D. Ferrari and D. Verma, "A scheme for real-time channel establishment in wide-area networks," *IEEE J. Sel. Areas Commun.*, vol. 8, no. 3, pp. 368–379, Apr. 1990.

[36] J. Hong, X. Tan, and D. Towsley, "A performance analysis of minimum laxity and earliest deadline scheduling in a real-time system," *IEEE Trans. Comput.*, vol. 38, no. 12, Dec. 1989.

[37] S. Shakkottai and R. Srikant, "Scheduling real-time traffic with deadlines over a wireless channel," *ACM/Baltzer Wireless Networks*, vol. 8, no. 1, pp. 13–26, Jan. 2002.

[38] K.-H. Teh, P.-Y. Kong, and S. Jiang, "Proactive earliest due date scheduling in wireless packet networks," *Proceedings of the International Conference on Communication Technology*, pp. 816–820, Apr. 2003.

[39] P.-Y. Kong and K.-H. Teh, "Performance of proactive earliest due date packet scheduling in wireless networks," *IEEE Trans. Veh. Technol.*, vol. 53, no. 4, pp. 1224–1234, July 2004.

[40] Y. Zhou, P. C. Yip, and H. Leung, "On the efficient prediction of fractal signals," *IEEE Trans. Signal Process.*, vol. 45, no. 7, pp. 1865–1868, July 1997.

[41] S. Choi and K. Shin, "A unified wireless LAN architecture for real-time and non-real-time communication services," *IEEE/ACM Trans. Network.*, vol. 8, no. 1, pp. 44–59, Feb. 2000.

[42] K. Nicholas, V. Jacobson, and L. Zhang, "A two-bit differentiated services architecture for the internet," Internet draft, Apr. 1999.

[43] C. Dovrolis, D. Stiliadis, and P. Rmamanathan, "Proportional differentiated services: delay differentiation and packet scheduling," *Proceedings of ACM SIGCOMM*, Aug. 1999.

[44] C. Dovrolis, "Proportional differentiated services for the internet," Ph.D. dissertation, University of Wisconsin–Madison, 2000.

[45] M. R. Jeong, K. Kakami, H. Morikawa, and T. Aoyama, "Wireless scheduler providing relative delay differentiation," *Proceedings of WPMC*, pp. 1050–1055, Nov. 2000.

[46] P.-Y. Kong, K. C. Chua, and B. Bensaou, "Compound QoS commitments for a wireless network with variable capacity," *Proceedings of the IEEE International Conference on Multimedia and Expo*, pp. 1232–1235, Aug. 2001.

[47] P.-Y. Kong, K. C. Chua, and B. Bensaou, "A novel scheduling scheme to share dropping ratio while guaranteeing a delay bound in a multicode-CDMA network," *IEEE/ACM Trans. Network.*, vol. 11, no. 6, pp. 994–1006, Dec. 2003.

[48] D. Li and P.-Y. Kong, "A scheme to provide proportionally differentiated end-to-end packet delay in wireless multi-hop ad hoc networks," *Proceedings of the 5th IFIP Networking Conference—Networking 2006*, Springer LNCS 3976, pp. 1–12, May 2006.

[49] P.-Y. Kong, D. Li, and Y. Zhang, "Quality of service in wireless multi-hop ad hoc networks: a cross-layer framework," *Wireless Quality-of-Service: Techniques, Standards, and Applications,* CRC Press, pp. 179–218, Oct. 2008.

[50] P.-Y. Kong, K. C. Chua, and B. Bensaou, "MultiCode-DRR: a packet scheduling algorithm for delay guarantee in a MultiCode-CDMA network," *IEEE Trans. Wireless Commun.*, vol. 4, no. 6, pp. 2694–2704, Nov. 2005.

CHAPTER 8

QUALITY OF SERVICE FOR MULTIMEDIA SERVICES

8.1 INTRODUCTION

Quality of service (QoS) is an engineering term that refers to the relative priorities provided to different users in a network, or to guaranteeing a certain level of performance to a data flow. This level of performance could be delay, delay jitter, packet loss, minimum bandwidth, and so on. Additional examples of levels of performance are described in this chapter. These parameters are also related to the types of traffic in the network. The types of traffic could be voice, video, and data traffic, or a mixture. This gives rise to multimedia traffic, which can be a combination of voice, video, and data. In numerous performance analyses of communication networks, the arrival process is often assumed to be Poisson distributed. This is applicable for modeling connection arrival processes. Such a Markovian process enables closed-form or analytically tractable solutions in the performance analyses. However, based on *measurements*, it was found that network technologies such as Ethernet local area network (LAN) and integrated services digital network (ISDN), application layer protocols such as world wide web (WWW), TELNET, and file transfer protocol (FTP), and signaling protocols such as common channel signaling network (CCSN) have *self-similar data traffic* [1,2]. Thus, the performance of communication networks with a new self-similar data traffic model should be studied rather than with the extensively used Poisson data traffic model at the packet level for these traffic models. Since self-similar data traffic is not Markovian, a closed-form solution is difficult,

---

Wireless Broadband Networks, By David Tung Chong Wong, Peng-Yong Kong, Ying-Chang Liang, Kee Chaing Chua, and Jon W. Mark

although more results are emerging in the literature. For example, the aggregate of heavy-tailed on/off data sources, with Pareto-distributed on and off periods, forms self-similar data traffic. The Pareto distribution is a heavy-tailed distribution because its tail distribution decays slower than that of an exponential distribution. Examples of sources that can be modeled as Pareto distribution include Ethernet traffic and the web browser [1]. Thus, traffic models that are derived from measurements are more realistic. Therefore, performance analysis should be done with these traffic models. These traffic models can also help in network planning where the end user's QoS is also very important.

Basically, traffic can be classified as delay-sensitivity and loss-sensitivity. If the traffic is *delay sensitive*, it is a real-time traffic. Otherwise, the traffic is a non-real-time traffic. On the other hand, if the traffic is *loss sensitive*, it cannot tolerate any loss in data. Otherwise, the traffic is not loss sensitive and it can tolerate some data loss. For example, voice traffic is delay sensitive but not loss sensitive, as it is real-time traffic but can tolerate some voice clippings. At the other end, data traffic is not delay sensitive but is loss sensitive, as it is generally not real time but cannot tolerate any data loss. Besides delay and loss sensitivity, an additional QoS parameter for video traffic is *delay jitter*, the variation in the delays of the packets received. This has an impact on the size of the video playout buffer, which ensures the smoothness of the playout of video scenes.

In this chapter a number of traffic models are first presented for voice, voice over IP, video, file transfer protocol (FTP), email, web browsing, and network gaming. For voice over IP (VoIP), the voice signal is packetized and transmitted through the Internet using the Internet protocol (IP). Video can be encoded using MPEG-1, MPEG-2, and MPEG-4 for storage or transmission. FTP is used to transfer files between two hosts linked through a network or the Internet. Email is electronic mail that is written and sent through a network or through the Internet. Web browsing uses software that allows the user to search for information on the Internet. Network gaming allows a number of users whose computers are connected through a network to play or compete with each other in a computer game.

In the next part of this chapter, the quality of service (QoS) requirements or relative QoS requirements for cellular systems, WiMax fixed broadband wireless access system, IEEE 802.11e wireless local area network (WLAN), and WiMedia wireless personal area network (WPAN) are presented. There are four traffic classes in universal mobile telecommunications systems (UMTSs), and there are four scheduling services in WiMax fixed broadband wireless access system. On the other hand, there are four access categories in both IEEE 802.11e enhanced distributed channel access (EDCA) MAC and WiMedia prioritized channel access (PCA) MAC. A universal mobile telecommunications system (UMTS) is a third generation (3G) cellular access technology. UMTS is specified by 3G partnership project (3GPP). Wideband code-division multiple access (WCDMA) in UMTS can support a data rate of up to 2 Mbps. Long-term evolution (LTE) is a 3.9G cellular access technology supported by 3GPP. It can support a downlink data rate of 100 Mbps and an uplink data rate of 50 Mbps. More details on LTE are provided in Chapter 9. WiMax is a wireless metropolitan area network (WMAN) that can support a raw data rate in excess of 120 Mbps. IEEE 802.11e is used to provide multiclass differentiation in WLANs of up to four access categories: IEEE 802.11b, IEEE 802.11a, IEEE 802.11g, and the upcoming IEEE 802.11n WLAN. IEEE 802.11b can support a data rate of up to 11 Mbps, and IEEE 802.11a and IEEE 802.11g can support a data rate of up to 54 Mbps. The upcoming IEEE 802.11n can support a data rate of up to 600 Mbps with multiple input, multiple output (MIMO) technology. MIMO means that multiple transmit and receive antennas are used. More details on IEEE 802.11e and IEEE 802.11n are given in Chapter 11. WiMedia can support a data rate of up to 480 Mbps. More details on WiMedia are given in Chapter 12.

In the last part of this chapter, the outage probability, which is also a QoS metric, is studied through a case study. A performance analysis of multiclass video services in a multirate DS-CDMA system is presented. This analysis can also be used to model voice and data traffics.

The chapter is organized as follows. In Section 8.1 we give an overview of QoS and the shift in traffic modeling from Poissonian models to self-similar models. In Section 8.2 we present the various traffic models and in Section 8.2.1, traffic models for voice traffic, including the classical voice traffic model and the voice over IP traffic model. Section 8.2.2 covers traffic models for video traffic, including Maglaris's model, Sen's model, the Markov modulated Poisson process (MMPP) model, and the fractional ARIMA model. In Section 8.2.3 we present traffic models for data traffic. The data traffic that are considered in this section are file transfer protocol and email. Other traffic models are presented in Section 8.2.4. These models include Web browsing and network gaming traffic. In Section 8.3 we present the quality of four types of wireless systems. The UMTS QoS requirements are presented in Section 8.3.1, and the four traffic classes in UMTSs are described. Four scheduling services and their key QoS parameters in WiMax are described in Section 8.3.2. Four traffic classes in IEEE 802.11e wireless local area network and its access category parameters are presented in Section 8.3.3. Similarly, four traffic classes in a WiMedia wireless personal area network and its access category parameters are presented in Section 8.3.4. Another QoS parameter of outage probability is considered in an example of multiclass video services in a multirate DS-CDMA system in Section 8.4. In Section 8.5 we summarize the salient aspects of the contents of the chapter, including extensive references to papers on modeling of traffic in the literature.

8.2 TRAFFIC MODELS

8.2.1 Voice Traffic

Two traffic models for voice traffic are presented in this section: a traditional voice traffic model and a voice over IP traffic model. The former model leads to easy closed-form analysis, whereas it is more difficult to get closed-form analysis using the latter.

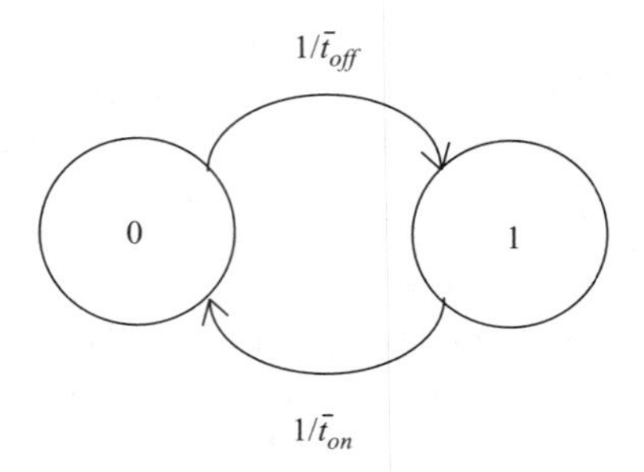

FIGURE 8.1 Classical voice traffic model.

8.2.1.1 Traditional Voice Traffic Model Traditionally, voice is transmitted from one person to another person through a circuit-switched network. The source model for the voice traffic can be represented by a two-state Markov chain as shown in Figure 8.1. In the figure, $\bar{t}_{\text{on}}$ and $\bar{t}_{\text{off}}$ are, respectively, the mean on and mean off periods. The voice traffic can be modeled by an exponential on/exponential off process as shown in Figure 8.2, where T is the packetization time. Let $\tilde{t}_{\text{on}}$ and $\tilde{t}_{\text{off}}$ denote, respectively, the random variables for the on and off periods. The probability density function (pdf) of the on and off periods is given, respectively, by

$$f_{\tilde{t}_{\text{on}}}(t) = \frac{1}{\bar{t}_{\text{on}}} e^{-t/\bar{t}_{\text{on}}} \tag{8.1}$$

and

$$f_{\tilde{t}_{\text{off}}}(t) = \frac{1}{\bar{t}_{\text{off}}} e^{-t/\bar{t}_{\text{off}}}. \tag{8.2}$$

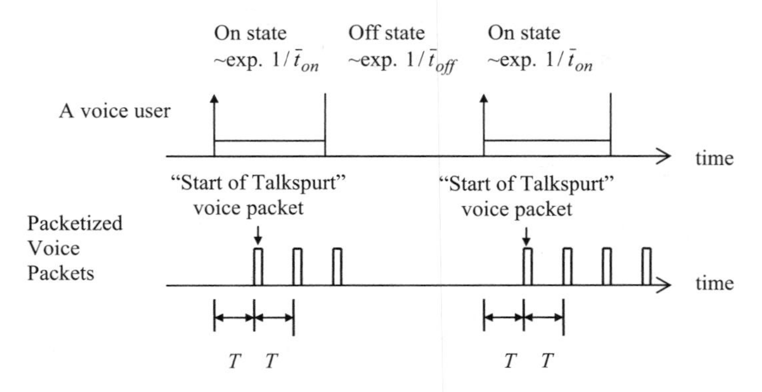

FIGURE 8.2 Exponential on/exponential off process for voice traffic.

Let p_{on} and p_{off} denote the probability in the on state and the probability in the off state, respectively. p_{on} and p_{off} are given, respectively, by

$$p_{\text{on}} = \frac{\bar{t}_{\text{on}}}{\bar{t}_{\text{on}} + \bar{t}_{\text{off}}} \tag{8.3}$$

and

$$p_{\text{off}} = \frac{\bar{t}_{\text{off}}}{\bar{t}_{\text{on}} + \bar{t}_{\text{off}}}. \tag{8.4}$$

8.2.1.2 Heavy-Tailed Voice over IP Traffic Model

For voice over IP (VoIP), the voice signal is packetized and transmitted through the Internet using Internet protocol (IP). The source model for the VoIP traffic can be represented by an on/off process. However, the on and off periods are not distributed exponentially but are Pareto distributed [3]. Let $\tilde{t}_{\text{on}}$ and $\tilde{t}_{\text{off}}$ denote, respectively, the random variables for the on and off periods. Their pdf's are given, respectively, by

$$f_{\tilde{t}_{\text{on}}} = \frac{a_{\text{on}} b_{\text{on}}^{\alpha_{\text{on}}}}{x^{\alpha_{\text{on}}+1}}, \qquad x \geq b_{\text{on}} \tag{8.5}$$

and

$$f_{\tilde{t}_{\text{off}}} = \frac{a_{\text{off}} b_{\text{off}}^{\alpha_{\text{off}}}}{x^{\alpha_{\text{off}}+1}}, \qquad x \geq b_{\text{off}}, \tag{8.6}$$

where a_{on}, b_{on}, and α_{on} are the constant, the scaling parameter, and the shaping parameter of the Pareto-distributed on period, respectively, and a_{off}, b_{off}, and α_{off} are the constant, the scaling parameter, and the shaping parameter of the Pareto-distributed off period, respectively.

8.2.2 Video Traffic

Video can be encoded using MPEG-1, MPEG-2, or MPEG-4 for storage or transmission. Four traffic models for video traffic are presented in this section. One is Maglaris's variable-bit-rate video traffic model, an other is Sen's variable-bit-rate video traffic model with scene changes, a third is the MMPP video traffic model, and the fourth model is a fractional ARIMA video traffic model. The former three models lead to easy closed-form analysis, whereas it is more difficult to perform closed-form analysis with the fourth model although it generates self-similar variable-bit-rate video traffic.

8.2.2.1 Maglaris's Video Traffic Model

A classical video traffic model is Maglaris's model, which can be represented by a one dimensional Markov chain with finite states as shown in Figure 8.3. Each state represents the discrete level of bit rate generated by a single source. Figure 8.4 shows a variable-bit-rate video source

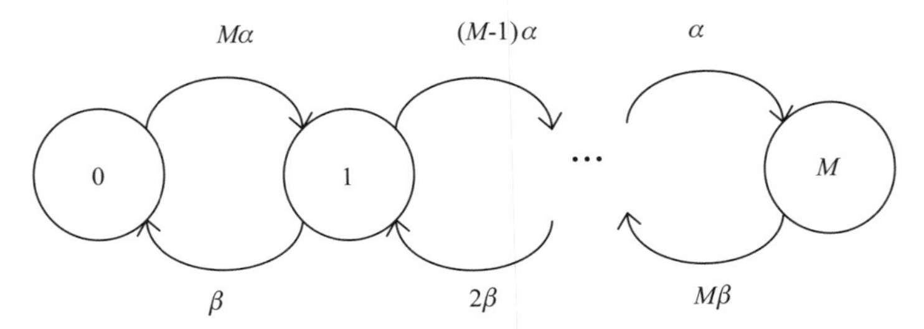

FIGURE 8.3 Classical Maglaris video traffic model.

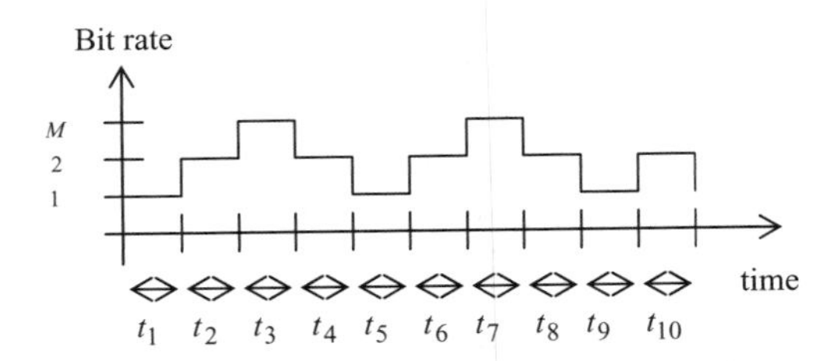

FIGURE 8.4 Discrete levels for the variable-bit-rate Maglaris video traffic model.

with discrete bit rate levels. We assume that the highest level is state M. If $M = 1$, the source is an on/off source. The bit rate fluctuation is represented in Figure 8.4 by a thin solid line. For ease of illustration, the time durations at each level for a source are shown to be equal. Each level can be modeled by a two-state minisource with an increase rate of α and a decrease rate of β. Thus, the continuous-time Markov chain for a single video source at state m has an increase rate of $(M - m)\alpha$ and a decrease rate of $m\beta$.

The steady-state probability of being in state m, denoted by P_m, is given by

$$P_m = \binom{M}{m} p^m (1-p)^{M-m}, \qquad m = 0, 1, 2, \ldots, M, \tag{8.7}$$

where

$$p = \frac{\alpha}{\alpha + \beta} \tag{8.8}$$

and its mean, second moment, and variance are Mp, $Mp[1 + (M - 1)p]$, and $Mp[1 - p]$, respectively. Note that P_m is a binomial distribution.

8.2.2.2 Sen's Video Traffic Model Another classical traffic model is Sen's model, which is an extension of Maglaris's model to account for scene changes. It is representable by a two-dimensional Markov chain, as shown in Figure 8.5. The horizontal state transitions are for low-bite-rate fluctuations, and the vertical state

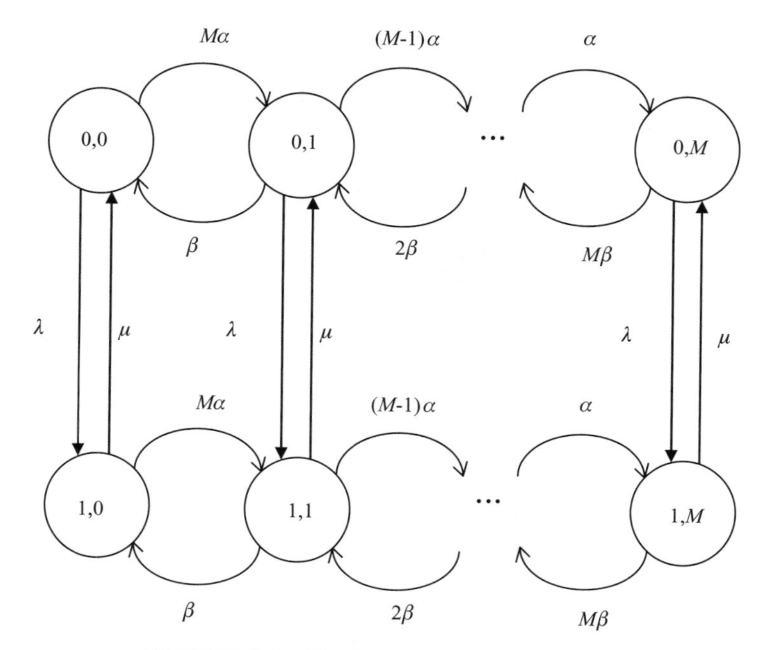

FIGURE 8.5 Classical Sen video traffic model.

transitions are for high-bit-rate transitions. Figure 8.6 shows a variable-bit-rate video source with discrete bit rate levels. Each level has a bit rate of R_l. The bit rate level depends on the sum of low- and high-bit-rate levels. The highest level for low-bit-rate fluctuation is M, and the level for high-bit-rate fluctuation is either zero or X. The low-bit-rate fluctuation is represented in Figure 8.6 by a thin solid line, and the high-bit-rate fluctuation is represented by a dashed line. The total bit rate is represented by a thick solid line. Thus, it has a highest bit rate of $(M + X)R_l$. For ease of illustration, the time durations at each level for a source are shown as being equal. Each

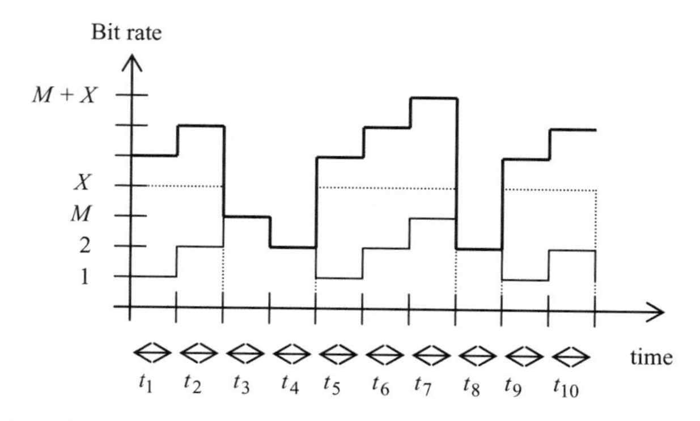

FIGURE 8.6 Discrete levels for the variable-bit-rate Sen video traffic model.

low-bit-rate level can be modeled by a two-state minisource with an increase rate of α and a decrease rate of β, while each high-bit-rate level can be modeled by a two-state minisource with an increase rate of λ and a decrease rate of μ. Thus, the horizontal low-bit-rate Markov chain at state (x, m) has an increase rate of $(M - m)\alpha$ and a decrease rate of $m\beta$ in the horizontal directions.

The steady-state probability of being in state (x, m), denoted by $P_{x,m}$, is given by

$$P_{x,m} = \binom{1}{x} q^x (1-q)^{1-x} \binom{M}{m} p^m (1-p)^{M-m}, \qquad m = 0, 1, 2, \ldots, M, \quad (8.9)$$

where

$$q = \frac{\lambda}{\lambda + \mu} \quad (8.10)$$

$$p = \frac{\alpha}{\alpha + \beta}. \quad (8.11)$$

Note that $P_{x,m}$ is a product of two binomial distributions.

8.2.2.3 Markov Modulated Poisson Process Video Traffic Model Another classical video traffic model is the M-state Markov modulated Poisson process (MMPP) model shown in Figure 8.7. The video arrival process is modeled as an M-state MMPP. Let h_i be the probability of being in phase i in steady state. Then we have

$$[h_1 h_2 \cdots h_M] = [h_1 h_2 \cdots h_M] \boldsymbol{H}\ , \quad (8.12)$$

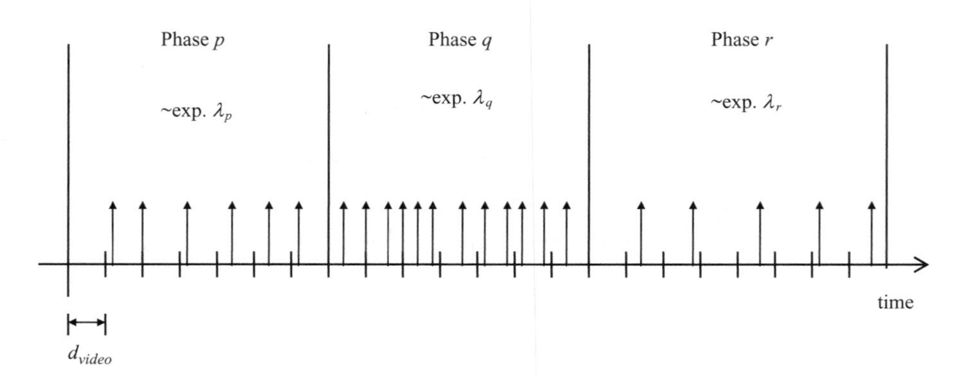

FIGURE 8.7 M-state Markov modulated Poisson process video traffic model.

where the $M \times M$ phase transition probability matrix in a slot time $\boldsymbol{H} = [h_{ij}]$, $(1 \leq i, j \leq M)$, is given by [4]

$$H = \begin{bmatrix} 9.998{\cdot}10^{-1} & 9.848{\cdot}10^{-5} & 5.820{\cdot}10^{-9} & 2.675{\cdot}10^{-13} & 1.054{\cdot}10^{-17} & 3.738{\cdot}10^{-22} & 1.227{\cdot}10^{-26} & 1.929{\cdot}10^{-31} \\ 1.688{\cdot}10^{-4} & 9.997{\cdot}10^{-1} & 1.182{\cdot}10^{-4} & 8.148{\cdot}10^{-9} & 4.280{\cdot}10^{-13} & 1.897{\cdot}10^{-17} & 7.475{\cdot}10^{-22} & 1.371{\cdot}10^{-26} \\ 1.694{\cdot}10^{-8} & 2.007{\cdot}10^{-4} & 9.997{\cdot}10^{-1} & 1.379{\cdot}10^{-4} & 1.086{\cdot}10^{-8} & 6.420{\cdot}10^{-13} & 3.162{\cdot}10^{-17} & 6.959{\cdot}10^{-22} \\ 1.039{\cdot}10^{-12} & 1.847{\cdot}10^{-8} & 1.840{\cdot}10^{-4} & 9.997{\cdot}10^{-1} & 1.576{\cdot}10^{-4} & 1.397{\cdot}10^{-8} & 9.172{\cdot}10^{-13} & 2.523{\cdot}10^{-17} \\ 4.346{\cdot}10^{-17} & 1.030{\cdot}10^{-12} & 1.539{\cdot}10^{-8} & 1.673{\cdot}10^{-4} & 9.997{\cdot}10^{-1} & 1.772{\cdot}10^{-4} & 1.746{\cdot}10^{-8} & 6.403{\cdot}10^{-13} \\ 1.309{\cdot}10^{-21} & 3.877{\cdot}10^{-17} & 7.724{\cdot}10^{-13} & 1.259{\cdot}10^{-8} & 1.505{\cdot}10^{-4} & 9.997{\cdot}10^{-1} & 1.969{\cdot}10^{-4} & 1.083{\cdot}10^{-8} \\ 2.920{\cdot}10^{-26} & 1.038{\cdot}10^{-21} & 2.585{\cdot}10^{-17} & 5.618{\cdot}10^{-13} & 1.007{\cdot}10^{-8} & 1.338{\cdot}10^{-4} & 9.997{\cdot}10^{-1} & 1.100{\cdot}10^{-4} \\ 4.885{\cdot}10^{-31} & 2.026{\cdot}10^{-26} & 6.059{\cdot}10^{-22} & 1.645{\cdot}10^{-17} & 3.932{\cdot}10^{-13} & 7.835{\cdot}10^{-9} & 1.171{\cdot}10^{-4} & 9.999{\cdot}10^{-1} \end{bmatrix}. \tag{8.13}$$

In each phase, the video arrival process is Poisson distributed with a mean video arrival rate λ_i, in a video slot d_{video}. For a Poisson arrival process with a mean rate λ, the probability of $\tilde{G}$ arrivals in any interval of length t is given by

$$\Pr[\tilde{G} = g] = \frac{(\lambda t)^g e^{-\lambda t}}{g!}. \tag{8.14}$$

Let $\tilde{C}$ be the number of video packets arriving during an interval of length t. For one video user, $\tilde{C}$ is the weighted sum of the Poisson arrival processes in each of the M phases. Its pdf is given by

$$\Pr[\tilde{C} = c] = \sum_{v=1}^{M} \frac{\left[\frac{\lambda_v}{d_{\text{video}}} t\right]^c e^{-(\lambda_v/d_{\text{video}})t}}{c!} h_v. \tag{8.15}$$

For two video users, $\tilde{C}$ is the weighted sum of the Poisson arrival processes in each of the M phases for users 1 and 2. From the property of Poisson processes, the sum of two Poisson processes with mean arrival rate λ_x and λ_y is also a Poisson process with a mean rate $\lambda_x + \lambda_y$. Then for two video users, the pdf of $\tilde{C}$ is given by

$$\Pr[\tilde{C} = c] = \sum_{u=1}^{M} \sum_{v=1}^{M} \frac{[[(\lambda_u + \lambda_v)/d_{\text{video}}]t]^c e^{-(\lambda_u+\lambda_v)/d_{\text{video}}t}}{c!} h_u h_v. \tag{8.16}$$

For M_{video} video users, $\tilde{C}$ is the weighted sum of the Poisson arrival processes in each of the M phases for users 1 to M_{video}.

Therefore, generalizing for M_{video} video users, the pdf of $\tilde{C}$ is given by

$$\Pr[\tilde{C} = c] = \sum_{i_1=1}^{M} \sum_{i_2=1}^{M} \cdots \sum_{i_{M_{\text{video}}}=1}^{M} \left[\frac{\sum_{l=1}^{M_{\text{video}}} \lambda_{i_l}}{d_{\text{video}}} t\right]^c \frac{e^{-\left(\sum_{l=1}^{M_{\text{video}}} \lambda_{i_l}/d_{\text{video}}\right)t}}{c!} \prod_{l=1}^{M_{\text{video}}} h_{i_l}. \tag{8.17}$$

8.2.2.4 Fractional ARIMA Video Traffic Model This model uses a fractional ARIMA(0,d,0) process to generate the background sequence, where the zeros indicate that no autoregressive (AR) and moving average (MA) parameters are specified [5,6]. d is the fractional differencing parameter and is equal to $H - 0.5$, where H is the Hurst parameter. The process X_k has Gaussian marginals with zero mean and variance v_0 and its autocorrelation function is given by

$$\rho_k = \frac{d(1+d)\cdots(k-1+d)}{(1-d)(2-d)\cdots(k-d)}. \tag{8.18}$$

X_0 is chosen from the normal distribution $N(0, v_0)$. The background process, $\{X_k\}$, is generated using Hosking's algorithm, which requires $o(n^2)$ computation time. Each X_k is Gaussian distributed with mean m_k and variance v_k, which are given recursively by iterating the following for $k = 1, \ldots, n$:

$$N_k = \rho_k - \sum_{j=1}^{k-1} \phi_{k-1,j}\rho_{k-j}, \tag{8.19}$$

$$D_k = \frac{D_{k-1} - N_{k-1}^2}{D_{k-1}}, \tag{8.20}$$

$$\phi_{kk} = \frac{N_k}{D_k}, \tag{8.21}$$

$$\phi_{kj} = \phi_{k-1,j} - \phi_{kk}\phi_{k-1,k-j}, \quad j = 1, \ldots, k-1, \tag{8.22}$$

$$m_k = \sum_{j=1}^{k} \phi_{kj} X_{k-j}, \tag{8.23}$$

$$v_k = (1 - \phi_{kk}^2)v_{k-1}. \tag{8.24}$$

The initial values of N_0 and D_0 are zero and 1 respectively. The number of bits per frame is represented by the foreground sequence, $\{Y_k\}$, which is generated by mapping each point as

$$Y_k = F_{\Gamma/P}^{-1}(F_N(X_k)), \tag{8.25}$$

where F_N is the standard normal cumulative distribution function and $F_{\Gamma/P}^{-1}$ is the inverse cumulative distribution function of a hybrid Gamma/Pareto cumulative distribution function. This hybrid distribution consists of a concatenation of a Gamma distribution for the left tail and a Pareto distribution for the right tail of the empirical distribution. The Gamma distribution has a pdf given by [5]

$$f_\Gamma(x) = e^{-\lambda x}\frac{\lambda(\lambda x)^{s-1}}{\Gamma(s)}, \tag{8.26}$$

where s and λ are the shape and scale parameters and can be determined from the mean and variance of the empirical trace. The Pareto pdf is given by [5]

$$f_P(x) = \frac{ak^a}{x^{a+1}}, \qquad x \geq k, \tag{8.27}$$

with a cumulative distribution function of

$$F_P(x) = 1 - \left(\frac{k}{x}\right)^a. \tag{8.28}$$

The parameters that are needed in this model are $(\mu_\Gamma, \sigma_\Gamma, m_T, H)$, where μ_Γ and σ_Γ are the mean and variance of the Gamma distribution and m_T is the slope of the line that best fits the tail of the Pareto distribution.

8.2.3 Data Traffic

Two traffic models for data traffic are presented in this section. One is a file transfer protocol traffic model; the other is an email traffic model.

8.2.3.1 File Transfer Protocol File transfer protocol (FTP) is used to transfer files between two hosts linked through a network or the Internet. The FTP traffic can be modeled by a Pareto on/Weibull off process, as shown in Figure 8.8, where T is the packetization time. In the figure, $\bar{t}_{\text{on}}$ and $\bar{t}_{\text{off}}$ are, respectively, the mean on and off periods. Let $\tilde{t}_{\text{on}}$ and $\tilde{t}_{\text{off}}$ denote, respectively, the random variables for the on and off periods. The pdfs of the on and off periods are given, respectively, by

$$f_{\tilde{t}_{\text{on}}}(t) = \frac{\alpha_{\text{on}}\beta_{\text{on}}^{\alpha_{\text{on}}}}{(\beta_{\text{on}} + t)^{\alpha_{\text{on}}+1}} \tag{8.29}$$

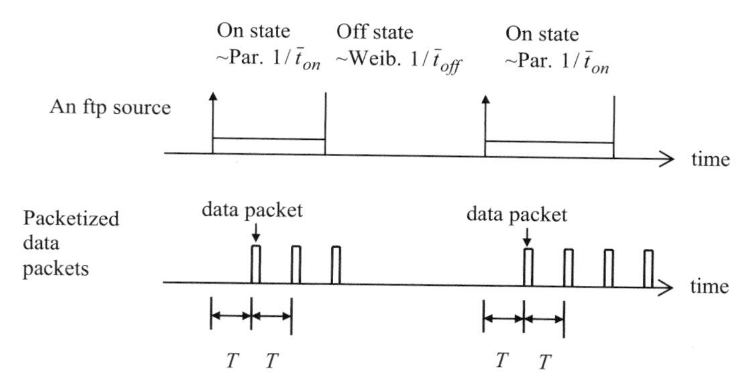

FIGURE 8.8 File transfer protocol traffic model.

and

$$f_{\tilde{t}_{\mathrm{off}}}(t) = \alpha_{\mathrm{off}}\beta_{\mathrm{off}}t^{\beta_{\mathrm{off}}-1}e^{-\alpha_{\mathrm{off}}t^{\beta_{\mathrm{off}}}}, \qquad \alpha_{\mathrm{off}} > 0 \text{ and } \beta_{\mathrm{off}} \geq 0, \tag{8.30}$$

where α_{on} is the intensity of the Noah effect (i.e., have high variability or infinite variance) or shaping parameter, β_{on} is the scaling parameter for the on period, α_{off} is the shaping parameter, and β_{off} is the scaling parameter for the off period. Assuming a Pareto-distributed on period, its cumulative distribution function (CDF) is given by

$$1 - F_{\tilde{t}_{\mathrm{on}}}(t) = \frac{\beta_{\mathrm{on}}^{\alpha_{\mathrm{on}}}}{(\beta_{\mathrm{on}} + t)^{\alpha_{\mathrm{on}}}}, \qquad 1 < \alpha_{\mathrm{on}} < 2, \tag{8.31}$$

and β_{on} is given by

$$\beta_{\mathrm{on}} = \bar{t}_{\mathrm{on}}\,(\alpha_{\mathrm{on}} - 1)\,. \tag{8.32}$$

The aggregate of heavy-tailed on/off traffic such as the Pareto or Weibull distribution forms self-similar traffic.

8.2.3.2 Email Email is electronic mail that is written and sent through a network or through the Internet. Email traffic can be modeled by a Weibull on/Pareto off process as shown in Figure 8.9, where T is the packetization time. In the figure, $\bar{t}_{\mathrm{on}}$ and $\bar{t}_{\mathrm{off}}$ are, respectively, the mean on and off periods. Let $\tilde{t}_{\mathrm{on}}$ and $\tilde{t}_{\mathrm{off}}$ denote, respectively, the random variables for the on and off periods. The pdfs of the on and off periods are given, respectively, by

$$f_{\tilde{t}_{\mathrm{on}}}(t) = \alpha_{\mathrm{on}}\beta_{\mathrm{on}}t^{\beta_{\mathrm{on}}-1}e^{-\alpha_{\mathrm{on}}t^{\beta_{\mathrm{on}}}}, \quad \alpha_{\mathrm{on}} > 0, \quad \text{and} \quad \beta_{\mathrm{on}} \geq 0 \tag{8.33}$$

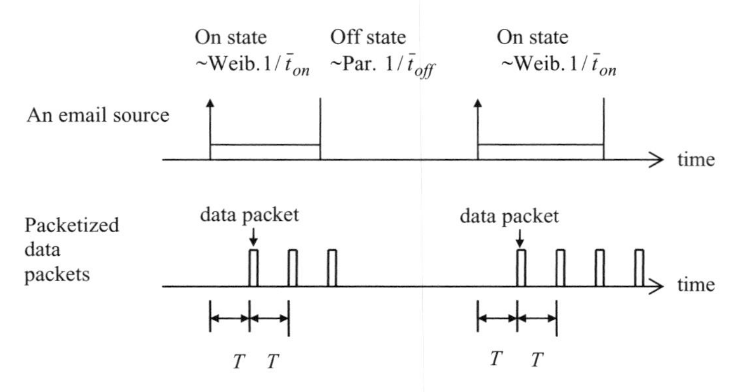

FIGURE 8.9 Email and web browsing traffic model.

and

$$f_{\tilde{t}_{\text{off}}}(t) = \frac{\alpha_{\text{off}} \beta_{\text{off}}^{\alpha_{\text{off}}}}{(\beta_{\text{off}} + t)^{\alpha_{\text{off}}+1}}, \tag{8.34}$$

where α_{on} is the shaping parameter β_{on} is the scaling parameter for the on period, α_{off} is the shaping parameter, and β_{off} is the scaling parameter for the off period. Assuming a Pareto-distributed off period, its CDF is given by

$$1 - F_{\tilde{t}_{\text{off}}}(t) = \frac{\beta_{\text{off}}^{\alpha_{\text{off}}}}{(\beta_{\text{off}} + t)^{\alpha_{\text{off}}}}, \qquad 1 < \alpha_{\text{off}} < 2, \tag{8.35}$$

and β_{on} is given by

$$\beta_{\text{off}} = \bar{t}_{\text{off}} \left(\alpha_{\text{off}} - 1\right). \tag{8.36}$$

8.2.4 Other Traffic

Two other traffic models are presented in this section. One is a web browsing traffic model, while the other is a network gaming traffic model.

8.2.4.1 Web-Browsing Traffic

Web browsing uses software that allows the user to search for information on the Internet. The web browsing traffic can be modeled by a Pareto on/Pareto off process as shown in Figure 8.10, or a Weibull on/Pareto off process as shown in Figure 8.9, where T is the packetization time. In the figure, $\bar{t}_{\text{on}}$ and $\bar{t}_{\text{off}}$ are, respectively, the mean on and off periods. Let $\tilde{t}_{\text{on}}$ and $\tilde{t}_{\text{off}}$ denote, respectively, the random variables for the on and off periods. The pdfs of the on period for Pareto or Weibull distribution are given, respectively, by (8.33) and (8.29), and the pdf of the off period for Pareto distribution is given by (8.34).

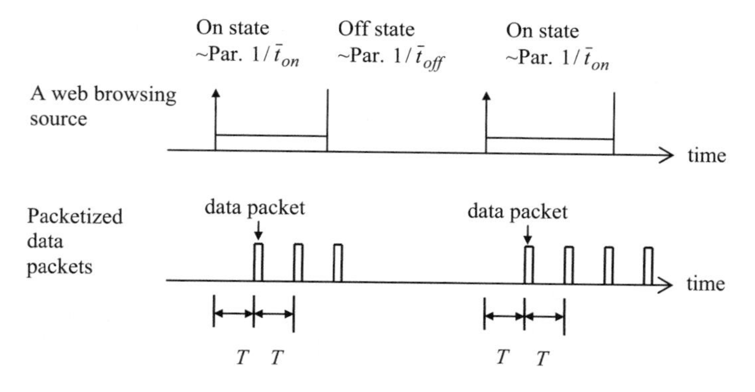

FIGURE 8.10 Web browsing traffic model.

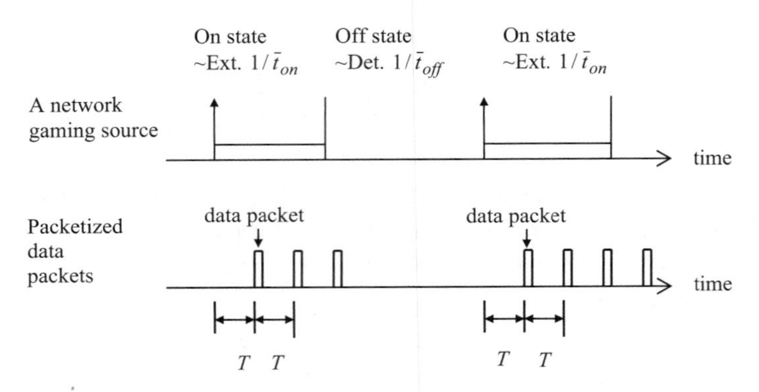

FIGURE 8.11 Network gaming traffic model.

8.2.4.2 Network Gaming Traffic Network gaming allows a number of users whose computers are connected through a network to play or compete with each other in a computer game. Both the network gaming service and client traffic can be modeled by an extreme value function on/deterministic off process, as shown in Figure 8.11. In the figure, $\bar{t}_{\text{on}}$ and $\bar{t}_{\text{off}}$ are, respectively, the mean on and off periods. Let $\tilde{t}_{\text{on}}$ and $\tilde{t}_{\text{off}}$ denote, respectively, the random variables for the on and off periods. The pdf of the on and off periods for the server per client are given, respectively, by [7]

$$f_{\tilde{t}_{\text{on}}}(t) = \frac{1}{b} e^{-(x-a)/b} e^{-e^{-(x-a)/b}}, \qquad b > 0 \tag{8.37}$$

and

$$f_{\tilde{t}_{\text{off}}}(t) = \bar{t}_{\text{off}}, \tag{8.38}$$

where

$$a = 34.5 + 4.2n, \tag{8.39}$$

$$b = 9 + 3n, \tag{8.40}$$

$$\bar{t}_{\text{off}} = 60 \text{ ms}. \tag{8.41}$$

and n is the number of players. Similarly, the pdf of the on and off periods for the client are given, respectively, by (8.37) and (8.38), with $a = 41$, $b = 6$, and $\bar{t}_{\text{off}} = 50$ ms.

8.3 QUALITY OF SERVICE IN WIRELESS SYSTEMS

8.3.1 Universal Mobile Telecommunications System Traffic

Universal mobile telecommunications systems (UMTSs) are third-generation (3G) mobile phone systems. The most common system is based on wideband code-division multiple access (WCDMA). The underlying standard is standardized by 3GPP. A WCDMA system can deliver a data rate of up to 2 Mbps, with geographical coverage that includes indoor and low-range outdoor, urban outdoor, and rural outdoor. A UMTS defines four traffic classes: (1) conversational traffic, (2) streaming traffic, (3) interactive traffic, and (4) background traffic. The UMTS QoS requirements are shown in Table 8.1.

8.3.1.1 Conversational Traffic Conversational traffic is real-time delay-sensitive traffic with stringent low delay. Voice, video gaming, and video telephony traffic are a few examples. Voice can be modeled by an exponential on/exponential off source for circuit-switched traffic.

TABLE 8.1 UMTS QoS Requirements

				Key Performance Parameters Values and Target		
Medium	Application	Degree of Symmetry	Data Rate (kbps)	End-to-end One-Way Delay	Delay Variation Within a Call	Information Loss
Audio	Conversational voice	Two-way	4–13	<150 ms preferred; <400 ms limit	<1 ms	<3% FER
Video	Videophone	Two-way	32–384	<150 ms preferred; <400 ms limit; lip-synch. <100 ms	—	<1% FER
Data	Web browsing–HTML	Primarily one-way	—	<4 s per page	—	Zero
Data	Email (server access)	Primarily one-way	—	<4 s	—	Zero
Data	Bulk data transfer/retrieval	Primarily one-way	—	<10 s	—	Zero

8.3.1.2 Streaming Traffic Streaming traffic is real-time streaming multimedia traffic with a preserved time relation between the information entities of the stream. Video on demand, multimedia, and webcast are a few examples. Video can be modeled by Sen's model or the fractional ARIMA model for video traffic.

8.3.1.3 Interactive Traffic Interactive traffic is non-real-time loss-sensitive traffic with a delay response better than that of background traffic. However, the response time should not be too inconsistent or too long or in an unordered fashion. Web browsing, network gaming, and database access are a few examples. Web browsing can be modeled by a Pareto-distributed on/off source or a Weibull-distributed on/pareto-distributed off source.

8.3.1.4 Background Traffic Background traffic is non-real-time loss-sensitive traffic with data integrity and reliable data transfer. Email, FTP, and short message service (SMS) are some examples. Email can be modeled by a Weibull-distributed on/Pareto-distributed off source, while FTP can be modeled by a Pareto-distributed on/Weibull-distributed off source.

8.3.1.5 UMTS QoS Parameters/Attributes UMTS defines the following QoS parameters and attributes:

- Maximum bit rate in kbps
- Guaranteed bit rate in kbps
- Delivery order in terms of yes/no
- Maximum service data unit (SDU) in bytes
- SDU format information in terms of bits
- SDU error ratio
- Residual bit error ratio
- Delivery of erroneous SDU in terms of yes/no
- Transfer delay in milliseconds
- Traffic-handling priority
- Allocation–retention priority
- Source statistics descriptor in terms of speech or unknown
- Signaling indication in terms of yes/no

The maximum bit rate is the maximum number of bits delivered by UMTS and to UMTS at a service access point (SAP) within a period of time, divided by the duration of the period. A token bucket algorithm with a token rate equal to the maximum bit rate and a bucket size equal to the maximum SDU size enforces that the traffic is conformant to the maximum bit rate. This bit rate is also the upper limit that an application can accept or provide. All UMTS bearer service attributes may be fulfilled for traffic up to this bit rate. The guaranteed bit rate is the guaranteed number of bits delivered at a SAP within a period of time, divided by the duration of the period. A token bucket algorithm with a token rate equal to the guaranteed bit rate and a bucket size equal to the maximum SDU size enforces that the traffic is conformant to the guaranteed bit rate. The UMTS bearer service attributes are guaranteed for traffic up to this bit rate. The attributes could be delay and reliability. If the traffic exceeds this bit rate, the UMTS bearer service attributes are not guaranteed. Delivery order is used to indicate whether the UMTS bearer will provide in-sequence SDU delivery. The maximum SDU size is a negotiated QoS that the network must satisfy. The SDU format information is a list of possible exact sizes of the SDUs. SDU error ratio is the fraction of SDUs that are lost or detected as erroneous. This ratio is defined only for conforming traffic. Residual bit error ratio is used to indicate the undetected bit error ratio in the SDUs delivered. If error detection is not requested, the residual bit error ratio is the bit error ratio in the SDUs delivered. Delivery of erroneous SDU is used to deliver or discard SDUs detected erroneously. The transfer delay is used to indicate the maximum delay of the 95th percentile of the distribution of delay for all SDUs delivered during the lifetime of a bearer service. This delay for an SDU is the time from a request to transfer an SDU at a transmitting SAP to its delivery at the receiving SAP. The traffic-handling priority is used to specify the relative priority in the handling of all SDUs belonging to the UMTS bearer compared to the SDUs of the other UMTS bearers. The allocation–retention priority is used to specify the priority for allocation and retention of the UMTS bearer compared to those of the other UMTS bearers. This attribute is a subscription attribute. The source statistics descriptor is used to specify the characteristics of the source SDUs submitted. The signaling indication is used to indicate the signaling nature of the SDUs submitted. The defined UMTS bearer attributes and the relevancy for each bearer traffic class are shown in Table 8.2.

TABLE 8.2 UMTS QoS Attributes Defined for Each Traffic Class

Attribute	Conversation Class	Streaming Class	Interactive Class	Background Class
Maximum bit rate	×	×	×	×
Delivery order	×	×	×	×
Maximum SDU size	×	×	×	×
SDU format information	×	×		
SDU error ratio	×	×	×	×
Residual bit error ratio	×	×	×	×
Delivery of erroneous SDUs	×	×	×	×
Transfer delay	×	×		
Guaranteed bit rate	×	×		
Traffic-handling priority			×	
Allocation–retention priority	×	×	×	×
Source statistics descriptor	×	×		
Signaling indication			×	

8.3.2 WiMax Scheduling Services

As a wireless system, WiMax is for fixed broadband wireless access. It consists of at least one base station (BTS) and a number of subscriber stations (SSs). In some cases it may include repeaters. The architecture for the basic WiMax system can be point to multipoint or multipoint to multipoint. The raw data rates are in excess of 120 Mbps. The range of a WiMax system is up to several kilometers. As a fixed broadband wireless access system, WiMax defines four scheduling services: (1) unsolicited grant service, (2) real-time polling service, (3) non-real-time polling service, and (4) best effort service.

8.3.2.1 Unsolicited Grant Service Unsolicited grant service (UGS) is designed to handle real-time data streams consisting of fixed-length packets generated at periodic intervals such as T1/E1 and voice over IP without silence suppression. The mandatory QoS service flow parameters for UGS are maximum sustained traffic rate, maximum latency, tolerated jitter, and request/transmission policy. If the minimum reserved traffic rate parameter is present, it should have the same value as that of the maximum sustained traffic rate parameter. The maximum sustained traffic rate is the peak information rate of the service. The value of the maximum latency specifies the maximum latency between the reception of a packet by the base station or subscriber station on its network interface and the forwarding of the packet to its RF interface. The jitter tolerance defines the maximum delay variation (jitter) for the connection. The value in the request/transmission policy parameter provides the capability to specify certain attributes for the associated flow.

8.3.2.2 Real-Time Polling Service Real-time polling service (rtPS) is designed to handle real-time data streams consisting of variable-length data packets generated at periodic intervals such as the moving pictures experts group (MPEG) video. The mandatory QoS service flow parameters for rtPS are minimum reserved traffic rate, maximum sustained traffic rate, maximum latency, and request/transmission policy. The minimum reserved traffic rate is the minimum information rate reserved for this service flow.

8.3.2.3 Non-Real-Time Polling Service Non-real-time polling service (nrtPS) is designed to handle delay-tolerant data streams consisting of variable-length data packets with a required minimum data rate such as file transfer protocol. The mandatory QoS service flow parameters for nrtPS are minimum reserved traffic rate, maximum sustained traffic rate, traffic priority, and request/transmission policy. The value in the traffic priority parameter specifies the priority assigned to a service flow.

8.3.2.4 Best Effort Service Best effort (BE) service is designed to handle delay-tolerant data streams for which no minimum service level is required and may be handled on a space-available basis.

TABLE 8.3 Access Category Parameters for IEEE 802.11e

Access Category	AC Number	AC Designation	CW_{min}	CW_{max}	AIFSN	TXOP Limit (ms)
AC_BK	0	Background	aCW_{min}	aCW_{max}	7	0
AC_BE	1	Best effort	aCW_{min}	aCW_{max}	3	0
AC_VI	2	Video	$\frac{aCW_{min}+1}{2}-1$	aCW_{min}	2	3
AC_VO	3	Voice	$\frac{aCW_{min}+1}{4}-1$	$\frac{aCW_{min}+1}{2}-1$	2	1.5

8.3.3 Access Categories in IEEE 802.11e and 802.11n Draft Wireless Local Area Networks

The IEEE 802.11e standard specifies the modifications to the IEEE 802.11 standard for wireless local area networks. The coverage or range of WLAN is up to about 100 m. The IEEE 802.11 standard allows only a single traffic class. Thus, the modifications in the IEEE 802.11e standard allow eight priority classes with four access categories. The four access categories for the IEEE 802.11e standard in its enhanced distributed channel access (EDCA) medium access control (MAC) are (1) voice traffic, (2) video traffic, (3) best effort traffic, and (4) background traffic.

The EDCA MAC is based on carrier-sense multiple access with collision avoidance (CSMA/CA). The traffics in EDCA are given relative QoS priorities for access to the channel, with voice traffic having the highest priority and background traffic having the lowest priority. The shorter the arbitration interframe space (AIFS) and the smaller the minimum and maximum contention window, the higher the priority of the access categories. The AIFS is determined from the AIFS number (AIFSN). The access category parameters are shown in Table 8.3.

The IEEE 802.11n draft standard specifies a number of MAC enhancements to increase the user throughput. These MAC enhancements include frame aggregation, reverse direction protocol, enhanced block acknowledgments, transmission modes, and reduced interframe space. These techniques can increase user throughput, delay, and jitter QoSs. IEEE 802.11n can support a data rate of up to 600 Mbps with MIMO technology. More details of these enhanced MAC techniques are presented in Chapter 11.

8.3.4 Access Categories in WiMedia Wireless Personal Area Networks

WiMedia standard specifies the physical (PHY) and MAC for a wireless personal area network. Its coverage or range is only 10 m. It can support a data rate of up to 480 Mbps. The PHY is based on multiband OFDM alliance (MBOA); the MAC consists mainly of the distributed reservation protocol (DRP) and prioritized channel access (PCA). Basically, PCA MAC is similar to the IEEE 802.11e EDCA MAC based on CSMA/CA. Similarly, WiMedia WPAN specifies four access categories

TABLE 8.4 Access Category Parameters for WiMedia

Access Category	AC Number	AC Designation	CW_{min}	CW_{max}	AIFSN	TXOP Limit (μs)
AC_BK	0	Background	15	1023	7	512
AC_BE	1	Best effort	15	1023	4	512
AC_VI	2	Video	7	511	2	1024
AC_VO	3	Voice	3	255	1	256

in its PCA MAC: (1) voice traffic, (2) video traffic, (3) best effort traffic, and (4) background traffic.

They are also given relative QoS priorities for access to the channel, with voice traffic having the highest priority and background traffic having the lowest priority. The shorter the arbitration interframe space (AIFS) and the smaller the minimum and maximum contention window, the higher the priority of the access categories. The AIFS is determined from the AIFS number (AIFSN). The access category parameters are shown in Table 8.4.

The WiMedia MAC specifications also define an enhanced MAC technique for improving user throughput. This technique uses burst-mode packet transmissions with minimum interframe spaces (MIFSs) between packets when the access of the channel is successful. The physical-layer convergence protocol (PLCP) preamble of the first packet is a standard PLCP preamble; the PLCP preambles of subsequent packets are burst PLCP preambles, which are shorter than the standard PLCP preamble. This cuts down the collision times for transmitting each packet and improves the header-to-payload overheads. Thus, user throughput can be increased.

8.4 OUTAGE PROBABILITY FOR VIDEO SERVICES IN A MULTIRATE DS-CDMA SYSTEM

Besides delay, delay jitter, and packet loss, outage probability in a wireless network is also a QoS metric for multimedia traffic. If the outage probability is high due to fading or mobility, some packets may be lost. If it is voice traffic, some voice clippings can be tolerated. On the other hand, if it is data traffic, automatic repeat request can be used to retransmit the lost packets.

In this section, the system capacity in the uplink of a multirate DS-CDMA system with Sen's video traffic model having *scene changes* is derived. We generalize the analytical results in [8] for low-bit-rate *multiclass* services, which are modeled by a one-dimensional Markov chain (MC), to low- and high-bit-rate *multiclass video* services, modeled by a two-dimensional MC. Each user uses a combination of low- and high-bit-rate spreading codes. Here, the probabilities of bit error rate QoS for different classes are formulated in terms of the number of users of different classes, the number of low- and high-bit-rate active spreading codes, the intracell received powers for all classes, the intercell interference for all classes, the low- and high-bit-rate spreading gains for all classes, the bit energy-to-interference ratio requirements for all classes, and the background noise. Although we are considering multiclass video, the model presented here is general enough to account for voice and data traffic as well by assuming that there is no high-bit-rate transition in the vertical direction and that the low-bit-rate transitions in the horizontal direction are only between two states. Furthermore, it can be approximated that the probability of being in the on state for Pareto or Weibull distribution is equal to its mean on period divided by the sum of its mean on and off periods.

We consider *VBR multiclass video* traffic in the uplink of a wireless cellular multirate DS-CDMA network with *video scene changes*. The cell site (base station) supports K classes of video services that can originate from the mobile users in the cell. The following system parameters are used throughout this section:

K	number of traffic classes
W	spread-spectrum bandwidth
$R_{i,l}$	transmission rate of class i traffic using one low-bit-rate class i spreading code, $i = 1, 2, \ldots, K$
$R_{i,h}$	transmission rate of class i traffic using one high-bit-rate class i spreading code, $i = 1, 2, \ldots, K$
$G_{i,l}$	spreading gain of class i traffic using low-bit-rate spreading codes
$G_{i,h}$	spreading gain of class i traffic using high-bit-rate spreading codes
n_i	number of class i users per sector
$\mathrm{BER}_{i,l}$	bit error rate of class i using low-bit-rate spreading codes
$\mathrm{BER}_{i,h}$	bit error rate of class i using high-bit-rate spreading code
$\mathrm{SIR}_{i,l}$	signal-to-interference ratio of class i using low-bit-rate spreading codes
$\mathrm{SIR}_{i,h}$	signal-to-interference ratio of class i using high-bit-rate spreading codes
$\mathrm{BER}^*_{i,l}$	bit error rate requirement of class i using low-bit rate spreading codes
$\mathrm{BER}^*_{i,h}$	bit error rate requirement of class i using high-bit-rate spreading codes
$\mathrm{SIR}^*_{i,l}$	signal-to-interference ratio requirement of class i using low-bit-rate spreading codes
$\mathrm{SIR}^*_{i,h}$	signal-to-interference ratio requirement of class i using high-bit-rate spreading codes
$\gamma_{i,l}$	E_b/I_0 requirement of class i traffic using low-bit-rate spreading codes
$\gamma_{i,h}$	E_b/I_0 requirement of class i traffic using high-bit-rate spreading codes
$S_{i,l}$	received power of class i traffic using low-bit-rate spreading codes
$S_{i,h}$	received power of class i traffic using high-bit-rate spreading codes
$I_{i,l}$	intercell interference power of class i traffic using low-bit-rate spreading codes
$I_{i,h}$	intercell interference power of class i traffic using high-bit-rate spreading codes
η	power spectral density of ambient noise
ρ_i	density of class i users per unit area
$\psi_{ij,l}$	random variable for the number of active low-bit-rate spreading code used by a user j belonging to class i traffic

$\psi_{ij,h}$ random variable for the number of active high-bit-rate spreading code used by user j belonging to class i traffic
M_i maximum number of active low-bit-rate spreading codes used by a class i user
α_i increase rate of a two-state low-bit-rate minisource
β_i decrease rate of a two-state low-bit-rate minisource

The following assumptions are made to facilitate the analytical formulation:

- The transmission rates of other classes are integer multiples of that for the class with the basic rate.
- The processing gain, $G_{i,l}$, for class i users a using low-bit-rate spreading code are given by $W/R_{i,l}$.
- The processing gain, $G_{i,h}$, for class i users using a high-bit-rate spreading code are given by $W/R_{i,h}$.
- The system is made up of hexagonal cells.
- The mobile users have omnidirectional antennas.
- The base station has three sectors in each cell.
- The sectorization in the cells is perfect.
- Users are uniformly distributed in each cell.
- There are equal numbers of users from each class in every cell.
- There is perfect power control in each cell.
- The spreading gain can be varied for a different traffic class.
- The spreading gain is the same within the same class.
- The channel is modeled as a combination of path loss and lognormal shadowing, represented by $r^{-4} \cdot 10^{\varepsilon/10}$, where r is the normalized distance between the mobile and the serving base station and ε is a Gaussian random variable with zero mean and variance σ^2. The path loss exponent, which is normally determined from measurements and is in the range 2 to 5, is assumed to be 4 in this section.

We are concerned with the uplink system capacity of a multirate DS-CDMA system with VBR video traffic having scene changes in terms of the number of users, n_i, that can be supported for the ith class. The capacity region for K classes is derived by considering the outage probability in terms of the signal/interference ratio (SIR) specification. These probabilities are expressed in terms of the number of class i users, the number of active low- and high-bit-rate spreading codes, the intracell received powers for K classes, the intercell interference for K classes, the low- and high-bit-rate spreading gains for K classes, the E_b/I_0 requirements for K classes, and the thermal noise.

The capacity of the K-class system is defined by $(n_1, \ldots, n_i, \ldots, n_K)$. The aim here is to determine the maximum number of users for the K classes that are allowable in the system while maintaining the required QoS.

From Sen's model [6–8], a VBR video source can be modeled by a two-dimensional continuous-time MC with finite states as shown in Figure 8.12. Each state (x, m) represents the combined discrete level of low and high bit rates that are generated by a single source. The combined data rate of each source is $mR_{i,l} + xR_{i,h}$, where $R_{i,l}$ is the low for user i using one low-bit-rate spreading code and $R_{i,h}$ is the high bit rate for user i using one high-bit-rate spreading code. That is, we assume that each level of low or high bit rate uses one low- or high-bit-rate spreading code for a class i user. Each low-bit-rate level is modeled by a two-state minisource with an increase rate of α_i and a decrease rate of β_i. Thus, the two-dimensional continuous-time MC for a single video source at state (x, m) has an increase rate of $(M_i - m)\alpha_i$ and a decrease rate of $m\beta_i$ for low-bit-rate fluctuations, where M_i is the highest level in the low-bit-rate states, and this is also the maximum number of active spreading codes used by a class i user for low-bit-rate fluctuations. Each high-bit-rate fluctuation is modeled by a two-state MC with an increase rate of λ_i and a decrease rate of μ_i. There are only two states in the high-bit-rate fluctuation. $x \in 0, 1$ represents the high-bit-rate state that the video source is in. State 0 $(x = 0)$ means that there are no high-bit-rate fluctuations, only low-bit-rate fluctuations; state 1 $(x = 1)$ means that there are both low- and high-bit-rate fluctuations. If the high-bit-rate fluctuation has only one state (state 0), it reduces to Maglaris's model. Furthermore, if $M_i = 1$, the source is an on/off source. The steady-state probability of being in state m, denoted by P_m, is given by a binomial distribution with parameters (M_i, p_i, m), where $m = 0, 1, \ldots, M_i$ and $p_i = \alpha_i/(\alpha_i + \beta_i)$, and its first two moments and variance are $M_i p_i$, $M_i p_i[1 + (M_i - 1)p_i]$, and $M_i p_i[1 - p_i]$, respectively. The steady-state probability of being in state x, denoted by Q_x, obeys a Bernoulli distribution with parameters (q_i, x), where $x = 0,1$ and $q_i = \lambda_i/(\lambda_i + \mu_i)$, and its first two moments and variance are q_i, q_i, and $q_i[1 - q_i]$, respectively. Next, let $\gamma_{i,l}$ denote the E_b/I_0 for class i using low-bit-rate spreading codes. It is given by

$$\gamma_{i,l} = \frac{G_{i,l}}{\sum_{j=1}^{n_i-1}\left(\psi_{ij,l} + \psi_{ij,h}\frac{S_{i,h}}{S_{i,l}}\right) + \sum_{\substack{k=1\\k\neq i}}^{K}\sum_{j=1}^{n_k}\left(\psi_{kj,l}\frac{S_{k,l}}{S_{i,l}} + \psi_{kj,h}\frac{S_{k,h}}{S_{i,l}}\right) + \sum_{k=1}^{K}\left(\frac{I_{k,l}}{S_{i,l}} + \frac{I_{k,h}}{S_{i,l}}\right) + \frac{\eta}{S_{i,l}}},$$

$$i = 1, 2, \ldots, K, \tag{8.42}$$

where $\psi_{ij,l} \in \{0, 1, 2, \ldots, m, \ldots, M_i\}$ is a binomial random variable indicating the number of active low-bit-rate spreading codes used by the jth user of class i, while $\psi_{ij,h} \in \{0, 1\}$ is a Bernoulli random variable indicating the number of active high-bit-rate spreading codes used by the jth user of class i. The probability that m active low-bit-rate spreading codes are used by a video source, denoted by $Pr[\psi_{ij,l} = m]$, is given by $\Pr[\psi_{ij,l} = m] = P_m,\ m = 0, 1, 2, \ldots, M_i$, while the probability that x active high-bit-rate spreading code is used by a video source, denoted by $Pr[\psi_{ij,h} = x]$, is given by $\Pr[\psi_{ij,h} = x] = Q_x,\ x = 0,1$. The numerator in the right-hand side of (8.42) is the class i processing gain with low-bit-rate spreading gain. In the denominator, the first and second terms are due to the intracell interference from other users

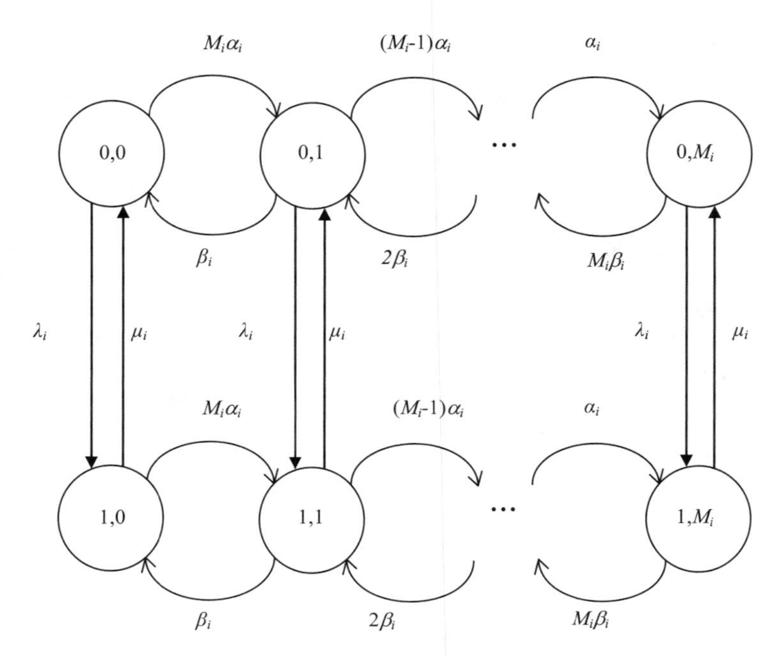

FIGURE 8.12 Classical Sen multiclass video traffic model.

in class i with low- and high-bit-rate spreading codes, respectively; the third and fourth terms are due to the intracell interference from users from other classes with low- and high-bit-rate spreading codes, respectively; the fifth and sixth terms are due to the intercell interference from all classes with low- and high-bit-rate spreading codes, respectively; and the last term is due to background noise. The spreading gain is assumed to be variable for different traffic classes, while the low- or high-bit-rate spreading gain is assumed to be the same within the same class having the corresponding low or high bit rate. Similarly, let $\gamma_{i,h}$ denote the E_b/I_0 for class i using high-bit-rate spreading codes. It is given by

$$\gamma_{i,h} = \frac{G_{i,h}}{\left[\sum_{j=1}^{n_i-1}\left(\psi_{ij,l}\frac{S_{i,l}}{S_{i,h}} + \psi_{ij,h}\right) + \sum_{\substack{k=1\\k\neq i}}^{K}\sum_{j=1}^{n_k}\left(\psi_{kj,l}\frac{S_{k,l}}{S_{i,h}} + \psi_{kj,h}\frac{S_{k,h}}{S_{i,h}}\right) + \sum_{k=1}^{K}\left(\frac{I_{k,l}}{S_{i,h}} + \frac{I_{k,h}}{S_{i,h}}\right) + \frac{\eta}{S_{i,h}}\right]},$$

$$i = 1, 2, \ldots, K. \tag{8.43}$$

To solve for the outage probabilities with low- and high-bit-rate spreading codes, the power ratios $S_{i,h}/S_{i,l}$, $S_{i,l}/S_{k,l}$, $S_{i,h}/S_{k,h}$, and $S_{k,h}/S_{i,l}$ must be determined.

Rearranging (8.42) and (8.43), we have

$$M_i p_i(n_i - 1)S_{i,l} + q_i(n_i - 1)S_{i,h} + \sum_{k=1,k\neq i}^{K} M_k p_k n_k S_{k,l} + \sum_{k=1,k\neq i}^{K} q_k n_k S_{k,h} - \frac{S_{i,l}G_{i,l}}{\gamma_{i,l}}$$

$$= -\sum_{k=1}^{K}(I_{k,l} + I_{k,h}) - \eta \tag{8.44}$$

and

$$M_i p_i(n_i - 1)S_{i,l} + q_i(n_i - 1)S_{i,h} + \sum_{k=1,k\neq i}^{K} M_k p_k n_k S_{k,l} + \sum_{k=1,k\neq i}^{K} q_k n_k S_{k,h}$$

$$-\frac{S_{i,h}G_{i,h}}{\gamma_{i,h}} = -\sum_{k=1}^{K}(I_{k,l} + I_{k,h}) - \eta. \tag{8.45}$$

Manipulation of (8.44) and (8.45) yields the power ratio

$$\frac{S_{i,h}}{S_{i,l}} = \frac{G_{i,l}/\gamma_{i,l}}{G_{i,h}/\gamma_{i,h}}. \tag{8.46}$$

Manipulation of (8.44) yields the power ratio

$$\frac{S_{i,l}}{S_{k,l}} = \frac{\varpi_k}{\varpi_i}, \qquad i, k \in \{1, 2, \ldots, K\}, \quad i \neq k, \tag{8.47}$$

where $\varpi_i = (G_{i,l}/\gamma_{i,l})[1 + q_i/(G_{i,h}/\gamma_{i,h})] + M_i p_i$. Similarly, manipulation of (8.45) yields

$$\frac{S_{i,h}}{S_{k,h}} = \frac{\theta_k}{\theta_i}, \qquad i, k \in \{1, 2, \ldots, K\}, \quad i \neq k, \tag{8.48}$$

where $\theta_i = (G_{i,h}/\gamma_{i,h})[1 + M_i p_i/(G_{i,l}/\gamma_{i,l})] + q_i$.

Using the results from (8.45) and (8.47) and manipulating, we have

$$\frac{S_{k,h}}{S_{i,l}} = \frac{S_{i,h}/S_{i,l}}{S_{k,h}/S_{i,h}} = \frac{\varpi_i}{\theta_k}, \qquad i \neq k. \tag{8.49}$$

To solve for the outage probabilities with low- and high-bit-rate spreading codes, the upper bounds of the means and variances of the power ratios $I_{i,l}/S_{i,l}$, $I_{i,h}/S_{i,h}$, $I_{k,l}/S_{i,l}$, $I_{k,h}/S_{i,h}$, $I_{k,l}/S_{i,h}$, and $I_{k,h}/S_{i,l}$ must also be determined. From [9], the intercell interference-to-signal ratio for the jth user in class i, with interference power I_{ij} and received signal power S_i, is given by

$$\frac{I_{ij}}{S_i} = \left(\frac{r_m}{r_d}\right)^4 \cdot 10^{(\varepsilon_d - \varepsilon_m)/10}, \tag{8.50}$$

where r_d is the distance between the intercell mobile that causes interference and the intracell base station, r_m is the distance between the intercell mobile and its own base station, and ε_d and ε_m are Gaussian random variables with zero mean and standard deviation σ. Since ε_d and ε_m are independent, $\varepsilon_d - \varepsilon_m$ is a Gaussian random variable with zero mean and variance $2\sigma^2$. The mean and variance of $I_{i,l}/S_{i,l}$ are upper bounded by [9]

$$E\left[I_{i,l}/S_{i,l}\right] = M_i p_i \rho_i \iint f\left(r_m/r_d\right) dA \leq \mu_{ii,l} \tag{8.51}$$

and

$$\begin{aligned} \operatorname{Var}\left[I_{i,l}/S_{i,l}\right] &\leq \rho_i \iint \left[M_i p_i[1 + (M_i - 1)p_i] g\left(r_m/r_d\right) - (M_i p_i)^2 f^2\left(r_m/r_d\right)\right] dA \\ &= \sigma_{ii,l}^2, \end{aligned} \tag{8.52}$$

where $\mu_{ii,l}$ is the upper bound on the mean of $I_{i,l}/S_{i,l}$, $\sigma_{ii,l}^2$ is the upper bound on the variance of $I_{i,l}/S_{i,l}$, and ρ_i is the density of class i users per unit area given by $\rho_i = 2n_i/\sqrt{3}$:

$$f\left(r_m/r_d\right) = \left(r_m/r_d\right)^4 e^{(\sigma \ln 10/10)^2} \left[1 - Q\left(40 \log(r_d/r_m)/\sqrt{2\sigma^2} - \sqrt{2\sigma^2} \ln 10/10\right)\right], \tag{8.53}$$

$$g\left(r_m/r_d\right) = \left(r_m/r_d\right)^8 e^{(\sigma \ln 10/5)^2} \left[1 - Q\left(40 \log(r_d/r_m)/\sqrt{2\sigma^2} - \sqrt{2\sigma^2} \ln 10/5\right)\right], \tag{8.54}$$

and $Q(y) = (1/\sqrt{2\pi}) \int_y^\infty e^{-x^2/2} dx$. The mean and variance of $I_{i,h}/S_{,i,h}$ are upper bounded by [3]

$$E\left[I_{i,h}/S_{i,h}\right] \leq q_i \rho_i \iint f\left(r_m/r_d\right) dA = \mu_{ii,h} \tag{8.55}$$

and

$$\operatorname{Var}\left[I_{i,h}/S_{i,h}\right] \leq \rho_i = \sigma_{ii,h}^2, \tag{8.56}$$

where $\mu_{ii,h}$ is the upper bound on the mean of $I_{i,h}/S_{i,h}$, and $\sigma_{ii,h}^2$ is the upper bound on the variance of $I_{i,h}/S_{i,h}$. The random variable, $I_{k,l}/S_{i,l}$, can be expressed as

$$\frac{I_{k,l}}{S_{i,l}} = \frac{I_{1,l}}{S_{1,l}} \frac{S_{k,l}}{S_{i,l}} \left(\frac{I_{k,l}}{S_{k,l}} / \frac{I_{1,l}}{S_{1,l}}\right), \qquad i, k = 1, 2, \ldots, K. \tag{8.57}$$

Thus, the mean and variance of $I_{k,l}/S_{i,l}$ satisfy the inequalities

$$E\left[I_{k,l}/S_{i,l}\right] \leq \left[S_{k,l} M_k p_k \rho_k / \left(S_{i,l} M_1 p_1 \rho_1\right)\right] \mu_{11,l} \tag{8.58}$$

and

$$\mathrm{Var}\left[I_{k,l}/S_{i,l}\right] \leq \left(S_{k,l}/S_{i,l}\right)^2 \sigma_{kk,l}^2. \tag{8.59}$$

Similarly, the random variable, $I_{k,h}/S_{i,h}$, can be expressed as

$$\frac{I_{k,h}}{S_{i,h}} = \frac{I_{1,h}}{S_{1,h}} \frac{S_{k,h}}{S_{i,h}} \left(\frac{I_{k,h}}{S_{k,h}} \Big/ \frac{I_{1,h}}{S_{1,h}} \right), \qquad i, k = 1, 2, \ldots, K. \tag{8.60}$$

Thus, the mean and variance of $I_{k,h}/S_{i,h}$ satisfy the inequalities:

$$E\left[I_{k,h}/S_{i,h}\right] \leq \left(S_{k,h} q_k \rho_k / S_{i,h} q_1 \rho_1\right) \mu_{11,h} \tag{8.61}$$

and

$$\mathrm{Var}\left[I_{k,h}/S_{i,h}\right] \leq \left(S_{k,h}/S_{i,h}\right)^2 \sigma_{kk,h}^2. \tag{8.62}$$

On the other hand, the random variable, $I_{k,l}/S_{i,h}$, can be expressed as

$$\frac{I_{k,l}}{S_{i,h}} = \frac{S_{k,l}}{S_{i,h}} \frac{I_{k,l}}{S_{k,l}}, \qquad i, k = 1, 2, \ldots, K. \tag{8.63}$$

Thus, the mean and variance of $I_{k,l}/S_{i,h}$ satisfy the inequalities:

$$\begin{aligned} E\left[I_{k,l}/S_{i,h}\right] &\leq \left(S_{k,l}/S_{i,h}\right) \mu_{kk,l} \\ &= \left(S_{k,l} M_k p_k \rho_k / S_{i,h} M_1 p_1 \rho_1\right) \mu_{11,l} \end{aligned} \tag{8.64}$$

and

$$\mathrm{Var}\left[I_{k,l}/S_{i,h}\right] \leq \left(S_{k,l}/S_{i,h}\right)^2 \sigma_{kk,l}^2. \tag{8.65}$$

Similarly, the random variable, $I_{k,h}/S_{i,l}$, can be expressed as

$$I_{k,h}/S_{i,l} = \left(S_{k,h}/S_{i,l}\right)\left(I_{k,h}/S_{k,h}\right), \qquad i, k = 1, 2, \ldots, K. \tag{8.66}$$

Thus, the mean and variance of $I_{k,h}/S_{i,l}$ satisfy the inequalities

$$E\left[I_{k,h}/S_{i,l}\right] \leq \left(S_{k,h}/S_{i,l}\right) \mu_{kk,h} = \left[S_{k,h} q_k \rho_k / \left(S_{i,l} q_1 \rho_1\right)\right] \mu_{11,h} \tag{8.67}$$

and

$$\mathrm{Var}\left[I_{k,h}/S_{i,l}\right] \leq \left(S_{k,h}/S_{i,l}\right)^2 \sigma_{kk,h}^2. \tag{8.68}$$

Let $\mathrm{BER}_{i,l}^*$ and $\mathrm{BER}_{i,h}^*$ denote the BER requirement for class i users with low- and high-bit-rate spreading codes, respectively. Furthermore, let $\mathrm{SIR}_{i,l}^*$ and $\mathrm{SIR}_{i,h}^*$ denote the SIR requirement for class i users with low- and high-bit-rate spreading codes, respectively. The system capacity is defined as the maximum $(n_1, \ldots, n_i, \ldots, n_K)$ that can be supported such that the $\mathrm{SIR}_{i,l}$ achieved is greater than or equal to the $\mathrm{SIR}_{i,l}^*$ required and the $\mathrm{SIR}_{i,h}$ achieved is greater than or equal to the required $\mathrm{SIR}_{i,h}^*$ 99% of the time for all classes. That is, the outage probabilities are defined as

$$\begin{aligned}\Pr[\mathrm{BER}_{i,l} \geq \mathrm{BER}_{i,l}^*] &= \Pr[\mathrm{SIR}_{i,l} \leq \mathrm{SIR}_{i,l}^*] \\ &= \Pr\left[\sum_{j=1}^{n_i-1}\left(\psi_{ij,l} + \psi_{ij,h}\frac{S_{i,h}}{S_{i,l}}\right) + \sum_{k=1,k\neq i}^{K}\sum_{j=1}^{n_k}\right. \\ &\left.\times\left(\psi_{kj,l}\frac{S_{k,l}}{S_{i,l}} + \psi_{kj,h}\frac{S_{k,h}}{S_{i,l}}\right) + \sum_{k=1}^{K}\left(\frac{I_{k,l}}{S_{i,l}} + \frac{I_{k,h}}{S_{i,l}}\right) \geq \delta_{i,l}\right]\end{aligned} \tag{8.69}$$

and

$$\begin{aligned}\Pr[\mathrm{BER}_{i,h} \geq \mathrm{BER}_{i,h}^*] &= \Pr[\mathrm{SIR}_{i,h} \leq \mathrm{SIR}_{i,h}^*] \\ &= \Pr\left[\sum_{j=1}^{n_i-1}\left(\psi_{ij,l}\frac{S_{i,l}}{S_{i,h}} + \psi_{ij,h}\right) + \sum_{k=1,k\neq i}^{K}\sum_{j=1}^{n_k}\right. \\ &\left.\times\left(\psi_{kj,l}\frac{S_{k,l}}{S_{i,h}} + \psi_{kj,h}\frac{S_{k,h}}{S_{i,h}}\right) + \sum_{k=1}^{K}\left(\frac{I_{k,l}}{S_{i,h}} + \frac{I_{k,h}}{S_{i,h}}\right) \geq \delta_{i,h}\right],\end{aligned} \tag{8.70}$$

where $\delta_{i,l} = G_{i,l}/\gamma_{i,l} - \eta/S_{i,l}$ and $\delta_{i,h} = G_{i,h}/\gamma_{i,h} - \eta/S_{i,h}$, $i = 1, 2, \ldots, K$.

The probability that l_i active low-bit-rate spreading codes are used by n_i class i sources, denoted by $\Pr[\phi_{i,l} = l_i]$, is given by a binomial distribution with parameters $(M_i n_i, p_i, l_i)$, where $l_i = 0, 1, \ldots, M_i n_i$, while the probability that h_i active high-bit-rate spreading codes are used by n_i class i sources, denoted by $\Pr[\phi_{i,h} = h_i]$, is given by a binomial distribution with parameters (n_i, q_i, h_i), where $h_i = 0, 1, \ldots, n_i$.

Using the central limit approximation and solving (8.69) by conditioning on the active low- and high-bit-rate spreading codes used and then unconditioning the probability in (8.69) by summing up all cases for the numbers of active low- and high-bit-rate spreading codes used for all classes and multiplying by all the corresponding binomial probabilities of active low- and high-bit-rate spreading codes used, we have

$$
\begin{aligned}
&\Pr[\mathrm{BER}_{i,l} \geq \mathrm{BER}^*_{i,l}] \\
&= \sum_{l_1=0}^{M_1 n_1} \cdots \sum_{l_i=0}^{M_i(n_i-1)} \cdots \sum_{l_K=0}^{M_K n_K} \sum_{h_1=0}^{n_1} \cdots \sum_{h_i=0}^{n_i-1} \cdots \sum_{h_K=0}^{n_K} \Pr\left[\sum_{k=1}^{K}\left(\frac{I_{k,l}}{S_{i,l}} + \frac{I_{k,h}}{S_{i,l}}\right) \geq \delta_{i,l} - l_i \right. \\
&\quad - \sum_{\substack{k=1\\k\neq i}}^{K} l_k \frac{S_{k,l}}{S_{i,l}} - \sum_{k=1}^{K} h_k \frac{S_{k,h}}{S_{i,l}} \left| \sum \psi_{1j,l} = l_1, \sum \psi_{2j,l} = l_2, \ldots, \sum \psi_{Kj,l} = l_K, \right. \\
&\quad \left. \times \sum \psi_{1j,h} = h_1, \sum \psi_{2j,h} = h_2, \ldots, \sum \psi_{Kj,h} = h_K \right] \\
&\quad \times \prod_{k=1}^{K} \left(\Pr\left[\sum \psi_{kj,l} = l_k\right] \Pr\left[\sum \psi_{kj,h} = h_k\right] \right) \\
&= \sum_{l_1=0}^{M_1 n_1} \cdots \sum_{l_i=0}^{M_i(n_i-1)} \cdots \sum_{l_K=0}^{M_K n_K} \sum_{h_1=0}^{n_1} \cdots \sum_{h_i=0}^{n_i-1} \cdots \sum_{h_K=0}^{n_K} Q\left(\frac{\delta_{i,l} - \mu_{i,l}}{\sigma_{i,l}}\right) \binom{M_i(n_i-1)}{l_i} \\
&\quad \times p_i^{l_i} (1-p_i)^{M_i(n_i-1)-l_i} \binom{n_i - 1}{h_i} q_i^{h_i} (1-q_i)^{n_i-1-h_i} \\
&\quad \times \prod_{\substack{k=1\\k\neq i}}^{K} \binom{M_k n_k}{l_k} p_k^{l_k} (1-p_k)^{M_k n_k - l_k} \binom{n_k}{h_k} q_k^{h_k} (1-q_k)^{n_k - h_k},
\end{aligned} \tag{8.71}
$$

and similarly from (8.70), we have

$$
\begin{aligned}
&\Pr[\mathrm{BER}_{i,h} \geq \mathrm{BER}^*_{i,h}] \\
&= \sum_{l_1=0}^{M_1 n_1} \cdots \sum_{l_i=0}^{M_i(n_i-1)} \cdots \sum_{l_K=0}^{M_K n_K} \sum_{h_1=0}^{n_1} \cdots \sum_{h_i=0}^{n_i-1} \cdots \sum_{h_K=0}^{n_K} \Pr\left[\sum_{k=1}^{K}\left(\frac{I_{k,l}}{S_{i,h}} + \frac{I_{k,h}}{S_{i,h}}\right) \geq \delta_{i,h} \right. \\
&\quad - \sum_{k=1}^{K} l_k \frac{S_{k,l}}{S_{i,h}} - h_i - \sum_{\substack{k=1\\k\neq i}}^{K} h_k \frac{S_{k,h}}{S_{i,h}} \left| \sum \psi_{1j,l} = l_1, \sum \psi_{2j,l} = l_2, \ldots, \sum \psi_{Kj,l} = l_K, \right. \\
&\quad \left. \times \sum \psi_{1j,h} = h_1, \sum \psi_{2j,h} = h_2, \ldots, \sum \psi_{Kj,h} = h_K \right] \\
&\quad \times \prod_{k=1}^{K} \left(\Pr\left[\sum \psi_{kj,l} = l_k\right] \Pr\left[\sum \psi_{kj,h} = h_k\right] \right)
\end{aligned}
$$

$$
= \sum_{l_1=0}^{M_1 n_1} \cdots \sum_{l_i=0}^{M_i(n_i-1)} \cdots \sum_{l_K=0}^{M_K n_K} \sum_{h_1=0}^{n_1} \cdots \sum_{h_i=0}^{n_i-1} \cdots \sum_{h_K=0}^{n_K} Q\left(\frac{\delta_{i,h} - \mu_{i,h}}{\sigma_{i,h}}\right) \binom{M_i(n_i-1)}{l_i}
$$

$$
\times p_i^{l_i}(1-p_i)^{M_i(n_i-1)-l_i} \binom{n_i - 1}{h_i} q_i^{h_i}(1-q_i)^{n_i-1-h_i}
$$

$$
\times \prod_{\substack{k=1 \\ k \neq i}}^{K} \binom{M_k n_k}{l_k} p_k^{l_k}(1-p_k)^{M_k n_k - l_k} \binom{n_k}{h_k} q_k^{h_k}(1-q_k)^{n_k - h_k}, \tag{8.72}
$$

where

$$
\mu_{i,l} = l_i + \sum_{\substack{k=1 \\ k \neq i}}^{K} \left(\frac{S_{k,l}}{S_{i,l}}\right) l_k + \sum_{k=1}^{K} \left(\frac{S_{k,h}}{S_{i,l}}\right) h_k + \left(\frac{M_i p_i \rho_i}{M_1 p_1 \rho_1}\right) \mu_{11,l}
$$

$$
+ \sum_{\substack{k=1 \\ k \neq i}}^{K} \left(\frac{S_{k,l} M_k p_k \rho_k}{S_{i,l} M_1 p_1 \rho_1}\right) \mu_{11,l} + \sum_{k=1}^{K} \left(\frac{S_{k,h} q_k \rho_k}{S_{i,l} q_1 \rho_1}\right) \mu_{11,h}, \tag{8.73}
$$

$$
\mu_{i,h} = \sum_{k=1}^{K} \left(\frac{S_{k,l}}{S_{i,h}}\right) l_k + h_i + \sum_{\substack{k=1 \\ k \neq i}}^{K} \left(\frac{S_{k,h}}{S_{i,h}}\right) h_k + \frac{q_i \rho_i}{q_1 \rho_1} \mu_{11,h} + \sum_{\substack{k=1 \\ k \neq i}}^{K} \left(\frac{S_{k,h} q_k \rho_k}{S_{i,h} q_1 \rho_1}\right) \mu_{11,h}
$$

$$
+ \sum_{k=1}^{K} \left(\frac{S_{k,l} M_k p_k \rho_k}{S_{i,h} M_1 p_1 \rho_1}\right) \mu_{11,l}, \tag{8.74}
$$

$$
\sigma_{i,l} = \sqrt{\sigma_{ii,l}^2 + \sum_{\substack{k=1 \\ k \neq i}}^{K} \left(\frac{S_{k,l}}{S_{i,l}}\right)^2 \sigma_{kk,l}^2 + \sum_{k=1}^{K} \left(\frac{S_{k,h}}{S_{i,l}}\right)^2 \sigma_{kk,h}^2}, \tag{8.75}
$$

$$
\sigma_{i,h} = \sqrt{\sigma_{ii,h}^2 + \sum_{\substack{k=1 \\ k \neq i}}^{K} \left(\frac{S_{k,h}}{S_{i,h}}\right)^2 \sigma_{kk,h}^2 + \sum_{k=1}^{K} \left(\frac{S_{k,l}}{S_{i,h}}\right)^2 \sigma_{kk,l}^2}, \tag{8.76}
$$

and $i = 1, 2, \ldots, K$.

SUMMARY

The first part of this chapter covers the traffic models for voice, voice over IP, video, file transfer protocol, email, web browsing, and network gaming. Some of these models are suitable for analytical formulation but may not represent the self-similar characteristics of measured traffic, while others are for self-similar traffic but may only be suitable for simulations. Other traffic models can be found in references [10–25]. The quality of service (QoS) requirements or relative QoS requirements for UMTS systems, WiMax, IEEE 802.11e WLAN, and WiMedia WPAN, are presented in Section 8.3. UMTSs have absolute QoS requirements for four classes of traffic, and WiMax has specific QoS parameters for its four classes of scheduling services. On the other hand, IEEE 802.11e MAC and WiMedia MAC have four relative classes of access categories. More details of some of these systems are presented in Part II of the book. The performance analysis of multiclass video services in a DS-CDMA system is presented in the last section of this chapter. This model is general enough to capture voice and data traffic as well by setting some appropriate parameters of the video source model for modeling voice and data traffic sources.

REFERENCES

[1] W. Stallings, *High-Speed Networks and Internet: Performance and Quality of Service*, 2nd ed., Prentice Hall, Upper Saddle River, NJ, 2001.

[2] A. Jamalipour, *The Wireless Mobile Internet: Architectures, Protocols and Services*, Wiley, Hoboken, NJ, 2003.

[3] O. I. Sheluhin, S. M. Smolskiy, and A. V. Osin, *Self-Similar Processes in Telecommunications*, Wiley, Hoboken, NJ, 2007.

[4] G.-L. Wu and J. W. Mark, "Computational methods for performance evaluation of a statistical multiplexer supporting bursty traffic," *IEEE/ACM Trans. Network.*, vol. 4, no. 3, pp. 386–397, June 1996.

[5] M. W. Garrett and W. Willinger, "Analysis, modeling and generation of self-similar VBR video traffic," *IEEE SIGCOMM 1994*, pp. 269–280, 1994.

[6] M. R. Izquierdo and D. S. Reeves, "A survey of statistical source models for variable-bit-rate compressed video," *Multimedia Syst.*, vol. 7, pp. 199–213, 1999.

[7] J. Farber, "Traffic modeling for fast action network games," *Multimedia Tools and Applications 2004, Conference CD-ROM*, 2004.

[8] T. C. Wong, J. W. Mark, K. C. Chua, and B. Kannan, "Performance analysis of variable bit rate multiclass services in the uplink of wideband CDMA," *IEEE International Conference on Communications 2003*, pp. 363–367, May 2003.

[9] R. Vannithamby and E. S. Sousa, "Performance of multi-rate data traffic using variable spreading gain in the reverse link under wideband CDMA," *IEEE Vehicular Technology Conference, Conference Record*, pp. 1155–1159, 2000.

[10] H. Michiel and K. Laevens, "Teletraffic engineering in a broad-band era," *Proc. IEEE*, vol. 85, no. 12, pp. 2007–2033, Dec. 1997.

[11] G. D. Stamoulis, M. E. Anagnostou, and A. D. Georgantas, "Traffic source models for ATM networks: a survey," *Comput. Commun.*, vol. 17, no. 6, pp. 428–438, June 1994.

[12] B. Ryu, "Modeling and simulation of broadband satellite networks: II. Traffic modeling," *IEEE Commun. Mag.*, pp. 48–56, July 1999.

[13] N. Jefferies, A. Munro, J. M. Irvine, and S. Hope, "Modelling mobile multimedia services," *Electron. Commun. Eng. J.*, pp. 271–279, Dec. 2000.

[14] V. S. Frost and B. Melamed, "Traffic modeling for telcommunications networks," *IEEE Commun. Mag.*, pp. 271–279, Mar. 1994.

[15] A. Adas, "Traffic models in broadband networks," *IEEE Commun. Mag.*, pp. 70–81, July 1997.

[16] L. Muscariello, M. Mellia, M. Meo, M. A. Marsan, and R. L. Cigno, "Markov models of internet traffic and a new hierarchical MMPP model," *Comput. Commun.*, vol. 28, pp. 1835–1851, 2005.

[17] K. Park and W. Willinger, *Self-Similar Network Traffic and Performance Evaluation*, Wiley-Interscience, New York, 2000.

[18] A. Erramilli, M. Roughan, D. Veitch, and W. Willinger, "Self-similar traffic and network dynamics," *Proc. IEEE*, vol. 90, no. 5, pp. 800–819, May 2002.

[19] B. Tsybakov and N. D. Georganas, "Self-similar processes in communications networks," *IEEE Trans. Inf. Theory*, vol. 44, no. 5, pp. 1713–1725, Sept. 1998.

[20] Z. Sahinoglu and S. Takinay, "On multimedia networks: self-similar traffic and network performance," *IEEE Commun. Mag.*, pp. 48–52, Jan. 1999.

[21] P. Salvador and R. Valadas, "Multiscale fitting procedure using Markov modulated poisson processes," *Telecommun. Syst.*, vol. 23, no. 1–2, pp. 123–148, 2003.

[22] J. Beran, R. Sherman, M. S. Taqqu, and W. Willinger, "Long-range dependence in variable-bit-rate video traffic," *IEEE Trans. Commun.*, vol. 43, no. 2–3–4, Feb.–Mar.–Apr., pp. 1566–1579, 1995.

[23] O. Rose, "Simple and efficient models for variable bit rate MPEG video traffic," *Perform. Eval.*, vol. 30, pp. 69–85, 1997.

[24] M. S. Borella, "Source models for network game traffic," *Comput. Commun.*, vol. 23, pp. 403–410, 2000.

[25] A. Abdennour, "VBR video traffic modeling and synthetic data generation using GA-optimized Volterra filters," *Int. J. Network Manage.*, vol. 17, pp. 231–241, 2007.

PART II

SYSTEMS FOR WIRELESS BROADBAND NETWORKS

CHAPTER 9

LONG-TERM-EVOLUTION CELLULAR NETWORKS

9.1 INTRODUCTION

Cellular systems are widely deployed in the world today. A mobile subscriber is connected to the core network through the base station that is serving his or her mobile phone. One of the most common cellular systems in use is the second-generation (2G) global system for mobile communications (GSM). The GSM cellular access network can support both voice and data with a data rate of up to 9.6 kbps. Most first-generation (1G) cellular systems are no longer in use. An evolved cellular system from GSM is the 2.5G general packet radio service (GPRS) cellular network. GPRS can support a data rate of up to 384 kbps. Further evolution from the GSM and GPRS is the third-generation (3G) wideband code-division multiple access (WCDMA) in universal mobile telecommunications systems (UMTSs). WCDMA can support a data rate of up to 2 Mbps. UMTS is specified by the 3G partnership project (3GPP). Another enhancement on top of WCDMA is the 3.5G high-speed packet access (HSPA). HSPA can support a downlink data rate of up to 14.4 Mbps and an uplink data rate of 5.76 Mbps. Downlink represents transmission in the direction from the base station to the mobile user, while uplink represents transmission in the direction from the mobile user to the base station.

Long term-evolution (LTE) in 3GPP is the latest technology in cellular networks that is being standardized. It is considered as a 3.9G cellular system. The name LTE

Wireless Broadband Networks, By David Tung Chong Wong, Peng-Yong Kong, Ying-Chang Liang, Kee Chaing Chua, and Jon W. Mark

comes from the evolved universal terrestrial radio access network (E-UTRAN). It is based on an all-Internet protocol (all-IP) framework that is not limited by past design. The peak data rate is also increased, with the downlink having 100 Mbps and the uplink having 50 Mbps. LTE uses multiple-input, multiple-output (MIMO) technique (i.e., it uses multiple antennas). The downlink is based on orthogonal frequency-division multiple access (OFDMA) at the physical layer, while the uplink is based on single carrier frequency-division multiple access (SC-FDMA). Each radio frame is 10 ms long and can be divided into subframes each of 1 ms duration. Each subframe can be further divided into two slots, each of 0.5 ms duration. LTE also uses transmission power control (TPC) and adaptive modulation and coding (AMC). Access to the network is dependent on the physical layer, medium access control, radio link control, packet data convergence protocol, and radio resource control. Tight interaction among these building blocks is necessary to transmit a packet efficiently with low latency as well as without transmission errors. Both automatic repeat request (ARQ) and hybrid ARQ (HARQ) are used to combat transmission errors. HARQ with soft combining is used at the medium access control and the physical layer. The medium access control is used for signaling control, while the physical layer is used for retention of transmission blocks and soft combining. Soft combining makes use of the old transmission block and combines with the new transmission block to make a decision for decoding. The ARQ is needed when the HARQ fails to deliver error-free transmission blocks.

Mobility management is also needed in LTE for the mobile user to move from one cell to another cell. Soft handoff, which is used in code-division multiple access (CDMA) cellular networks, is not used in LTE. Instead, hard handoff is used in LTE. *Soft handoff* allows a user to be connected to several base stations during handoff before being handed off from the source base station to the target base station. *Hard handoff* causes a mobile to break off its connection from a source base station before connecting to the target base station. Handoff in LTE is also initiated by the network (i.e., network-controlled handoff).

Radio resource management is also needed to make efficient use of the resources available to meet radio resource requirements. The functions of radio resource management include radio bearer control, radio admission control, connection mobility control, dynamic resource allocation and packet scheduling, intercell interference coordination, load balancing, and interradio access technology radio resource management.

The quality of service in LTE is greatly improved by having a high peak data rate and low latency. As mentioned earlier, the peak data rate in the downlink is 100 Mbps, while the peak data rate in the uplink is 50 Mbps. The peak data rate in 3G cellular networks much as wideband CDMA is at most only 2 Mbps. Thus, there is a tremendous increase in peak data rate, which means that the packet transfer delay is much shorter for the same amount of payload, or many more packets can be transferred in the same amount of time.

There are also many applications for LTE. These applications include web browsing, file transfer protocol, video streaming, music streaming, voice over IP, network gaming, real-time video, push to talk, and push to view. More details on these applications are described in this chapter.

A succinct introduction to LTE is given in Section 9.1. In Section 9.2 we present the LTE network architecture. The physical layer for LTE is explained in detail in Section 9.3, and layer 2, including the medium access control (MAC) for LTE, is explained in Section 9.4. Both the downlink and uplink are described. In Section 9.5 we describe the mobility resource management for handoff in LTE, and radio resource management in LTE is discussed in Section 9.6. Security in LTE is described briefly in Section 9.7. The qualities of service in LTE are described in Section 9.8, and the applications of LTE are presented in Section 9.9. Finally, the salient features of this chapter are summarized at the end of this chapter.

To facilitate concise presentation, acronyms used throughout the chapter are tabulated in Table 9.1.

9.2 NETWORK ARCHITECTURE

The LTE network architecture is shown in Figure 9.1. The user equipment (UE) or mobile phone is connected wirelessly to the enhanced node B (eNB) or base station. In 3G networks, the base station is known simply as node B (NB). The eNBs can communicate with each other using a X2 interface, as shown by the long dashed lines, while the eNBs can communicate with the mobility management entity (MME) in the control plane and/or system architecture evolved (SAE) in the user plane using a S1 interface, as shown by short dashed lines. The MME/SAE gateway is called the *evolved packet core* (EPC).

9.3 PHYSICAL LAYER

In many countries, 3G cellular services have been deployed, including, for example, CDMA2000 and wideband CDMA. Solutions to enhancing the performances of these systems, such as high-speed downlink packet access (HSDPA), high-speed uplink packet access (HSUPA), and multimedia broadcast multicast service (MBMS), have also been standardized. These solutions are usually called *midterm evolutions* of 3G systems since they are based fundamentally on the same multiple-access technologies as those of basic 3G systems. Over the years, 3.9G LTE, which uses completely different multiple access schemes, has been under standardization. In particular, OFDMA and SC-FDMA have been adopted as the physical-layer air interfaces for LTE downlink and uplink, respectively. LTE provides a major advance in cellular technology; it meets the needs for high-speed data transmission, supporting multimedia broadcasting, as well as high-capacity voice users. The technologies developed in LTE are also considered to be important solutions for 4G cellular systems.

TABLE 9.1 Acronyms

1G	first generation	NAS	nonaccess stratum
2G	second generation	NB	node B
3G	third generation	NRT	non-real time
4G	fourth generation	OFDM	orthogonal frequency-division multiplexing
3GPP	third-generation partnership project	OFDMA	orthogonal frequency-division multiple access
ACK	acknowledgment	PAPR	peak-to-average power ratio
AMC	adaptive modulation and coding	PBCH	physical broadcast channel
ARQ	automatic repeat request	PCCH	paging control channel
BCCH	broadcast control channel	PCH	paging channel
BCH	broadcast channel	PDCP	packet data convergence protocol
CCCH	common control channel	PDSCH	physical downlink shared data channel
CDMA	code-division multiple access	PDU	protocol data unit
CMC	connection mobility control	PMCH	physical multicast channel
CP	cyclic prefix	PRACH	physical random access channel
CQI	channel quality indication	PS	packet scheduling
CRC	cyclic redundancy check	PTT	push to talk
DCCH	dedicated control channel	PTV	push to view
DFT	discrete Fourier transform	PUCCH	physical uplink control channel
DL-SCH	downlink shared channel	PUSCH	physical uplink shared data channel
DRA	dynamic resource allocation	QAM	quadrature amplitude modulation
DTCH	dedicated traffic channel	QoS	quality of service
eNB	enhanced node B	QPSK	quadrature phase shift keying
EPC	evolved packet core	RAC	radio admission control
		RACH	random access channel
		RA-RNTI	random access radio network temporary identifier
E-UTRAN	evolved universal terrestrial radio access network	RAT	radio access technology
		RB	resource block
FDD	frequency-division duplex	RBC	radio bearer control
FSTD	frequency-switched transmitting diversity		random access channel
FTP	file transfer protocol		random access radio network temporary identifier

TABLE 9.1 (*Continued*)

FTT	fast Fourier transform		resource block
GPRS	general packet radio service	RLC	radio link control
GSM	global system for mobile communications	ROHC	robust header compression
HARQ	hybrid ARQ	RRC	radio resource control
HFN	hyper frame number	RRM	radio resource management
HSDPA	high-speed downlink packet access	RT	real time
HSPA	high-speed packet access	RTD	round-trip delay
HSUPA	high-speed uplink packet access	SAE	system architecture evolved
ICIC	intercell interference coordination	SC-FDMA	single-carrier frequency-division multiple access
IFFT	inverse FFT	SDU	service data unit
IP	Internet protocol	SFBC	space-frequency block code
LB	load balancing	SFN	single-frequency network
LTE	long-term evolution	TDD	time-division duplex
MAC	medium access control	TPC	transmission power control
MBMS	multimedia broadcast multicast service	TSTD	time-switched transmit diversity
MCCH	multicast control channel	TTI	transmission time interval
MCH	multicast channel	UE	user equipment
MIMO	multiple-input, multiple-output	UL-SCH	uplink shared channel
MME	mobility management entity	UMTS	universal mobile telecommunications systems
MTCH	multicast traffic channel	VoIP	voice over IP
NACK	negative acknowledgment	WCDMA	wideband code-division multiple access

The major design targets for LTE are as follows:

- The peak data rate is 100 Mbps for downlink and 50 Mbps for uplink over a 20-MHz bandwidth.
- LTE provides communications with a terminal, moving up even to 500 km/h.
- The cell coverage is around 5 km when performance is met. However, cell coverage of up to 100 km is not precluded.
- LTE supports scalable bandwidths of 1.25, 2.5, 5, 10, 15, and 20 MHz.

LTE supports both time-division duplex (TDD) and frequency-division duplex (FDD) for uplink and downlink separation. Furthermore, adaptive link adaptation,

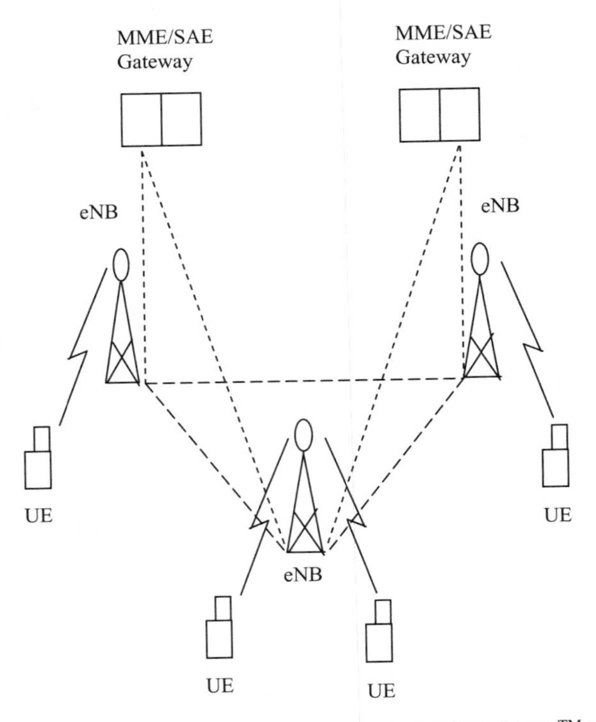

FIGURE 9.1 Network architecture for LTE. (From [4] © 2007.) 3GPP™ TSs and TRs are the property of ARIB, ATIS, ETSI, CCSA, TTA, and TCC, who jointly own the copyright. They are subject to further modifications and are therefore provided to you "as is" for information purposes only. Further use is strictly prohibited.

time–frequency scheduling, and MIMO antenna systems are incorporated in LTE. In this section we review the key issues related to the LTE physical layer.

9.3.1 LTE Downlink

9.3.1.1 Frame Structure Figure 9.2 shows the frame structure of LTE downlink transmission. Each radio frame is of 10 ms duration and consists of 10 subframes. Each subframe consists of two time slots of 0.5 ms each. The basic time unit is $T_s = 1/30{,}720$ ms. In one slot, there are seven or six OFDM symbols, depending on whether short or long cyclic prefix (CP) length is used (see Figure 9.2). The short CP length is 5.21 μs for the first OFDM symbol and 4.69 μs for the remaining six OFDM symbols; the long CP length is 16.67 μs for all six OFDM symbols. The long CP structure is used to support MBMS with a single-frequency network (SFN) in the downlink. Table 9.2 summarizes the OFDM parameters associated with LTE downlink.

LTE downlink physical resource can be visualized as a time–frequency grid, as shown in Figure 9.3. In this figure, one *resource element* corresponds to one OFDM subcarrier during one OFDM symbol duration. The available downlink bandwidth

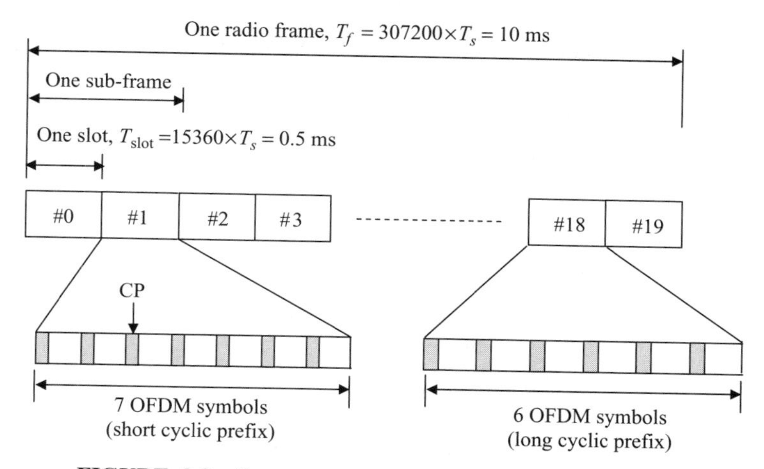

FIGURE 9.2 Frame structure of LTE downlink transmission.

consists of N_{BW} subcarriers with an OFDM subcarrier spacing of $\Delta f = 15\,\text{kHz}$. In the frequency domain, $N_{RB} = 12$ contiguous subcarriers of one slot are grouped to form a resource block (RB). The number of subcarriers in one RB is the same for different bandwidths.

9.3.1.2 Physical Downlink Channels Next, we provide more details on the downlink reference signals (Table 9.3). At the terminal, in order to reconstruct the signals transmitted from the base station, the terminal needs to estimate the downlink channel. This can be done by using reference signals embedded in the time–frequency OFDM grid.

As illustrated in Figure 9.4, the reference signals are inserted within the first- and third-from-last OFDM symbols of each slot, with a frequency spacing of six carriers. In addition, the separation in the frequency domain between the reference signals at the first symbol and the third from the last symbol is three subcarriers. The terminal has to perform interpolation over multiple reference signals to have channel information for the entire frequency–time grid. For one resource block, the terminal can use reference signals from other resource blocks and/or previous slots to improve the quality of the estimates. However, interpolation quality in the frequency or time domain depends heavily on the characteristics of the channel in the corresponding domain.

9.3.2 LTE Uplink

9.3.2.1 Frame Structure LTE uplink transmission is based on single-carrier FDMA (SC-FDMA). Compared to OFDMA, SC-FDMA has low PAPR, which is a desired feature for the user terminals since low PARP brings a great benefit in terms of power consumption. The block diagram of SC-FDMA signal generation is shown

TABLE 9.2 3.9G LTE Downlink OFDM Parameters

Transmission bandwidth (MHz)	1.25	2.5	5	10	15	20
Subframe duration				0.5 ms		
Subcarrier spacing				15 kHz		
Sampling frequency (MHz)	1.92	3.84	7.68	15.36	23.04	30.72
FFT size	128	256	512	1024	1536	2048
OFDM symbols per slot				7 for short CP, 6 for long CP		
CP length (μs/samples)						
short	$(5.21/10) \times 1$ $(4.69/9) \times 6$	$(5.21/20) \times 1$ $(4.69/18) \times 6$	$(5.21/40) \times 1$ $(4.69/36) \times 6$	$(5.21/80) \times 1$ $(4.69/72) \times 6$	$(5.21/120) \times 1$ $(4.69/108) \times 6$	$(5.21/160) \times 1$ $(4.69/144) \times 6$
long	16.67/32	16.67/64	16.67/128	16.67/256	16.67/384	16.67/512

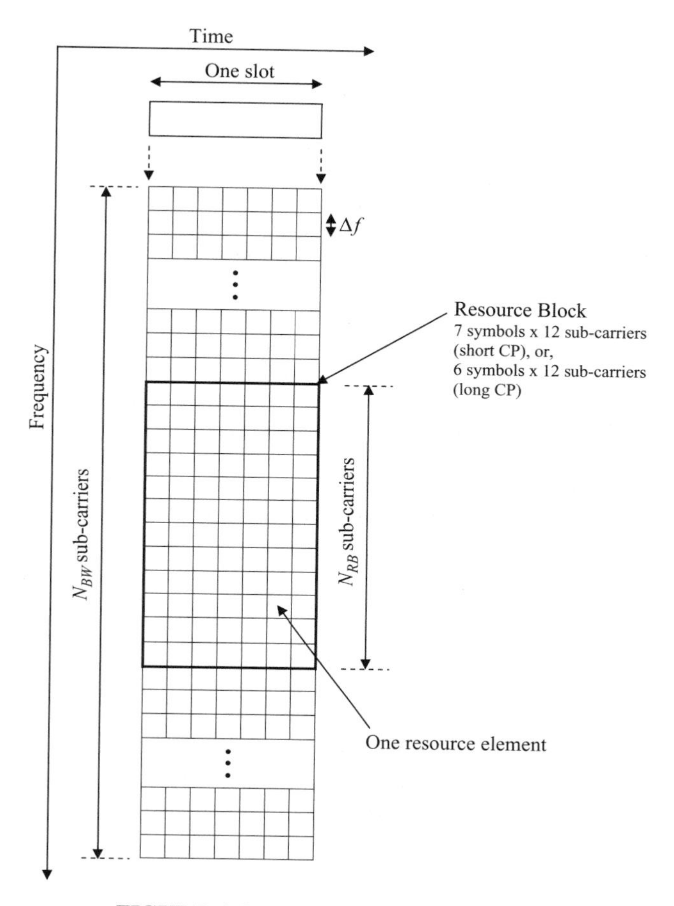

FIGURE 9.3 LTE downlink resource grid.

in Figure 9.5. The frequency mapping block in Figure 9.5 determines which part of the available frequency band can be used for transmission. Basically, there are two possible strategies for frequency allocation: *localized* and *distributed* allocations. However, in the uplink transmission, only the localized method is deployed, that is, the frequency mapping block in Figure 9.5 maps the output of the M-point DFT block to consecutive inputs of the N-point IFFT block. The localized SC-FDMA achieves multiuser diversity through frequency-domain scheduling.

The frame structure of the LTE uplink transmission is the same as that of the downlink transmission (see Figure 9.2); that is, one radio frame consists of 10 subframes, and each consists of two slots. Each slot accommodates seven SC-FDMA symbols (with short CP) or six SC-FDMA symbols (with long CP). Table 9.4 summarizes the parameters for the uplink frame structure. Uplink resource block allocation for different users is illustrated in Figure 9.6, in which each user is assigned consecutive

TABLE 9.3 Summary of the Physical Downlink Channels Defined in 3.9G LTE

Channel/Signal	Description
Reference signal	Reference signal is used for measuring channel quality to assist in time–frequency scheduling, channel estimation for coherent demodulation, and synchronization.
Synchronization signal	Synchronization signal is designed for cell search. Two types of synchronization signals are defined: primary and secondary.
Physical broadcast channel (PBCH)	PBCH is used to broadcast the system- and cell-specific control information.
Downlink layer 1/layer 2 (L1/L2) control channel	This channel is used to carry the control information for PDSCH (physical downlink shared data channel) and physical uplink shared data channel (PUSCH), such as channel allocation, link adaptation, and hybrid ARQ (HARQ).
Physical downlink shared data channel (PDSCH)	PDSCH is used for the transmission of user data and higher-layer signaling.
Physical multicast channel (PMCH)	PMCH is used for multimedia multicast/broadcast service (MBMS) with soft combining in a single-frequency network (SFN).

RB(s). Similar to the downlink transmission, one RB is defined to comprise 12 subcarriers over one slot of 0.5 ms in duration.

Another possible allocation with *frequency hopping* is shown in Figure 9.7. Frequency hopping means that a user is allocated RBs from one slot to another that do not occupy the same set of subcarriers. Frequency hopping provides frequency diversity with a condition that the frequency separation is equal to or greater than the channel coherence bandwidth.

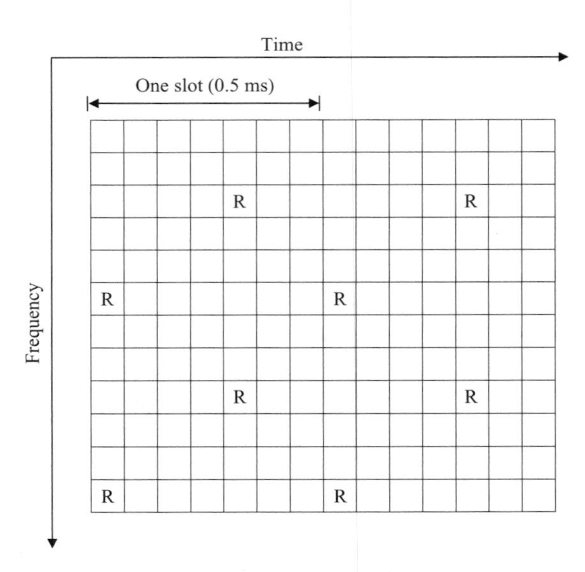

FIGURE 9.4 LTE downlink reference signals.

TABLE 9.4 Parameters for Uplink Frame Structure

Configuration	Number of Symbols	Cyclic Prefix in Samples	Cyclic Prefix Length
Short cyclic prefix	7	160 for first symbol, 144 for other symbols	5.21 μs for first symbol, 4.69 μs for other symbols
Long cyclic prefix	6	512	16.7 μs

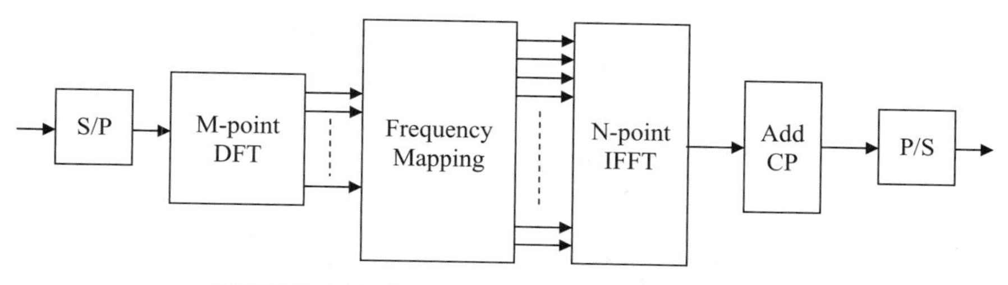

FIGURE 9.5 Structure of an SC-FDMA transmitter.

9.3.2.2 Physical Uplink Channels The following physical channels are defined for the LTE uplink transmission [2]:

- *Physical uplink shared channel* (PUSCH). User data are carried on this channel.
- *Physical uplink control channel* (PUCCH). This channel carries uplink control information: channel quality indication (CQI), ACK/NACK information related to data packets received in the downlink, and HARQ and uplink scheduling requests. Note that the user equipment (UE) uses PUCCH only when it does not have any data to transmit on PUSCH. If the UE has data to transmit on PUSCH, it would multiplex the control information with data on PUSCH.
- *Physical random access channel* (PRACH). This channel is used for initial link connection without timing synchronization. It consists of a cyclic prefix, a preamble, and a guard time, during which no signal is transmitted. The channel occupies a 1.08-MHz transmission bandwidth, which is equivalent to six RBs (72 contiguous subcarriers). The preambles are derived from Zadoff–Chu sequences.

Again, we provide more details on uplink reference signals. Uplink reference signals are used for two different purposes. On the one hand, they are used for channel estimation in the base station to demodulate control and data signals from the terminal. This type of reference signal is called a *demodulation* reference signal. The reference signals also provide channel quality information as a basis for uplink

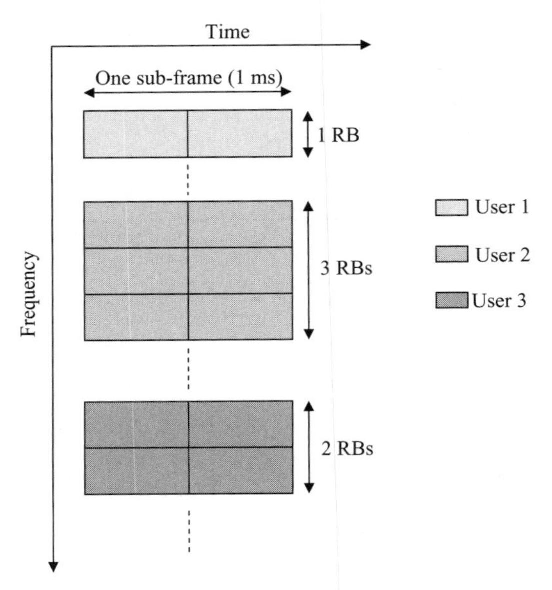

FIGURE 9.6 Uplink resource allocation.

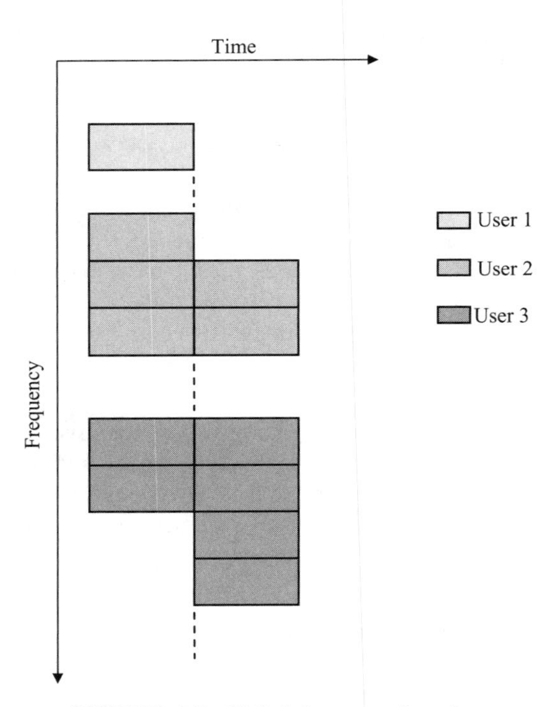

FIGURE 9.7 Uplink frequency hopping.

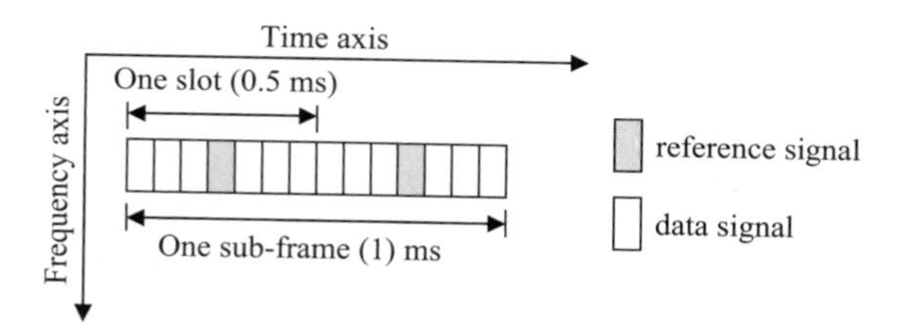

FIGURE 9.8 Uplink demodulation reference signals.

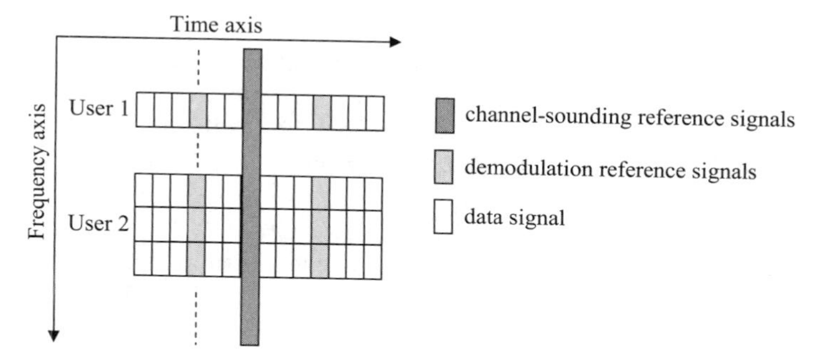

FIGURE 9.9 Channel-sounding reference signals.

channel-dependent scheduling decision making at the base station. The latter function is referred to as *channel sounding*.

The demodulation reference signals of a user are transmitted on the fourth symbol of each uplink slot over a bandwidth equal to the one granted to the user. These signals are taken from Zadoff–Chu sequences. The allocation for this reference signal is illustrated in Figure 9.8. The demodulation reference signals provide channel information only on the bandwidth at which they are transmitted. This information is not enough for the base station to determine the channel-dependent uplink scheduling. This operation requires channel information on a wide range of frequencies. The wideband channel-sounding reference signals are used to supply the necessary information for the base station. The allocation for these reference signals is illustrated in Figure 9.9. The channel-sounding reference signals are also taken from Zadoff–Chu sequences. They typically do not need to transmit as often as the demodulation reference signals. Furthermore, the channel-sounding reference signals may even need to be transmitted by terminals that have not been assigned any uplink resource for data transmission.

9.3.3 Advanced Features

9.3.3.1 Time-Frequency Domain Scheduling In LTE, time-frequency domain channel-dependent scheduling is supported for both uplink (PUSCH) and downlink (PDSCH). This is achieved through acquiring the CQIs of the respective channels.

By doing so, multiuser diversity gain is achieved. In both links, the minimum allocation is one RB. Furthermore, in downlink, discontinuous RB allocation is allowed, but in uplink, only continuous RB allocation is supported, for the purpose of achieving low PAPR.

9.3.3.2 Link Adaptation In LTE, downlink supports QPSK, 16-QAM, and 64-QAM modulations. In uplink, QPSK and 16-QAM are supported, with 64-QAM being optional. Based on CQI and QoS requirement, LTE supports various link adaptations, including transmission power control, adaptive modulation, and coding as well as adaptive bandwidth allocation. Further, LTE supports HARQ.

9.3.3.3 MBMS LTE supports MBMS through single-frequency networks (SFN). That is, all cells provide the broadcasting using the same set of resources (time and frequency). The signals coming from the surrounding cells are treated as multiple delayed versions (multiple signals); thus, through using long CP, the broadcast signals can easily be decoded by the user terminals.

9.3.3.4 MIMO MIMO is one of the main features supported by LTE, for the purpose of achieving its design targets. In fact, to achieve a 100-Mbps peak data rate in downlink transmission over a 20-MHz bandwidth, two independent data streams are multiplexed through 2×2 MIMO or 4×2 MIMO systems.

In LTE downlink, the supportable antenna configurations are 4×2, 2×2, 1×2, and 1×1. In uplink, the antenna configurations are 1×2 and 1×1. The baseline systems achieving a 100-Mbps downlink peak data rate and a 50-Mbps uplink peak data rate are 2×2 MIMO for downlink and 1×2 SIMO for uplink. When 4×4 MIMO systems are used, four data streams can be supported, achieving a 300-Mbps downlink peak data rate over a 20-MHz bandwidth.

Spatial diversity is exploited using MIMO, especially for control channels and MBMS. For example, in downlink, delay diversity is used for MBMS; space–frequency block code (SFBC) and frequency-switched transmit diversity (FSTD) are designed for control channels. In the uplink, time-switched transmit diversity (TSTD) is designed for PRACH. Finally, adaptive beamforming is also supported in LTE to achieve increased coverage.

9.4 MEDIUM ACCESS CONTROL SCHEDULING

Scheduling of system resources in LTE cellular networks is performed by the medium access control (MAC) protocol. In this section, the MAC functionality is described in detail. Essentially, the MAC does scheduling through coordination not only of the link layer but also the physical layer and radio resource control. The MAC has to be designed for both the downlink and uplink in LTE cellular networks. The logical channels have to be mapped to the transport channels for both the downlink and the uplink. Furthermore, random access channels are needed for initial access. LTE specifies two types of random access procedures, as explained in this section.

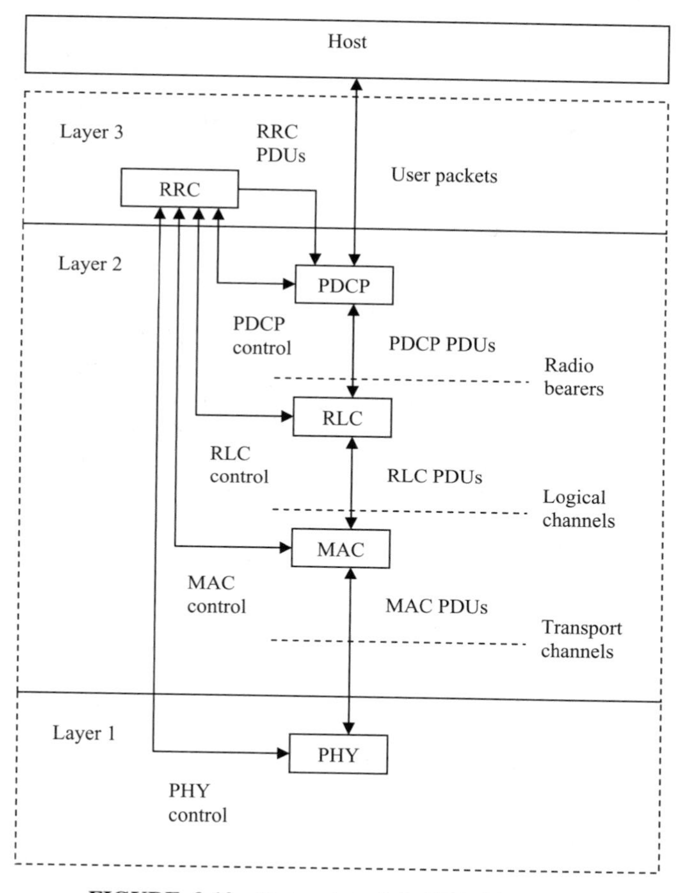

FIGURE 9.10 Protocol stack in LTE. (From [3].)

The MAC controls access to a shared medium or radio spectrum. The MAC decides who will transmit how much data information at what time and at what rate. Thus, this definition of MAC function consists essentially of the following in LTE, as shown in Figure 9.10:

- Radio resource control (RRC)
- Packet data convergence protocol (PDCP)
- Radio link control (RLC)
- MAC

Above this stack is the host; below this stack is the physical layer. IP packets from the host are packets into the PDCP protocol data unit (PDU) with PDCP headers. These are then passed down to the RLC and packed together with RLC headers to form RLC PDUs. These, in turn, are passed to the MAC layer as MAC service data units (SDUs), and MAC headers and paddings are added to form MAC PDUs. These are then passed to the physical layer as transport blocks. Each transport block is a subframe of 1 ms, and each subframe consists of two slots, each of 0.5 ms. Ten subframes form one radio frame of 10 ms. The 1-ms subframe length adopted is to achieve a short round-trip delay (RTD).

9.4.1 Downlink

The MAC function includes hybrid automatic repeat request (HARQ), mapping between the logical channels and transport channels, downlink scheduling, format selection, and measurements for the RRC. HARQ is used to provide robustness against transmission errors using retransmissions with soft combining at the physical layer. Soft combining uses the previous transport block to combine with the current transport block to improve the probability of decoding the transport block. Multiple parallel HARQ processes can run to retry several transport blocks. When the transport block fails the cyclic redundancy check (CRC), the MAC sends a negative acknowledgment (NACK) to the transmitter. The transmitter sends the transport block with a different puncturing code. This retransmitted transport block is then soft-combined with the previous transmitted transport block(s). When the transport block is decoded correctly, the MAC will send an acknowledgment (ACK) to the transmitter. For the downlink in LTE, an asynchronous HARQ is used. Thus, the MAC performs the signaling while the physical layer performs the retention of the transmission blocks and soft combining.

The logical channels are above the MAC, whereas the transport channels are below the MAC. The logical channels represent the data transfer services that are offered by the MAC; the transport channels represent the data transfer services that are offered by the physical layer. Logical channels are defined by what type of information they carry, whereas transport channels are defined by how information is carried. There are two types of logical channels: control channels and traffic channels. *Control channels* are used for control plane information, and *traffic channels* are used for user plane information.

The control channels offered by the MAC are as follows:

- Broadcast control channel (BCCH)
- Paging control channel (PCCH)
- Common control channel (CCCH)
- Multicast control channel (MCCH)
- Dedicated control channel (DCCH)

The BCCH is a common channel for the downlink. It is used by the network to broadcast system information to the UEs that are in the radio cell. The PCCH is also a common channel for the downlink. However, it is used to transfer paging information to the UEs that are in the radio cell. The CCCH is a special type of transport channel.

It is used for communications between the UEs and LTE when no RRC connection is available. The MCCH is used for transmission of multimedia broadcast and multicast service (MBMS) information from the network to one or more of the UEs. The DCCH is a point-to-point bidirectional channel that supports control information between a UE and the network.

The traffic control channels offered by the MAC are as follows:

- Dedicated traffic channel (DTCH)
- Multicast traffic channel (MTCH)

The DTCH is a point-to-point bidirectional data or application-level signaling channel that is used between a UE and the network. The MTCH is a point-to-multipoint data channel. It is used for transmission of data traffic from the network to one or more UEs.

The transport channels offered by the physical layer are as follows:

- Paging channel (PCH)
- Broadcast channel (BCH)
- Downlink shared channel (DL-SCH)
- Multicast channel (MCH)

The PCH is associated with the PCCH logical channel; the BCH is associated with the BCCH logical channel. The DL-SCH is used to transport user control or data traffic in the downlink. The MCH is associated with the MBMS user of transport control information. The mapping between downlink logical channels and downlink transport channels is shown in Figure 9.11.

The formation selection is specified in the information of each of the next transport blocks at the eNB. The measurements regarding local status and conditions from the local physical layer to the RRC are coordinated by the MAC. The RRC sends the measurement back to the eNB through control messages. The eNB informs the RRC on the physical-layer modulation and configuration settings that it controls. Measurements are used by the MAC for downlink scheduling. The downlink scheduler determines which terminal(s) will receive DL-SCH transmission and the resources in a dynamic manner. The data rates and radio conditions at the UE are utilized by the eNB. If the data rate is low, more time slots are needed to transmit the data information. On the other hand, if the data rate is high, fewer time slots are needed to transmit the same amount of data information.

The functions of the RLC are segmentation and reassembly, acknowledgment policy, in-sequence delivery, and duplication detection. In many instances, unpacking an RLC PDU into RLC SDUs and packing of two parts of two RLC PDUs into one RLC SDU may be possible. There are three acknowledgment policies: transparent mode, acknowledged mode, and unacknowledged mode. No RLC header is used for the first mode, while an RLC header is used for each of the latter two modes. The transparent mode is used when the PDU sizes are known (e.g., broadcasting system

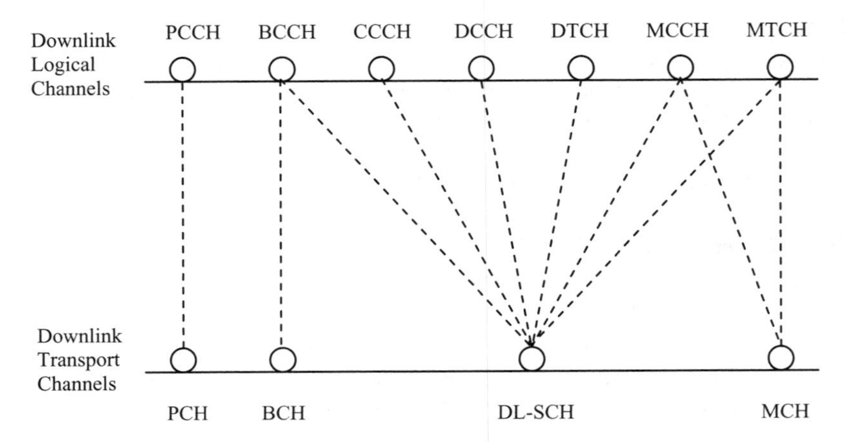

FIGURE 9.11 Mapping between downlink logical channels and downlink transport channels. (From [4] © 2007.) 3GPP™ TSs and TRs are the property of ARIB, ATIS, ETSI, CCSA, TTA, and TCC, who jointly own the copyright. They are subject to further modifications and are therefore provided to you "as is" for information purposes only. Further use is strictly prohibited.

information). The acknowledged mode is used for non-real-time (NRT) traffic, and the unacknowledged mode is used for real-time (RT) traffic. A RLC SDU has ARQ, too. This is used when the HARQ fails to deliver error-free transmission blocks.

The functions of the PDCP span both the user and control planes. In the user plane, the PDCP functions include decryption, robust header compression (ROHC) and decompression, and transfer of user data. In the control plane, the PDCP functions include decryption and integrity protection. ROHC is designed to support wireless links with higher error rates and longer round-trip times.

9.4.2 Uplink

The main differences between the uplink and the downlink are that the peak rate of the uplink is only half that of the downlink; the access in the uplink is granted by the eNB, the use of random access for initial transmission, and some changes in the logical and transport channels.

The functions of the PDCP in the uplink are symmetric to those in the downlink. The PDCP functions include header compression and encryption. The functions of the RLC in the uplink are also symmetric to those in the downlink. The RLC functions include adding RLC header, segmentation, and concatenation into transmission blocks.

The MAC function includes hybrid automatic repeat request (HARQ), mapping between the logical channels and transport channels, uplink scheduling, and format selection. HARQ in the downlink is synchronous. There is a fixed time after the uplink transmission when the NACK/ACK is transmitted on the downlink from the eNB.

The control channels offered by the MAC are as follows:

- Common control channel (CCCH)
- Dedicated control channel (DCCH)

The traffic control channel offered by the MAC is as follows:

- Dedicated traffic channel (DTCH)

The transport channels offered by the physical layer are as follows:

- Random access channel (RACH)
- Uplink shared channel (UL-SCH)

The RACH is a specific transport channel that supports limited control information. It is used during the early phases of communication establishment as well as RCC state change. The UL-SCH is used to transport user control or data traffic in the uplink. The mapping between uplink logical channels and uplink transport channels are shown in Figure 9.12. All UL-SCH MAC transmissions must be scheduled by the uplink scheduler. The UE sends an uplink scheduling request to the eNB, and the eNB replies with a grant. Uplink scheduling information is transmitted on the downlink physical layer control channel. It includes the transport format, resource allocation, HARQ information related to the DL-SCH, and NACK/ACK in reply to the uplink transmission. The transport format includes modulation and coding for the control channels. The modulation and coding determine the data rate and thus the capacity of a transmission block. The RACH provides a way for devices that

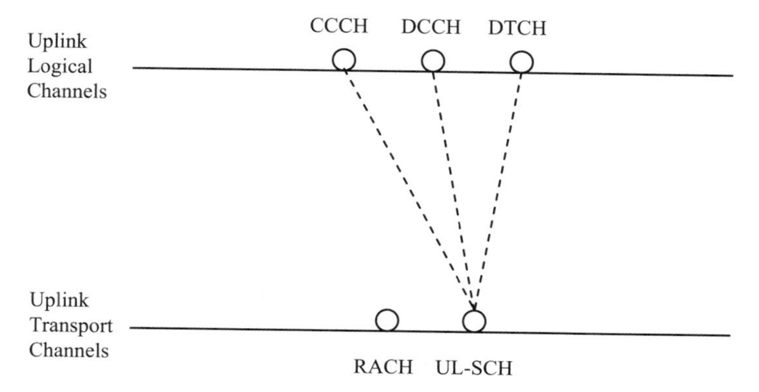

FIGURE 9.12 Mapping between uplink logical channels and uplink transport channels. (From [4] © 2007.) 3GPP™ TSs and TRs are the property of ARIB, ATIS, ETSI, CCSA, TTA, and TCC, who jointly own the copyright. They are subject to further modifications and are therefore provided to you "as is" for information purposes only. Further use is strictly prohibited.

are not connected to the eNB to have initial access. It can also be used for the handoff procedure.

The random access procedure is used for the following cases [4]:

- Initial access when the UE is not connected
- Initial access after a radio link failure
- Handoff requiring random access procedure
- Downlink or uplink data arrival when the UE is connected after the uplink physical layer has lost synchronization, which may be due to a power-save operation
- Uplink data arrival when no dedicated scheduling request channels are available

There are two types of random access procedures. One is contention based; the other is non-contention based. The contention-based random access procedure is shown in Figure 9.13; the non-contention-based random access procedure is shown in Figure 9.14. Let us first consider the contention-based random access procedure. The UE sends a random access preamble to the eNB using CDMA-like coding to allow simultaneous transmissions to be decoded. The eNB responds with a random access response sent on the DL-SCH within a window of a few transmission time intervals (TTIs). The response conveys at least a random access preamble, the timing alignment information, the initial uplink grant, and the assignment of a temporary cell radio network temporary identifier (C-RNTI). One or more UEs may be addressed in one response message. The UE will then schedule transmission to the eNB using HARQ and the RLC transparent mode on UL-SCH. It conveys at least the UE identifier. Finally, the eNB will send a contention resolution to the UE to end the random access procedure.

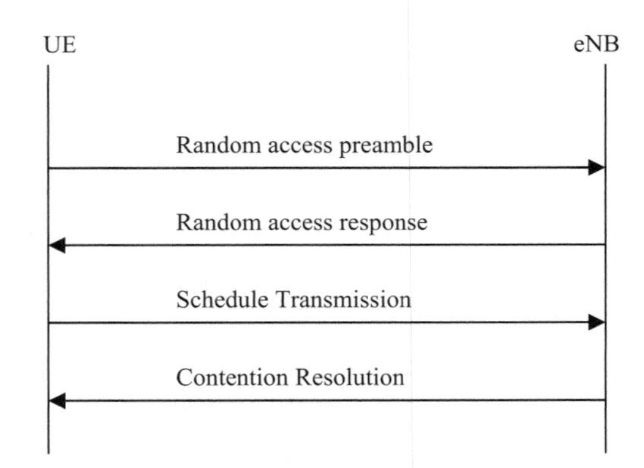

FIGURE 9.13 Contention-based random access procedure. (From [4] © 2007.) 3GPP™ TSs and TRs are the property of ARIB, ATIS, ETSI, CCSA, TTA, and TCC, who jointly own the copyright. They are subject to further modifications and are therefore provided to you "as is" for information purposes only. Further use is strictly prohibited.

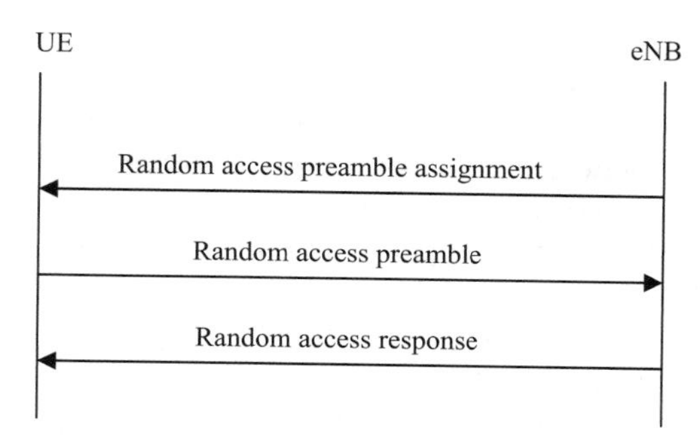

FIGURE 9.14 Non-contention-based random access procedure. (From [4] © 2007.) 3GPP™ TSs and TRs are the property of ARIB, ATIS, ETSI, CCSA, TTA, and TCC, who jointly own the copyright. They are subject to further modifications and are therefore provided to you "as is" for information purposes only. Further use is strictly prohibited.

Next, let us consider the non-contention-based random access procedure. The eNB assigns the 6-bit preamble code to the UE. The UE transmits the assigned preamble to the eNB. The eNB will then send a random access response, which is the same as that in the contention-based random access procedure. In addition, the response contains the random access preamble identifier if it is addressed to the random access radio network temporary identifier (RA-RNTI) on the layer 1/layer 2 control channel. One or more UEs may be addressed in one response message.

9.5 MOBILITY RESOURCE MANAGEMENT

Mobility resource management in LTE cellular networks is very different from that in 3G cellular networks. 3G cellular networks based on wideband code-division multiple access (WCDMA) use soft handoff. That is, an UE or mobile can be connected to two more NBs or base stations during handoff before being handed off from the source base station to the target base station. LTE uses a hard handoff for mobility management. That is, an UE or mobile will break its connection with a source eNB or base station before connecting to a target eNB or base station.

An UE can be in an idle mode or an active mode. An UE in an idle mode means that it is in a power-saving mode and is not transmitting or receiving packets. An UE in an active mode means that it is registered and is connected to its eNB with packet transmissions.

For handoff in an active mode, measurements are made between the UE and the source eNB [4]. The source eNB will make a decision for handoff; that is, it is a network-controlled handoff. The source eNB will request for a handoff to the target eNB. The target eNB will check its admission control to see if it can admit the call. If it can admit the call, it will respond with a handoff request acknowledgment. Then the source eNB will inform the UE of the handoff. The UE will detach itself from the old cell and synchronize itself with the new cell. The packets are buffered in the source eNB and forwarded to the target eNB. The target eNB will buffer the packets from the source eNB. After the synchronization and the necessary uplink allocation, the UE will confirm the handoff to the target eNB. The target eNB will do a path switch to the MME, and the MME will send an update request to the serving gateway in the user plane. The serving gateway will switch the downlink path and confirm the update by responding to the MME in the user plane. The MME will send an acknowledgment for the path switch to the target eNB. The target eNB will inform the source eNB to release the resource, and the source eNB will flush its downlink buffer and continue to deliver the in-transit packets to the target eNB. After this, the source eNB will release the resources for the connection, and the handoff is completed.

9.6 RADIO RESOURCE MANAGEMENT

Radio resource management (RRM) ensures efficient use of the radio resource available and provides mechanisms to enable LTE to meet radio resource requirements. RRM provides the means to assign, reassign, and release radio resources for both single cell and multicell functions [4]. The functions of RRM are as follows:

- Radio bearer control (RBC)
- Radio admission control (RAC)
- Connection mobility control (CMC)
- Dynamic resource allocation (DRA) and packet scheduling (PS)
- Intercell interference coordination (ICIC)
- Load balancing (LB)
- Interradio access technology (RAT) radio resource management (RRM)

The configuration of radio resources is needed for the establishment, maintenance, and release of radio bearers. A radio bearer is the service provided by layer 2 to transfer user data between the UE and LTE. When a radio bearer for a service is being set up, RBC considers the overall resources in LTE, the quality-of-service (QoS) requirement of in-progress sessions, and the QoS requirement of the new service. RBC is also responsible for the maintenance of radio bearers of in-progress sessions due to handoff or mobility or other reasons. When a session terminates or hands off, RBC is also involved in the release of radio resources.

The purpose of RAC is to decide whether to admit or reject the establishment of requests for new radio bearers. RAC's admission decision is based on the overall resources in LTE, the QoS requirements, the priority levels, the QoS provided by in-progress sessions, and the QoS requirement of the new radio bearer request. The objective of the RAC is to ensure high radio resource utilization and protect existing QoS for in-progress sessions. High radio resource utilization is achieved by admitting radio bearer requests when radio resources are available, while the existing QoS for in-progress sessions are protected by rejecting radio bearer requests if they cannot be admitted.

CMC manages the radio resources due to handoff. Handoff can be in an idle or active mode. In the idle mode, the cell reselection algorithms are controlled by thresholds and hysteresis values. These values define the best cell and/or determine when the UE should select a new cell. LTE broadcasts these parameters that configure the UE measurement and reporting procedures. In the active mode, handoff decisions may be based on UE and eNB measurements. Other factors that affect the decisions include neighbor cell load, traffic distribution, transport and hardware resources, and operator-defined policies.

DRA is located in the eNB. The purpose of DRA or PS is to allocate and deallocate resources to users and control plane packets. These resources include buffer, processing resources, and resource block(s). DRA's subtasks also include the selection of a radio bearer whose packets are to be scheduled and managing resources such as power levels or specific resource blocks used. Intercell interference coordination may also be taken into account. PS considers the QoS requirements of the radio bearers, the channel quality information for the UEs, buffer status, and interference.

The job of ICIC is to manage radio resources so that intercell interference is kept under control. ICIC is a multicell RRM function that considers information from multiple cells. The information includes resource usage status and traffic load situation. The job of LB is to handle uneven distribution of the traffic load over multiple cells. The objectives of LB are to distribute the traffic load evenly so that radio resources remain highly utilized, to ensure that the QoS of in-progress sessions are maintained and call-dropping probabilities are kept sufficiently small. The algorithms used in LB are to handoff or make cell reselection decisions such that the traffic load is redistributed from highly loaded cells to underutilized cells.

Inter-RAT RRM is to manage radio resources due to handoff between RATs. The handoff decision considers the resource situations of the RATs, UE capabilities, and operator policies involved.

9.7 SECURITY

Security in LTE applies the following principles. The eNB keys are cryptographically separated from the EPC keys used for nonaccess stratum (NAS) protection [4]. Thus, it is impossible to use the eNB key to figure out an EPC key [4]. The keys are derived in the EPC/UE from key material that was generated by an NAS (EPC/UE)-level AKA procedure. The eNB keys are sent from the EPC to the eNB when the UE is entering the active state in LTE [4]. Key material for the eNB keys is sent between the eNBs during intra-LTE mobility during the active state in LTE [4]. A sequence number is used as input to the ciphering and integrity protection [4]. Ciphering is the process of encrypting the message, while integrity protection is to ensure that the message sent by the source to the destination has not been modified or tampered with. A given sequence number must be used only once for a given eNB key, except for identical retransmission [4]. The same sequence number can be used for both ciphering and integrity protection [4]. A hyper frame number (HFN) due to an overflow counter mechanism is used in the eNB and UE to limit the actual number of sequence number bits needed to be sent over the wireless link [4]. The HFN needs to be synchronized between the UE and the eNB [4].

9.8 QUALITY OF SERVICE

The quality of service (QoS) in LTE is greatly improved by having a high peak data rate and low latency. The peak data rate in the downlink is 100 Mbps; the peak data rate in the uplink is 50 Mbps. The interruption times for handoff between LTE and WCDMA and between LTE and GSM are both 500 and 300 ms for non-real-time and real-time traffic, respectively. In the control plane, the transition time or latency from idle state to active state is less than 100 ms, while the transition time from dormant state to active state is less than 50 ms. The radio access network (RAN) latency in the user plane is less than 5 ms one way. Each transport block in LTE is a subframe of 1 ms. The 1 ms subframe length adopted is to achieve a short RTD. LTE supports end-to-end QoS control, which is optimized for a wide variety of packet services, such as VoIP, real-time gaming, real-time video, and data services [5].

The SAE bearer service layered architecture is shown in Figure 9.15. The SAE bearer service is used to provide edge-to-edge QoS for service data flows [4]. The SAE bearer is made up of two parts: a SAE radio bearer and a SAE access bearer. The

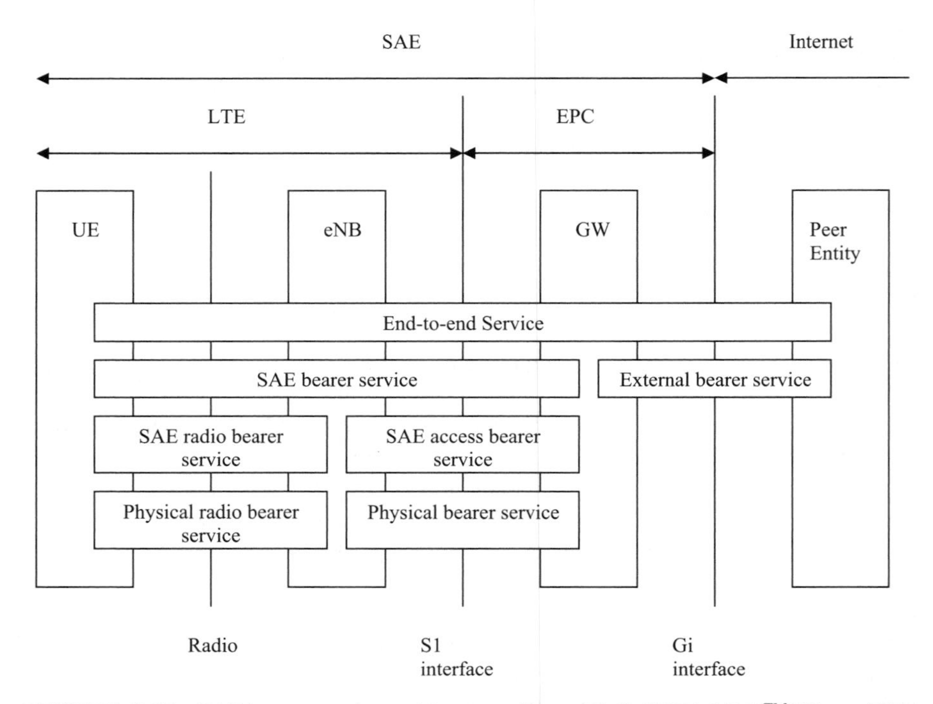

FIGURE 9.15 SAE bearer service architecture. (From [4] © 2007.) 3GPP™ TSs and TRs are the property of ARIB, ATIS, ETSI, CCSA, TTA, and TCC, who jointly own the copyright. They are subject to further modifications and are therefore provided to you "as is" for information purposes only. Further use is strictly prohibited.

SAE radio bearer provides transport of the SAE bearer service data units between the UE and the eNB. This is done based on the SAE QoS profile associated with each SAE bearer. The SAE access bearer service provides transport of the SAE bearer service data units between the eNB and the serving gateway. This is done based on the SAE QoS profile associated with each SAE bearer. The mappings between an SAE bearer and an SAE radio bearer and between an SAE radio bearer and a logical channel are both one-to-one.

9.9 APPLICATIONS

A good cellular system is also dependent on the types of applications that it can support. LTE is expected to support different types of applications. These applications include the following:

- Web browsing
- File transfer protocol (FTP)
- Video streaming
- Music streaming
- Voice over IP (VoIP)
- Network gaming
- Real-time video
- Push to talk (PTT)
- Push to view (PTV)

The following applications are applicable to LTE. Web browsing uses software that allows the user to search for information on the Internet. Files and pictures can be transferred from a source UE to a destination UE very quickly using LTE, due to its high data rate and low latency using file transfer protocol (FTP). FTP is used to transfer files between two hosts linked through LTE and the core network. Video streaming and music streaming are the continuous transmission of video or music, respectively, to the end UE connected through the eNB. The voice signal is packetized and transmitted through the Internet using Internet protocol (IP) for voice over IP (VoIP). Interactive network gaming connects different players together to play games with each other through LTE and the core network. Real-time video allows interaction between two users talking to each other "face to face." PTT is a way to send transmissions instantaneously to other users in LTE, emulating walkie-talkie communications. PTV allows a user to take a picture and send it instantaneously to other users in LTE.

SUMMARY

In this chapter we have focused mainly on the physical layer and the medium access control in LTE. Both the downlink and uplink are described and explained. The downlink is based on OFDMA, while the uplink is based on SC-FDMA. The relationship between the MAC and the RRC, and PDCP and RLC, are described and explained. Furthermore, the peak data rate for the downlink is 100 Mbps, while that for the uplink is 50 Mbps. These high data rates in both links improve the quality of service in LTE. Some related issues on network architecture, mobility resource management, radio resource management, qualities of service, and applications are also discussed and explained.

LTE is designed in such a way that it is not tied down by backward compatibility. It is not backwardly compatible with GSM, GPRS, WCDMA, and HSPA. It is also based on a new "all-IP" framework. In summary, LTE is a new 3.9G cellular technology to watch in the future. LTE will also give a smooth introduction to future 4G systems. The next phase of standardization in cellular technology is an extension of LTE: LTE-advanced, which will be the next 4G cellular technology to look for following LTE.

REFERENCES

[1] J. Zyren, "Overview of the 3GPP long term evolution physical layer," online, white paper, Freescale Semiconductor, July 2007.

[2] K. Higuchi, "Introduction of evolved UTRA and UTRAN," seminar at the Institute for Infocomm Research, Singapore, March 25, 2008.

[3] T. Godfrey, "Long-term evolution protocol: how the standard impacts media access control," available online on August 16, 2007, Freescale Semiconductors, June 26, 2007.

[4] *3GPP TS 36.300 V8.2.0* 2007–2009.

[5] "3GPP long term evolution," online, Qualcomm, San Diego, CA, Jan. 2008.

CHAPTER 10

WIRELESS BROADBAND NETWORKING WITH WiMAX

10.1 INTRODUCTION

In this chapter we first provide an overview of WiMAX, which is an emerging technology for wireless broadband networking. We compare WiMAX to its competing technologies. For WiMAX, there are three modes of operations: point-to-multipoint, mesh, and multihop relay. For the three modes, we also provide their MAC frame structures and operational details.

10.2 WiMAX OVERVIEW

Worldwide interoperability for microwave access (WiMAX) is a wireless broadband telecommunication system for metropolitan areas. In other words, WiMAX aims to provide high-speed wireless access in the range of Mbps over a long distance in the range of kilometers. WiMAX has formally been described by the WiMAX Forum [1] as "a standards-based technology enabling the delivery of last mile wireless broadband access as an alternative to cable and DSL." The forum was formed in June 2001 to promote and certify conformance and interoperability of the wireless broadband products based on the harmonized IEEE 802.16 standard. Interoperability can help in reducing deployment cost where certified WiMAX equipment from all manufacturers can work together seamlessly ([2,3]).

Wireless Broadband Networks, By David Tung Chong Wong, Peng-Yong Kong, Ying-Chang Liang, Kee Chaing Chua, and Jon W. Mark

The IEEE 802.16 standard has been developed by the working group, which targets the design of high-speed, high-bandwidth, and high-capacity standards for both fixed and mobile wireless broadband networks. The fixed and mobile wireless network standards are IEEE 802.16d [4] and IEEE 802.16e [5], respectively. The products that conform to IEEE 802.16d and IEEE 802.16e are commonly called fixed WiMAX and mobile WiMAX, respectively. *Fixed WiMAX* enables broadband access to homes and businesses, whereas *mobile WiMAX* offers the full mobility of cellular networks at true broadband speeds.

An alternative to the IEEE 802.16 standard has been developed in Europe by the broadband radio access networks (BRAN) group of the European Telecommunications Standards Institute (ETSI). The European standard is called the high-performance radio metropolitan area network (HIPERMAN). ETSI HIPERMAN defines only one physical layer with 256-carrier OFDM. The HIPERMAN physical layer is a mandatory mode in the IEEE 802.16 standard that has additionally defined two other physical layers, single carrier (SC) and 2048-carrier OFDMA. ETSI HIPERMAN has been harmonized with and has become a subset of the IEEE 802.16 standard. Therefore, products that conform to ETSI HIPERMAN also conform to WiMAX.

Another wireless broadband access technology, known as wireless broadband (WiBro), has been developed in South Korea. WiBro is designed with a 2048-carrier OFDMA air interface for operation with 8.75-MHz channel size over the 2.3-GHz frequency band, while IEEE 802.16 supports different channel sizes up to 20 MHz and a flexible service frequency band within the range 2 to 11 GHz. WiBro has been adopted as a specific subset of the IEEE 802.16 standard. Thus, WiBro products also conform to WiMAX.

Given the harmonization of IEEE 802.16 with ETSI HIPERMAN and WiBro, they are not competitors with each other, so we focus on IEEE 802.16 while referring to the technical aspects of the current WiMAX. At present, WiMAX supports two modes of operation: the point-to-multipoint (PMP) mode and the mesh mode [Figure 10.1(a) and (b)]. In the figure as in well as in the rest of this chapter, subscriber stations (SSs) represent a network node that may or may not be the end-user node, and may be static or mobile. This avoids additional definition of the term *mobile station* (MS) and allows interchangeable use of SS and MS.

In the PMP mode, all SSs are connected to a base station (BS) within a single hop. In addition, the BS is connected to the back-haul Internet, thus acting as the gateway nodes for all SSs. The PMP mode requires all SSs to be located within the radio range of the BS. In the mesh mode, not all SSs need to be within the radio range of the BS. This is because an SS can act as a router for other SSs that are not within the radio range of the BS. As such, a WiMAX mesh is indeed a multihop wireless network where new SSs can be added between the BS and an SS so that coverage areas of the BS can be improved and transmission capacity may be increased with an improved signal quality. The benefits of multihop networking as shown in a WiMAX mesh have motivated new activity in the IEEE 802.16 working group to introduce the multihop networking idea to the PMP mode. In this new activity, a multihop relay

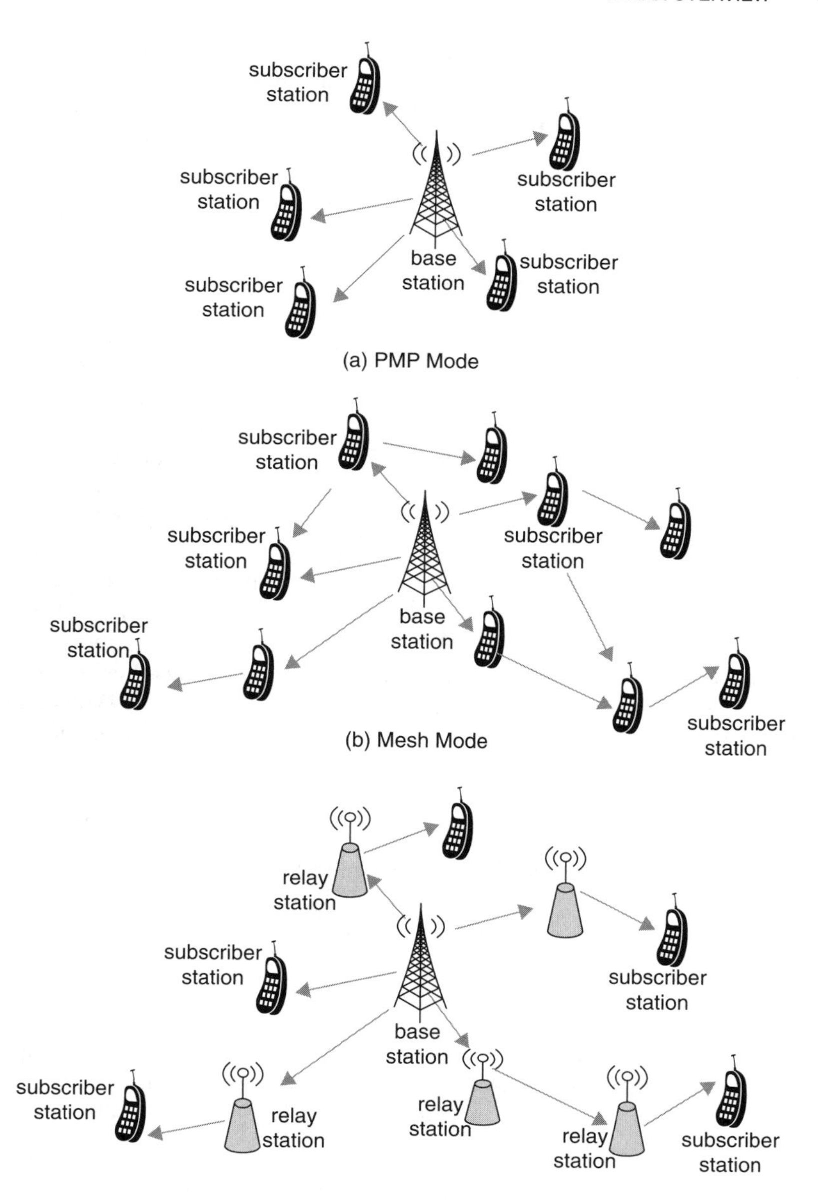

FIGURE 10.1 PMP, mesh, and multihop relay modes of operation.

network [Figure 10.1(c)] can be formed by adding a static or mobile relay station (RS) between the BS and an SS. In the context of a multihop relay network, the RS is not an SS but a simplified version of the BS where unnecessary functions are removed to save cost. We present details of the PMP mode, mesh mode, and multihop relay mode in Sections 10.5 to 10.7, respectively.

10.3 COMPETING TECHNOLOGIES

Fixed WiMAX has been described as a competitor to Wi-Fi [6] because both can be used for wireless networking connecting end users to the Internet, with WiMAX providing a much higher data rate over a much longer distance than Wi-Fi. Also, WiMAX uses both licensed and unlicensed frequency bands, whereas Wi-Fi uses only the unlicensed frequency band. Compared to Wi-Fi, which generally offers only besteffort service, WiMAX can provide QoS, which makes it more suitable for VoIP, IPTV, and other emerging multimedia applications. On the other hand, the existing widespread deployment of Wi-Fi hotspots may provide sufficient coverage for users to render subscription to WiMAX unnecessary. Practically, fixed WiMAX and Wi-Fi may not compete with each other directly, and WiMAX can complement Wi-Fi in providing cost-efficient back-haul networking to interconnect Wi-Fi hotspots.

As shown in Table 10.1 mobile WiMAX, has several competing technologies, primarily the incumbent 3G cellular network technologies that have evolved to provide broadband data rates. Compared to the emerging WiMAX, these incumbent 3G cellular networks have the advantage of having existing infrastructure that can

TABLE 10.1 Competing Technologies of Mobile WiMAX

		Data Rate (Mbps)		
Technology	Air Interface	Downlink	Uplink	Remark
WiMAX	OFDM, OFDMA	75	75	20-MHz channel bandwidth
UMTS-TDD	TD-CDMA	16	16	5-MHz channel bandwidth
3GPP LTE	OFDM, OFDMA	100	50	20-MHz channel bandwidth
CDMA2000/ EVDO	FD-CDMA	3.1	1.8	$N \times 1.25$-MHz channel bandwidth, where $N \leq 4$ is the number of channels being bundled together
3GPP2 ultra mobile broadband	OFDMA	275	75	20-MHz channel bandwidth
MBWA	OFDMA	1	1	120 to 350-kmph mobility speed

be upgraded to offer broadband wireless access with incremental cost. As such, in a fully mobile environment, users can seamlessly fall back to the existing systems when they move out of the coverage area of the upgraded equipment.

There are two major 3G technologies, UMTS and CDMA2000. For UMTS, UMTS-TDD, currently proposed to use TD-CDMA as the air interface, aims to provide a data rate of 16 Mbps in both downlink and uplink over a channel bandwidth of 5 MHz. Further, UMTS will move to 4G following the route of 3GPP LTE, where OFDMA and OFDM may be used as the air interfaces to achieve data rates of 100 and 50 Mbps for downlink and uplink, respectively, over a channel bandwidth of 20 MHz. For CDMA2000, EVDO has currently been proposed using FD-CDMA as the air interface to provide data rates of 3.1 and 1.8 Mbps, respectively, for downlink and uplink over a channel bandwidth of 5 MHz. Similar to UMTS, CDMA2000 will also move to 4G. CDMA2000's route to 4G will follow 3GPP2 ultra mobile broadband, which may use OFDMA as the air interface to offer a data rate of 275 and 75 Mbps for downlink and uplink, respectively, over a channel bandwidth of 20 MHz.

Apart from the 3G and 4G cellular network technologies, mobile broadband wireless access (MBWA), which is currently being developed by the IEEE 802.20 working group, may post a challenge to the current WiMAX. MBWA, similar to WiMAX, aims to provide wireless high-speed connectivity to a mobile user at high-speed mobility (120 to 350 kmph). At such high speeds, MBWA uses OFDMA as the air interface and will provide a data rate of 1 Mbps. MBWA may be adopted as part of the future WiMAX.

10.4 OVERVIEW OF THE PHYSICAL LAYER

Figure 10.2 illustrates the evolution of the IEEE 802.16 physical layer. The first IEEE 802.16 standard was completed in December 2001. This version defined only the fixed PMP mode, with a requirement of line-of-sight (LOS) transmissions employing the single carrier air interface within the frequency band 10 to 66 GHz. IEEE 802.16 can offer data rates up to 134 Mbps but cannot operate in a non-line-of-sight (NLOS) environment, due to the high carrier frequency used. In a subsequent version, IEEE

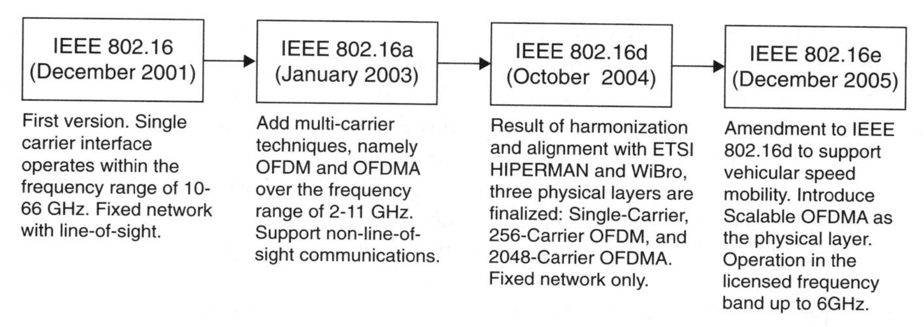

FIGURE 10.2 Evolution of IEEE 802.16 WiMAX physical layer technologies.

802.16a has defined operations in a NLOS environment in the frequency band 2 to 11 GHz. This version was completed in January 2003 and is capable of supporting data rates up to 75 Mbps using OFMA and OFDMA air interfaces. After harmonizing with ETSI HIPERMAN's OFDM air interface and WiBro's OFDMA air interface, IEEE 802.16d was finalized in October 2004 with all three physical interfaces: single carrier, 256-carrier OFDM, and 2048-carrier OFDMA. While IEEE 802.16d is for fixed network, the subsequent IEEE 802.16e amendment, completed in December 2005, supports mobility. In IEEE 802.16e, scalable OFDMA was introduced to further improve spectrum efficiency.

Since OFDM is the basis for both OFDMA and scalable OFDMA, which are the more advanced physical layers compared to the single carrier, it is discussed briefly in the rest of this section. Details of OFDM are provided in Chapter 1.

10.4.1 OFDM, OFDMA, and Scalable OFDMA

OFDM is a multicarrier modulation scheme that uses a large number of closelyspaced orthogonal subcarriers for simultaneous transmissions. With simultaneous transmissions, when each narrow subcarrier is modulated with a conventional modulation scheme such as QPSK at a low symbol rate, the aggregated data rate can be much higher, as it is a wideband communication.

The orthogonality in OFDM is important in making sure that there is no crosstalk between two adjacent subcarriers, and hence eliminating the need for intercarrier guard bands. This can greatly improve the spectrum efficiency since almost the entire frequency band can be utilized for data transmission.

Although orthogonality is important, maintaining it requires very accurate frequency synchronization between the receiver and the transmitter. Orthogonality can be destroyed when subcarriers suffer from frequency offsets that are caused by any Doppler shift due to node movement. These frequency offsets cannot easily be compensated in the presence of multipath propagation impairments because reflections at different propagation paths may appear with different strengths at various frequency offsets. The loss of perfect orthogonality leads to intercarrier interference, and the level of interference can worsen with increased node speed. Thus, there is always a supported maximum speed. In WiMAX, the maximum speed is 120 kmph.

Recall from above that OFDM has a low symbol rate and thus relatively long symbol duration compared to the channel coherent time. Hence, compared to conventional single-carrier modulation, OFDM suffers less intersymbol interference caused by multipath propagation impairment. The low intersymbol interference can be potentially eliminated by inserting a guard interval between two adjacent OFDM symbols. Inserting a guard interval will not be economical if OFDM has short symbol durations, because the guard interval can take up a large percentage of the symbol duration. Figure 10.3 shows the time structure of an OFDM symbol where the guard interval T_g precedes the OFDM waveform time duration T_b. Practically, T_b is the actual useful symbol time to transmit data. The guard interval is used to transmit the cyclic prefix, which is the last T_g duration of the intended OFDM symbol.

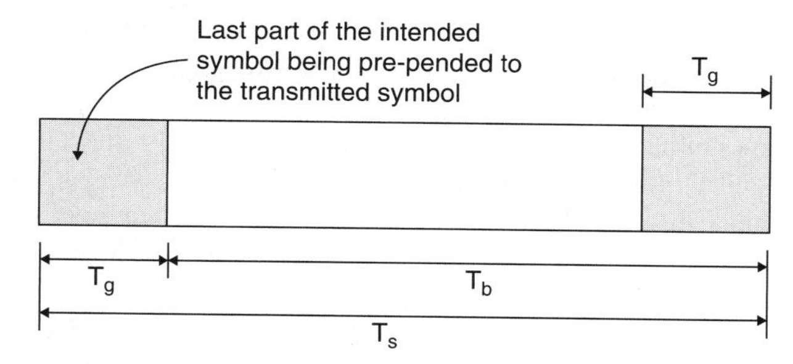

FIGURE 10.3 Time structure of an OFDM symbol.

A wireless link may suffer from significant performance degradation due to short-term multipath fading, long-term shadowing, and Doppler shift and time-dispersive effects. To improve wireless link resilience against these several propagation impairments, channel information acquired by the receiver can be fed back to the transmitter. Based on the feedback information, adaptive modulation and channel coding, as well as dynamic power control, may be performed on the subcarriers. Table 10.2 shows the set of predefined modulation and coding schemes that have been standardized in IEEE 802.16 WiMAX and used in the link adaptation algorithm [7].

The adaptive control described above can be applied monotonically on all subcarriers, or selectively (multitonically) only on the subcarrier affected. In the latter case, if a particular subcarrier suffers from bad signal quality, the subcarrier can be disabled or made to run slower by applying more robust modulation or channel coding. In this case, a bit-loading algorithm is needed to determine the configuration of each subcarrier in response to its instantaneous channel condition.

OFDM may not be considered a multiple access technique because its primary function is to transfer data robustly over a communication channel using a sequence of OFDM symbols. In practice, OFDM can be combined with various common multiple access techniques, such as TDMA, FDMA and CDMA, to support multiple users. A combination of OFDM and FDMA leads to OFDMA. In OFDMA, different

TABLE 10.2 Adaptive Modulation and Coding Schemes in IEEE 802.16 WiMAX

Modulation Scheme	Bytes per Subchannel	Bytes per 10-ms MAC Frame with 50% Asymmetry
QPSK 1/2	6	36
QPSK 3/4	9	54
16-QAM 1/2	12	72
16-QAM 3/4	18	108
64-QAM 1/2	18	108
64-QAM 2/3	24	144
64-QAM 3/4	27	162

OFDM subcarriers are assigned to different users. As such, OFDMA may support differentiated QoS by allocating different numbers of subcarriers to different users. By allocating subcarriers to achieve QoS, OFDMA may avoid complex packet scheduling algorithms or medium access control schemes.

In WiMAX, OFDMA divides the total number of carriers into N_g groups, each with N_c carriers. This will form N_c subchannels, where each subchannel takes a carrier from the N_g groups. For example, in the 2048-carrier OFDMA, $N_g = 48$ and $N_c = 32$ for the downlink. For the uplink, $N_g = 53$ and $N_c = 32$.

In 2048-carrier OFDMA, the FFT size is fixed at 2048. Since WiMAX supports variable channel size, ranging from 1.25 to 20 MHz, the subcarrier spacing will vary when the channel size changes. However, if the subcarrier spacing is not fixed, Doppler shift of a moving node can affect the signal quality negatively. Thus, scalable OFDMA [8] is proposed such that the subcarrier spacing is always kept constant while FFT size can be changed based on the channel size. In mobile WiMAX, 1024-carrier Scalable OFDMA has been adopted. Thus, when a node travels through the network, it may receive signal through 128-carrier OFDMA, 512-carrier OFDMA, and 1024-carrier OFDMA, depending on the channel size.

10.5 PMP MODE

Generally, PMP communication refers to communication over any media with a root station that can broadcast on a single frequency through a trunk-and-branch structure to its leaf stations. The leaf stations can unicast on a single common frequency through the same trunk-and-branch structure to the root station. In the trunk-and-branch structure, the leaf nodes may not communicate directly with each other, and thus there is no peer-to-peer communication. In the PMP communications, the common frequency requirement may be relaxed in cases where separate and unique frequencies are used between each root station/leaf station pair. In the forward communication direction, a central antenna or antenna array at the root station broadcasts to several receiving antennas at the leaf stations. On the other hand, the system uses a form of multiple access schemes to allow for transmissions in the reverse direction. The communication channel can be shared between the forward and reverse directions through time-division duplexing (TDD) or frequency-division duplexing (FDD).

In WiMAX, the root station and leaf station are the BS and SS, respectively. As illustrated in Figure 10.1(a), signal from the BS can reach all its SSs directly. Hence, PMP in WiMAX is architecturally similar to a cellular network, where one cellular tower can coordinate the incoming and outgoing signals to and from multiple mobile phones. While the forward direction communication is broadcast in nature, unicast transmission from BS can be performed where each message that is directed to a particular SS is indicated by the SS's connection identifier (CID). Although all SSs may receive all broadcast messages, an SS will process only those packets that are intended for it, or those that are intended explicitly for all SSs.

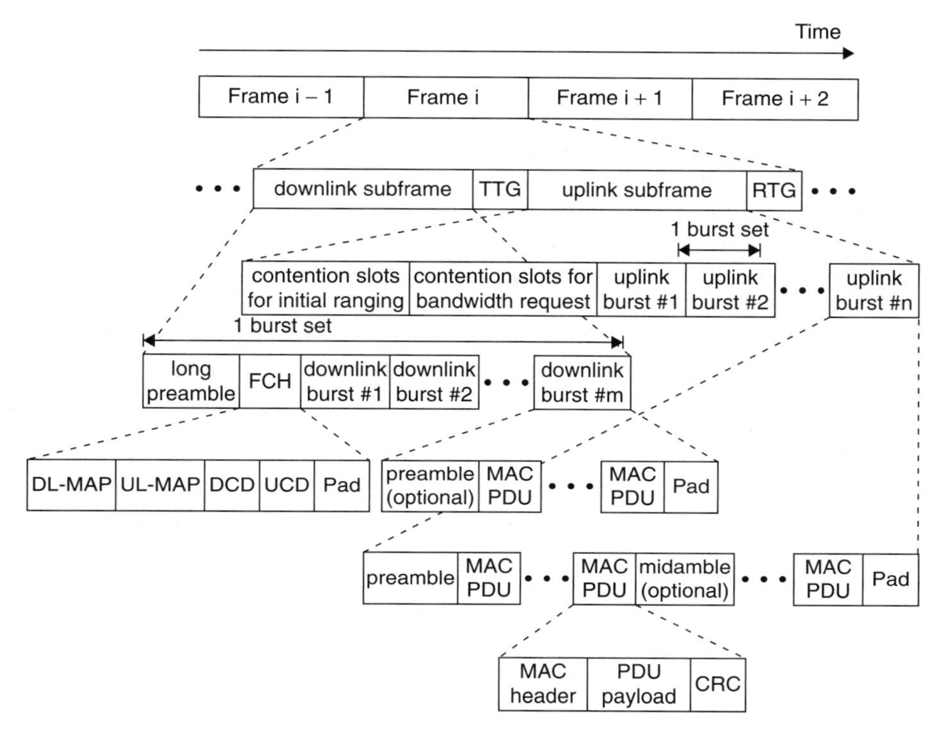

FIGURE 10.4 MAC frame structure with TDD for WiMAX PMP mode.

10.5.1 MAC Frame Structure and Operation

WiMAX PMP adopts a frame-based transmission where each channel is made up of periodic time frames called MAC *frames*. Figure 10.4 depicts the MAC frame structure for the OFDM physical layer operating in the TDD mode. The MAC frame length is fixed but configurable to be between 2.5 and 20 ms. Each MAC frame is further divided into a downlink subframe and an uplink subframe, separated by a transmit–receive transition gap (TTG) and a receive–transmit transition gap (RTG). Specifically, TTG is required between the downlink and uplink subframes, and it allows the BS to turn its operation around from transmission to reception. Similarly, RTG is required between the uplink and downlink subframes, and it enables the BS to change its operation from reception to transmission. Both TTG and RTG should not exceed 100 μs [9].

Usually, the downlink subframe consists of only a single burst set, which may contain multiple downlink bursts. As illustrated in Figure 10.4, the downlink subframe starts with a long preamble of two OFDM symbols. The long preamble is used for synchronization, and it is followed by a frame control header (FCH) of one OFDM symbol. The FCH is modulated with the most robust modulation and coding scheme, binary phase-shift keying (BPSK) with coding rate 1/2. The FCH is followed by one or multiple downlink bursts. The first downlink burst contains all the broadcast MAC management messages, which include, but are not limited to, downlink channel descriptor (DCD), uplink channel descriptor (UCD), downlink map (DL-MAP), and uplink map (UL-MAP). Both DCD and UCD define the physical channel characteristics for downlink and uplink, respectively. In practice, the channel characteristics are given as a set of burst profiles that can be used, where each burst profile is a combination of parameters that describe the transmission properties, such as modulation and coding schemes. In the specification [4], each burst profile is identified through an assigned code, called downlink interval usage code (DIUC) and uplink interval usage code (UIUC). DL-MAP defines the burst start times, but not duration, for each individual downlink burst and its associated DIUC. UL-MAP indicates the start time and duration of time intervals granted to different SSs for their uplink transmissions, as well as the associated UIUC.

In the downlink subframe, besides the first downlink burst, the other downlink bursts begin with an optional preamble to enhance the synchronization and channel estimation. Following the preamble, a number of MAC protocol data units (PDUs) are packaged into a single downlink burst. These MAC PDUs may be associated with different service flows, connections, or SSs, but all of these PDUs are encoded and modulated using the same burst profile. Each MAC PDU consists of a MAC header, PDU payload, and cyclic redundancy check code (CRC). The MAC header may carry information for automatic repeat request (ARQ), packing, and fragmentation. Packing is the process of combining multiple MAC service data units (SDUs) from the higher protocol layer into a single MAC PDU. On the other hand, fragmentation is the process of dividing a MAC SDU into multiple MAC PDUs.

The uplink subframe consists of three contention slots for initial ranging, 10 contention slots for bandwidth request, and one or multiple slots of an uplink burst set. The purpose of initial ranging is to facilitate the SSs' entry into the system by providing power control, frequency offset adjustment, time offset correction, and basic management request. The bandwidth request contention slots are used for the SSs to transmit their bandwidth request messages. Contention resolution in the slots is achieved through binary exponential backoff, where the start and end of the contention window range are announced periodically in UDC.

Different from the downlink burst set, which may consist of multiple downlink bursts, each uplink burst set has only one uplink burst. The uplink burst structure is similar to the downlink burst. One noticeable difference is that each uplink burst may have optional midambles inserted periodically to enhance synchronization and channel estimation. In both downlink burst and uplink burst, the size of each burst is an integer number of the OFDM symbol length to exactly match the OFDM symbol and burst boundaries. To form an integer number of OFDM symbols, unused bytes in the burst payload might be padded by the bytes 0xFF.

As illustrated in Figure 10.5, the BS maintains a downlink packet queue for each SS, and the queue is used to hold SDUs from the higher layer. During the process of node entry into system, each SS negotiates its QoS requirement with the BS. Based on the QoS negotiated and downlink queue length, the downlink scheduler at the BS schedules the next SDU to be transmitted. Several SDUs may be packed to form a

PDU, and multiple PDUs can be packed further to form a burst for transmission. The time boundaries for all the PDUs and bursts transmitted in a downlink subframe are indicated in the preceding DL-MAP, as depicted in Figure 10.4.

For uplink transmission, the actual packet queues reside at respective SS. When there is SDU awaiting uplink transmission in the queue, SS sends a bandwidth request to the BS. The bandwidth request can be sent through contention slots in an uplink subframe or unicast polling. In unicast polling, the BS allocates a polled uplink connection to an SS, to which the SS may respond by transmitting its bandwidth request. The BS can also issue broadcast polls to all SSs, but this may lead to collisions when there are multiple simultaneous bandwidth requests in response to the broadcast. SS can also piggyback its bandwidth requests on a PDU. At the BS, the bandwidth requests received are used to estimate the backlog size of each actual uplink queue at the SS. Based on the estimation, uplink virtual queues are formed at the BS for the respective SS as illustrated in Figure 10.5. Based on the length of the uplink virtual queue and QoS requirement, the uplink scheduler at the BS allocates time intervals to different SSs for uplink transmissions. The allocation outcomes are called grants in the UL-MAP, which indicates the start and end times for scheduled uplink bursts. For SS with a grant, its local scheduler will decide which SDU from which actual uplink queue is transmitted next within the allocated time interval.

In Figure 10.5, there are two schedulers at the BS and one scheduler at each SS, but the IEEE 802.16 standard does not specify the scheduling algorithm. Instead, the standard defines support for four different types of QoS:

1. *Unsolicited grant service* (UGS). In UGS, applications that generate fixed-size data packets on a periodic basis, such as voice over IP (VoIP), are supported. An

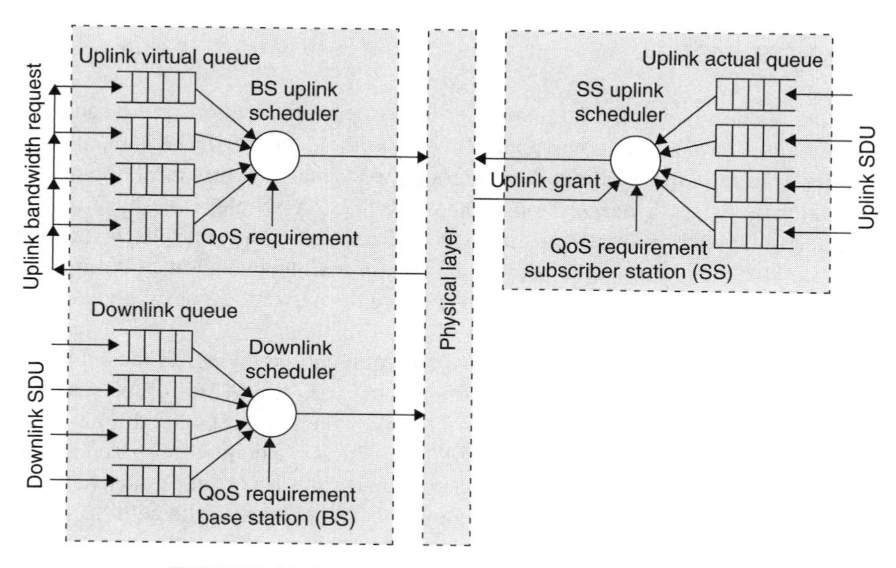

FIGURE 10.5 MAC operation in WiMAX PMP mode.

UGS service flow may request for a fixed amount of bandwidth on a periodic basis during registration of the uplink channel of the SS. Once the registration is done, bandwidth is never requested explicitly.

2. *Real-time polling service* (rt-PS). In rt-PS, real-time applications that generate variable-size data packets on a periodic basis are supported.
3. *Non-real-time polling service* (nrt-PS). In nrt-PS, non-real-time applications that generate variable-size data packets on a periodic basis are supported.
4. *Best efforts* (BE). In BE, all other types of non-real-time applications are supported without a minimum bandwidth guaranteed.

WiMAX networks are already available for deployment. In fact, during Indonesia's tsunami catastrophe in 2005, Intel deployed its WiMAX network in Acheh to provide communication support for the relief effort.

In PMP mode, research on QoS support [10–14] is crucial due to the different types of service flows the system is targeted to serve, but no scheduling algorithm has been standardized. In [12], the uplink and downlink packet scheduling algorithms have been proposed for WiMAX PMP mode. The main objective of the algorithm proposed is to provide delay and bandwidth guarantee, as well as maintaining fairness. The proposed algorithms have been implement and evaluated through simulation using QualNet. Unfortunately, the simulation assumes the physical layer of IEEE 802.11 instead of the WiMAX's OFDM physical layer [15]. The similar IEEE 802.11 physical layer assumption is made in WiMAX study in [16]. Here, the IEEE 802.16 MAC is implemented in GloMoSim with simplified signaling messages, fixed transmission-time-interval allocations, and no QoS differentiation.

10.6 MESH MODE

Mesh mode communication refers to peer-to-peer communication where each node can communicate with another node directly without going through the BS, via multihop routing or forwarding. A wireless network that supports mesh mode communication is called a *wireless mesh network* [17]. A wireless mesh network is characterized by dynamic self-organization, self-configuration, and self-correction to enable flexible integration, quick deployment, easy maintenance, low cost, high scalability, and reliable services, as well as to enhance the network capacity, connectivity, and throughput.

There are two standardized mesh mode communications, WiMAX mesh [18] and Wi-Fi mesh. The most fundamental difference between the MACs of WiMAX and Wi-Fi is that WiMAX mesh is based on TDMA, which is a scheduled time-slotted system, whereas Wi-Fi is based on CSMA/CA, which is a contention-based system. In WiMAX mesh, all data transmissions are carried out according to a schedule to prevent collision. The transmission scheduling is achieved with the help of a three-way handshaking mechanism that involves exchange of bandwidth request, grant, and grant confirmation between two communicating nodes. The three-way shaking mechanism is described in detail later in this section. In Wi-Fi mesh, each node tries to reserve in a distributed manner the channel for transmission and will perform exponential backoff if it senses the channel busy. To avoid collision, Wi-Fi mesh uses a two-way-handshaking mechanism that involves exchange of RTS and CTS signaling between two communicating nodes.

WiMAX mesh specifies a separate channel in the time domain for control messages, whereas Wi-Fi mesh does not. By separating the control channel from the data channel, WiMAX mesh ensures that scheduled transmissions in the data channel will not be affected by contention-based transmissions of bandwidth request, grant, grant confirmation, and other control messages in the control channel. Because there is no separate data channel, RTS and CTS control messages in Wi-Fi mesh may collide with data transmissions, leading to deterioration in service flow QoS.

In WiMAX mesh, each successful three-way handshaking may allow for isochronous allocation of multiple time slots in contiguous MAC frames. This feature is useful in reducing control overhead while scheduling for constant-bit-rate traffic, which requires fixed time slot allocation in the entire connection lifetime. In Wi-Fi, the absence of such a feature results in each packet requiring to go through the two-way handshaking process to reserve the channel before transmission.

WiMAX mesh is topology-aware but Wi-Fi mesh is not. WiMAX mesh utilizes up to three-hop neighborhood information in its resource allocation and transmission scheduling algorithms, thus making it more robust against the presence of hidden and exposed nodes. In the contrast, Wi-Fi mesh has no notion of topology awareness because only single hop information is available.

10.6.1 MAC Frame Structure and Operation

WiMAX mesh adopts OFDM as the physical layer and TDMA with a periodic MAC frame structure as illustrated in Figure 10.6. Each MAC frame consists of a control subframe and a data subframe. The length of the control subframe is defined as MSH-CTRL-LEN $\times$ 7 OFDM symbols, where MSH-CTRL-LEN is a 4-bit parameter in the network descriptor information element (IE). By fixing the control subframe length, the remaining part of a MAC frame is the data subframe. Each data subframe is divided into 256 time slots, where each time slot is a basic unit for resource allocation. Unlike the WiMAX PMP mode, the data subframe in mesh mode is not divided into downlink and uplink.

Each control subframe is divided into multiple (up to 16) time slots, for signaling message transmission using the most robust modulation and coding schemes. The length of each time slot is seven OFDM symbols, part of which is used as guard time. There are two types of control subframe, network control subframe and scheduling control subframe. *Network control subframe* is used to create and maintain cohesion between different nodes in the network. *Scheduling control subframe* is used to facilitate time slot allocation for transmissions in the data subframe. Network control subframe appears less frequently than scheduling control subframe. Specifically, there is only one network control subframe in every scheduling frame MAC frame, where a scheduling frame has a value in multiples of 4.

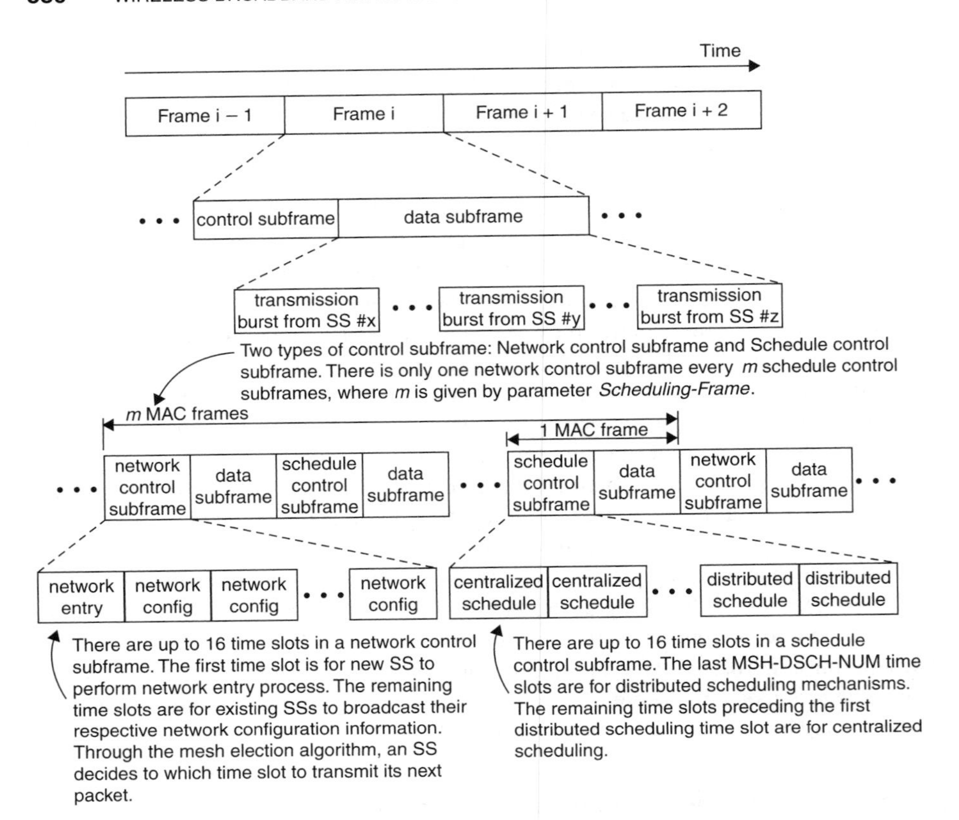

FIGURE 10.6 MAC frame structure for WiMAX mesh mode.

In a network control subframe, the first time slot is for network entry, followed by up to 15 network configuration time slots. In the network entry time slot, a new node may gain synchronization and initiate entry to the network. In the network configuration time slots, each SS announces its basic information to its neighboring nodes.

In a scheduling control subframe, the first (MSH-CTRL-LEN/MSH-DSCH-NUM) time slots are for centralized scheduling messages; the remaining MSH-DSCH-NUM time slots are for distributed scheduling messages, since WiMAX mesh supports both centralized and distributed scheduling paradigms. In centralized scheduling, a mesh BS as depicted in Figure 10.1(b) centrally performs the time slot allocation. In distributed scheduling, each transmitter–receiver pair requests and allocates time slots locally. In WiMAX mesh, distributed scheduling can be further classified into coordinated distributed scheduling and uncoordinated distributed scheduling. For *coordinated distributed scheduling*, the request–grant–confirmation three-way handshaking takes place in the scheduling control subframe. For *uncoordinated distributed scheduling*, the three-way handshaking is carried out opportunistically in the unused

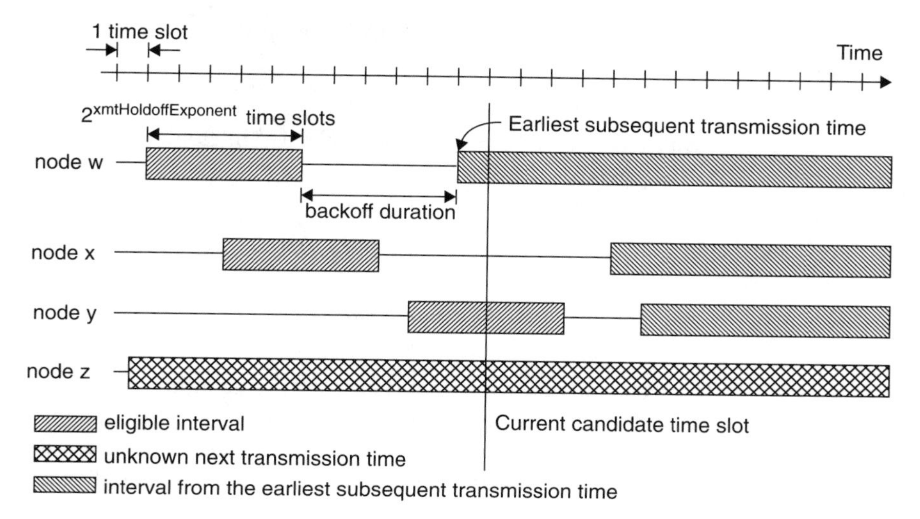

FIGURE 10.7 Eligible nodes to compete using the mesh election algorithm for transmission in the current time slot.

time slots in the data subframe. Thus, the performance of uncoordinated distributed scheduling is less predictable than that of coordinated distributed scheduling. We focus on coordinated distributed scheduling in this chapter.

In coordinated distributed scheduling, when an SS has a packet to send to another SS, it will transmit a bandwidth request to the SS in a time slot in the scheduling control subframe. The time slot is identified using the mesh election algorithm, which ideally allows for only one node (winner) to transmit in each time slot without collision. The algorithm is performed in a distributed manner at each node and an SS identifies itself as the winner for the current time slot if its node ID produces the highest value from a pseudorandom mixing function. The function takes in as inputs the sequence number of the current time slot and a set of node IDs for all the nodes that are eligible to transmit in the time slot. As illustrated in Figure 10.7, the set of eligible competing nodes includes all neighboring nodes that satisfy any of the following three conditions [19]:

1. The current time slot falls within the node's NextXmtTime interval. Here, NextXmtTime is determined as follows:

$$2^{\text{XmtHoldoffExponent}} \times \text{NextXmtXm} < \text{NextXmtTime} \leq 2^{\text{XmtHoldoffExponent}} \times (\text{NextXmtXm} + 1),$$

   where XmtHoldoffExponent and NextXmtXm of a node are, respectively, 3- and 5-bit parameters broadcast in the node's distributed scheduling control messages.

2. The current time slot occurs after the node's EarliestSubsequentXmtTime, which is determined as follows:

$$\text{EarliestSubsequentXmtTime} = \text{NextXmtXm} + \text{XmtHoldoffTime},$$

where XmtHoldoffTime is given by $2^{\text{XmtHoldoffExponent}+4}$. In practice, XmtHoldoffTime is the backoff duration of the node after it has transmitted in a time slot.
3. The node's NextXmtTime interval is unknown.

A transmitter node sends it bandwidth request after winning a time slot through the mesh election algorithm described above. Upon receiving the bandwidth request, the receiver node will allocate the first group of available time slots in the data subframe for transmission. The outcome of this data subframe time slot allocation will be announced by the receiver node as a grant. All the grants are transmitted in the control subframe after the node wins in the mesh election algorithm. In response to the grant, the transmitter node will send a grant confirmation in the control subframe following the same mesh election algorithm. Usually, the data packet will be transmitted in the data subframe only after successful transmission of the grant confirmation.

10.6.2 Research Activities

Research activities for WiMAX mesh is not as intense as that of WiMAX PMP mode, probably due to the lower complexity and higher maturity of WiMAX PMP. Furthermore, there is also no available WiMAX mesh simulation model in the public domain as well as commercial network simulation package. Commercial network simulation tools such as QualNet [20] and OPNET [21] have no WiMAX mesh simulation module. Currently, QualNet and OPNET support only WiMAX PMP mode, not mesh mode. Similar situation exists for the public-domain network simulation tools, such as GloMoSim [22] and NS2 [23], which are more popular with academic researchers, due to their license-free usage.

Published research works in WiMAX mesh have been related to optimization in route setup at the MAC layer and mesh scheduling mechanisms. While multihop routing is traditionally a network layer issue, routing at the WiMAX MAC layer ensures that every SS is able to obtain a default route to the BS upon successfully joining the mesh network. Given the default route, communication between the SS and its BS can be activated immediately after joining the network, without the need for any additional BS discovery process or any specialized ad hoc routing protocol. In the WiMAX mesh MAC, a newly arrived SS finds its default route to the BS by selecting a sponsoring node, which is the node that accepts the new SS into the network. Each sponsoring node has a route to the BS through its own sponsoring node when it entered the network earlier. Thus, each new SS is actually a leaf node added to the existing tree topology where the sponsoring node is its parent.

In [24] and [25], the main design goal is to derive an efficient mesh tree construction algorithm, which can provide high utilization network resources. To achieve the goal, [24] and [25] propose to select as a sponsoring node the node that can provide a route to the BS and incur the least interference cost from its surrounding neighbors. Here, any node that has already joined the mesh network can become a sponsoring node, and the level of interference is quantified in terms of a blocking metric. Evaluation has been conducted using Matlab simulation for throughput performance comparison against the original IEEE 802.16 standard, in which a default random selection of sponsoring node is used.

As described earlier in this section, transmission scheduling mechanisms in WiMAX mesh can be classified into (1) centralized scheduling, (2) uncoordinated distributed scheduling, and (3) coordinated distributed scheduling. Optimization of centralized scheduling was investigated in [24] and [25], where the authors proposed an interference-aware scheduling algorithm to exploit concurrent transmission opportunity to achieve high spectral utilization and hence high system throughput. In these works, the authors have evaluated the enhanced scheduling scheme against a basic scheduling algorithm with no spatial reuse in a linear chain topology and random mesh topology, and the results show that the enhanced scheme outperforms the basic scheme.

For the coordinated distributed scheduling mechanism, [26] has provided an analytical and simulation performance evaluation. Specifically, [26] has provided a good explanation on the operation of the mesh election algorithm, which is required by each node to determine the time slot to transmit it request, grant, and grant confirmation messages in the scheduling control subframe. This three-way handshaking process of sending bandwidth request, receiving grant, and acknowledging using grant confirmation ensures that once the time slot allocation has been confirmed for use by a sending node, no other node may utilize the allocated slots, and this will ensure collision-free data transmission as well as avoiding the hidden and exposed node problem. This three-way handshaking process has been modeled and analyzed in great detail in [26], such that the connection setup time for a node can be estimated based on the expected interval between successful control messages. The system parameters are the total node number, holdoff exponent value, and network topology. Logically, when the total number of nodes is increased or when using a larger holdoff exponent value, the expected time interval between control messages increases. From this study, holdoff exponent configuration has a significant impact on system performance. To meet the QoS of real-time traffic, such traffic flows can be configured to use a smaller holdoff exponent value so that these flows have a better chance of accessing the channel than that of non-real-time traffic flows.

While Cao et al. [26] focus on distributed coordinated scheduling, Redana and Lott [27] compare the throughput for both centralized and coordinated distributed scheduling. Based on the analytical and NS2 simulation results, it has been shown that centralized scheduling outperforms coordinated distributed scheduling. However, the evaluation results are derived from simple one- and two-hop scenarios, which may not hold in a generic WiMAX mesh network that has a greater hop count.

10.7 MULTIHOP RELAY MODE

WiMAX PMP and WiMAX mesh are not compatible in the sense that their MAC frame structures and operations cannot coexist. Thus, the benefits of multihop communications in WiMAX mesh are not readily available to WiMAX PMP. In view of this problem, in 2006 the IEEE 802.16j Working Group began to draft a specification for multihop relay mode operation that is backward-compatible to WiMAX PMP, as illustrated Figure 10.1(c). The backward compatibility means that the existing WiMAX PMP SSs will function normally in enhanced multihop relay networks. However, some modifications to the WiMAX PMP BS must be made to allow communication with the RS and support aggregation of traffic from multiple RSs.

To obtain the best possible network performance, the characteristics of RSs and their placement must be chosen carefully. There are three types of RSs: fixed, nomadic and mobile [28]. *Fixed RS* is installed permanently at a fixed location. *Nomadic RS* is intended to function from a location that is fixed for a period of time comparable to a user session. Finally, *mobile RS* is fully mobile and may be installed on moving vehicles, such as a bus, train, or ferry. In some cases, an SS may also act as a RS.

The multihop relay mode is significantly different from WiMAX mesh in the sense that it does not support peer-to-peer communications between two SSs. Similar to WiMAX mesh, the multihop relay mode is expected to bring about coverage extension and capacity improvement to WiMAX PMP. As of late 2007, the specification is still being developed and the first draft is undergoing letter balloting [29]. Hence, the detail presented in this section, although up to date, may differ from the final specification.

10.7.1 MAC Frame Structure and Operation

A main challenge in designing the MAC frame structure for multihop relay mode is to make it compatible with that of WiMAX PMP. Let an access link be defined as a radio link that originates or terminates at an SS. Also, let a relay link be defined as a radio link between the BS and an RS or between two adjacent RSs. Then, when all the access links and relay links on a given path do not interfere with each other, the links can be active concurrently, and thus the existing WiMAX PMP frame structure illustrated in Figure 10.4 can be used. However, this is not the case in practice since the BS and RSs can be within radio range of each other. Therefore, a new MAC frame structure needs to be designed for multihop relay operation. In the design, two types of multihop relay operations are considered: transparent mode and nontransparent mode. In the *transparent mode*, the RS does not transmit

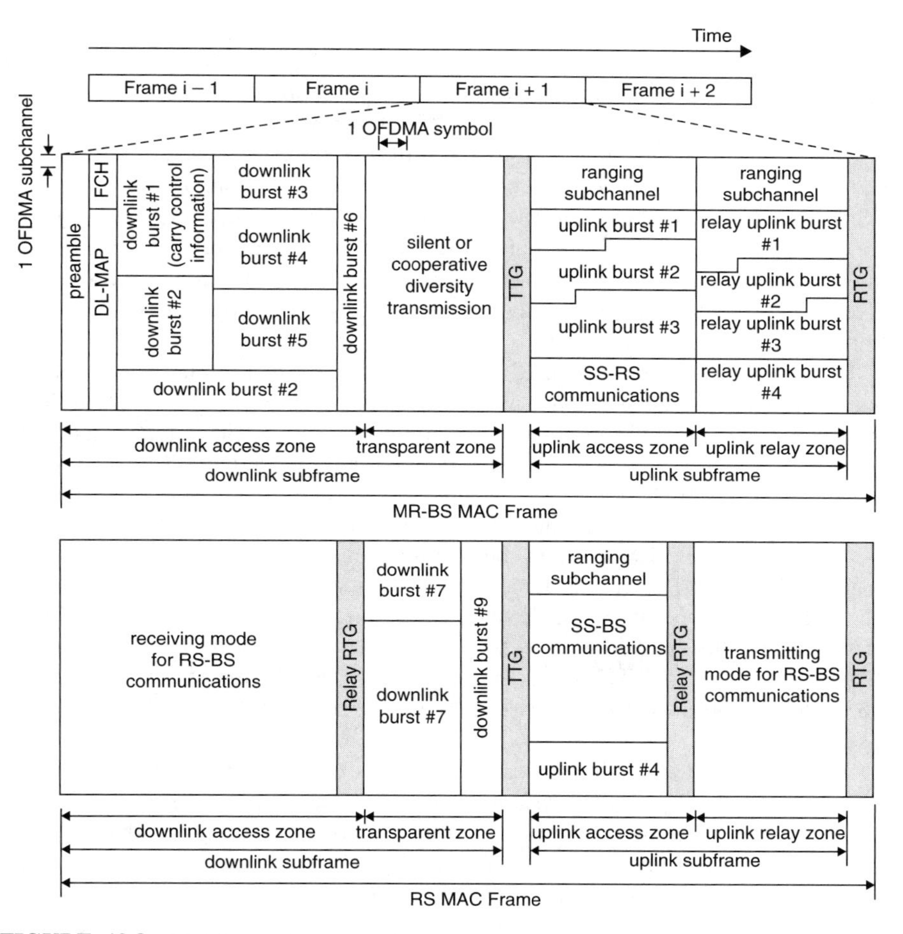

FIGURE 10.8 Multihop relay MAC frame structure in transparent mode with OFDMA physical layer.

a downlink frame-start preamble, FCH, DL-MAP, UL-MAP, DCD, and UCD. On the other hand, a *nontransparent* RS may transmit a downlink frame-start preamble, FCH, DL-MAP, UL-MAP, DCD, and UCD. Since a transparent RS does not transmit the control messages, all nodes must be within the radio range of the BS, although data transmissions may go through multiple hops.

Figure 10.8 shows an example of the multihop relay MAC frame structure in transparent mode with an OFDMA physical layer. Compared to WiMAX PMP (Figure 10.4) and WiMAX mesh (Figure 10.6), each of which has only one MAC frame structure for all nodes in the systems, multihop relay mode defines different MAC frame structures for the BS and RS: MR-BS MAC frame and RS MAC frame. The MR-BS (RS) frame is the MAC frame structure used for downlink transmission and uplink reception by the BS (RS).

In both the transparent MR-BS and RS MAC frames, each of the downlink subframe and uplink subframe is divided into zones. There are three types of zones: access, relay, and transparent. The *access zone* is a portion of the subframe in which the BS or RS transmits to an SS or receives from an SS. The *relay zone* is a portion of the subframe in which the BS or RS transmits to or receives from other RSs. Thus, a relay zone may be utilized for either transmission or reception, but the BS and any RS will not be required to switch between transmission and reception within a relay zone. The *transparent zone* is for an RS to transmit to its subordinate RS given that the BS may transmit concurrently with the RS in the transparent zone for cooperative diversity. It is not necessary to have all the three types of zones in a single MAC frame or subframe.

In the transparent MR-BS MAC frame, the downlink subframe includes at least one downlink access zone and may include one transparent zone. The uplink subframe may include an uplink access zone and an uplink relay zone. The ranging subchannel in the access zone is for an RS to perform initial ranging and an SS for all types of ranging operations. The ranging subchannel in the relay zone is for RSs to perform all types of ranging operations other than initial ranging.

In the transparent RS MAC frame, the downlink subframe includes one access zone for the BS-to-RS and SS transmission, and may include one transparent zone. The uplink subframe may include one access zone and one relay zone. In the RS MAC frame illustrated in Figure 10.8, if the RS switches from transmission to reception mode, an R-TTG (relay TTG) must be inserted. On the other hand, if the RS switches from reception to transmission mode, an R-RTG (relay RTG) must be inserted.

The transparent operation described above is mainly for two-hop relaying to improve link capacity. Nontransparent operation is needed when there are more than two hops. There are two ways of supporting multihop relaying in nontransparent mode. The first method is to group multiple MAC frames in a time domain into a repeatable multiframe structure with periodic allocation of relay zones. The second method is to create multiple relay zones within a single MAC frame. With either of the two methods, the BS and RSs are then assigned to transmit, receive, or idle in each of the relay zones. Figure 10.9 shows an example of the nontransparent multihop relay MAC frame structure using the second method.

In Figure 10.9, the MR-BS MAC frame has at least one downlink access zone and may include one or more downlink relay zones. When there are multiple downlink relay zones, only the first zone is required to include an R-FCH (relay FCH) and an R-MAP (relay MAP). As illustrated in Figure 10.9, the subchannel allocation in a downlink relay zone can be exactly the same as that in its preceding access zone.

In the nontransparent RS MAC frame structure, the downlink subframe start time must be aligned with that of the MR-BS MAC frame. Also, the RS transmits its frame-start preamble time aligned with its superordinate station's frame-start preamble. In addition to the preamble, RS also transmits FCH, DL-MAP, UL-MAP, DCD, and UCD, as illustrated in Figure 10.9. However, the content of these control messages in the RS MAC frame may be different from that in the MR-BS MAC frame.

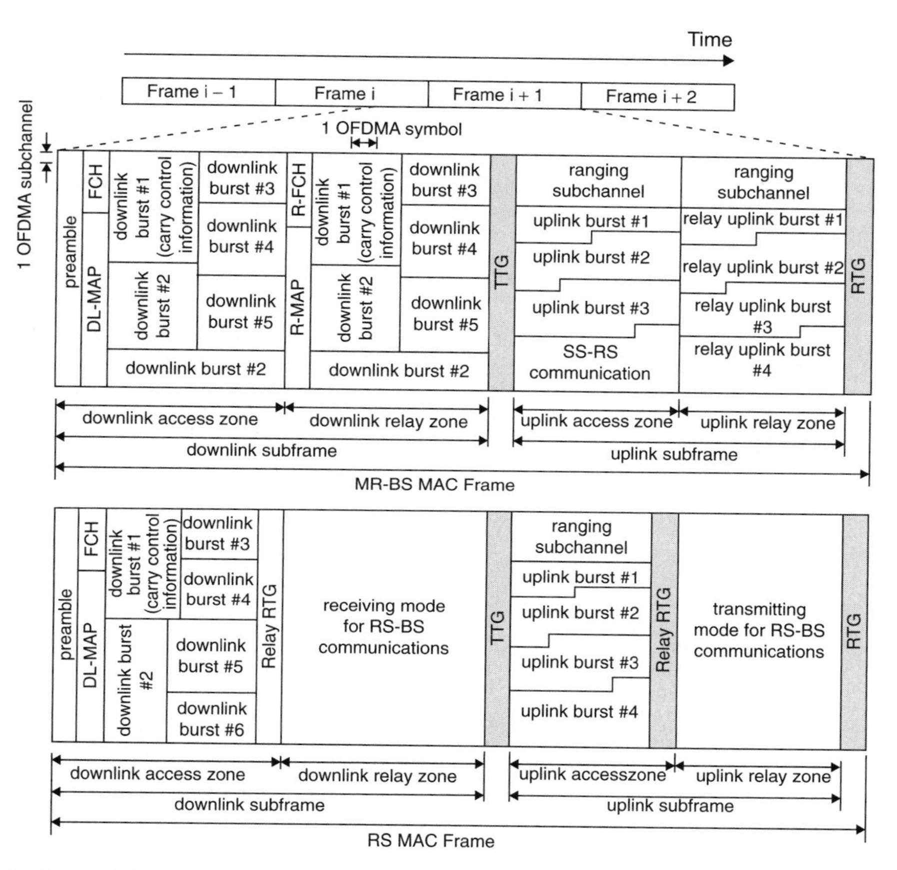

FIGURE 10.9 MAC frame structure in nontransparent mode using the single-frame method for multihop relaying.

SUMMARY

WiMAX plays an important role in wireless broadband networking. In this chapter, the three operation modes of WiMAX have been presented. Although both mesh mode and multihop relay mode, support multihop communications, only the multirelay mode is compatible to the PMP mode. Thus, the multihop relay mode which is currently being standardized by the IEEE 802.16j working group, deserves attentions.

REFERENCES

[1] WiMAX Forum, http://www.wimaxforum.org.

[2] F. Ohrtman, *WiMAX Handbook: Building 802.16 Wireless Networks*, McGraw-Hill, New York, May 2005.

[3] V. Gunasekaran and F. C. Harmantzis, "Affordable infrastructure for deploying WiMAX systems: mesh v. non mesh," *IEEE Vehicular Technology Conference*, May 2005.

[4] IEEE Std. 802.16-2004, "Part 16: Air interface for fixed broadband wireless access systems," *IEEE 802.16 Working Group Document*, Oct. 2004.

[5] IEEE Std 802.16e-2005, "Part 16: Air interface for fixed and mobile broadband wireless access systems, Amendment 2: Physical and medium access control layers for combined fixed and mobile operation in licensed bands, and corrigendum 1," *IEEE 802.16 Working Group Document*, Feb. 2006.

[6] Wi-Fi Alliance Knowledge Center, http://www.wi-fi.org/knowledge_center_overview.php.

[7] S. Ramachandran, C. W. Bostian, and S. F. Midkiff, "Link adaptation algorithm for IEEE 802.16," *Proceedings of the IEEE Wireless Communications and Networking Conference*, pp. 1466–1471, Mar. 2005.

[8] H. Yaghoobi, "Scalable OFDMA physical layer in IEEE 802.16 WirelessMAN," *Intel Technol. J.*, vol. 8, no. 3, pp. 201–212, Aug. 2004.

[9] C. Hoymann, "Analysis and performance evaluation of the OFDM-based metropolitan area network IEEE 802.16," *Comput. Networks*, vol. 49, no. 3, pp. 341–363, Oct. 2005.

[10] K. Wongthavarawat and A. Ganz, "Packet scheduling for QoS support in IEEE 802.16 broadband wireless access systems," *Int. J. Commun. Syst.*, vol. 16, no. 1, pp. 81–96, Feb. 2003.

[11] K. Wongthavarawat and A. Ganz, "IEEE 802.16 based last mile broadband wireless military networks with quality of service support," *IEEE Military Communications Conference (MILCOM)*, pp. 779–784, 2003.

[12] S. Maheshwari, "An efficient QoS scheduling architecture for IEEE 802.16 wireless MANs," Master of Technology thesis, Indian Institute of Technology–Bombay, 2005.

[13] M. Hawa and D. W. Petr, "Quality of service scheduling in cable and broadband wireless access systems," *IEEE International Workshop on Quality of Service*, pp. 247–255, May 2002.

[14] G. Chu, D. Wang, and S. Mei, "A QoS architecture for the MAC protocol of IEEE 802.16 BWA system," *IEEE International Conference on Communications, Circuits and System*, pp. 435–439, June 2002.

[15] S. Ramachandran, C. W. Bostian, and S. F. Midkiff, "Performance evaluation of IEEE 802.16 for broadband wireless access," *OPNETWORK*, Aug. 2002.

[16] B. Louazel, "Implementation of IEEE 802.16a in Glomosim/QualNet," Master of Engineering thesis, Dublin City University, Aug. 2004.

[17] I. F. Akyldiz, X. Wang, and W. Wang, "Wireless mesh networks: a survey," *Elsevier Comput. Networks J.*, vol. 47, no. 4, pp. 445–487, Mar. 2005.

[18] Y. Zhang, K.-S. Tan, P.-Y. Kong, J. Zheng, and M. Fujise, "IEEE 802.16 WiMAX mesh networking," *Wireless Mesh Networking: Architectures, Protocols and Standards*, Auerbach Publications, New York, pp. 425–466, Dec. 2006.

[19] D. Beyer, N. van Waes, and C. Eklund, "Tutorial: 802.16 MAC layer mesh extension overview," *IEEE 802.16 Working Group Document*, Mar. 2002, http://www.ieee802.org/16/tga/contrib/S80216a-02_30.pdf.

[20] Scalable Network Technologies, http://www.sclable-networks.com.

[21] OPNET Technologies Inc., http://www.opnet.com.

[22] GloMoSim, http://pcl.cs.ucla.edu/projects/glomosim.

[23] The Network Simulator, http://www.isi.edu/nsnam/ns.

[24] H.-Y. Wei, S. Ganguly, R. Izmailov, and Z. J. Haas, "Interference-aware IEEE 802.16 WiMAX mesh networks," *IEEE Vehicular Technology Conference*, May 2005.

[25] J. Tao, F. Liu, Z. Zeng, and Z. Lin, "Throughput enhancement in WiMAX mesh networks using concurrent transmissions," *IEEE International Conference on Wireless Communications, Networking and Mobile Computing* (*WCNC*), pp. 871–874, Sept. 2005.

[26] M. Cao, W. Ma, Q. Zhang, X. Wang, and W. Zhu, "Modelling and performance analysis of the distributed scheduler for IEEE 802.16 mesh mode," *ACM MobiHoc*, pp. 78–89, 2005.

[27] S. Redana and M. Lott, "Performance analysis of IEEE 802.16a in mesh operation mode," *IST SUMMIT*, June 2004.

[28] J. Sydir, "IEEE 802.16 Broadband Wireless Access Working Group: harmonized contribution on 802.16j (mobile multihop relay) usage models," *IEEE 802.16j Working Group Document 802.16j-06015*, Sept. 2006.

[29] IEEE Std. P802.16j/D1, "Part 16: Air interface for fixed and mobile broadband wireless access systems: multihop relay specification," *IEEE 802.16 Working Document*, Aug. 2007.

CHAPTER 11

WIRELESS LOCAL AREA NETWORKS

11.1 INTRODUCTION

A wireless local area network (WLAN) is a wireless network that allows two or more users to communicate with each other at relatively high speed compared to that offered by a cellular network. It is similar to Ethernet except that Ethernet is wired, whereas WLAN is wireless and supports user mobility. It is very commonly used in the office and home environments. The most prominent WLAN under deployment today is the IEEE 802.11 WLAN. The IEEE 802.11 standard has been evolved further into the following five types: IEEE 802.11b, 802.11a, 802.11g, 802.11n, and 802.11s. The upcoming IEEE 802.11n will provide higher throughput, and the upcoming IEEE 802.11s will support mesh networking.

IEEE 802.11b can operate up to 11 Mbps, while IEEE 802.11a can operate up to 54 Mbps. Both standards were published in 1999. The former operates in the 2.4-GHz unlicensed band; the latter, which operates in the 5-GHz band, has not really taken off. IEEE 802.11b uses direct-sequence spread spectrum (DSSS), while IEEE 802.11a uses orthogonal frequency-division multiplexing (OFDM) as the multiple-access technology. The IEEE 802.11g standard, released in 2003, which operates in the 2.4-GHz band and uses OFDM as multiple-access technology to achieve a data rate of up to 54 Mbps, has been gaining popularity.

A new task group, known as IEEE 802.11n, is developing the IEEE 802.11n standard, expected to be released in 2008. This new standard, which uses a combination of

Wireless Broadband Networks, By David Tung Chong Wong, Peng-Yong Kong, Ying-Chang Liang, Kee Chaing Chua, and Jon W. Mark

OFDM and multiple-input, multiple-output (MIMO) techniques to enhance diversity gain, aims to achieve a data rate of up to 600 Mbps. The goal is to design a medium access control (MAC) protocol that can deliver a user throughput of more than 100 Mbps at the MAC layer. Because of the ratio of the payload-to-header overheads, collisions of the access scheme, backoff algorithms of the access scheme, and inter-frame spaces (IFSs), the useful throughput will be lower than the physical data rate. But increasing the physical layer bit rate without enhancing the MAC techniques will have minimal gain in useful throughput. Thus, enhanced MAC techniques are needed to push the useful throughput beyond 100 Mbps. Some of these techniques include frame aggregation, enhanced block acknowledgments, reverse direction protocol, different transmission modes, and reduced IFS (RIFS). The main idea in frame aggregation is to transmit multiple frames within a short time period, thereby cutting down the channel access time and IFSs, which increase the payload/header ratio. Implicit block acknowledgment removes the need to transmit a special frame to request block acknowledgment, while compressed block acknowledgment cuts down its bitmap size to shorten its transmission frame. Reverse direction protocol allows bidirectional transmission of frames when the source station gains access to the channel. This improves throughput by cutting down on the channel access time of the destination station and reduces turnaround delay and jitter. There are three transmission modes: one that allows high-throughput stations to act as legacy (IEEE 802.11b/a/g) stations, one that allows high-throughput stations to talk to legacy stations, and one that allows high-throughput stations to talk only to each other. RIFS minimizes the IFSs between frames transmitted and thus improves the payload/overhead ratio. This, in turn, helps to increase user throughput.

The IEEE 802.11e standard, which allows for relative quality of service (QoS) between multiple classes of traffic, supports up to four access categories with eight traffic classes. IEEE 802.11b/a/g stations and IEEE 802.11n stations can be coupled together with IEEE 802.11e. IEEE 802.11e allows for relative QoS between multiple classes of traffic. This relative QoS is achieved using four access categories with different arbitration IFSs and different minimum and maximum contention window sizes. Handoff, or roaming from one access point to another access point in the current IEEE 802.11, takes a long time. Hence, a task group in IEEE 802.11r is looking into ways to improve this roaming time. There are also many applications for using IEEE 802.11n. These applications include voice over IP (VoIP), video streaming, music streaming, video conferencing, IPTV, interactive gaming, large file and picture transfer, and data backup and storage.

The IEEE 802.11s is another standard being established for wireless mesh networking. The IEEE 802.11s MAC uses the IEEE 802.11e EDCA MAC as a mandatory baseline. This is compatible with legacy stations and is easy to implement. Four main areas of MAC enhancements in IEEE 802.11s are mesh deterministic access (MDA), common channel framework (CCF), intramesh congestion control (IMCC), and power savings.

The balance of this chapter is organized as follow. In Section 11.2 we present an infrastructure-based network architecture, an ad hoc network architecture, and a wireless mesh network architecture for IEEE 802.11. The physical layer for IEEE

802.11n is explained in detail in Section 11.3, while some of the MAC enhancements for IEEE 802.11e, IEEE 802.11n, and IEEE 802.11s are explained in Section 11.4. These enhanced MAC techniques include frame aggregation, reverse direction protocol, enhanced block acknowledgments, different transmission modes, and reduced interframe space for IEEE 802.11n. Enhanced MAC techniques for IEEE 802.11s include MDA, CCF, IMCC, and power savings. The saturated throughputs and delays of IEEE 802.11e EDCA MAC with IEEE 802.11n are derived and some numerical results are presented in this section as well. In Section 11.5 we describe the mobility resource management for handoff or roaming in IEEE 802.11. The methods by which IEEE 802.11e and IEEE 802.11n provide quality of service are explained in Section 11.6. In Section 11.7 we present the applications in IEEE 802.11n. The chapter is summarized at the end of this chapter.

To facilitate concise presentation, acronyms and mathematical symbols used throughout the chapter are tabulated in Tables 11.1 and 11.2, respectively.

11.2 NETWORK ARCHITECTURES

IEEE 802.11 has three network architectures: an infrastructure network architecture, an ad hoc network architecture, and a wireless mesh network architecture, as shown in Figure 11.1. In the infrastructure network architecture, stations are connected directly to an access point (AP) connected to the distribution system. Thus, it is like a two-level multibranch tree network topology. A group consisting of an AP and mobile stations is known as a basic service set (BSS). In the ad hoc network architecture, stations are connected to each other directly in an ad hoc manner without an AP. This is like a mesh network topology, or is sometimes known as *peer-to-peer network topology*. This mode of operation is also known as an *independent BSS* (IBSS). In the wireless mesh network topology, the distribution system can be a wireless mesh network among the access points. The wireless links between mesh access points are shown in dashed lines in Figure 11.1(c).

11.3 PHYSICAL LAYER OF IEEE 802.11N

IEEE 802.11n supports high-throughput (HT) transmission for wireless local area networks. It provides improved performance over IEEE 802.11a and IEEE 802.11g by adopting MIMO-OFDM as the physical-layer air interface. The frequency bands used are 5 and/or 2.4 GHz, with the mandatory transmission bandwidth being 20 MHz. Thus, the IEEE 802.11n network is backwardly compatible with legacy devices, which include IEEE 802.11a and IEEE 802.11g devices. An optional transmission on 40-MHz bandwidth is supported to achieve further throughput enhancement. In particular, transmission of one to four spatial streams has been defined for operation in the 40-MHz bandwidth. IEEE 802.11n supports a shorter guard interval (cyclic prefix) than IEEE 802.11a, and makes more efficient use of the available

TABLE 11.1 Acronyms

Acronym	Meaning
AC	access category
ACK	acknowledgment
AIFS	arbitration IFS
A-MPDU	aggregate MAC protocol data unit
A-MSDU	aggregate MAC service data unit
AP	access point
APSD	automatic power-save delivery
ATIM	announcement traffic indication message
BACK	block acknowledgment
BAR	block acknowledgment request
BCC	binary convolutional code
BPSK	binary phase-shift keying
BSS	basic service set
CCF	common channel framework
CRC	cyclic redundancy check
CSD	cyclic shift
CSI	channel state information
CSMA/CA	carrier-sense multiple access with collision avoidance
CTS	clear to send
DCF	distributed coordination function
DSSS	direct-sequence spread spectrum
ESS	extended service set
FCS	frame check sequence
FEC	forward error correction
GI	guard interval
HT	high throughput
HWMP	hybrid wireless mesh protocol
IBSS	independent BSS
ID	identity
IDFT	inverse discrete Fourier transform
IFFT	inverse fast Fourier transform
IFS	interframe space
IMCC	intramesh congestion control
IP	Internet protocol
LDPC	low-density parity check
MAC	medium access control
MCS	modulation and coding set
MDA	mesh deterministic access
MIMO	multiple-input, multiple-output
MSDU	MAC service data unit
OFDM	orthogonal frequency-division multiplexing
PAD	padding
PAPR	peak/average power ratio
PHY	physical layer
PLCP	physical-layer convergence protocol
PPDU	physical protocol data unit
PSDU	PHY service data unit
PSMP	power-save multipoll
QAM	quadrature amplitude modulation
QoS	quality of service
QPSK	quadrature phase-shift keying
RA-OLSR	radio-aware optimized link-state routing
RIFS	reduced IFS
RM-AODV	radio metric ad hoc on-demand distance vector
RTS	request to send
SF	subframe
SIFS	short IFS
SMPS	spatial multiplexing power save
SSID	service set ID
STBC	space–time block codes
TC	traffic category
TIM IE	traffic indication message information element
TS	traffic stream
TXOP	transmit opportunity
VoIP	voice over IP
WEP	wired equivalent privacy
WLAN	wireless local area network

TABLE 11.2 Mathematical Symbols

Symbol	Description
$a(i,t)$	random process representing the backoff stage j, $j = 0,1,\ldots,L_{\text{i,retry}}$, at time t
$b_{i,j,k}$	stationary distribution of the Markov chain
$b(i,t)$	random process representing the value of the backoff counter at time t
D_i	class i packet delay, $i = 0,\ldots,N-1$
D_k	coverage distance, $k = 1,\ldots,K$
$E(B_i)$	total number of busy slots that a packet encounters during backoff stages
$E(X_i)$	total number of idle slots that a packet encounters during backoff stages
i	priority class
K	number of generic data rate
L	length of the payload, including tail bits and pad bits
L^*	length of the longest frame in a collision
$L_{i,\text{ retry}}$	retry limit
n_i	number of class i stations
N	number of priority in a given station
$E(N_{i,\text{retry}})$	average number of retries for the ith priority class
p_b	probability that the channel is busy, happens when at least one station transmits during a time slot
P_k	probability of being in each region, within concentric circles of radii D_{k-1} and D_k, using direct transmission at data rate R_k, $k = 1,\ldots,K$
R_k	data rate, $k = 1,\ldots,K$
S	total actual saturated throughput
S_i	actual saturated throughput for the ith priority class
t	time
τ_i	probability that a station in the ith priority class transmits during a generic time slot
$\mathrm{T}_{\text{ACK},k}$	time to transmit an acknowledgment with data rate R_k
$T_{\text{AIFS}(i)}$	arbitration interframe space (AIFS) time for a class i station
T_c	average time that the channel has a collision
$T_{\text{CTS},k}$	time to transmit a clear-to-send (CTS) frame with data rate R_k
T_{DIFS}	DIFS time
$T_{E(L),\,k}$	time to transmit the average payload of average length, $E(L)$, with data rate R_k
$T_{E(L^*),k}$	time to transmit a payload with length $E(L^*)$ at data rate R_k
$T_{H,k}$	time to transmit the header (including PLCP preamble, physical layer header, MAC header) at data rate R_k for the MAC header
$T_{\text{RTS},k}$	time to transmit a request-to-send (RTS) frame with data rate R_k
T_s	average time required to transmit a packet successfully

(Continued)

TABLE 11.2 Mathematical Symbols (*Continued*)

Symbol	Description	Symbol	Description
p_i	probability that a transmitted frame collides or probability that a station in a backoff stage for the *i*th class senses the channel is busy	T_{SIFS}	SIFS time
p_s	probability that a successful transmission occurs in a time slot	$W_{i,j}$	maximum backoff counter value in backoff stage *j* for class *i*
$p_{s,i}$	probability that a successful transmission occurs in a time slot for the *i*th priority class	X_i	total number of backoff slots that a packet encounters without the case when the backoff counter freezes, for the *i*th priority class

subcarriers for data transmission. With these new features, 802.11n is capable of supporting data rates up to 600 Mbps.

The modulation schemes used in the IEEE 802.11n PHY include BPSK, QPSK, 16-QAM, and 64-QAM. Convolutional coding is chosen as the mandatory coding scheme with a code rate of 1/2, 2/3, 3/4, or 5/6 for forward error correction (FEC). Low-density parity check (LDPC) codes are defined as an optional feature to achieve enhanced performance.

With MIMO antenna systems, IEEE 802.11n uses spatial multiplexing as the basic scheme to achieve high throughput. Furthermore, transmit beamforming and space–time block codes (STBCs) are adopted as the optimal features related to multiple antennas.

11.3.1 802.11n Transmitter

The IEEE 802.11n transmitter block diagram is illustrated in Figure 11.2. Some of the parameters used frequently in an IEEE 802.11n transmitter are defined in Table 11.3. The scrambler randomizes the input data bits to avoid the occurrence of long zeros or ones. The scrambler's output is fed to the encoder parser, which generates outputs to N_{ES} FEC encoders. When binary convolutional code (BCC) is used for FEC and the PHY data rate is less than 300 Mbps, only one BCC encoder is used ($N_{\text{ES}} = 1$). If the PHY data rate exceeds 300 Mbps, two BCC encoders are used ($N_{\text{ES}} = 2$). Code rates 1/2, 2/3, 3/4, and 5/6 are supported. If LDPC is used for FEC, $N_{\text{ES}} = 1$. The stream parser splits the outputs from the FEC encoders into N_{SS} spatial streams, where N_{SS} can be 1, 2, 3, or 4. The coded data bits in each spatial stream are interleaved and then mapped onto the constellation points based on the modulation scheme used for each spatial stream, which is defined by the modulation and coding sets (MCSs). If LDPC is used for FEC, no interleaving is performed. The output from the constellation mapper is a string of complex numbers. This string is divided into groups of N_{D} complex symbols, where each group is associated with

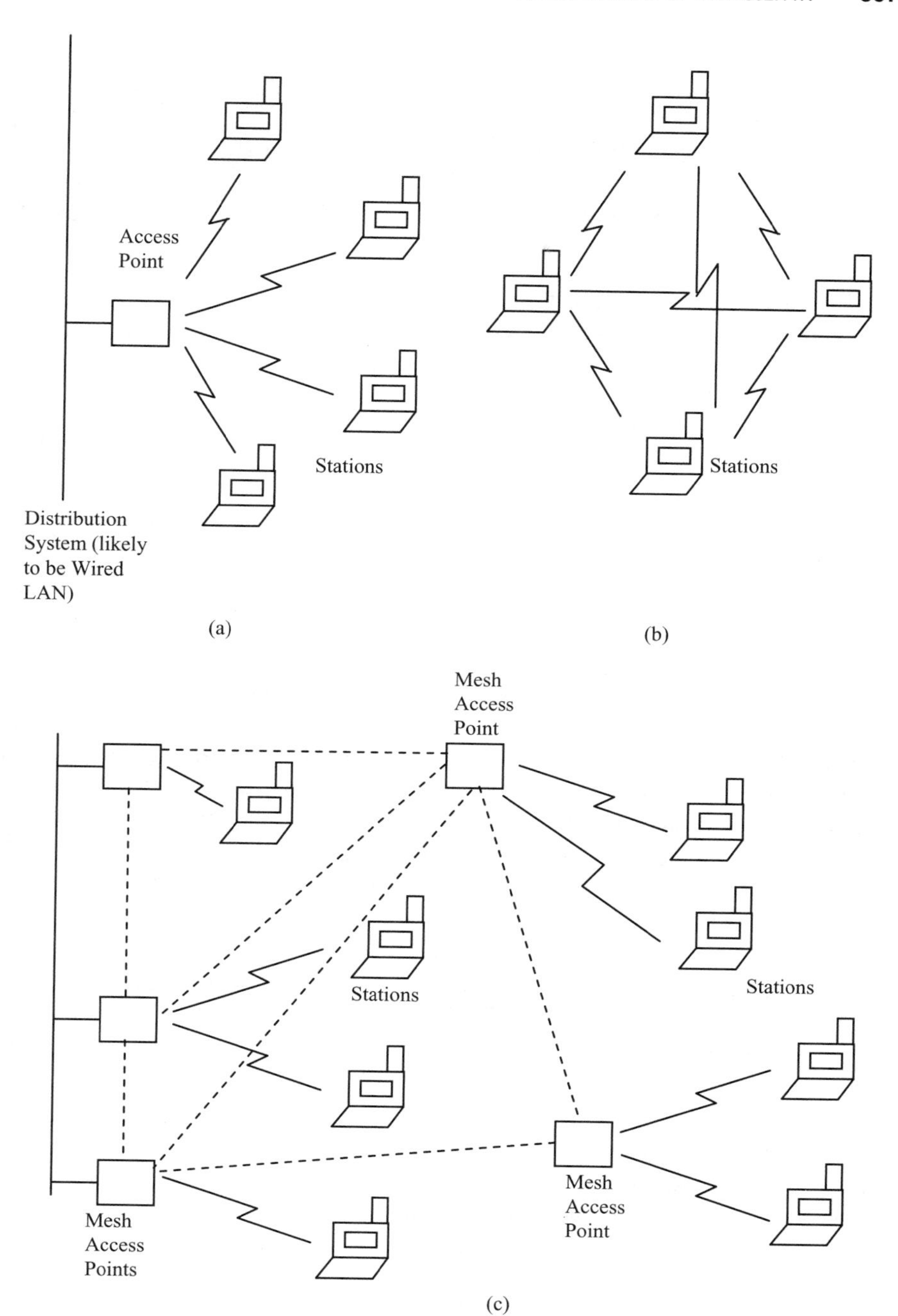

FIGURE 11.1 Network architectures for IEEE 802.11 WLAN with (a) infrastructure mode, (b) ad hoc mode, and (c) wireless mesh mode.

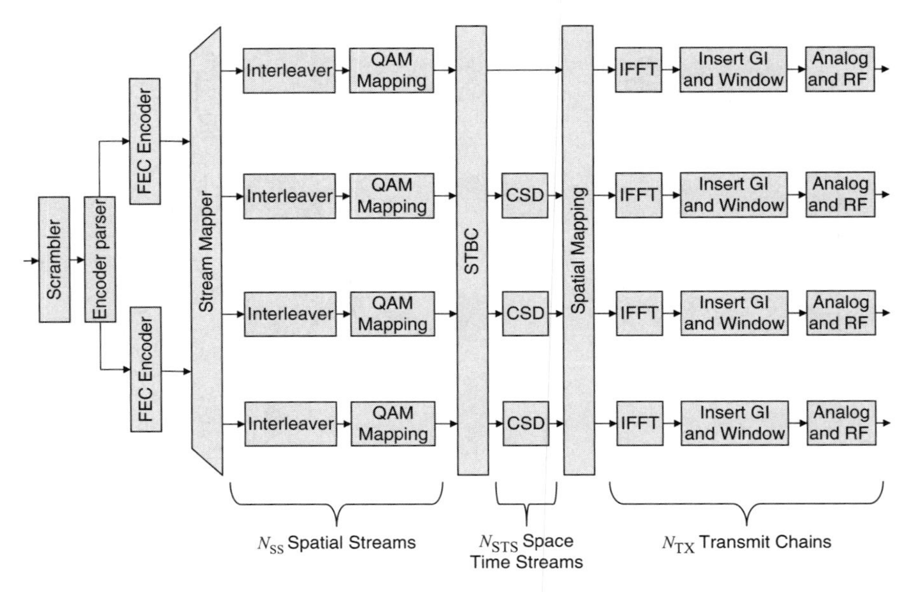

FIGURE 11.2 Block diagram of an IEEE 802.11n transmitter.

one OFDM symbol. If STBC is used, the N_{SS} OFDM symbols from N_{SS} spatial streams are spread in space and time to obtain N_{STS} OFDM symbols; that is, STBC is applied independently to each subcarrier. If STBC is not used, $N_{STS} = N_{SS}$. The spatial mapper maps N_{STS} space–time streams to N_{TX} transmit chains.

The spatial mapper has the provision to scale and/or rotate the output vector from the constellation mapper or STBC. Spatial mapping can be any one of the following: (1) direct mapping, (2) spatial expansion, and (3) beamforming. Following the spatial

TABLE 11.3 Frequently Used Parameters in an IEEE 802.11n Transmitter

Symbol	Explanation
N_{ES}	Number of BCC encoders for the data field $N_{ES} = 1$ if data rate is less than 300 Mbps and BCC is used; $N_{ES} = 2$ if data rate is 300 Mbps and above and BCC is used; $N_{ES} = 1$ if LDPC is used
N_{SS}	Number of spatial streams
N_{STS}	Number of space–time streams
N_{ESS}	Number of extension spatial streams
N_{TX}	Number of transmit chains
N_{CBPS}	Number of coded bits per symbol
$N_{CBPSS}(i)$	Number of coded bits per symbol for spatial stream i
N_{DBPS}	Number of data bits per symbol
N_{BPSC}	Number of coded bits per single carrier
$N_{BPSCS}(i)$	Number of coded bits per single carrier for spatial stream i

mapper, each transmit chain consists of inverse discrete Fourier transformation (IDFT), cyclic shift (CSD) insertion, guard interval (GI) insertion, and windowing.

Based on the bandwidth used by the devices in an IEEE 802.11n network, the operational modes can be classified as follows:

- *Legacy mode*. The operation is similar to IEEE 802.11a/b/g. The channel bandwidth used is 20 MHz.
- *Duplicate legacy mode*. In this mode, the devices use a 40-MHz channel bandwidth, but the same data are transmitted in the upper and lower regions of the 20-MHz channel. To reduce the peak/average power ratio (PAPR), the data in the upper part of the channel are rotated by 90°.
- *High-throughput* (HT) *mode*. HT mode is available for both 20- and 40-MHz channels. In this mode, supporting one and two spatial streams is mandatory. A maximum of four spatial streams is supported. This mode also includes the HT duplicate mode.
- *40-MHz upper mode*. In this mode, only the upper 20 MHz of the 40-MHz channel is used.
- *40-MHz lower mode*. In this mode, only the lower 20 MHz of the 40-MHz channel is used.

11.3.2 OFDM Parameters

Table 11.4 lists the OFDM parameters used in the IEEE 802.11n system. The number of subcarriers and the sampling frequency depend on the channel bandwidth used for transmission. The IFFT size is 64 for the 20-MHz channel bandwidth and 128 for the 40-MHz channel bandwidth. The subcarrier spacing in the 20- and 40-MHz channels is 312.5 kHz, and the FFT/IFFT period is 3.2 μs. As shown in Figure 11.3, for the 20-MHz channel the four pilot subcarriers are located at subcarrier indices −21,−7, 7, and 21. For the 40-MHz [except for modulation and coding scheme (MCS) 32 and non-HT format] the pilot subcarrier indices are −53, −25, −11, 11, 25, and 53.

11.3.3 802.11n Date Rates

The data rates supported in an IEEE 802.11n network range from 6.5 to 600 Mbps. Table 11.5 lists the selected data rates supported with different code rates, modulation, number of subcarriers, number of spatial streams, and different guard intervals. The data rates in the table are given when the same modulation format is used in all the spatial streams.

11.3.4 PLCP Layer

The physical protocol data unit (PPDU) of IEEE 802.11n consists of the physical layer convergence protocol (PLCP) preamble, PLCP header, and PHY service data

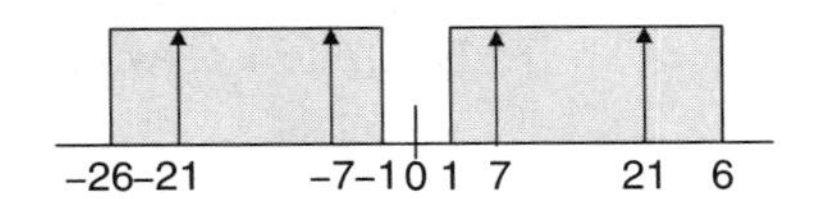

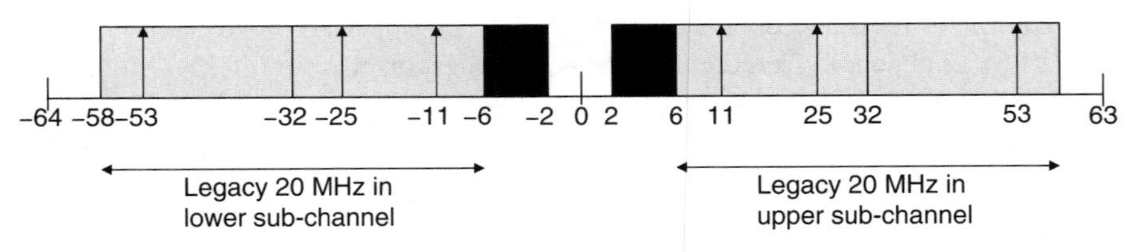

FIGURE 11.3 Pilot subcarrier allocation in IEEE 802.11n.

TABLE 11.4 OFDM Parameters Used in IEEE 802.11n

Parameter	Non-HT 20 MHz	HT 20 MHz	HT 40 MHz	MCS 32 and Non-HT 40 MHz
Total number of subcarriers, N_{FFT}	64	64	128	128
Number of data subcarriers, N_{D}	48	52	108	48
Number of pilot subcarriers, N_{P}	4	4	6	4
Number of guard and null subcarriers, $N_{\mathrm{G}} + N_{\mathrm{NULL}}$	12	8	14	24
Sampling frequency, f_s	20 MHz	20 MHz	40 MHz	40 MHz
Subcarrier spacing, $\Delta_f (= f_s/N_{\mathrm{FFT}})$	312.5 kHz	312.5 kHz	312.5 kHz	312.5 kHz
FFT/IFFT period, $T_{\mathrm{FFT}} (= 1/\Delta_F)$	3.2 μs	3.2 μs	3.2 μs	3.2 μs
Guard interval duration, $T_{\mathrm{GI}} = T_{\mathrm{FFT}}/4$	0.8 μs	0.8 μs	0.8 μs	0.8 μs
Double GI, $T_{\mathrm{GI2}} = T_{\mathrm{FFT}}/2$	1.6 μs	1.6 μs	1.6 μs	1.6 μs
Short guard interval, $T_{\mathrm{GIS}} = T_{\mathrm{FFT}}/8$	N/A[a]	0.4 μs	0.4 μs	0.4 μs[b]
Symbol interval, $T_{\mathrm{SYM}} (= T_{\mathrm{FFT}} + T_{\mathrm{GI}})$	4 μs	4 μs	4 μs	4 μs
Short GI symbol interval, $T_{\mathrm{SYMS}} (= T_{\mathrm{FFT}} + T_{\mathrm{GIS}})$	N/A	3.6 μs	3.6 μs	3.6 μs[b]

[a]N/A, not applicable.
[b]Not applicable to non-HT formats.

TABLE 11.5 Selected Data Rates Supported in IEEE 802.11n

Modulation	Coding Rate	Bandwidth MHz	Number of Data Subcarriers	Number of Spatial Streams	Data Rate (Mbps) for GI = 800 ns	Data Rate (Mbps) for GI = 400 ns
BPSK	1/2	20	52	1	6.5	7.2
64-QAM	5/6	20	52	1	65	72.2
BPSK	1/2	40	108	1	13.5	15
64-QAM	5/6	40	108	1	135	150
BPSK	1/2	20	52	2	13	14.4
64-QAM	5/6	20	52	2	130	144
BPSK	1/2	40	108	2	27	30
64-QAM	5/6	40	108	2	270	300
BPSK	1/2	20	52	3	19.5	21.7
64-QAM	5/6	20	52	3	195	216.7
BPSK	1/2	40	108	3	40.5	45
64-QAM	5/6	40	108	3	405	450
BPSK	1/2	20	52	4	26	28.9
64-QAM	5/6	20	52	4	260	288.9
BPSK	1/2	40	108	4	54	60
64-QAM	5/6	40	108	4	540	600

unit (PSDU). The frame formats support three modes of operation:

1. *Legacy mode*. In this mode both the legacy and MIMO-OFDM systems shall coexist. At the transmitter, only one transmit antenna is used. Receive diversity is exploited in this mode.
2. *Mixed mode*. In this mode, the MIMO-OFDM transmitter should be able to generate legacy packets for legacy systems and HT packets for MIMO-OFDM devices.
3. *Greenfield mode*. In this mode, the transmission is only between MIMO-OFDM systems. There is no protection for MIMO-OFDM systems from the legacy devices. When a greenfield device is transmitting, the legacy systems will not transmit, but use the carrier-sensing mechanism in the physical layer to avoid collision. Receivers enabled in this mode should be able to decode packets from the legacy, the mixed mode, and the greenfield mode transmitters.

The legacy mode is a non-HT mode, whereas the mixed and greenfield modes are HT modes. The elements to be used in the frame formats are tabulated in Table 11.6. The frame formats for the forgoing three modes when LDPC is not used for FEC are shown in Figure 11.4.

11.3.4.1 PLCP Preamble and Header The PLCP preamble for legacy mode consists of two parts, L-STF and L-LTF. L-STF consists of 10 identical short symbols each of duration 0.8μs. L-LTF consists of an extended cyclic prefix of duration 1.6 μs followed by two identical long symbols of duration 3.2 μs. L-SIG contains information about the rate and length of payload.

TABLE 11.6 Elements of an IEEE 802.11n PLCP Packet

Element	Description
L-STF	Non-HT short training field
L-LTF	Non-HT long training field
L-SIG	Non-HT signal field
HT-SIG	HT signal field
HT-STF	HT short training field
HT-GF-STF	HT-greenfield short training field
HT-LTF1	First HT long training field (data HT-LTF)
HT-LTFs	Additional HT long training field (data HT-LTFs and extension HT-LTFs)
Data	Data field includes the PSDU (PHY service data unit)

The preamble for mixed mode consists of legacy preamble. This takes care of the compatibility with legacy devices. The HT-SIG contains the information about the modulation scheme used, channel bandwidth, length of payload, coding details, information abut GI, number of HT-LTFs, and tail bits for the encoder. The duration of HT-SIG is 0.8 μs. HT-STF is similar to L-STF and is of duration 4 μs. Following HT-STF is a sequence of HT-LTFs, which are used to obtain better automatic gain control (AGC) and MIMO channel estimation. The duration of one HT-LTF is 4 μs. The number of HT-LTFs is decided by the antenna configuration and presence of STBC.

In greenfield mode, the preamble and header are intended only for MIMO-OFDM systems. There is no legacy preamble. HT-GF-STF, which is of duration 8 μs, is

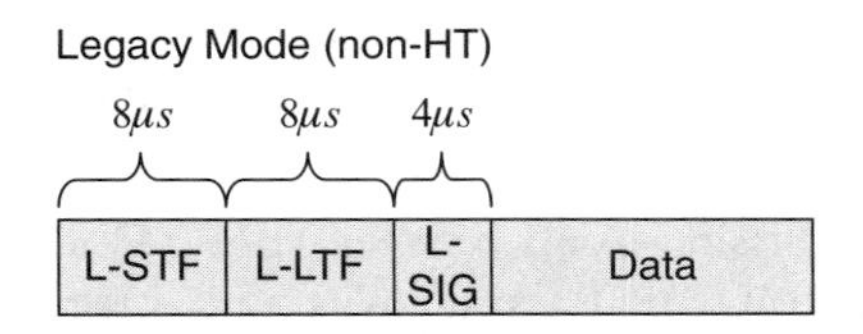

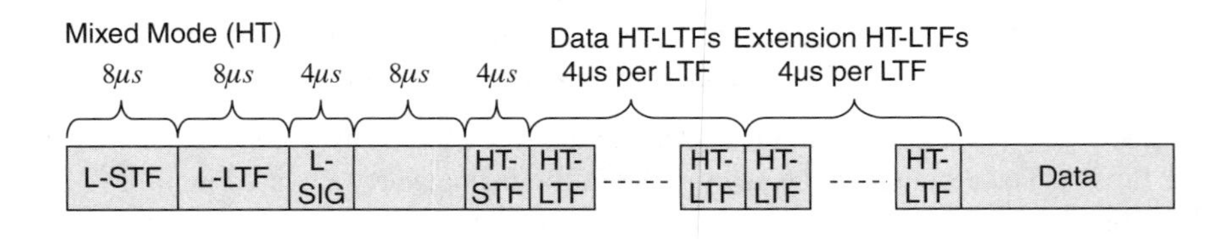

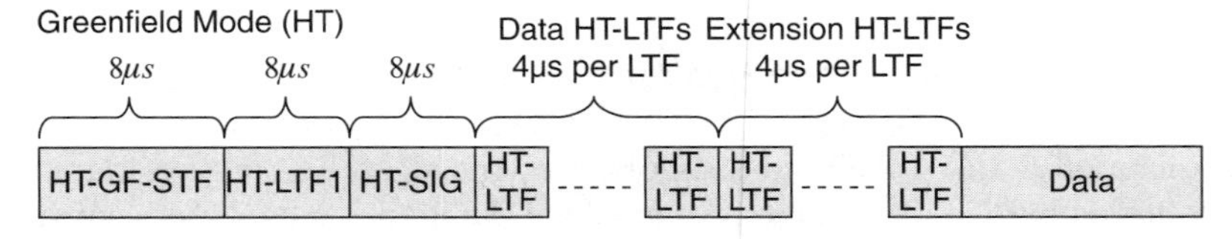

FIGURE 11.4 PPDU frame formats.

TABLE 11.7 IEEE 802.11n MIMO Configurations

Number of Transmit Antennas, N_{TX}	Number of Receive Antennas, N_{RX}	Maximum N_{STS} Without STBC	Maximum N_{STS} With STBC
1	1	1	—
2	1	1	2
3	1	1	2
3	2	2	3
4	1	1	2
4	2	2	4

SERVICE 16 bits	Scrambled PSDU	$6 \times N_{ES}$ Tail bits	Pad bits

FIGURE 11.5 PSDU format.

used for AGC convergence, timing acquisition, and coarse frequency acquisition. HT-LTF1 is twice as long as HT-LTF (i.e., HT-LTF1 is of duration 8 μs).

11.3.4.2 PSDU The PSDU of IEEE 802.11n consists of 16 service bits, scrambled data payload, followed by $6N_{ES}$ tail bits and pad bits. The structure of PSDU is shown in Figure 11.5.

11.3.5 Channel Coding Schemes

In IEEE 802.11n, BCC is used as the mandatory channel coding scheme. The use of LDPC is optional. The BCC used in IEEE 802.11n achieves the coding rates of 1/2, 3/4, 2/3, and 5/6. The BCC encoder is brought back to the all-zero state by inserting six zero bits in the tail of the encoder's input. The use of LDPC code can provide enhanced performance. LDPC code uses the code rates supported by BCC.

11.3.6 MIMO Configurations

Spatial multiplexing is the mandatory scheme supported by IEEE 802.11n for achieving high throughput. Space–time coding and transmit beamforming are optional features related to MIMO configurations. The MIMO configuration and the corresponding maximum available spatial streams are tabulated in Table 11.7.

11.3.6.1 Space–Time Block Coding STBC is applied to each OFDM subcarrier independently. For $N_{TX} = 2$ and $N_{RX} = 1$, Alamouti codes are used. Table 11.8 shows STBC schemes designed for other antenna configurations.

TABLE 11.8 STBC schemes

Transmit Antenna	T1	T2
	$N_{TX} = 3$ and $N_{RX} = 2$	
1	s_1	s_2
2	$-s_2^*$	s_1^*
3	s_3	s_4
	$N_{TX} = 4$ and $N_{RX} = 2$	
1	s_1	s_2
2	$-s_2^*$	s_1^*
3	s_3	s_4
4	$-s_4^*$	s_3^*

11.3.6.2 Transmit Beamforming IEEE 802.11n supports closed-loop transmit beamforming (singular-value decomposition-based eigen-beamforming), which is especially useful when there are more transmit chains than space–time streams.

To support transmit beamforming, the following issues have been detailed in IEEE 802.11n specifications: (1) support for sounding the channel, (2) support for asymmetric MCSs, and (3) channel state information (CSI) feedback support. Antenna array calibration and various CSI feedback schemes (compressed and uncompressed) have also been defined.

11.3.6.3 Spatial Expansion When there are more transmit chains than space–time streams, spatial expansion can also be used to map the space–time streams to the transmit chains. This is done by introducing a precoding matrix between these two blocks. Let V and H_{act} be, respectively, the precoding matrix and the actual channel transfer function vector between the transmitter and the receiver; after precoding, an effective channel $H_{eff} = H_{act}V$ is generated.

There are three ways to generate the precoding matrix. The first method is to apply cyclic shifts across the transmit antennas. This can be done either in the time domain or equivalently, in the frequency domain. The second scheme involves two layers of operations: The first operation is orthogonal spreading, and the second operation is cyclic shift across the transmit antennas. The third scheme is to use a beamforming steering matrix. Whereas the third scheme requires the CSI, the first two schemes do not.

11.4 MEDIUM ACCESS CONTROL

11.4.1 IEEE 802.11 DCF MAC

The 802.11 DCF MAC [1] uses the CSMA/CA MAC protocol described in Section 4.3. To recap and to make this chapter self-contained, the CSMA/CA MAC protocol from Section 4.3 is summarized as follows. There are two access methods in CSMA/CA MAC: the basic access method and the request to send/clear to send (RTS/CTS) access method. The basic access method is a two-way handshaking, while the RTS/CTS access method is a four-way handshaking. In the former access method, the source station sends its frame to the destination station as the data transmission phase. After receiving the frame correctly, the destination station sends an acknowledgment to the source station in the acknowledgment phase. Thus, this process completes the two-way handshaking. In the latter access method, the source station sends a RTS frame to the destination station. If the destination station receives the RTS frame correctly and is available for reception, it replies with a CTS frame. Then the source station sends its data frame to the destination station. Upon receiving the data frame correctly, the destination station acknowledges receipt of the data frame with an acknowledgment frame. This completes the four-way handshaking. If the payload is below a certain threshold, the basic access method is used; otherwise, the RTS/CTS access method is used.

The DCF MAC works as follows (see Section 4.3):

- If the channel is idle for more than a distributed coordination function interframe space–time, a station can transmit immediately.
- If the channel is busy, the station will generate a random backoff period. This is selected uniformly from zero to the current contention window size minus one. The backoff counter decrements by one if the channel is idle for each time slot and freezes if the channel is sensed to be busy. The backoff counter is reactivated to count down when the channel is sensed idle for more than a distributed coordination function interframe space–time. At the initial backoff stage, the current contention window size is set at the minimum contention window size.
- If the backoff counter reaches zero, the station will attempt to transmit its frame. If it is successful, the destination station will send an acknowledgment after a short interframe space, and the current contention window size is reset to the minimum contention window size. If it is not successful, it will increase the current contention window size by doubling it only until a maximum contention window size is reached in the next backoff stage and a new random backoff period is selected as before.
- This process repeats itself until the frame is transmitted successfully or until the maximum retry limit is reached. If the frame is still not transmitted successfully, it is dropped.
- If a station does not receive an acknowledgment within an acknowledgment timeout period after a frame is transmitted, it will continue to attempt to retransmission of the frame according to the backoff algorithm.
- In the RTS/CTS access method, if a station does not receive a CTS frame within a CTS timeout period after sending an RTS frame, it will attempt to retransmit the frame according to the RTS/CTS access method and the backoff algorithm.

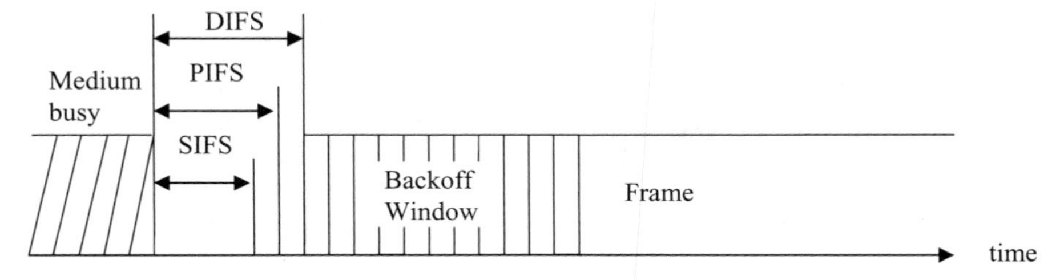

FIGURE 11.6 Channel access in IEEE 802.11 DCF MAC.

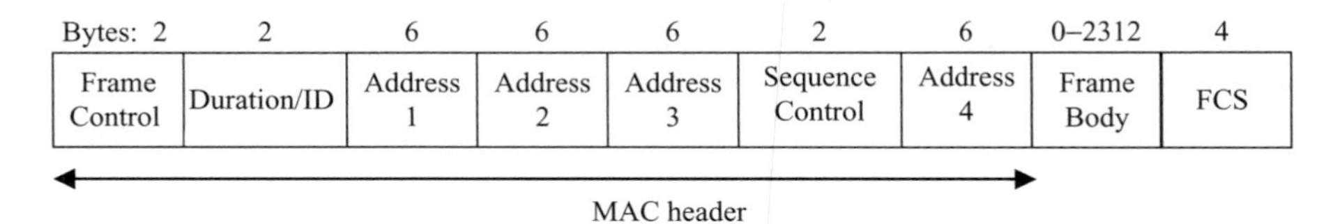

FIGURE 11.7 IEEE 802.11 MAC frame format.

Figure 11.6 shows the channel access in IEEE 802.11 DCF MAC. IEEE 802.11b has data rates of 1, 2, 5.5, and 11 Mbps, while IEEE 802.11a and IEEE 802.11g have identical data rates of 6, 9, 12, 18, 24, 36, 48, and 54 Mbps.

11.4.1.1 MAC Frame Format of IEEE 802.11 The MAC frame format of IEEE 802.11 is shown in Figure 11.7. The frame control indicates the type of frames, such as control frame, management frame, or data frame. It also provides control. Control information includes whether a frame is to or from an AP, fragmentation information, and privacy information. The duration/ID indicates the time the channel will be allocated for a successful MAC frame transmission. The addresses indicate the source, destination, transmit, and receive stations, depending on the case under consideration. The sequence control is used for fragmentation and reassembly with a 4-bit fragment number subfield and for numbering frames sent between a transmit and a receive station with a 12-bit sequence number subfield. The frame body contains a MAC service data unit (MSDU) or a fragment of an MSDU. The frame check sequence (FCS) is a 32-bit cyclic redundancy check (CRC).

11.4.1.2 IEEE 802.11 Security The security architecture and protocol in IEEE 802.11 is called *wired equivalent privacy* (WEP). WEP is responsible for providing authentication, confidentiality, and data integrity. Authentication is the process of verifying that a station or user is in fact who it claims to be. Confidentiality is achieved by sharing a secret key on how to encrypt and decrypt messages. The secret key is the method or algorithm or cipher that the mobile station and the access point use to decrypt messages. Integrity ensures that the message sent by the source to the destination has not been modified or tampered with.

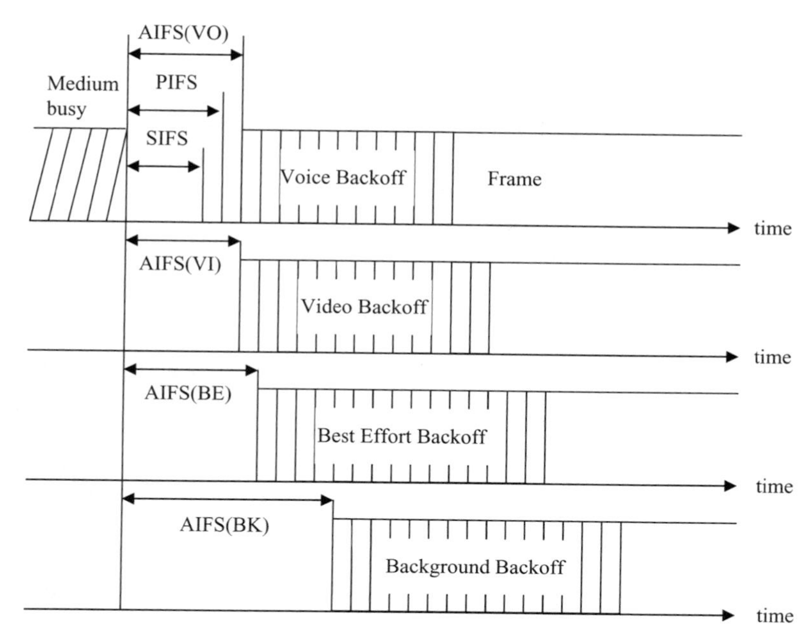

FIGURE 11.8 Channel access in IEEE 802.11e EDCA MAC.

11.4.2 IEEE 802.11e EDCA MAC

The contention-based IEEE 802.11e [2] uses carrier-sense multiple access with collision avoidance (CSMA/CA) similar to that in Section 4.3. The main differences are that it allows for multiple classes and supports the transmission of several data frames at one go with block acknowledgment. There are eight priority classes, and they are mapped into four access categories (ACs). The four access categories are for background, best effort, video, and voice traffic. The channel access for these traffics is differentiated by using different arbitration interframe spaces (AIFSs) and the minimum and maximum contention window sizes. The shorter the AIFS, the higher the priority for these access categories. Figure 11.8 shows the channel access for IEEE 802.11e. Table 11.9 shows the AIFSN for background, best effort, video, and voice

TABLE 11.9 AIFSN and Minimum and Maximum Contention Window Sizes for IEEE 802.11e EDCA MAC

Traffic Class	AIFSN	CW_{min}	CW_{max}
Background	7	aCW_{min}	aCW_{max}
Best effort	3	aCW_{min}	aCW_{max}
Video	2	$\frac{aCW_{min}+1}{2}-1$	aCW_{min}
Voice	2	$\frac{aCW_{min}+1}{4}-1$	$\frac{aCW_{min}+1}{2}-1$

traffic. The larger the AIFSN value, the longer the AIFS. A key element in IEEE 802.11e, like IEEE 802.11, is that retransmissions are controlled by using a backoff counter. The initial value in the backoff counter is set between zero and the minimum contention window size. The backoff counter value will decrement if the channel is idle and will freeze if the channel is busy. If the backoff counter value is zero, the station will attempt to transmit its packet. If it is not successful, the new contention window size will be doubled that of the previous window size plus one. The backoff counter will then randomly choose a value in the range [0, new contention window size]. After each backoff stage, if packet transmission is unsuccessful, the backoff and retransmission process is repeated until the retry limit is reached. When the retry limit is reached, the packet will be discarded if its transmission is still not successful. If the packet transmission is successful at any backoff stage or when the retry limit is reached, a new packet will be selected for new transmission at backoff stage 0. The minimum and maximum contention window sizes for differentiating priority in each class are also shown in Table 11.9. The smaller the minimum and maximum contention window sizes, coupled with the AIFSs, the shorter the channel access time and the higher the share of throughput.

11.4.2.1 MAC Frame Format of IEEE 802.11e The MAC frame format of IEEE 802.11e is shown in Figure 11.9. The frame control indicates the type of frame, such as control frame, management frame, or data frame. It also provides control information and indication that more data frames are buffered for the station at the AP. Control information includes the types of acknowledgment policy (e.g., normal acknowledgment, no acknowledgment, no explicit acknowledgment block acknowledgment), the transmit opportunity (TXOP) limit, queue size, transmit duration requested, and buffer status information. The duration/ID indicates the time the channel will be allocated for a successful MAC frame transmission. The addresses indicate the source, destination, transmitting, and receiving stations, depending on the case under consideration. The sequence control is used for fragmentation and reassembly with a 4-bit fragment number subfield and for numbering frames sent between a transmitting and a receiving station with a 12-bit sequence number subfield. The QoS control field is a 16-bit field that identifies the traffic category (TC) or traffic stream (TS) to which the frame belongs and various other QoS-related information about the frame that varies by frame type and subtype. Note that the QoS control field is the main difference between IEEE 802.11 MAC frame format and IEEE 802.11e MAC frame formation. The frame body contains a MSDU or a fragment of an MSDU. The FCS is a 32-bit CRC.

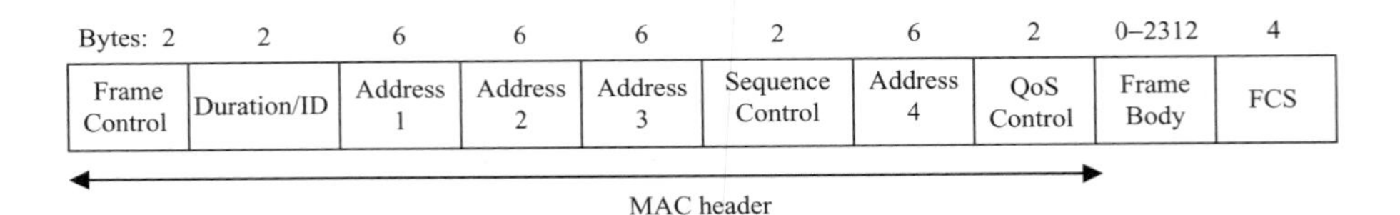

FIGURE 11.9 IEEE 802.11e MAC frame format.

TABLE 11.10 Supported Data Rates for a 800-ns Guard Interval

S/No.	Data Rates (Mbps)
1	6.5, 13, 19.5, 26, 39, 52, 58.5, 65
2	13, 26, 39, 52, 78, 104, 117, 130
3	19.5, 39, 58.5, 78, 117, 156, 175.5, 195
4	26, 52, 78, 104,156, 208, 234, 260
5	13.5, 27, 40.5, 54, 81, 108, 121.5, 135
6	27, 54, 81, 108, 162, 216, 243, 270
7	40.5, 81, 121.5, 162, 243, 324, 264.5, 405
8	54, 108, 162, 216, 324, 432, 486, 540
9	6
10	39, 52, 65, 58.5, 78, 97.5
11	52, 65, 65, 78, 91, 91, 104, 78, 97.5, 97.5, 117, 136.5, 136.5, 156
12	65, 78, 91, 78, 91, 104, 117, 104, 117, 130, 130, 143, 97.5, 117, 136.5, 117, 136.5, 156, 175.5, 156, 175.5, 195, 195, 214.5
13	81, 108, 135, 121.5, 162, 202.5
14	108, 135, 135, 162, 189, 189, 216, 162, 202.5, 202.5, 243, 283.5, 283.5, 324
15	135, 162, 189, 162, 189, 216, 243, 216, 243, 270, 270, 297, 202.5, 243, 283.5, 243, 283.5, 324, 364.5, 324, 364.5, 405, 405, 445.5

11.4.3 IEEE 802.11n Draft MAC

The goal of IEEE 802.11n is to increase the data rate to up to 600 Mbps. However, it is still rate adaptive and it can support multirate with 800- and 400-ns guard intervals, as shown in Tables 11.10 and 11.11, respectively. A few of them have the same or repeated data rates, but may have a different modulation scheme. In this section, enhanced MAC techniques for IEEE 802.11n are presented. These techniques include frame aggregation, reverse direction protocol, enhanced block acknowledgment, transmission modes, and reduced interframe space.

11.4.3.1 MAC Frame Format of IEEE 802.11n The MAC frame format of IEEE 802.11n is shown in Figure 11.10. The frame control indicates the type of frame, such as control frame, management frame, or data frame. It also provides control information and an indication that more data frames are buffered for the station at the AP. Control information includes block acknowledgment, block acknowledgment request, power-save poll, request to send (RTS), clear to send (CTS), and acknowledgment (ACK). The duration/ID indicates the time the channel will be allocated for a successful MAC frame transmission. The addresses indicate the source, destination, transmitting, and receiving stations, depending on the case under consideration. The sequence control is used for fragmentation and reassembly with a 4-bit fragment number subfield and for numbering frames sent between transmitting and receiving stations with a 12-bit sequence number subfield. The QoS control field is a 16-bit field that identifies the TC or TS to which the frame belongs and various other QoS-related information about the frame, which varies by frame type and subtype. The HT control field is for high-throughput station transmission. It includes

TABLE 11.11 Supported Data Rates for a 400-ns Guard Interval

S/No.	Data Rates (Mbps)
1	7.2, 14.4, 21.7, 28.8, 43.3, 57.8, 65, 72.2
2	14.4, 28.9, 43.3, 57.8, 86.7, 115.6, 130, 144.4
3	21.7, 43.3, 65, 86.7, 130, 173.3, 195, 216.7
4	28.9, 57.8, 86.7, 115.6, 173.3, 231.1, 260, 288.9
5	15, 30, 45, 60, 90, 120, 135, 150
6	30, 60, 90, 120, 180, 240, 270, 300
7	45, 90, 135, 180, 270, 360, 405, 450
8	60, 120, 180, 240, 360, 480, 540, 600
9	6.7
10	43.3, 57.8, 72.2, 65, 86.7, 108.7
11	57.8, 72.2, 72.2, 86.7, 101.1, 101.1, 115.6, 86.7, 108.3, 108.3, 130, 151.7, 151.7, 173.3
12	72.2, 86.7, 101.1, 86.7, 101.1, 115.6, 130, 115.6, 130, 144.4, 144.4, 158.9, 108.3, 130, 151.7, 130, 151.7, 173, 195, 173.3, 195, 216.7, 216.7, 238.3
13	90, 120, 150, 135, 180, 225
14	120, 150, 150, 180, 210, 210, 240, 180, 225, 225, 270, 315, 315, 360
15	150, 180, 210, 180, 210, 240, 270, 240, 270, 300, 300, 330, 225, 270, 315, 270, 315, 360, 405, 360, 405, 450,4 50, 495

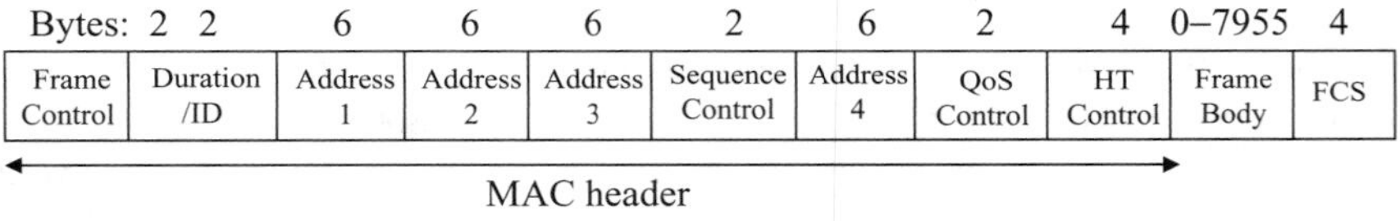

Bytes:	2	2	6	6	6	2	6	2	4	0–7955	4
	Frame Control	Duration /ID	Address 1	Address 2	Address 3	Sequence Control	Address 4	QoS Control	HT Control	Frame Body	FCS

MAC header

FIGURE 11.10 IEEE 802.11n MAC frame format.

link adaptation, position calibration, sequence calibration, channel state information (CSI) and steering, and reverse direction data flow grant. The QoS control field and the HT control field are the main difference between the IEEE 802.11 and IEEE 802.11n MAC frame formats. The frame body contains an aggregated MAC service data unit (A-MSDU), a MSDU, or a fragment of an MSDU. The maximum frame body in IEEE 802.11n is 7955 bytes, whereas those in IEEE 802.11 and 802.11e are both 2312 bytes. The FCS is a 32-bit CRC.

11.4.3.2 Frame Aggregation Figure 11.11 shows the data transfer for legacy IEEE 802.11, frame aggregation using an aggregate MAC service data unit (A-MSDU), and frame aggregation using an aggregate MAC protocol data unit (A-MPDU). First, let us consider the case of a legacy IEEE 802.11 WLAN. When the channel becomes idle, station 0 can try to access the channel and will experience an access delay before transmitting the data packet [including physical layer preamble, physical layer header, MAC header, MSDU, and frame check sequence (FCS)]. The access delay includes the interframe spaces (IFSs) and backoff time and could possibly include other stations' packet transmissions. Station 1 acknowledges the receipt

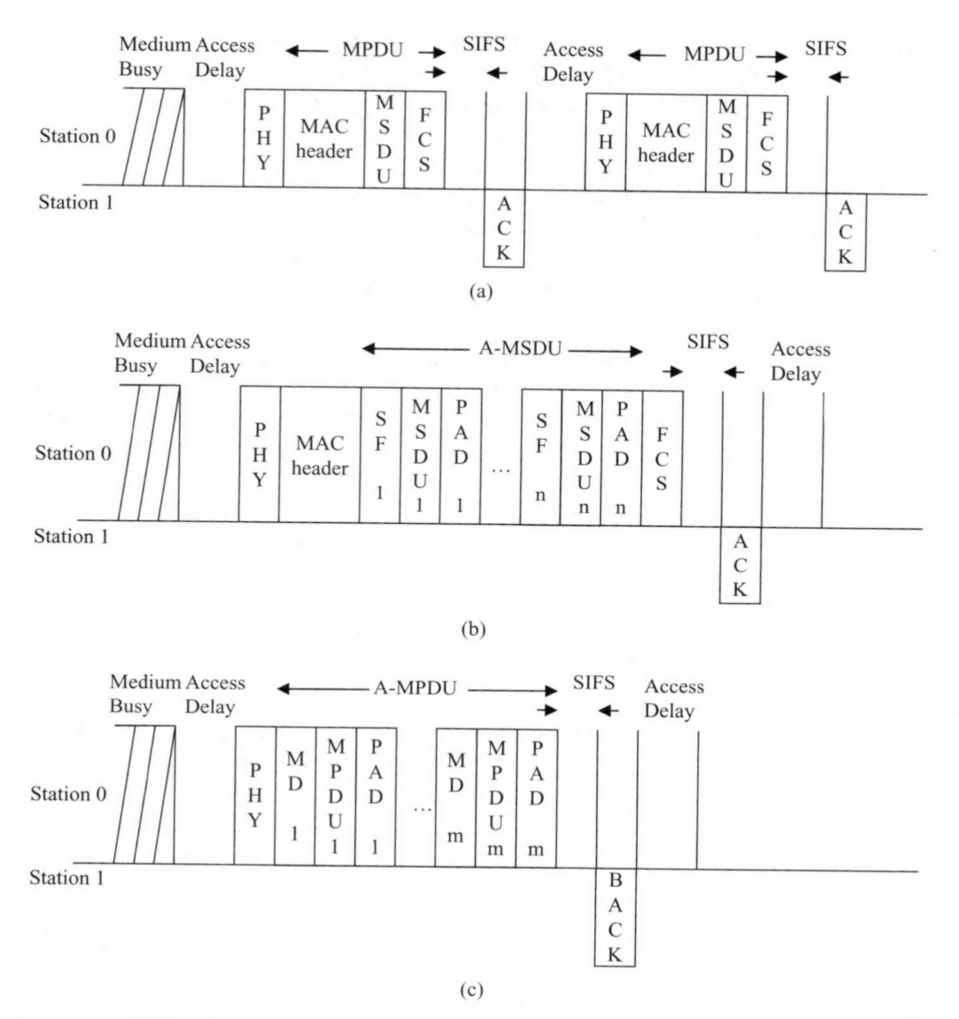

FIGURE 11.11 Data transfer for (a) legacy IEEE 802.11, (b) IEEE 802.11n frame aggregation using A-MSDU, and (c) IEEE 802.11n frame aggregation using A-MPDU.

of the data packet after a short IFS (SIFS). Next, let us consider frame aggregation of MSDUs using A-MSDU. Similarly, there is an access delay followed by the data packet transmission. The main difference here is that there are multiple MSDUs encapsulated by subframe (SF) headers and paddings (PADs). These are encapsulated by the physical layer preamble, physical layer header, MAC header, and FCS. Similarly, station 1 acknowledges the receipt of the aggregated MSDUs after a SIFS. Using the A-MSDU technique, the maximum size of the MAC frame is increased from the legacy length of 2304 bytes to 7955 bytes. Another frame aggregation technique used in IEEE 802.11n is the aggregation of MPDUs known as A-MPDU. A

MPDU consists of the MAC header, payload, or MSDU and FCS. The main difference here is that multiple MPDUs are concatenated together, with each MPDU encapsulated by a MPDU delimiter (MD) and padding. The physical layer preamble and physical layer header are transmitted before the A-MPDU. After a SIFS, the A-MPDU is block-acknowledged for each MPDU. Using the A-MPDU technique, the maximum size of the MAC frame is increased from the legacy length of 2304 bytes to 65535 bytes. Note that the maximum MPDU length that can be transported using the A-MPDU technique is 4095 bytes [3]. The minimum time between the start of adjacent MPDUs within an A-MPDU can be {0, 1/4, 1/2, 1, 2, 4, 8, 16} μs.

11.4.3.3 Block Acknowledgment In an implicit block acknowledgment request (BAR) when using frame aggregation, the source station may exclude a BAR frame, and the high-throughput destination station that receives the data frames will consider this an implicit BAR. The destination station should reply with a block acknowledgment (BACK). When the source station does not receive a BACK frame after transmitting the frames, it will transmit a BAR frame to get the destination to reply with a BACK frame. However, if the destination station cannot receive all the frames, it will not send a BACK frame.

When sending a BACK frame between two high-throughput stations, a compressed BACK frame can be used. It reduces the bitmap size in the IEEE 802.11e BACK frame from 128 bytes to 8 bytes. This increases the network efficiency.

11.4.3.4 Reverse Direction Protocol The reverse direction protocol allows the destination station to transmit a data frame immediately after the source station has transmitted its data frame and a BAR frame. This protocol cuts down access delay for the destination station and improves turnaround time for the destination station to respond to the source station. This is particularly important for voice application, which is a bidirectional traffic. Figure 11.12 shows the data transfer between two stations using request to send/clear to send (RTS/CTS). Station 0 first sends an RTS frame. The physical layer preamble, physical layer header, and MD and PAD for

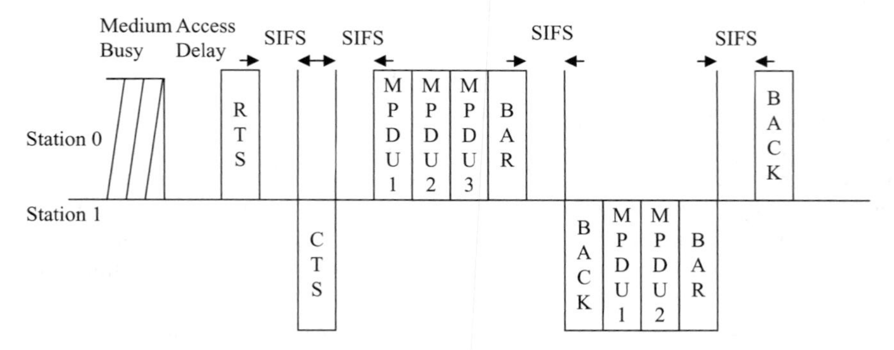

FIGURE 11.12 Data transfer using reverse direction protocol.

each MPDU data frame are not shown in the figure, for simplicity. Station 0 responds with a CTS frame, including a request for reverse direction data flow. Station 0 will transmit its MPDU data frames with a BAR frame at the end. Three MPDU data frames are sent by station 0 in this example. The reverse direction data flow is granted by a grant that is piggybacked on the MPDU frame or the BAR frame. Station 1 will block-acknowledge the data frames from station 0 by a BA frame. If a grant is granted by station 0, station 1 can transmit its MPDU data frames and a BAR frame in the reverse direction to station 0. 2 MPDU data frames are transmitted by station 1 in this example. Station 0 will block-acknowledge the data frame from station 1 by a BACK frame.

11.4.3.5 Transmission Modes There are three transmission modes for the PLCP formats in the IEEE 802.11n draft: the legacy (or non-HT) mode, mixed (or HT-mixed) mode, and Greenfield mode. The legacy mode is used to act as the legacy stations, such as 802.11a, 802.11b, 802.11g, or even 802.11 stations. The mixed mode is used when transmitting in a mixed or legacy stations and high-throughput (HT) stations such as 802.11n stations. The Greenfield mode is used when all the stations are high-throughput stations, such as 802.11n stations, and this gives the best performance.

11.4.3.6 Reduced Interframe Space RIFS reduces the amount of time between data frames and thus can improve network efficiency. However, this can be used only in the Greenfield mode, where all stations are high-throughput stations.

11.4.3.7 Power Savings IEEE 802.11n has extended the power savings of the IEEE 802.11 and 802.11e MACs. Two methods are used to extend the power savings of IEEE 802.11n MAC and IEEE 802.11e MAC. One is spatial multiplexing power save (SMPS), and the other is power-save multipoll (PSMP). The former allows an IEEE 802.11n station, not the access point (AP), to power down to just using one of its radios. The latter extends the automatic power save delivery (APSD) mechanism defined in IEEE 802.11e. PSMP's scheduling mechanism reduces the contention between stations and the AP and between stations. This reduces the backoff time and the number of retries for packet transmission. Thus, it improves the power savings in the stations.

11.4.4 IEEE 802.11s Draft MAC

IEEE 802.11s MAC uses IEEE 802.11e EDCA MAC as a mandatory baseline. This is compatible with legacy stations and is easy to implement. Four main areas of MAC enhancements in IEEE 802.11s are mesh deterministic access (MDA), common channel framework (CCF), intramesh congestion control (IMCC), and power savings. Mesh deterministic access is a distributed reservation protocol. Thus, it is suitable for real-time traffic. Slots are reserved for transmissions using the MDA protocol. Traffics using the EDCA protocol are transmitted between MDA reserved slots. A data frame cannot be transmitted if the time required to transmit the data frame,

SIFSs, and acknowledgment before the beginning of the next MDA reserved slot is insufficient. CCF enables a pair of source and destination to transmit RTS/CTS frames on a common channel and then transmits the data frame on another channel after a switching time. Thus, RTSs/CTSs can be sent by multiple source–destination pairs on the common channel after each other and their corresponding data frames can be sent through orthogonal channels with the data frames overlapping each other in time but differing in the frequency channels. IMCC is achieved by using local congestion monitoring, congestion control signaling, and local rate control. Power savings use the existing mechanism but with some modifications. These modifications, made to the APSD, include a traffic indication message information element (TIM IE) in the beacon, and new rules for the announcement traffic indication message (ATIM) window and ATIM frames. The default routing scheme in IEEE 802.11s is the hybrid wireless mesh protocol (HWMP), while the optional routing scheme is the radio-aware optimized link state routing (RA-OLSR). The former utilizes the flexibility of an on-demand route discovery with efficient proactive routing to a mesh portal. The mesh portal is the point where the MSDU exits and enters a mesh WLAN. On-demand routing is based on radio metric ad hoc on-demand distance vector (RM-AODV) routing, while the proactive routing is based on tree-based routing. RS-OLSR proactively maintains a link state for routing. The changes in link state are communicated to neighborhood nodes.

11.4.4.1 MAC Frame Format of IEEE 802.11s The MAC frame format of IEEE 802.11s is shown in Figure 11.13. The frame control indicates the type of frame, such as control frame, management frame, or data frame. It also provides control information and indication more data frames are buffered for the station at the AP. Control information includes the types of acknowledgment policy (e.g., normal acknowledgment, no acknowledgment, no explicit acknowledgment, block acknowledgment), the transmit opportunity (TXOP) limit, queue size, transmit

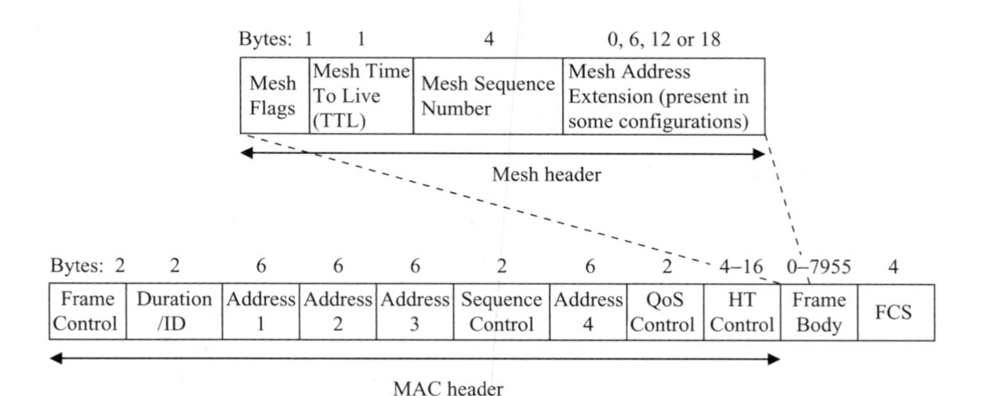

FIGURE 11.13 IEEE 802.11s MAC frame format. (From [4].)

duration requested, and buffer status information. The duration/ID indicates the time the channel will be allocated for a successful MAC frame transmission. The addresses indicate the source, destination, transmitting, and receiving stations depending on the case. The sequence control is used for fragmentation and reassembly with a 4-bit fragment number subfield and for numbering frames sent between a transmitting and a receiving station with a 12-bit sequence number subfield. The QoS control field is a 16-bit field that identifies the TC or TS to which the frame belongs and various other QoS-related information about the frame, which varies by frame type and subtype. The HT control field is for high-throughput station transmission. The mesh header field encapsulated in the frame body is for mesh flags, time to live, mesh sequence number, and up to three optional mesh addresses. The mesh flags, time to live, and mesh sequence number are always present in the mesh header, but the three additional mesh addresses are optional. The mesh header field and the HT control field are the main differences between the IEEE 802.11e and IEEE 802.11s MAC frame formats. The frame body contains a MSDU or a fragment of an MSDU. The maximum frame body in IEEE 802.11s is 7955 bytes, whereas those in IEEE 802.11 and 802.11e are both 2312 bytes and that in IEEE 802.11n is 7955 bytes. Thus, the maximum body is the same in IEEE 802.11s and IEEE 802.11n. The FCS is a 32-bit CRC.

11.4.5 IEEE 802.11e EDCA MAC Analysis with IEEE 802.11n

The analysis here assumes the same AIFS for all access categories. However, it still accounts for the variation of the contention window size and the backoff process as in [5]. We assume that there are K generic data rates, denoted by R_k, where $k = 1, \dots, K$ and $R_1 > R_2 > \cdots > R_K$, and their corresponding coverage distances are denoted by D_k, $D_1 < D_2 < \cdots < D_K$. We also assume that all stations are distributed uniformly in a coverage area of radius R, where $R = D_K$; the access point is located at the center of the circle. Let us consider the case of direct transmissions. Stations within the circle of radius D_1 use data rate R_1 to communicate with the AP, while stations within concentric circles of radii D_{k-1} and D_k use data rate R_k, $k = 2, \dots, K$. Let $P_k, k = 1, \dots, K$, denote the probability of being in each of these regions using direct transmission at data rate R_k. P_k is given by

$$P_k = \begin{cases} \dfrac{\pi D_k^2}{\pi R^2}, & k = 1 \\ \dfrac{\pi (D_k^2 - D_{k-1}^2)}{\pi R^2}, & k = 2, \dots, K. \end{cases} \tag{11.1}$$

Let N be the number of the priority in a given station, where the ith priority class is given by the index $i = 0, 1, \dots, N - 1$. Let $a(i, t)$ be a random process representing the backoff stage j, $j = 0, 1, \dots, L_{i,\text{retry}}$, at time t, where $L_{i,\text{retry}}$ is the retry limit if the packet transmission starts at backoff stage 0. Let $b(i, t)$ be a random process representing the value of the backoff counter at time t. The value of the backoff counter $b(i, t)$ is chosen uniformly from the range $(0, 1, \ldots, W_{i,j})$, where $W_{i,j} = 2W_{i,j-1} + 1$ for $j \leq m_i \leq L_{i,\text{retry}}$ and $W_{i,j} = \text{CW}_{\max,i}$ if $m_i < j \leq L_{i,\text{retry}}$.

$$W_{i,j} = \begin{cases} (W_i + 1)2^j - 1, & (0 \leq j \leq m_i \\ (W_i + 1)2^{m_i} - 1, & m_i < j \leq L_{i,\text{retry}}, \end{cases} \tag{11.2}$$

where $W_i = W_{i,o}$. Let p_i denote the probability that a transmitted frame collides; it is also the probability that a station in a backoff stage for the ith class senses the channel to be busy. The discrete-time Markov chain for the state transition diagram for the ith priority class is shown in Figure 11.14.

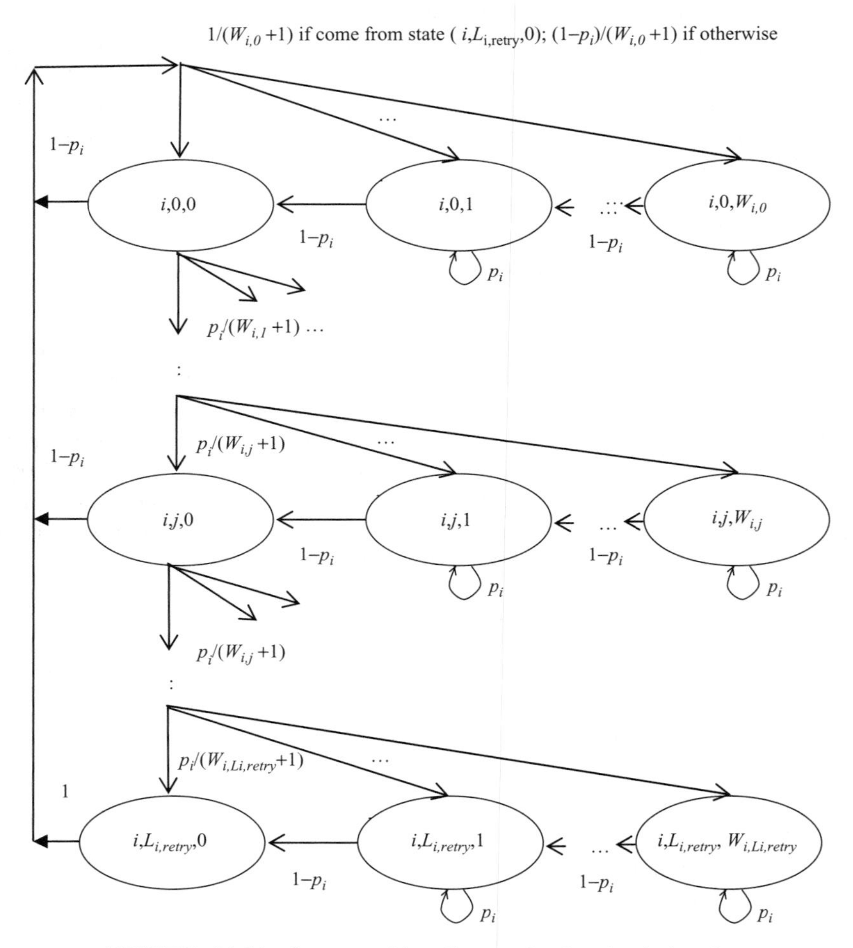

FIGURE 11.14 State transition diagram for the ith priority class.

Let $b_{i,j,k} = \lim t \to \infty \Pr[a(i,t) = j, b(i,t) = k]$ be the stationary distribution of the Markov chain. The non-null transition probabilities are listed as follows:

$$\Pr[(i,0,k)|(i,j,0) = \frac{1-p_i}{W_{i,0}+1}, 0 \le k \le W_{i,0}, 0 \le j < L_{i,\mathrm{retry}}$$

$$\Pr[(i,0,k)|(i,j,0) = \frac{1}{W_{i,0}+1}, 0 \le k \le W_{i,0}, j = L_{i,\mathrm{retry}}$$

$$\Pr[(i,j,k)|(i,j-1,0) = \frac{p_i}{W_{i,j}+1}, 0 \le k \le W_{i,j}, 1 \le j \le L_{i,\mathrm{retry}}$$

$$\Pr[(i,j,k)|(i,j,k) = p_i, 1 \le k \le W_{i,j}, 0 \le j \le L_{i,\mathrm{retry}}$$

$$\Pr[(i,j,k)|(i,j,k+1) = 1 - p_i, 0 \le k \le W_{i,j} - 1, 0 \le j \le L_{i,\mathrm{retry}}.$$

The first and second equations above represent the transition probability for each state in backoff stage 0, while the third equation represents the transition probability for each state in backoff stage j. The fourth equation represents the self-transition probability in a state due to the channel being busy, while the fifth equation represents the backoff counter decrementing by one, as the channel is not busy. In steady state we can derive the following relationships through chain regularities:

$$b_{i,j,0} = p_i^j b_{i,0,0}, k = 0, 0 \le j \le L_{i,\mathrm{retry}} \tag{11.3}$$

$$b_{i,j,k} = \frac{W_{i,j} - k + 1}{W_{i,j} + 1} \frac{p_i^j}{1 - p_i} b_{i,0,0}, 1 \le k \le W_{i,j} - 1, 0 \le j \le L_{i,\mathrm{retry}}. \tag{11.4}$$

Using (11.2) and the following identity,

$$\sum_{k=1}^{W_{i,j}} \frac{W_{i,j} - k + 1}{W_{i,j} + 1} = \frac{W_{i,j}}{2}, \tag{11.5}$$

we have

$$\sum_{k=1}^{W_{i,j}} \frac{W_{i,j} - k + 1}{W_{i,j} + 1} = \begin{cases} \frac{(W_i + 1)2^j}{2} - \frac{1}{2}, & 0 \le j \le m_i \\ \frac{(W_i + 1)2^{m_i}}{2} - \frac{1}{2}, & m_i < j \le L_{i,\mathrm{retry}} \end{cases} \tag{11.6}$$

By the property of total probability,

$$\sum_{j=0}^{L_{i,\mathrm{retry}}} \sum_{k=0}^{W_{i,j}} b_{i,j,k} = 1. \tag{11.7}$$

From (11.7), we have

$$b_{i,0,0} = \begin{cases} \dfrac{2(1-p_i)^2(1-2p_i)}{\left\{(1-2p_i)^2\left[1-p_i^{L_{i,\text{retry}}+1}\right]+(W_i+1)(1-p_i)\left[1-(2p_i)^{L_{i,\text{retry}}+1}\right]\right\}}, & m_i \geq L_{i,\text{retry}}, \\ \dfrac{2(1-p_i)^2(1-2p_i)}{\left\{\begin{array}{l}(1-2p_i)^2\left[1-p_i^{L_{i,\text{retry}}+1}\right]+(W_i+1)(1-p_i)\left[1-(2p_i)^{m_i+1}\right] \\ +2^{m_i}p_i^{m_i+1}(1-2p_i)\left(1-p_i^{L_{i,\text{retry}}-m_i}\right)\end{array}\right\}}, & m_i < L_{i,\text{retry}}. \end{cases} \tag{11.8}$$

A station transmits when its backoff counter reaches zero; that is, the station is at any of the states $\{i, j, 0\}$, where $0 \leq j \leq L_{i,\text{retry}}$. The probability that a station in the ith priority class transmits during a generic time slot, denoted by τ_i, is given by

$$\tau_i = \sum_{j=0}^{L_{i,\text{retry}}} b_{i,j,0} = \sum_{j=0}^{L_{i,\text{retry}}} p_i^j b_{i,0,0} = \frac{1-p_i^{L_{i,\text{retry}}+1}}{1-p_i} b_{i,0,0}. \tag{11.9}$$

Substituting (11.8) into (11.9), we have

$$\tau_i = \begin{cases} \dfrac{2(1-p_i)(1-2p_i)(1-p_i^{L_{i,\text{retry}}+1})}{(1-2p_i)^2[1-p_i^{L_{i,\text{retry}}+1}]+(W_i+1)(1-p_i)[1-(2p_i)^{L_{i,\text{retry}}+1}]}, & m_i \geq L_{i,\text{retry}} \\ \dfrac{2(1-p_i)(1-2p_i)(1-p_i^{L_{i,\text{retry}}+1})}{\left\{\begin{array}{l}(1-2p_i)^2[1-p_i^{L_{i,\text{retry}}+1}]+(W_i+1)(1-p_i)[1-(2p_i)^{m_i+1}] \\ +2^{m_i}p_i^{m_i+1}(1-2p_i)(1-p_i^{L_{i,\text{retry}}-m_i})\end{array}\right\}}, & m_i < L_{i,\text{retry}}. \end{cases} \tag{11.10}$$

A transmitted frame collides when one or more stations transmit during a time slot. The probability that a station in the backoff stage for the ith priority class senses the channel as busy, denoted by p_i, is given by

$$p_i = 1 - (1-\tau_i)^{n_i-1} \prod_{h=0,h\neq i}^{N-1} (1-\tau_h)^{n_h}, \tag{11.11}$$

where n_i is the number of class i stations. τ_i and p_i can be solved numerically.

The probability that the channel is busy happens when at least one station transmits during a time slot, denoted by p_b, and is given by

$$p_b = 1 - \prod_{h=0}^{N-1} (1 - \tau_h)^{n_h}. \tag{11.12}$$

The probability that a successful transmission occurs in a time slot for the ith priority class, denoted by $p_{s,i}$, is given by

$$p_{s,i} = n_i \tau_i (1 - \tau_i)^{n_i - 1} \prod_{h=0, h \neq i}^{N-1} (1 - \tau_h)^{n_h}, \qquad i = 0, 1, \ldots, N-1. \tag{11.13}$$

The probability that a successful transmission occurs in a time slot, denoted by p_s, is given by

$$p_s = \sum_{i=0}^{N-1} p_{s,i} = (1 - p_b) \sum_{i=0}^{N-1} \frac{n_i \tau_i}{1 - \tau_i}. \tag{11.14}$$

The probability that the channel is idle for a time slot is $1 - p_b$. The probability that the channel is neither idle nor successful for a time slot is $[1 - (1 - p_b) - p_s] = p_b - p_s$.

Let $T_{E(L),k}$, T_s, and T_c denote the time to transmit the average payload of average length $E(L)$, with data rate R_k, the average time required to transmit a packet successfully and the average time that the channel has a collision, respectively. The actual saturated throughput for the ith priority class, denoted by S_i, is given by

$$\begin{aligned} S_i &= \sum_{k=1}^{K} P_k \frac{E(\text{payload transmission time in a time slot for class } i)}{E(\text{length of a time slot})} R_k \\ &= \sum_{k=1}^{K} P_k \frac{p_{s,i} E(L)}{(1 - p_b)\delta + p_s T_s + (p_b - p_s) T_c} \\ &= \frac{p_{s,i} E(L)}{(1 - p_b)\delta + p_s T_s + (p_b - p_s) T_c}. \end{aligned} \tag{11.15}$$

Note that $E(L) = T_{E(L),k} R_k$. The total actual saturated throughput, denoted by S, is given by

$$S = \sum_{i=0}^{N-1} S_i. \tag{11.16}$$

Let $T_{H,k}$, L, $T_{E(L),k}$, L^*, $T_{E(L^*),k}$, T_{SIFS}, T_{DIFS}, $T_{\text{ACK},k}$, $T_{\text{RTS},k}$, $T_{\text{CTS},k}$, and $T_{\text{AIFS}(i)}$ denote, respectively, the time to transmit the header (including PLCP preamble, physical layer header, MAC header) at data rate R_k for the MAC header, the length of the payload including tail bits and pad bits, the time to transmit a payload with length $E(L)$ at data rate R_k, the length of the longest frame in a collision, the time to transmit a payload with length $E(L^*)$ at data rate R_k, the SIFS time, the DIFS time, the time to transmit an acknowledgment with data rate R_k, the time to transmit a request-to-send (RTS) frame with data rate R_k, the time to transmit a clear-to-send (CTS) frame with data rate R_k, and an arbitration inter frame space (AIFS) time for a class i station. For the RTS/CTS access, T_s and T_c are given, respectively, by

$$T_s = \sum_{k=1}^{K} P_k \left(T_{\text{RTS},k} + T_{\text{CTS},k} + T_{H,k} + T_{E(L),k} + 3T_{\text{SIFS}} + T_{\text{ACK},k} + T_{\text{AIFS}(i)}\right) \tag{11.17}$$

$$T_c = \sum_{k=1}^{K} P_k \left(T_{\text{RTS},k} + T_{\text{SIFS}} + T_{\text{ACK},k} + T_{\text{DIFS}}\right). \tag{11.18}$$

Let X_i denote the total number of backoff slots that a packet encounters without the case when the backoff counter freezes for the ith priority class. The expected value of X_i, denoted by $E(X_i)$, is given by (similar to [5])

$$E(X_i) = \sum_{j=0}^{L_{i,\text{retry}}} \frac{p_i^j(1-p_i)}{1-p_i^{L_{i,\text{retry}}+1}} \sum_{h=0}^{j} \frac{W_{i,h}}{2}. \tag{11.19}$$

Using (11.2) and the identities

$$\sum_{j=0}^{n} jp^j = p\left[\frac{1-p^n}{(1-p)^2} - \frac{np^n}{1-p}\right] \tag{11.20}$$

and

$$\sum_{j=m}^{n} jp^j = \frac{p^{m+1}-p^{n+2}}{(1-p)^2} + \frac{mp^m-(n+1)p^{n+1}}{1-p}, \tag{11.21}$$

we have

$$E(X_i) = \begin{cases} \dfrac{(W_i+1)(1-p_i)\left[1-(2p_i)^{L_{i,\text{retry}}+1}\right]}{(1-2p_i)(1-p_i^{L_{i,\text{retry}}+1})} - \left(\dfrac{W_i}{2}+1\right) \\ \quad -\dfrac{p_i(1-p_i^{L_{i,\text{retry}}})}{2(1-p_i)(1-p_i^{L_{i,\text{retry}}+1})} + \dfrac{L_{i,\text{retry}}\,p_i^{L_{i,\text{retry}}+1}}{2(1-p_i^{L_{i,\text{retry}}+1})}, & m_i \geq L_{i,\text{retry}} \\ \dfrac{(W_i+1)(1-p_i)\left[1-(2p_i)^{m_i+1}\right]}{(1-2p_i)(1-p_i^{L_{i,\text{retry}}+1})} - \left(\dfrac{W_i}{2}+1\right) \\ \quad -\dfrac{p_i(1-p_i^{m_i})}{2(1-p_i)(1-p_i^{L_{i,\text{retry}}+1})} + \dfrac{m_i\,p_i^{m_i+1}}{2(1-p_i^{L_{i,\text{retry}}+1})} \\ \quad +\dfrac{\left[(W_i+1)2^{m_i} - \left(\dfrac{W_i+1}{2}\right)2^{m_i}m_i\right]p_i^{m_i+1}\left(1-p_i^{L_{i,\text{retry}}-m_i}\right)}{1-p_i^{L_{i,\text{retry}}+1}} \\ \quad +\dfrac{\left[(W_i+1)2^{m_i}-1\right]p_i\left[p_i^{m_i+1}-p_i^{L_{i,\text{retry}}+1}\right]}{2(1-p_i)(1-p_i^{L_{i,\text{retry}}+1})} \\ \quad +\dfrac{\left[(W_i+1)2^{m_i}-1\right]p_i\left[(m_i+1)p_i^{m_i} - \left(L_{i,\text{retry}}+1\right)p_i^{L_{i,\text{retry}}}\right]}{2(1-p_i^{L_{i,\text{retry}}+1})}, & m_i < L_{i,\text{retry}}. \end{cases} \tag{11.22}$$

Only successful transmissions are considered above.

Let B_i denote the total number of slots that an ith priority class packet encounters when the backoff counter freezes. The expected value of B_i, denoted by $E(B_i)$, is given by

$$E(B_i) = \frac{E(X_i)}{1-p_i}p_i. \tag{11.23}$$

$E(X_i)$ and $E(B_i)$ can be treated as the total number of idle and busy slots that a packet encounters during backoff stages, respectively. Let $E(N_{i,\text{retry}})$ denote the average number of retries for the ith priority class. It is given by

$$\begin{aligned} E(N_{i,\text{retry}}) &= \sum_{j=0}^{L_{i,\text{retry}}} \frac{j p_i^j (1-p_i)}{1-p_i^{L_{i,\text{retry}}+1}} \\ &= \frac{p_i(1-p_i^{L_{i,\text{retry}}})}{(1-p_i)(1-p_i^{L_{i,\text{retry}}+1})} - \frac{L_{i,\text{retry}}\,p_i^{L_{i,\text{retry}}+1}}{1-p_i^{L_{i,\text{retry}}+1}}. \end{aligned} \tag{11.24}$$

TABLE 11.12 Parameter Values Used for IEEE 802.11n and 802.11e MAC Based on IEEE 802.11n PHY

Symbol	Value	Symbol	Value
W_0	7	$W_{0,j}$	{7,15, 31, 63, 127, 255}
W_1	15	$W_{1,j}$	{15, 31, 63, 127, 255, 511}
m_1	6	Data rate, R_k	13.5, 27, 40.5, 54, 81, 108, 121.5, 135 Mbps
m_2	6	Distance, D_k	85, 78, 57, 55, 39, 32, 28, 21 m
$L_{0,\text{retry}}$	6	$T_{H,k}$	$24\ \mu\text{s} + 24 \cdot 8/R_k$
$L_{1,\text{retry}}$	6	$E(L) = E(L\ \cdot)$	7955 bytes
δ	9 μs	$T_{\text{RTS},\,k}$	$24\ \mu\text{s} + 28 \cdot 8/R_k$
T_{SIFS}	16 μs	$T_{\text{CTS},\,k}$	$24\ \mu\text{s} + 28 \cdot 8/R_k$
T_{DIFS}	32 μs	$T_{\text{ACK},\,k}$	$24\ \mu\text{s} + 28 \cdot 8/R_k$
$T_{\text{AIFS}(0)}$	34 μs	$T_{\text{AIFS}(1)}$	34 μs

Let D_i denote the class i packet delay. From [4] it is given by

$$E(D_i) = E(X_i)\delta + E(B_i)\left[\frac{p_s}{p_b}T_s + \frac{(p_b - p_s)}{P_b}T_c\right] + E(N_{i,\text{retry}})(T_c + T_o) + T_s, \tag{11.25}$$

where

$$T_o = \text{SIFS} + T_{\text{CTS timeout}}. \tag{11.26}$$

Here we present the result for the saturated throughput and delay with two classes ($N = 2$) as an illustrative example with a payload of 7955 bytes. The parameter values used in the numerical examples for IEEE 802.11n MAC and IEEE 802.11e MAC are tabulated in Table 11.12. Note that the data rates versus distances are used for illustration purposes, which can vary in reality. The RTS/CTS access is assumed in this section.

The saturated throughputs of class 0 and class 1 stations and the total saturated throughput for IEEE 802.11n are shown in Figure 11.15. Class 0 has a higher saturated throughput than that of class 1, as it has a higher priority. Figure 11.16 shows the saturated packet delays of class 0, D_0, and class 1, D_1, for IEEE 802.11n. Similarly, class 0 has a lower saturated delay compared to that of class 1. Note that class 1 has a higher priority due to smaller minimum and maximum contention window sizes of 7 and 255. Note that these results do not include the enhanced MAC features of IEEE 802.11n draft but catered only for a larger payload of 7955 bytes.

11.5 MOBILITY RESOURCE MANAGEMENT

A number of infrastructure network architectures of BSSs are cellular networks. Thus, stations in one of the BSSs can handoff to another BSS. Handoff in this case is better known as roaming in WLAN context. Handoff in IEEE 802.11 WLAN

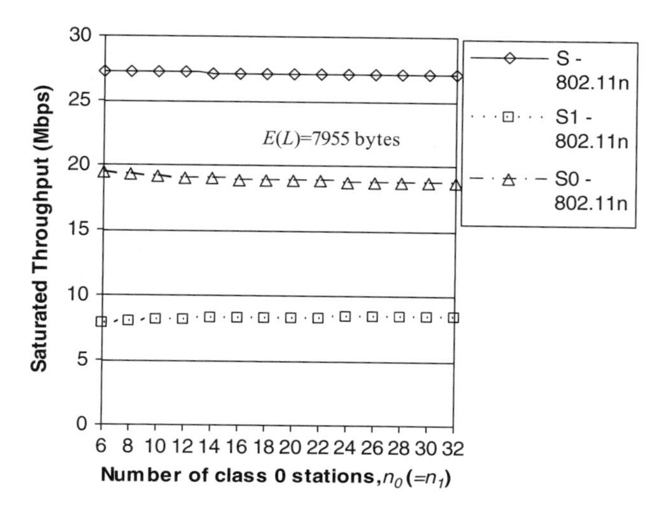

FIGURE 11.15 Saturated throughput for two classes.

uses hard handoff. That is, a connection is broken before a new connection is made or a "break-before-make" handoff. Handoff can be triggered by a change in signal strength between the station and AP. The BSS is identified by the address of the AP and is known simply as the service set ID (SSID). Before a station joins an AP, it has to perform either a passive scan or an active scan to get the capability information and information elements from the SSID. In a passive scan, a station will listen to all the channels for the different SSIDs advertised by the APs in these channels. In

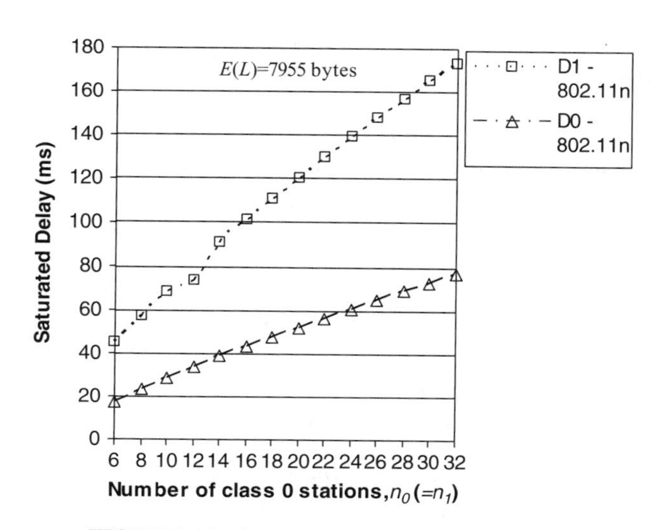

FIGURE 11.16 Saturated delay for two classes.

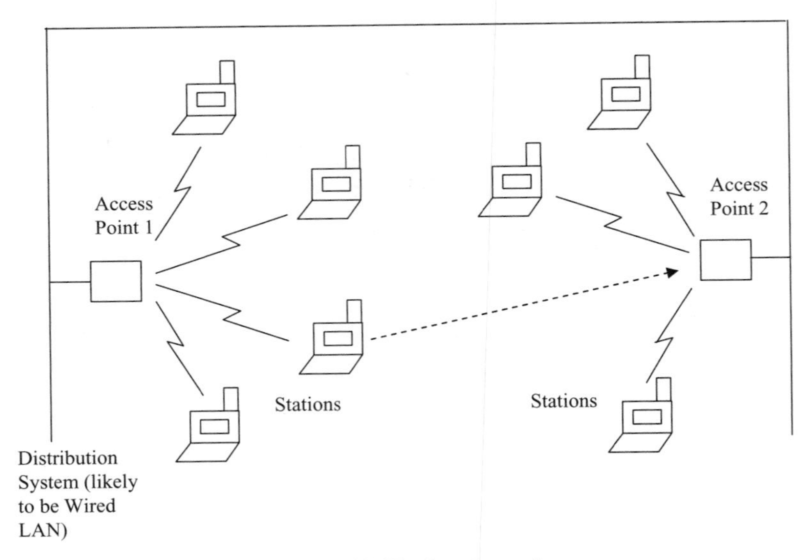

FIGURE 11.17 Local roaming.

an active scan, a station first tunes to the channel, listens and avoids collision as in CSMA/CA, and then sends a probe frame in that channel. It will listen for any probe response or beacon. The AP will reply with a probe response if the probe request matches the SSID of the AP. This is a directed probe. Any AP can reply if the probe request has a broadcast SSID. This is an undirected probe. We assume that the station has been authenticated with the AP before proceeding with the following. When the station has identified an AP, it can then send an association request frame to the AP or a reassociation request frame to the AP if it is already connected to another AP. The AP that the station wants to associate with will decide whether to accept or reject the association or reassociation using an association response frame or a reassociation response frame, accordingly. If the station roams within a set of APs with the same SSID or within the same extended service set (ESS), this is local roaming. Figure 11.17 shows an ESS with two APs. AP 1 has four stations connected to it, while AP 2 has three stations connected to it. One of the stations in AP 1 is shown moving to AP 2. It will proceed with the procedure above to do a local roaming or handoff. If the station roams between different ESSs, this is global roaming as shown in Figure 11.18. There are two ESSs. One of the ESSs has APs 1 and 2, while the other ESS has APs 3 and 4. One of the stations in AP 1 is shown moving to AP 3. It will proceed with the procedure above to do a global roaming or handoff. In this case, the IP address of the station changes. This can be taken care of using mobile IP as described in Section 5.7.

Handoff or roaming in the current IEEE 802.11 WLAN takes a long time. This aspect is being examined by a task group in IEEE 802.11r. The focus of this group is on fast secure roaming between APs within the same ESS.

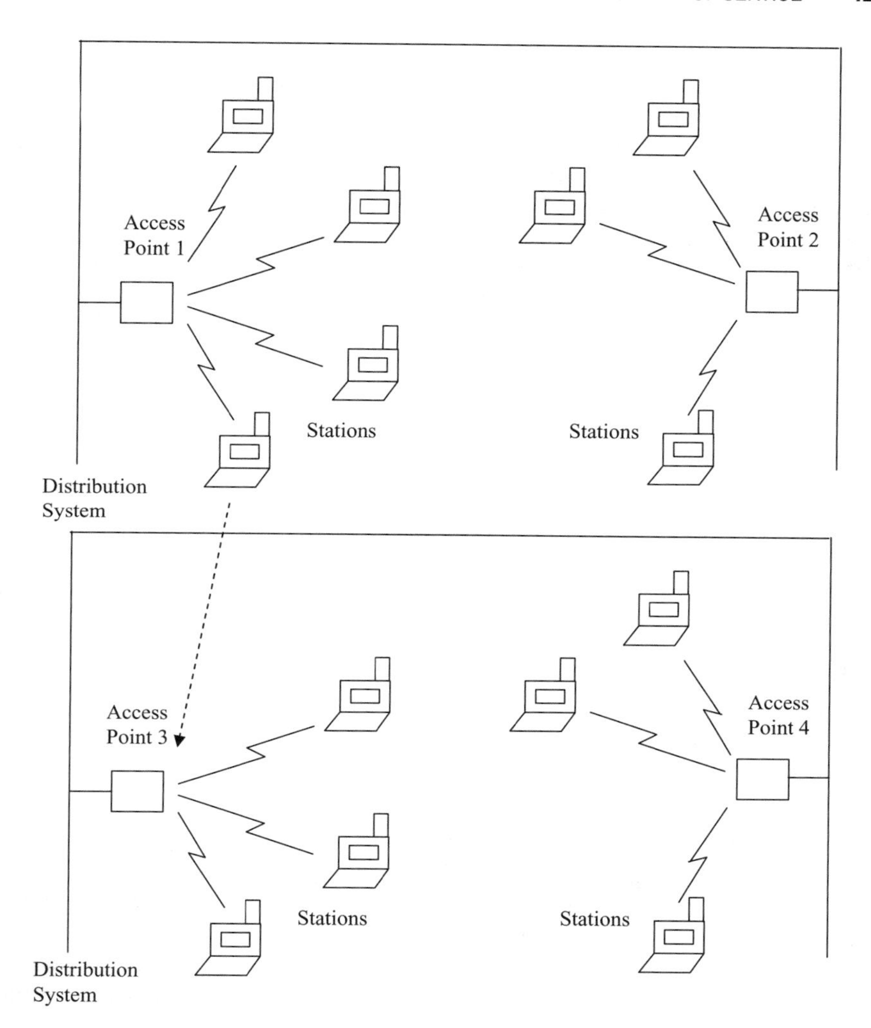

FIGURE 11.18 Global roaming.

11.6 QUALITY OF SERVICE

The qualities of service (QoSs) in IEEE 802.11e EDCA MAC are provided by its channel access methods. For real-time traffic, IEEE 802.11e HCCA MAC can be used. For less stringent QoS traffic, IEEE 802.11e EDCA MAC can be used. There are four types of access categories using IEEE 802.11e EDCA MAC. This gives rise to relative QoS, depending on the minimum and maximum contention window sizes and the arbitration interframe space (AIFS). In general, the smaller the minimum and maximum contention window sizes and the AIFS, the higher the priority of that access category. IEEE 802.11e EDCA MAC is allowed in IEEE 802.11n stations. That is, stations with quality of service differentiations are allowed in IEEE 802.11n MAC transmissions. Frame aggregation techniques, block acknowledgment techniques, reverse direction protocol, and RIFS together can help to improve throughput, delay, and jitter QoSs. If all the stations are high-throughput stations and they are operating in the Greenfield mode, this will improve the throughput tremendously. However, if there is even one station transmitting in a low data rate in a mixed mode, the total throughput of the stations will be severely reduced as the portion of time used to transmit by a low-rate station will be much higher that that by a high-rate station.

11.7 APPLICATIONS

There can be many applications using IEEE 802.11n, some of which are:

- Voice over IP (VoIP)
- Video streaming
- Music streaming
- Video conferencing
- IPTV
- Interactive gaming
- Large file and picture transfer
- Data backup and storage

The preceding applications apply to IEEE 802.11n. The voice signal is packetized and transmitted through the Internet using Internet protocol (IP) for voice over IP (VoIP). Video and music streaming are the continuous transmission of video or music, respectively, to the end user connected through the WLAN when using 802.11n. Video conferencing is the video and voice interaction among a number of users talking or discussing when connected together through a WLAN or using a WLAN as the "last mile." IPTV is the transmission of TV programs using IP. Interactive gaming connects different players together to play games with each other through the Internet. Files and pictures can be transferred from a source to a destination very quickly using IEEE 802.11n, due to its high user throughput. Similarly, it is also faster to backup and store data using IEEE 802.11n.

SUMMARY

We have focused mainly on the physical layer and the medium access control in IEEE 802.11n as well as IEEE 802.11e EDCA MAC. The enhanced MAC features in IEEE 802.11n are highlighted, and some of them are illustrated through diagrams. IEEE 802.11s is also described. Some related issues on network architecture, mobility radio management, qualities of service, and applications are discussed and explained. Besides Xiao's articles [5], there are other references for IEEE 802.11 MAC [6–9]. IEEE 802.11e MAC is also analyzed by Kong et al. [10]. Analysis and simulation on frame aggregation and reverse direction data flow protocol in IEEE 802.11n may be found in Lin and Wong's article [11].

IEEE 802.11a/b/g WLANs are already on the market. The next waves of WLAN technologies to watch in the market are IEEE 802.11n and IEEE 802.11s. IEEE 802.11n can support a data rate of up to 600 Mbps, while IEEE 802.11s can be used for wireless mesh networking. There is also an IEEE 802.11 very high throughput (VHT) study group that is looking at the feasibility of achieving gigabits per second MAC throughput. This would certainly be another WLAN direction to watch and participate in if a new IEEE 802.11 task group were formed.

REFERENCES

[1] IEEE 802.11, "Part 11: Wireless LAN medium access control (MAC) and physical layer (PYH) specifications," 1999.

[2] IEEE 802.11e, "Part 11: Wireless LAN medium access control (MAC) and physical layer (PYH) specifications," Nov. 2005.

[3] IEEE P802.11n/D2.07, "Part 11: Wireless LAN medium access control (MAC) and physical layer (PYH) specifications," Sept. 2007.

[4] IEEE P802.11s/D1.10, "Part 11: Wireless LAN medium access control (MAC) and physical layer (PYH) specifications," Mar. 2008.

[5] Y. Xiao, "Performance analysis of priority schemes for IEEE 802.11 and IEEE 802.11e wireless LANs," *IEEE Trans. Wireless Communi.*, vol. 4, no. 4, July 2005.

[6] G. Bianchi, "Performance analysis of the IEEE 802.11 distributed coordination function," *IEEE J. Sel. Areas Communi.*, vol. 18, no. 3, pp. 535–547, Mar. 2000.

[7] Y. C. Tay and K. C. Chua, "A capacity analysis of the IEEE 802.11 MAC protocol," *Wireless Networks*, vol. 7, pp. 159–171, 2001.

[8] E. Ziouva and T. Antonakopoulos, "CSMA/CA performance under high traffic conditions: throughput and delay analysis," *Comput. Communi.*, vol. 25, pp. 313–321, 2002.

[9] Y. Xiao and J. Rosdahl, "Throughput and delay limits of IEEE 802.11" *IEEE Communi. Lett.*, vol. 6, no. 8, Aug. 2002.

[10] Z. N. Kong, D. H. K. Tsang, and B. Bensaou, "Performance analysis of IEEE 802.11e contention-based channel access," *J. Sel. Areas Communi.*, vol. 22, no. 10, pp. 2095–2106, Dec. 2004.

[11] Y. Lin and V. W. S. Wong, "Frame aggregation and optimal frame size adaptation for IEEE 802.11n WLANs," *IEEE GLOBECOM 2006, Conference CD-ROM*, 2006.

CHAPTER 12

WIRELESS PERSONAL AREA NETWORKS

12.1 INTRODUCTION

Wireless personal area networks (WPANs) generally cover a small area with devices communicating among themselves. One such standardized WPAN is Bluetooth. It is also the IEEE 802.15.1 standard. Bluetooth can be configured with a master and up to seven slaves. Another standardized WPAN is IEEE 802.15.4, considered in Chapter 3. It is also known as ZigBee. However, the data rates in these WPANs are low. Bluetooth can support a data rate of up to 1 Mbps, while ZigBee can support a data rate of up to 250 kbps. Thus, another WPAN standard, known as WiMedia, is established recently in 2005. It is one of the two competing proposals in IEEE 802.15.3a. One of the proposals is based on direct-spread (DS) ultrawideband (UWB), while the other proposal is based on multiband orthogonal frequency-division multiplexing (MB-OFDM). IEEE 802.15.3a was disbanded after neither of the two groups could not get a majority vote for their proposals. The latter group joined WiMedia. WiMedia uses a multiband OFDM physical layer (PHY) and a distributed medium access control (MAC). The WiMedia specifications specify mainly only the PHY and the MAC. WiMedia has data rates of 53.3, 80, 106.7, 160, 200, 320, 400, and 480 Mbps. The first five data rates use quadrature phase-shift keying (QPSK) as the modulation method, while the latter three data rates use dual-carrier modulation (DCM). The coverage distance is about 10 m or less, depending on the data rates, distances between the transmitters and receivers, and channel conditions. In this chapter we focus on the

Wireless Broadband Networks, By David Tung Chong Wong, Peng-Yong Kong, Ying-Chang Liang, Kee Chaing Chua, and Jon W. Mark

high-data-rate WiMedia. One standardization use of WiMedia is in the wireless USB (WUSB) standard. WUSB uses a star topology with a host communicating with its devices. The WiMedia MAC has a superframe structure divided into 256 medium access slots (MASs). Each MAS is 256 μs, and each superframe is 65.536 ms. Three beacons can fit into one MAS. The superframe consists mainly of two parts. One is the beacon period, the other is the data transmission period. The beacon period is used by devices to transmit their beacons, and the data transmission period is used for actual data transfer. The WiMedia MAC has a reservation protocol for isochronous traffic and a contention-based protocol for nonisochronous traffic. The reservation protocol is known as the *distributed reservation protocol* (DRP), while the contention-based protocol is known as *prioritized channel access* (PCA). The distributed reservation protocol can be of the following type: hard, soft, private, and alien beacon period. In hard DRP, no one else except the device that reserves it can make use of its DRP transmission time. In soft DRP, other PCA devices can use the DRP transmission time if the owner is not using it. Private DRP is used by WUSB for data communications and data transfer. Alien beacon period DRP is used to protect nonoverlapping beacon periods when two groups of devices with different beacon periods merge. PCA is used in all transmission slots other than those occupied by DRPs and beacon period. PCA is similar to CSMA/CA used in IEEE 802.11e. Applications in WiMedia include connection of a personal computer or laptop to a printer, a storage device, a mobile phone, a MP3/4 player, a TV set, and so on. High-data-rate connectivity without wires is a big advantage of WiMedia and WUSB based on WiMedia.

In Section 12.2 we present a star topology network architecture for WiMedia as an example. The physical layer for WiMedia is explained in detail in Section 12.3, while the MAC for WiMedia is explained and analyzed in Section 12.4 together with WUSB. The approximate saturated throughputs and delays of WiMedia are derived, and extensive numerical results are presented in this section. In Section 12.5 we describe the merging protocol when two groups of WiMedia devices merge. In Section 12.6 we point out that no routing algorithm is specified in the WiMedia specifications, but routing algorithms can be applied in future applications due to the distributed MAC in WiMedia. The methods by which WiMedia provides qualities of service are explained in Section 12.7. In Section 12.8 we present the applications in WiMedia. The chapter is summarized at the end of this chapter.

12.2 NETWORK ARCHITECTURE

WiMedia can be used to connect different devices, such as mobile phone, printer, storage device, and MP3/4 player, to a laptop in a small wireless personal area network (WPAN) such as a piconet known in Bluetooth. A WPAN for WiMedia is shown in Figure 12.1. The connectivity between the laptop and the devices can be done using wireless USB. Wireless UWB makes use of a type of reservation known as the private distributed reservation protocol in WiMedia medium access control. The laptop can be connected to the local area network (LAN) via an access point. A

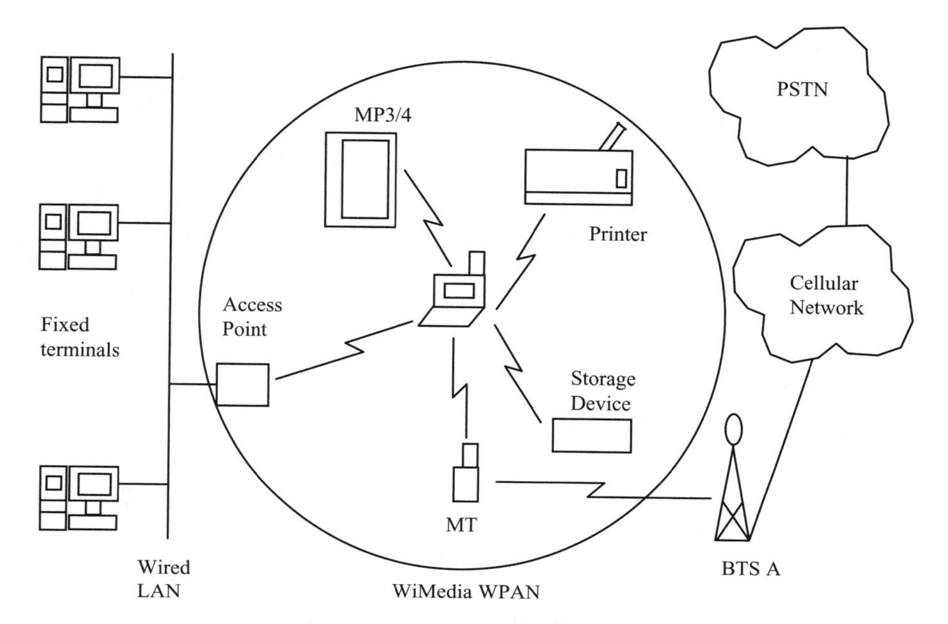

FIGURE 12.1 Star topology network architecture for WiMedia.

mobile phone can also be connected to a base station in a cellular network, which in turn is connected to a public-switched telephone network (PSTN). This example shows a star topology, but WiMedia does not need to be in this topology only. As it has a distributed medium access control, WiMedia can have other toplologies (e.g., those with mesh connectivity).

12.3 PHYSICAL LAYER

Multiband OFDM (MB-OFDM) is adopted as the physical layer air interface for WiMedia. Unlicensed UWB transmission is allowed in the 3.1- to 10.6-GHz band by the Federal Communications Commission (FCC). WiMedia divides this band into 14 subbands, each of which is 528 MHz wide. The center frequencies for these subbands are $f_i = (i - 1) \times 528 + 3432$ MHz, for $i = 1, \ldots, 14$. The subbands are divided further into four band groups of three bands each (band groups 1, 2, 3, and 4) and one band group of two bands (band group 5), as shown in Figure 12.2. Each beacon group is assigned a distinct hopping sequence among the two or three bands in the band group. This distinct hopping sequence is known as the *time frequency code* (TFC). An example of the hopping is shown in Figure 12.3. Each hop has a duration which includes one OFDM symbol plus a zero-padded suffix [1]. This pattern repeats every six OFDM symbols. The TFC and band group number in the

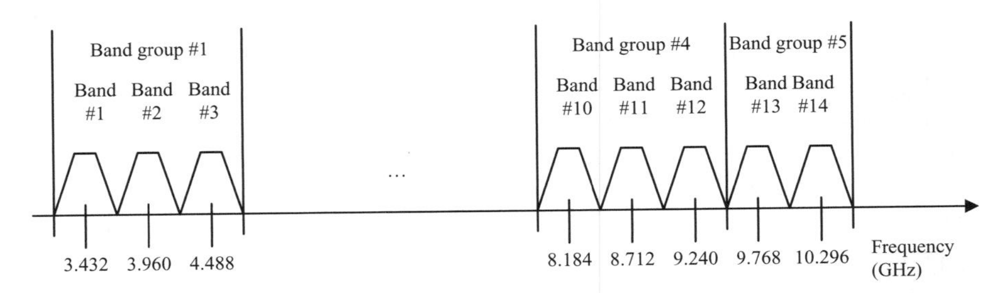

FIGURE 12.2 Subband division for the frequency range from 3.1 to 10.6 GHz.

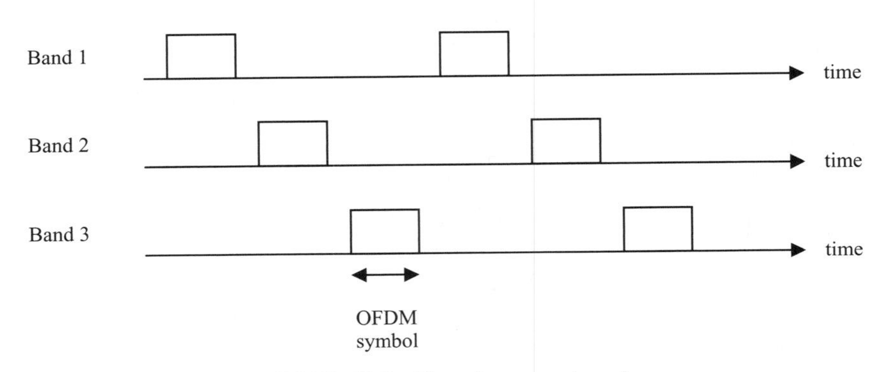

FIGURE 12.3 Time-frequency hopping.

beacon group are determined by the medium access control (MAC) depending on device capabilities and channel occupancy [2]. The TFCs for the five band groups are shown in Table 12.1. The numbers for each TFC in the band groups represent the hopping sequence.

12.3.1 MB-OFDM Transmitter

A functional block diagram of the MB-OFDM transmitter [3] is shown in Figure 12.4. The payload bits and headers are first encoded and possibly spread in a frequency-domain spreader (FDS). The output is then modulated and multiplexed with the frequency-domain preambles. The multiplexed output is then passed to the inverse fast Fourier transform (IFFT) to generate time-domain symbols, which are further multiplexed with time-domain preambles. Then zero suffix and guard interval (GI) are inserted, and the output is passed to a parallel-to-serial (P/S) converter and possibly to a time-domain spreader ((TDS). Finally, these samples are passed to a digital-to-analog converter (DAC) and up-converted to a carrier frequency f_c based on the TFC. It is pointed out that the OFDM used in WiMedia is based on a zero suffix instead of cyclic frequency as used in conventional OFDM.

TABLE 12.1 TFCs for Five Band Groups

Band Group Number	TFC Number	Band Number					
1	1	1	2	3	1	2	3
	2	1	3	2	1	3	2
	3	1	1	2	2	3	3
	4	1	1	3	3	2	2
	5	1	1	1	1	1	1
	6	2	2	2	2	2	2
	7	3	3	3	3	3	3
2	1	4	5	6	4	5	6
	2	4	6	5	4	6	5
	3	4	4	5	5	6	6
	4	4	4	6	6	5	5
	5	4	4	4	4	4	4
	6	5	5	5	5	5	5
	7	6	6	6	6	6	6
3	1	7	8	9	7	8	9
	2	7	9	8	7	9	8
	3	7	7	8	8	9	9
	4	7	7	9	9	8	8
	5	7	7	7	7	7	7
	6	8	8	8	8	8	8
	7	9	9	9	9	9	9
4	1	10	11	12	10	11	12
	2	10	12	11	10	12	11
	3	10	10	11	11	12	12
	4	10	10	12	12	11	11
	5	10	10	10	10	10	10
	6	11	11	11	11	11	11
	7	12	12	12	12	12	12
5	5	13	13	13	13	13	13
	6	14	14	14	14	14	14

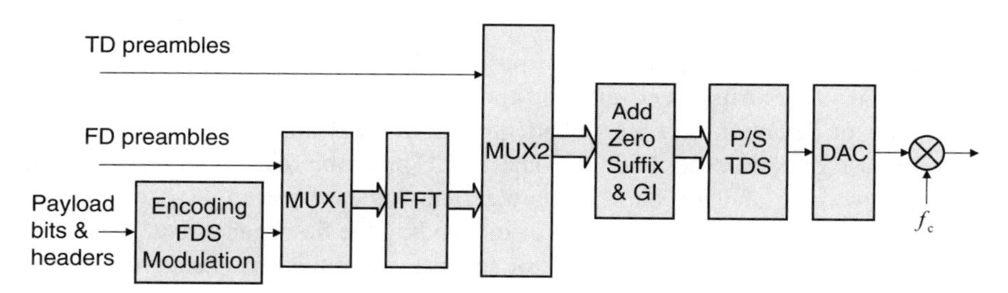

FIGURE 12.4 Block diagram of an MB-OFDM transmitter.

TABLE 12.2 MB-OFDM Parameters

Parameter	Value
Number of total subcarriers, N_{FFT}	128
Number of data subcarriers, N_{D}	100
Number of pilot subcarriers, N_{P}	12
Number of guard subcarriers, N_{G}	10
Number of null subcarriers, N_{NULL}	6
Sampling frequency, f_s	528 MHz
Subcarrier spacing, Δ_f (= f_s/N_{FFT})	4.125 MHz
FFT/IFFT period, T_{FFT} (= $1/\Delta_F$)	242.42 ns
Zero suffix duration, T_{ZP} (= $32/f_s$)	60.61 ns
Guard interval duration, T_{GI} (= $5/f_s$)	9.47 ns
OFDM symbol duration, T_{SYM} (= $T_{\mathrm{FFT}} + T_{\mathrm{ZP}} + T_{\mathrm{GI}}$)	312.5 ns
OFDM symbol rate, f_{SYM} (= $1/T_{\mathrm{SYM}}$)	3.2 MHz

12.3.2 MB-OFDM Parameters

Table 12.2 lists the key parameters related to the MB-OFDM system. The sampling frequency is 528 MHz and the IFFT size is 128. Of the 128 subcarriers, there are 100 data subcarriers, 12 pilot subcarriers, 10 guard subcarriers, and 6 NULL subcarriers. Both the guard subcarriers and the NULL subcarriers are allocated at the edge of the band for relaxing filter requirements [1]. Note that the dc value is also set to zero. The 12 pilot subcarriers are allocated uniformly across the band for the purpose of tracking the frequency and phase. The OFDM symbol interval is 312.5 ns, both IFFT and FFT with periods of 242.42 ns and a zero-padded suffix, and a guard interval period of 70.08 ns.

12.3.3 PLCP Layer

WiMedia has a physical protocol data unit (PPDU) which consists of the physical layer convergence protocol (PLCP) preamble, PLCP header, and the physical service data unit (PSDU). The PPDU frame structure is shown in Figure 12.5.

12.3.3.1 PLCP Preamble The purpose of the PLCP preamble is to assist in synchronization, carrier offset recovery, and channel estimation [1]. The PLCP preamble allows packet detection and piconet identification. A distinct PLCP preamble sequence is assigned to each TFC [1]. The PLCP preamble can be divided into the standard preamble and the burst preamble. The standard preamble is for sending standard transmission with one packet or for sending the first packet in a burst mode, while the burst preamble can be used for sending subsequent packets after the first packet transmission. The duration of the standard preamble is 9.375 μs, while the duration of the burst preamble is 5.625 μs. Minimum interframe spaces (MIFSs) can be used between packets that are transmitted in a burst mode, and an MIFS is shorter

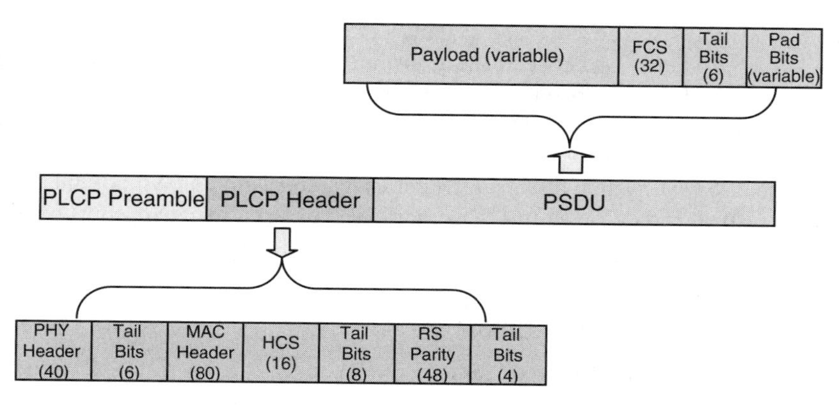

FIGURE 12.5 PPDU frame structure.

than a short interframe space (SIFS). As shown in Figure 12.4, the PLCP preamble consists of a time-domain (TD) portion and a frequency-domain (FD) portion.

12.3.3.2 PLCP Header The purpose of the PLCP header is to convey information on both the physical layer and the MAC so as to help in decoding the PSDU at the receiver. The PLCP header is transmitted with the lowest rate (53.3 Mbps). Both the PHY and MAC headers are embedded in the PLCP header. Reed–Solomon parity bits are also added to the PLCP header to increase the robustness of the PLCP header reception [1].

12.3.3.3 PSDU WiMedia provides eight PSDU data rates, ranging from 53.3 to 480 Mbps (Table 12.3). The first five data rates use quadrature phase-shift keying (QPSK) as the modulation method, and the latter three data rates use dual-carrier

TABLE 12.3 MB-OFDM PSDU Date Rates

Date Rate (Mbps)	Modulation	Coding Rate	2X FDS	2X TDS	Coded Bits/6 OFDM Symbols	Info Bits/6 OFDM Symbols
53.3	QPSK	$\frac{1}{3}$	Yes	Yes	300	100
80	QPSK	$\frac{1}{2}$	Yes	Yes	300	150
106.7	QPSK	$\frac{1}{3}$	No	Yes	600	200
160	QPSK	$\frac{1}{2}$	No	Yes	600	300
200	QPSK	$\frac{5}{8}$	No	Yes	600	375
320	DCM	$\frac{1}{2}$	No	No	1200	600
400	DCM	$\frac{5}{8}$	No	No	1200	750
480	DCM	$\frac{3}{4}$	No	No	1200	900

modulation (DCM). The coverage distance is about 10 m or less, depending on the data rates, distances between the transmitters and receivers, and the channel conditions. The lower the band group frequency, the longer the range.

The basic channel encoder is a rate-1/3 convolutional encoder. The other three rates (1/2, 3/4, and 5/8) are derived through puncturing. The coded and punctured bits are interleaved using a three-stage interleaver, consisting of symbol interleaver, tone interleaver, and cyclic shifter. The interleaved bits are then mapped using either Gray-mapped QPSK modulation or dual carrier modulation (DCM).

12.3.4 Dual Carrier Modulation

The basic idea behind DCM is as follows:

- To map four interleaved bits onto two 16-point symbols using two fixed but different mappings. This yields a 16-QAM-like constellation.
- To map the resulting two 16-point symbols onto two different IFFT tones separated by 50 tones.

By doing so, the same 4 bits of information are mapped onto two tones that are separated by at least 200 MHz; thus, frequency diversity can be achieved. This is because the probability that there is a deep fade on both tones is quite small. The DCM is applied to the modes at a rate of 320 Mbps or higher. The process for DCM is as follows. For four interleaved bits $\{u_{1,n},\ u_{2,n},\ u_{3,n},\ u_{4,n}\}$, we first convert them to bipolar symbols

$$x_{i,n} = 2u_{i,n} - 1 \tag{12.1}$$

for $i = 1, 2, 3, 4$. Then the outputs of DCM is given by

$$\begin{bmatrix} y_{1,n} \\ y_{2,n} \end{bmatrix} = \frac{1}{\sqrt{10}} \begin{bmatrix} 2 & 1 \\ 1 & -2 \end{bmatrix} \begin{bmatrix} x_{1,n} + jx_{3,n} \\ x_{2,n} + jx_{4,n} \end{bmatrix}. \tag{12.2}$$

$y_{1,n}$ and $y_{2,n}$ in fact take 16-QAM constellations. They are then transmitted over two subcarriers separated by 50 tones.

12.3.5 Time- and Frequency-Domain Spreading

To support lower data rates achieving diversity, frequency-domain spreading (FDS) and time-domain spreading (TDS) are applied. For FDS, the same symbol is transmitted over two subcarriers widely separated but within the same OFDM symbol. For TDS, the information contained in one OFDM symbol is also transmitted in the subsequent OFDM symbol. Thus, the spreading gain is 2. FDS is used for rates of 53.3 and 80 Mbps, while TDS is used for rates of 200 Mbps and lower.

12.4 MEDIUM ACCESS CONTROL

A recently standardized distributed medium access control (MAC) is the WiMedia MAC. It is formerly known as the multiband OFDM alliance (MBOA) MAC [1]. The WiMedia MAC uses a reservation-based protocol and a contention-based protocol. The reservation-based protocol is known as distributed reservation protocol (DRP), while the contention-based protocol is known as prioritized contention access (PCA).

12.4.1 WiMedia MAC Superframe Format

The superframe format is shown in Figure 12.6. Each superframe consists of two parts: a beacon period (BP) subframe and a data transmission subframe. The beacon period subframe consists of beacon slots for existing devices in the system. These beacons are packed at the beginning of the beacon period subframe after the N_{BSig} signaling beacon slots, while the latter slots and N_{BExt} extension beacon slots are used for new devices joining the system using a beacon collision resolution protocol. After successful entry into the system, the new devices are packed together. The number of signaling and extension beacon slots is assumed constant. When devices leave the system, a packing protocol is used to pack the remaining devices' beacon slots together. Thus, the beacon period is not fixed but varies according to the number of devices in the system. The beacon period subframe is assumed to be an integer number of medium access slots (MASs).

When a device starts up, it will scan the channels to detect for existing BPs. If a BP is detected, the device will associate itself with the BP. If a BP is not detected, the device will start its own BP by sending out the beacon period start time (BPST). The first two beacon slots in the BP are used for signaling. The device that starts the BP transmits its beacon in the third beacon slot. If there are a number of devices, they will occupy beacon slots after the first device's beacon. After these beacons, there are a number of extension beacon slots for new devices to join the BP. If a new device chooses a beacon slot that is beyond the BP, it will send a copy of its beacon to the signaling beacon slot to inform other devices to extend their BP. When two

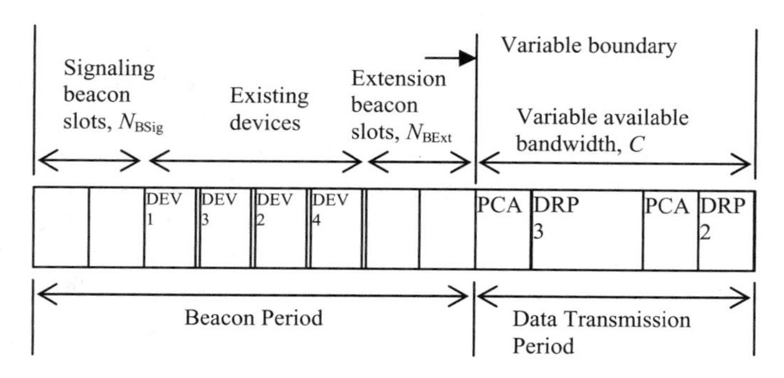

FIGURE 12.6 WiMedia MAC superframe format. (From [9] © 2006, IEEE.)

devices select the same beacon slot after the existing device or when their beacons in the signaling beacon slots collide, a beacon collision resolution rules is used to resolve the collision. Once the beacon slot is secured in the extension beacon slots, the device will move to the first beacon slot after the highest occupied beacon slots of the existing devices.

The data transmission period is used to transmit data packets whose data reservations are announced in its device beacon slot. This is called the distributed reservation protocol (DRP) [1]. DRP can be explicit, using command frames, or implicit, through the DRP information elements (IEs) in the beacons. The source device first sends a reservation request to destination device. Command frame can be sent using DRP or PCA, while DRP IE can be set to indicate the reservation request in the device's beacon. The destination device will check for the channel availability and reply to the source device using either a command frame or its DRP IE. If the reservation request cannot be accepted, the destination device can send the information on the available slots to the source device. Once the reservation request is accepted, the reservation is announced in the DRP IEs of the devices' beacons. The other devices are informed of the reservation by listening to these two devices' beacons, and the other devices do not transmit during the reserved DRP period. DRP transmission need not be in the same order as the devices in the beacon slots, and the DRP packets for each device also need not be transmitted immediately after other DRP packets. The number of data packets for each device to transmit is not fixed but can vary. Note that all devices announce their data reservations, and each device's beacon slot contains information on all other devices [1]. Since the beacon period varies, the available number of data packet slots for PCA transmissions, C, also varies, depending on the number of active devices in the system. In Figure 12.6, active devices 2 and 3 are transmitting DRP packets, while PCA devices 1 and 4 can transmit in the PCA periods. PCA is similar to IEEE 802.11e using CSMA/CA, except that sufficient time to transmit a packet must be available before the next DRP block or BP, and there is a timeout for retrying to transmit. The timeout for retrying has an effect similar to having a retry limit.

Hard and soft DRPs differ from each other in that no one else can use the reserved slots for hard DRPs, while PCA can use the reserved slots for soft DRPs if they are not used. Therefore, there are more time slots for PCA usage under soft DRPs than that under hard DRPs. In other words, soft DRPs allow PCA to share its reserved slots if they are not utilized, while hard DRPs allow no sharing of their reserved slots at all.

There are four traffic classes (access categories) in PCA, and one method of differentiating the priority of each class is based on its arbitration interframe space (AIFS). The shorter the AIFS and the smaller the minimum and maximum contention window sizes of the traffic class, the higher the priority. The shortest AIFS of the traffic classes contending for channel access will gain access to the channel faster. Figure 12.7 shows the channel access for IEEE 802.11e. Table 12.4 shows the AIFS for background, best effort, video, and voice traffic. A key element in PCA, like IEEE 802.11 and IEEE 802.11e, is the backoff counter. The initial value in the backoff counter is set between zero and the minimum contention window size. The backoff

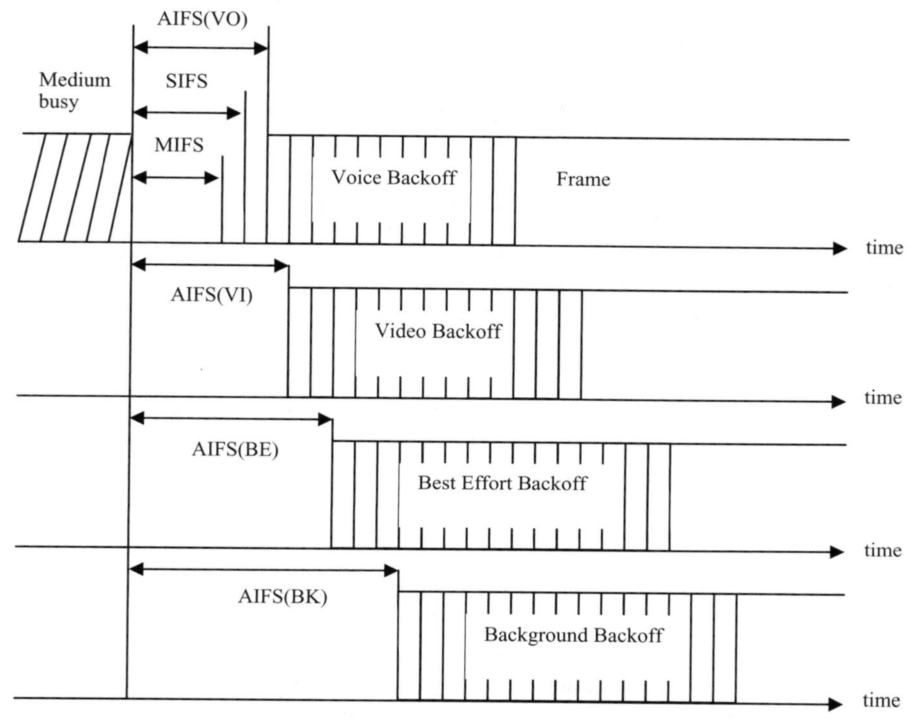

FIGURE 12.7 Channel access in WiMedia PCA MAC. (From [9] © 2006, IEEE.)

counter value will decrement if the channel is idle and will freeze if the channel is busy. If the backoff counter value is zero, the PCA device attempts to transmit its packet. If it is not successful, the new contention window size will be double that of the previous window size plus one. The backoff counter will then randomly choose a value in [0, new contention window size]. After each backoff stage, this is repeated if the attempt for transmission of a packet is not successful until the retry limit for the backoff stages is reached. At this stage, the packet will be discarded if

TABLE 12.4 AIFS and Minimum and Maximum Contention Window Sizes for PCA in WiMedia MAC

Traffic Class	AIFS (μs)	CW_{min}	CW_{max}
Background	73	15	1023
Best effort	46	15	1023
Video	28	7	511
Voice	19	3	255

its transmission is still not successful. If the packet transmission is successful at any backoff stage or when it is at the retry backoff stage, a new packet will be selected for new transmission at backoff stage 0. The minimum and maximum contention window sizes for differentiating priority in each class are also shown in Table 12.4.

12.4.2 Physical Layer Protocol Data Unit Structure

The PPDU structure for WiMedia is shown in Figure 12.8. The PLCP preamble has two types: the standard PLCP preamble and the burst PLCP preamble. The former is used for standard transmission and the latter is used for burst mode transmission. In standard transmission, there is only one packet. In burst mode transmission, multiple packets are transmitted when the device gets hold of the PCA channel. The first frame uses a standard PLCP preamble, while the subsequent packets use the burst PLCP header, which is shorter than the standard PLCP header. Minimum interframe spaces (MIFSs) are used to separate the packets rather than using short interframe space (SIFS). MIFS is shorter than SIFS. The PLCP preamble is followed by the PLCP header. Note that the MAC header in WiMedia is contained in the PLCP header. This is a big difference from the IEEE 802.11 WLAN MAC frame format. The PLCP header is followed by the physical layer data unit (PSDU). The PSDU consists of the frame payload of 0 to 4095 bytes. This is equivalent to the frame body in the IEEE 802.11 WLAN MAC frame format. The frame payload is followed by the frame check sequence (FCS), tail bits, and paddings. Note that the PLCP header is transmitted at 39.4 Mbps, while the PSDU is transmitted at one of the eight data rates of {53.3, 80, 106.7, 160, 200, 320, 400, 480} Mbps.

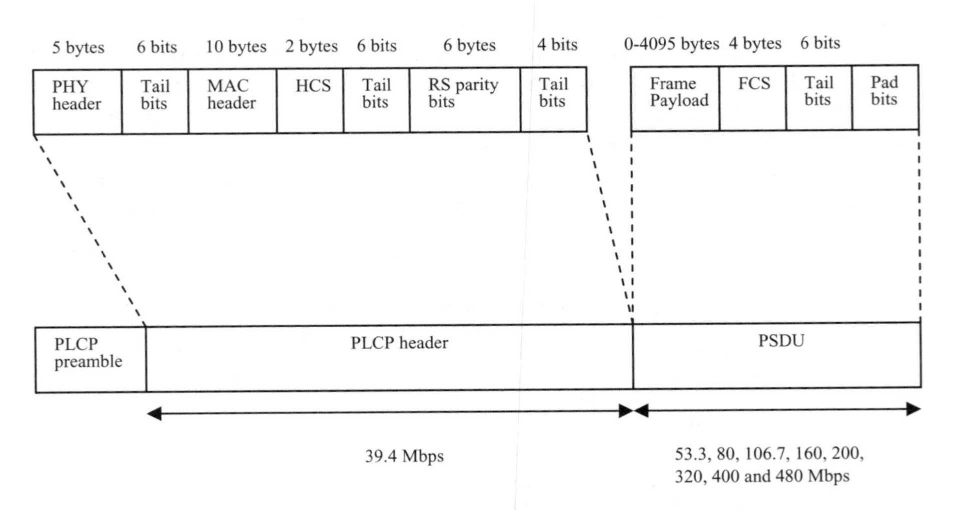

FIGURE 12.8 WiMedia PPDU structure.

12.4.3 Aggregation

A device can aggregate multiple MAC service data units (MSDUs) into one frame for transmission to another device. This cuts down on individual access times and acknowledgments, and interframe spaces, thereby increasing user throughput.

12.4.4 Acknowledgment Policies

There are three acknowledgment policies in WiMedia MAC: no acknowledgment, immediate acknowledgment, and block acknowledgment. No acknowledgment is suitable for real-time traffic. Immediate acknowledgment is the usual acknowledgment after a frame is transmitted. Block acknowledgment is used for acknowledging multiple frames.

12.4.5 Power-Saving Modes

The WiMedia MAC has two power-saving modes: active and hibernation. In the active mode, a device transmits its own beacon and listens to other beacons during the beacon period of a superframe. If it is not expecting any transmission in the superframe, it can go to sleep during the data transmission period of that superframe. In the hibernation mode, a device can go to sleep for a number of superframes. It does not participate in any beacon transmission or reception in the beacon periods in between the superframes when it is awake. Hibernation and the number of superframes that a device will hibernate are indicated in the hibernation mode IE in the beacon.

12.4.6 Security

WiMedia uses a four-way handshaking procedure for two devices to establish pairwise temporal keys (PTKs) and a secure relationship [1]. A device may solicit or distribute group temporal keys (GTKs) within a secure relationship [1]. The security mechanisms in WiMedia control the security operation of devices by setting appropriate security modes [1]. They allow devices to authenticate each other, to derive PTKs, and to establish secure relationships [1]. They also enable devices to solicit or distribute GTKs within established secure relationships [1].

12.4.7 WiMedia MAC Analysis

WiMedia MAC uses DRP and PCA to oversee access control. PCA in WiMedia is similar to IEEE 802.11e for multiple prioritized classes [4] using carrier-sense multiple access with collision avoidance (CSMA/CA). In traditional IEEE 802.11 wireless local area networks (WLANs), many of the vendors only implement the distributed coordination function (DCF) using CSMA/CA, not the point coordination function (PCF). Thus, saturated throughputs of IEEE 802.11 and IEEE 802.11e have been analyzed in the literature, assuming CSMA/CA as the access control protocol [5–7]. However, DRP in WiMedia is expected to be used for streaming and PCA is expected to be used in the presence of DRPs and the beacon period (BP). Furthermore, a device will not transmit if the time remaining to the start of the next DRP or BP is insufficient to transmit a packet for PCA transmission. Thus, this section extends the saturated throughput analysis of IEEE 802.11e to that of PCA in the presence of DRPs and BP in WiMedia MAC. The saturated delay is considered as well.

The approximate analyses to be presented in this section follow those in [8,9]. The analysis here has the same assumption of assuming the same AIFS for all access categories. However, it still accounts for the variation of the contention window size and the backoff process as in [8]. Nevertheless, the analytical results need to be verified by simulation results.

Let N be the number of the priority in a given PCA device, where the ith priority class is given by the index $i = 0, 1, \ldots, N - 1$. Let $a(i,t)$ be a random process representing the backoff stage j, $j = 0, 1, \ldots, L_{i,\text{retry}}$, at time t, where $L_{i,\text{retry}}$ is the retry limit. Let $b(i,t)$ be a random process representing the value of the backoff counter at time t. Let $c(i,t)$ be a random process representing the remaining time to the next DRP period or BP, which is insufficient to transmit a packet at time t, including the DRP period or BP. The value of the backoff counter $b(i,t)$ is chosen uniformly from the range of $[0, 1, \ldots, W_{i,j}]$, where $W_{i,j} = 2W_{i,j-1} + 1$ for $j \leq m_i \leq L_{i,\text{retry}}$ and $W_{i,j} = \text{CW}_{\max,i}$ if $m_i < j \leq L_{i,\text{retry}}$.

$$W_{i,j} = \begin{cases} (W_i + 1)2^j - 1, & 0 \leq j \leq m_i \\ (W_i + 1)2^{m_i} - 1, & m_i < j \leq L_{i,\text{retry}}, \end{cases} \tag{12.3}$$

where $W_i = W_{i,0}$. Let p_i denote the probability that a transmitted frame collides in a PCA period; it is also the probability that a device in a backoff stage for the ith class senses a busy channel in a PCA period. Let N_{BP} be the number of MASs occupied by the BP; it is given by

$$N_{\text{BP}} = \left\lceil \left(n + \sum_{i=0}^{N-1} n_i + N_{\text{BSig}} + N_{\text{BExt}} \right) \Big/ N_{\text{BS}} \right\rceil, \qquad n + \sum_{i=0}^{N-1} n_i \leq N_D, \tag{12.4}$$

where n is the number of hard DRP devices and n_i is the number of class i PCA devices, N_{BS} is the number of beacon slots equivalent to one MAS, and N_D is the maximum number of devices in a BP. The symbol $\lceil x \rceil$ denotes the smallest integer greater than or equal to x. Assuming that each hard DRP device uses one MAS, the number of MASs occupied by n hard DRP devices, denoted by N_{DRP}, is given by

$$N_{\text{DRP}} = n. \tag{12.5}$$

Let N_{SF} be the number of MASs in a superframe. The number of MASs for PCA, denoted by N_{PCA}, is therefore given by

$$N_{PCA} = N_{SF} - N_{BP} - N_{DRP}. \tag{12.6}$$

Furthermore, the probability of being in the BP, DRP, and PCA, denoted by P_{BP}, P_{DRP}, and P_{PCA}, respectively, are given by

$$P_{BP} = \frac{N_{BP}}{N_{SF}}, \tag{12.7}$$

$$P_{DRP} = \frac{N_{DRP}}{N_{SF}}, \tag{12.8}$$

$$P_{PCA} = \frac{N_{PCA}}{N_{SF}}. \tag{12.9}$$

Let C be the number of available MASs for DRP transmissions given by

$$C = C\,(n, n_0, \ldots, n_{N-1}) = N_{SF} - N_{BP}. \tag{12.10}$$

With the assumption that each hard DRP device uses one MAS, the number of combinations for the occupancy of the n DRP transmissions in C MASs is $\binom{C}{n}$, where $\binom{C}{n} = \frac{C!}{(C-n)!\,n!}$. Let y be the number of consecutive DRP MASs and weight with the length of y over the total DRP transmission slots, n, for each combination. Then summing the probability for all cases and combinations, it can be shown that the probability of the number of consecutive DRP MASs, denoted by P_y, is given by

$$P_y = \frac{(C-n+1)y}{n}\binom{C-y-1}{n-y}\Big/\binom{C}{n}, \qquad y = 1, \ldots, n. \tag{12.11}$$

We assume that all backoff counters freeze during these consecutive DRP transmissions and that a guard time is available at the end of the DRP block. Similarly, let x be the number of consecutive PCA MASs not used by DRP transmissions, and weight with the length of x over the total PCA transmission slots, $C - n$, for each combination. Then summing the probability for all cases and combinations, it can be shown that the probability of the number of consecutive PCA MASs, denoted by P_x, is given by

$$P_x = \frac{(n+1)x}{C-n}\binom{C-x-1}{C-n-x}\Big/\binom{C}{n}, \qquad x = 1, \ldots, C-n. \tag{12.12}$$

Let T_Q be the time required to transmit a PCA packet successfully, including header, payload, SIFS, ACK, SIFS, and guard time, T_G, at the point of decision. If the remaining time before the next DRP transmission or BP is greater than T_Q when the backoff counter is zero, the packet can attempt transmission in a PCA period;

otherwise, it will not be transmitted. Let r and q denote, respectively, the probability that the packet can and cannot be transmitted in a PCA period before the next DRP transmission or BP, when the backoff counter is zero. r is given by

$$r = \sum_{x=1}^{C-n} P_x \frac{\max(0,\ xT_{\text{MAS}} - T_Q)}{xT_{\text{MAS}}} P_{\text{PCA}}, \tag{12.13}$$

and q is given by

$$q = \left(1 - \sum_{x=1}^{C-n} P_x \frac{\max(0, xT_{\text{MAS}} - T_Q)}{xT_{\text{MAS}}}\right) P_{\text{PCA}}, \tag{12.14}$$

where T_{MAS} is the duration of a MAS . Note that $r + q = P_{\text{PCA}}$. Let δ denote a time slot in the backoff counter. The duration of the BP in terms of δ, denoted by D_{BP}, is given by

$$D_{\text{BP}} = \left\lfloor \frac{N_{\text{BP}} T_{\text{MAS}}}{\delta} \right\rfloor. \tag{12.15}$$

The duration of the number of consecutive DRP MASs in terms of δ, denoted by D_y, is given by

$$D_y = \left\lfloor \frac{yT_{\text{MAS}}}{\delta} \right\rfloor. \tag{12.16}$$

The average duration Q in terms of δ, that a packet cannot be transmitted in a PCA period before the next DRP transmission or BP when the backoff counter is zero is given by

$$Q = \left\lfloor \frac{\frac{1}{2} T_Q}{\delta} \right\rfloor, \tag{12.17}$$

where $\lfloor x \rfloor$ is the greatest integer smaller than or equal to x.

Let P_{D_y} be the probability of the number of consecutive DRP MASs, given by

$$P_{D_y} = P_y P_{\text{DRP}}, \tag{12.18}$$

and $\sum_{y=1}^{n} P_{D_y} = P_{\text{DRP}}$. Note that the probability q is associated with the next DRP transmission or the next BP when the backoff counter is zero. Therefore, no packet transmission in a PCA period is governed by two events: the next DRP transmission or the next BP when the backoff counter is zero. The duration of no packet transmission in a PCA period is thus either $Q + D_y$ or $Q + D_{\text{BP}}$. Thus, the probability of the duration $Q + D_y$ is given by

$$P_{Q+D_y} = qP_{Qy} + P_{D_y}, \tag{12.19}$$

and the probability of the duration $Q + D_{\text{BP}}$ is given by

$$P_{Q+\text{BP}} = qP_{Q,\text{BP}} + P_{\text{BP}}, \tag{12.20}$$

where

$$P_{Qy} = \frac{P_{D_y}}{P_{\text{BP}} + P_{\text{DRP}}}, \tag{12.21}$$

$$P_{Q,\text{BP}} = \frac{P_{\text{BP}}}{P_{\text{BP}} + P_{\text{DRP}}}. \tag{12.22}$$

Note that $\sum_{y=1}^{n} qP_{Q_y} + qP_{Q,\text{BP}} = q$ and $\sum_{y=1}^{n} P_{Q+D_y} + P_{Q+\text{BP}} = q + P_{\text{BP}} + P_{\text{DRP}}$.

The tridimensional random process $\{a(i,t), b(i,t), c(i,t)\}$ is a discrete-time Markov chain. Therefore, the state of each PCA device in the ith priority class is described by $\{i,j,k,l\}$, where $j \in \{0,1,\ldots,L_{i,\text{retry}}\}$ is the backoff stage, $k \in \{0,1,\ldots,W_{ij}\}$ in slots is the backoff delay, $l \in \{0,1,\ldots,D_y\}$ or $\{0,1,\ldots,D_{BP}\}$ is respectively the remaining time in the DRP period or BP, including the time that a PCA packet has sufficient time to transmit before each of these periods and takes values from $\{0,1,\ldots,Q+D_n\}$ if $D_n > D_{BP}$ or $\{0,1,\ldots,Q+D_{BP}\}$ if $D_n \leq D_{BP}$. Note that a time slot is smaller than a guard time.

The state transition diagram of the discrete-time Markov chain for the ith priority class is shown in Figure 12.9. For compactness, $L_{i,\text{retry}}$ is denoted in the figure as L. Let $b_{i,j,k,l} = \lim_{t\to\infty} \Pr[a(i,t) = j, b(i,t) = k, c(i,t) = l]$ be the stationary distribution

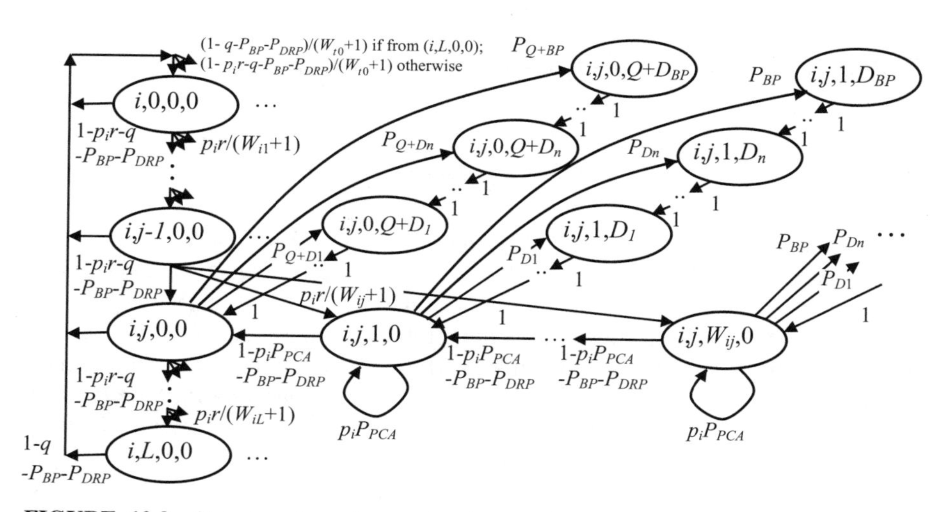

FIGURE 12.9 State transition diagram for the ith priority class using PCA in the presence of DRP transmissions and BP. (From [9] © 2006, IEEE.)

of the Markov chain. The nonnull transition probabilities are as follows:

$$\Pr[(i,0,k,0)|(i,j,0,0) = \frac{1-p_i r - q - P_{\text{BP}} - P_{\text{DRP}}}{W_{i,0}+1}, \quad 0 \le k \le W_{i,0}, \quad 0 \le j < L_{i,\text{retry}}, \quad l = 0$$

$$\Pr[(i,0,k,0)|(i,L_{i,\text{retry}},0,0) = \frac{1 - q - P_{\text{BP}} - P_{\text{DRP}}}{W_{i,0}+1}, \quad 0 \le k \le W_{i,0}, \quad j = L_{i,\text{retry}}, \quad l = 0$$

$$\Pr[(i,j,k,0)|(i,j,k,0) = p_i P_{\text{PCA}}, \quad 1 \le k \le W_{i,j}, \quad 0 \le j \le L_{i,\text{retry}}, \quad l = 0$$

$$\Pr[(i,j,k,0)|(i,j,k+1,0) = 1 - p_i P_{\text{PCA}} - P_{\text{BP}} - P_{\text{DRP}}, \quad 0 \le k \le W_{i,j} - 1, \quad 0 \le j \le L_{i,\text{retry}}, \quad l = 0$$

$$\Pr[(i,j,k,0)|(i,j-1,0,0) = \frac{p_i r}{W_{i,j}+1}, 0 \le k \le W_{i,j}, \quad 0 \le j \le L_{i,\text{retry}}, \quad l = 0$$

$$\Pr[(i,j,k,l)|(i,j,k,0) = P_{Q+\text{BP}}, k = 0, 0 \le j \le L_{i,\text{retry}}, \quad l = Q + D_{\text{BP}}$$

$$\Pr[(i,j,k,l)|(i,j,k,0) = P_{Q+D_y}, k = 0, \quad 0 \le j \le L_{i,\text{retry}}, \quad l = Q + D_y, \quad y = 1, \ldots, n$$

$$\Pr[(i,j,k,l)|(i,j,k,0) = P_{\text{BP}}, 1 \le k \le W_{i,j}, \quad 0 \le j \le L_{i,\text{retry}}, \quad l = D_{\text{BP}}$$

$$\Pr[(i,j,k,l)|(i,j,k,0) = P_{D_y}, 1 \le k \le W_{i,j}, \quad 0 \le j \le L_{i,\text{retry}}, \quad l = D_y, \quad y = 1, \ldots, n$$

$$\Pr[(i,j,k,l)|(i,j,k,l+1) = 1, k = 0, \quad 0 \le j \le L_{i,\text{retry}}, \quad l = 0, 1, \ldots, \max(Q + D_n, Q + D_{\text{BP}})$$

$$\Pr[(i,j,k,l)|(i,j,k,+1) = 1, 1 \le k \le W_{i,j}, \quad 0 \le j \le L_{i,\text{retry}}, \quad l = 0, 1, \ldots, \max(D_n, D_{\text{BP}})$$

The first equation above represents the transition probability to each state in backoff stage 0 from backoff stage $j, 0 \le j < L_{i,\text{retry}}$, while the second equation represents the transition probability to each state in backoff stage 0 from backoff stage $L_{i,\text{retry}}$. The third equation represents the self-transition probability in a state due to the channel being busy during a PCA period, while the fourth equation represents the backoff counter decrementing by one as the channel is not busy during a PCA period or it is not a DRP period or a BP. The fifth equation represents the transition probability to each state in backoff stage j. In this equation, there is sufficient time to transmit a packet before a BP or DRP transmission, but the channel is busy. That is why it is associated with the probability r. The sixth equation represents the transition probability to the $(Q + \text{BP})$ state due to the average time required to transmit a packet before a BP (Q) when the backoff counter is zero is insufficient and in the BP, while the seventh equation represents the transition probability to the (Q + DRP) state due to the average time required to transmit a packet before DRP transmissions (Q) when the backoff counter is zero is insufficient and the number of transmission packets is y. In these two equations, a packet cannot be transmitted before the BP or DRP transmissions. Thus, these equations are associated with probability q. The eighth equation represents the transition probability to the BP state due to a BP when the backoff counter is *not* zero, while the ninth equation represents the transition probability to the DRP state due to the DRP transmissions when the backoff counter is *not* zero and the number of transmission packets is represented by y. The tenth equation represents the remaining slot times that the Markov chain is in the BP or DRP plus Q when the backoff counter is zero, while the eleventh equation represents the remaining slot times that the Markov chain is in the BP or DRP when the backoff counter is *not* zero.

In steady state, we can derive the following relationships through chain regularities:

$$b_{i,j,0,0} = \left(\frac{p_i r}{1 - q - P_{\mathrm{BP}} - P_{\mathrm{DRP}}}\right)^j b_{i,0,0,0}, \quad 1 \le j \le L_{i,\mathrm{retry}}, \quad k = 0, \quad l = 0. \tag{12.23}$$

For $D_{\mathrm{BP}} > D_n$, we have (12.24), while for $D_{\mathrm{BP}} \le D_n$, we have (25), where $D_0 = 0$ and $s = p_i r/(1 - q - P_{\mathrm{BP}} - P_{\mathrm{DRP}})$.

$$b_{i,j,k,l} = b_{i,0,0,0}\frac{W_{i,j} - k + 1}{W_{i,j} + !}$$
$$\begin{cases} (s)^j \begin{cases} P_{Q+\mathrm{BP}}, & 0 \le j \le L_{i,\mathrm{retry}}, \ k = 0, \ l = Q + D_n + 1, \ldots, Q + D_{\mathrm{BP}} \\ \sum_{z=y}^{n} P_{Q+D_z} + P_{Q+\mathrm{BP}}, & 0 \le j \le L_{i,\mathrm{retry}}, \ k = 0, \\ \quad l = Q + D_{y-1} + 1, \ldots, Q + D_y, & y = 1, \ldots, n \\ \sum_{z=y}^{n} P_{Q+D_z} + P_{Q+\mathrm{BP}}, & 0 \le j \le L_{i,\mathrm{retry}}, \ k = 0, \\ \quad l = 1, \ldots, Q, & y = 1 \\ 1, & 0 \le j \le L_{i,\mathrm{retry}}, \ k = 0, \ l = 0 \end{cases} \\ \dfrac{1 - q - P_{\mathrm{BP}} - P_{\mathrm{DRP}}}{1 - p_i P_{\mathrm{PCA}} - P_{\mathrm{BP}} - P_{\mathrm{DRP}}}(s)^j \begin{cases} P_{\mathrm{BP}}, & 0 \le j \le L_{i,\mathrm{retry}}, \ 1 \le k \le W_{i,j}, \\ \quad l = D_n + 1, \ldots, D_{\mathrm{BP}} \\ \sum_{z=y}^{n} P_{D_z} + P_{\mathrm{BP}}, & 0 \le j \le L_{i,\mathrm{retry}}, \ 1 \le k \le W_{i,j}, \\ \quad l = D_{y-1} + 1, \ldots, D_y, & y = 1, \ldots, n \\ 1, & 0 \le j \le L_{i,\mathrm{retry}}, \ 1 \le k \le W_{i,j}, \ l = 0 \end{cases} \end{cases} \tag{12.24}$$

$$
b_{i,j,k,l} = b_{i,0,0,0}\frac{W_{i,j}-k+1}{W_{i,j}+1}
\begin{cases}
(s)^j \begin{cases}
\sum_{z=x}^{n} P_{Q+D_z}, \quad 0 \le j \le L_{i,\text{retry}}, \quad k=0, \quad l = Q+D_{x-1}+1,\ldots,Q+D_x, \\
\quad x = y+1,\ldots,n \\
\begin{cases}
\sum_{z=x}^{n} P_{Q+D_z}, \quad 0 \le j \le L_{i,\text{retry}}, \quad k=0, \\
\quad l = Q+D_{\text{BP}}+1,\ldots,Q+D_x, D_{\text{BP}} < D_x, \quad x=y \\
\sum_{z=x}^{n} P_{Q+D_z} + P_{Q+\text{BP}}, \quad 0 \le j \le L_{i,\text{retry}}, \quad k=0, \\
\quad l = Q+D_{x-1}+1,\ldots,Q+D_{\text{DB}}, D_{\text{BP}} \le D_x, \quad x=y
\end{cases} \\
\sum_{z=x}^{n} P_{Q+D_z} + P_{Q+\text{BP}}, \quad 0 \le j \le L_{i,\text{retry}}, \quad k=0, \\
\quad l = Q+D_{x-1}+1,\ldots,Q+D_x, \quad x=1,\ldots,y-1 \\
\sum_{z=x}^{n} P_{Q+D_z} + P_{Q+\text{BP}}, \quad 0 \le j \le L_{i,\text{retry}}, \quad k=0, \quad l=1,\ldots,Q, \quad x=1 \\
1, \quad 0 \le j \le L_{i,\text{retry}}, \quad k=0, \quad l=0
\end{cases} \\
\dfrac{1-q-P_{\text{BP}}-P_{\text{DRP}}}{1-p_i P_{\text{PCA}}-P_{\text{BP}}-P_{\text{DRP}}}(s)^j
\begin{cases}
\sum_{z=x}^{n} P_{D_z}, \quad 0 \le j \le L_{i,\text{retry}}, \quad 1 \le k \le W_{i,j}, \\
\quad l = D_{x-1}+1,\ldots,D_x, \quad x=y+1,\ldots,n \\
\begin{cases}
\sum_{z=x}^{n} P_{D_z}, \quad 0 \le j \le L_{i,\text{retry}}, \quad 1 \le k \le W_{i,j}, \\
\quad l = D_{\text{BP}}+1,\ldots,D_x, D_{\text{BP}} < D_x, \quad x=y \\
\sum_{z=x}^{n} P_{D_z} + P_{\text{BP}}, \quad 0 \le j \le L_{i,\text{retry}}, \quad 1 \le k \le W_{i,j}, \\
\quad l = D_{x-1}+1,\ldots,D_{\text{DB}}, D_{\text{BP}} \le D_x, \quad x=y
\end{cases} \\
\sum_{z=x}^{n} P_{D_z} + P_{\text{BP}}, \quad 0 \le j \le L_{i,\text{retry}}, \quad 1 \le k \le W_{i,j}, \\
\quad l = D_{x-1}+1,\ldots,D_x, \quad x=1,\ldots,y-1 \\
1, 0 \le j \le L_{i,\text{retry}}, \quad 1 \le k \le W_{i,j}, \quad l=0
\end{cases}
\end{cases}
\tag{12.25}
$$

Equations (12.24) and (12.25) show the steady-state probability in state (i,j,k,l) in the tridimensional Markov chain when the BP is larger than the transmission time of n hard DRPs and vice versa, respectively. Note that these steady-state probabilities are all expressed in terms of $b_{i,0,0,0}$. Using (12.3) and the identity

$$\sum_{k=1}^{W_{i,j}} \frac{W_{i,j}-k+1}{W_{i,j}+1} = \frac{W_{i,j}}{2}, \tag{12.26}$$

we have

$$\sum_{k=1}^{W_{i,j}} \frac{W_{i,j}-k+1}{W_{i,j}+1} = \begin{cases} \frac{(W_i+1)2^j}{2} - \frac{1}{2}, & 0 \le j \le m_i \\ \frac{(W_i+1)2^{m_i}}{2} - \frac{1}{2}, & m_i < j \le L_{i,\text{retry}}. \end{cases} \tag{12.27}$$

By the property of total probability for any Markov chain,

$$\sum_j \sum_k \sum_l b_{i,j,k,l} = 1, \tag{12.28}$$

we can solve for $b_{i,0,0,0}$. Similarly, we have two cases: $D_{\text{BP}} > D_n$ and $D_{\text{BP}} \le D_n$. Using (12.28), we have (12.29) and (12.30), respectively, for the two cases.

$$b_{i,0,0,0} = \begin{cases} \left[\frac{1-(s)^{L_{i,\text{retry}}+1}}{1-s} \left\{ 1 + \left(\sum_{z=1}^{n} P_{Q+D_z} + P_{Q+\text{BP}} \right) Q + \sum_{y=1}^{n} \left(\sum_{z=y}^{n} P_{Q+D_z} + P_{Q+\text{BP}} \right) (D_y - D_{y-1}) \right. \right. \\ \left. + P_{Q+\text{BP}}(D_{\text{BP}} - D_n) \right\} + \frac{1-q-P_{\text{BP}}-P_{\text{DRP}}}{1-p_i P_{\text{PCA}} - P_{\text{BP}} - P_{\text{DRP}}} \\ \times \left\{ (W_i+1) \frac{1-(2s)^{L_{i,\text{retry}}+1}}{2(1-2s)} - \frac{1-(s)^{L_{i,\text{retry}}+1}}{2(1-s)} \right\} \\ \left. \times \left\{ 1 + \sum_{y=1}^{n} \left(\sum_{z=y}^{n} P_{D_z} + P_{\text{BP}} \right) (D_y - D_{y-1}) + P_{\text{BP}}(D_{\text{BP}} - D_n) \right\} \right]^{-1}, & m_i > L_{i,\text{retry}} \\ \left[\frac{1-(s)^{L_{i,\text{retry}}+1}}{1-s} \left\{ 1 + \left(\sum_{z=1}^{n} P_{Q+D_z} + P_{Q+\text{BP}} \right) Q + \sum_{y=1}^{n} \left(\sum_{z=y}^{n} P_{Q+D_z} + P_{Q+\text{BP}} \right) (D_y - D_{y-1}) \right. \right. \\ \left. + P_{Q+\text{BP}}(D_{\text{BP}} - D_n) \right\} + \frac{1-q-P_{\text{BP}}-P_{\text{DRP}}}{1-p_i P_{\text{PCA}} - P_{\text{BP}} - P_{\text{DRP}}} \left\{ (W_i+1) \frac{1-(2s)^{m_i+1}}{2(1-2s)} \right. \\ \left. + (W_i+1) 2^{m_i} s^{m_i+1} \frac{1-(s)^{L_{i,\text{retry}}-m_i}}{2(1-s)} - \frac{1-(s)^{L_{i,\text{retry}}+1}}{2(1-s)} \right\} \\ \left. \times \left\{ 1 + \sum_{y=1}^{n} \left(\sum_{z=y}^{n} P_{D_z} + P_{\text{BP}} \right) (D_y - D_{y-1}) + P_{\text{BP}}(D_{\text{BP}} - D_n) \right\} \right]^{-1}, & m_i < L_{i,\text{retry}} \end{cases}$$

(12.29)

$$
b_{i,0,0,0} =
\begin{cases}
\Bigg[\frac{1-(s)^{L_{i,\mathrm{retry}}+1}}{1-s}\Bigg\{1+\left(\sum_{z=1}^{n}P_{Q+D_z}+P_{Q+\mathrm{BP}}\right)Q+\sum_{x=1}^{y-1}\left(\sum_{z=x}^{n}P_{Q+D_z}+P_{Q+\mathrm{BP}}\right)(D_x-D_{x-1}) \\
+\sum_{x=y}^{y}\left\{\left(\sum_{z=x}^{n}P_{Q+D_z}+P_{Q+\mathrm{BP}}\right)(D_{\mathrm{BP}}-D_{x-1})+\left(\sum_{z=x}^{n}P_{Q+D_z}\right)(D_x-D_{\mathrm{BP}})\right\} \\
+\sum_{x=y+1}^{n}\left(\sum_{z=x}^{n}P_{Q+D_z}\right)(D_x-D_{x-1})\Bigg\}+\frac{1-q-P_{\mathrm{BP}}-P_{\mathrm{DRP}}}{1-p_iP_{\mathrm{PCA}}-P_{\mathrm{BP}}-P_{\mathrm{DRP}}} \\
\times\left\{(W_i+1)\frac{1-(2s)^{L_{i,\mathrm{retry}}+1}}{2(1-2s)}-\frac{1-(s)^{L_{i,\mathrm{retry}}+1}}{2(1-s)}\right\}\Bigg\{1+\sum_{x=1}^{y-1}\left(\sum_{z=x}^{n}P_{D_z}+P_{\mathrm{BP}}\right)(D_x-D_{x-1}) \\
+\sum_{x=y}^{y}\left\{\left(\sum_{z=x}^{n}P_{D_z}+P_{\mathrm{BP}}\right)\times(D_{\mathrm{BP}}-D_{x-1})+\left(\sum_{z=x}^{n}P_{D_z}\right)(D_x-D_{\mathrm{BP}})\right\} \\
+\sum_{x=y+1}^{n}\left(\sum_{z=x}^{n}P_{D_z}\right)\times(D_x-D_{x-1})\Bigg\}\Bigg]^{-1}, \qquad m_i \geq L_{i,\mathrm{retry}} \\
\Bigg[\frac{1-(s)^{L_{i,\mathrm{retry}}+1}}{1-s}\Bigg\{1+\left(\sum_{z=1}^{n}P_{Q+D_z}+P_{Q+\mathrm{BP}}\right)Q+\sum_{x=1}^{y-1}\left(\sum_{z=x}^{n}P_{Q+D_z}+P_{Q+\mathrm{BP}}\right)(D_x-D_{x-1}) \\
+\sum_{x=y}^{y}\left\{\left(\sum_{z=x}^{n}P_{Q+D_z}+P_{Q+\mathrm{BP}}\right)\times(D_{\mathrm{BP}}-D_{x-1})+\left(\sum_{z=x}^{n}P_{Q+D_z}\right)(D_x-D_{\mathrm{BP}})\right\} \\
+\sum_{x=y+1}^{n}\left(\sum_{z=x}^{n}P_{Q+D_z}\right)(D_x-D_{x-1})\Bigg\}+\frac{1-q-P_{\mathrm{BP}}-P_{\mathrm{DRP}}}{1-p_iP_{\mathrm{PCA}}-P_{\mathrm{BP}}-P_{\mathrm{DRP}}} \\
\times\left\{(W_i+1)\frac{1-(2s)^{m_i+1}}{2(1-2s)}+(W_i+1)2^{m_i}s^{m_i+1}\frac{1-(s)^{L_{i,\mathrm{retry}}-m_i}}{2(1-s)}-\frac{1-(s)^{L_{i,\mathrm{retry}}+1}}{2(1-s)}\right\} \\
\times\Bigg\{1+\sum_{x=1}^{y-1}\left(\sum_{z=x}^{n}P_{D_z}+P_{\mathrm{BP}}\right)(D_x-D_{x-1})+\sum_{x=y}^{y}\left\{\left(\sum_{z=x}^{n}P_{D_z}+P_{\mathrm{BP}}\right)(D_{\mathrm{BP}}-D_{x-1})\right. \\
\left.+\left(\sum_{z=x}^{n}P_{D_z}\right)(D_x-D_{\mathrm{BP}})\right\}+\sum_{x=y+1}^{n}\left(\sum_{z=x}^{n}P_{D_z}\right)(D_x-D_{x-1})\Bigg\}\Bigg]^{-1}, \qquad m_i < L_{i,\mathrm{retry}}
\end{cases}
\tag{12.30}
$$

A PCA device transmits when its backoff counter reaches zero and there is no DRP or BP transmission ($k = 0$ and $l = 0$); that is, the PCA device is at any of the states $\{i, j, 0, 0\}$, where $0 \leq j \leq L_{i,\mathrm{retry}}$. The probability that a PCA device in the ith priority class transmits during a generic time slot, denoted by τ_i, is given by

$$
\tau_i = \sum_{j=0}^{L_{i,\mathrm{retry}}} b_{i,j,0,0} = \sum_{j=0}^{L_{i,\mathrm{retry}}} (s)^j b_{i,0,0,0} = \frac{\left(1-(s)^{L_{i,\mathrm{retry}}+1}\right) b_{i,0,0,0}}{1-s}. \tag{12.31}
$$

A transmitted frame collides when one or more PCA devices transmit during a time slot in a PCA period. The probability that a PCA device in the backoff stage for the ith priority class senses the channel busy conditioned on a PCA period, denoted by

p_i, is given by

$$p_i = 1 - (1 - \tau_i)^{n_i - 1} \prod_{h=0, h \neq i}^{N-1} (1 - \tau_h)^{n_h}. \tag{12.32}$$

τ_i and p_i can be solved numerically.

The probability that the channel is busy in a PCA period happens when at least one PCA device transmits during a time slot, denoted by p_b, and is given by

$$p_b = 1 - \prod_{h=0}^{N-1} (1 - \tau_h)^{n_h}. \tag{12.33}$$

The probability that a successful PCA transmission occurs in a time slot for the ith priority class, denoted by $p_{s,i}$, is given by

$$p_{s,i} = n_i \tau_i (1 - \tau_i)^{n_i - 1} \prod_{h=0, h \neq i}^{N-1} (1 - \tau_h)^{n_h}, \qquad i = 0, 1, \ldots, N - 1. \tag{12.34}$$

The probability that a successful transmission occurs in a time slot, denoted by p_s, is given by

$$p_s = \sum_{i=0}^{N-1} p_{s,i} = (1 - p_b) \sum_{i=0}^{N-1} \frac{n_i \tau_i}{1 - \tau_i}. \tag{12.35}$$

The probability that the channel is idle for a time slot is $1 - p_b$. The probability that the channel is neither idle nor successful for a time slot is $[1 - (1 - p_b) - p_s] = p_b - p_s$.

Let $T_{E(L)}$, T_s, and T_c denote the time to transmit the average payload, the average time required to transmit a PCA packet successfully, and the average time that the PCA channel has a collision, respectively. The fraction of PCA periods in a superframe, denoted by F_{PCA}, is given by

$$F_{\text{PCA}} = \frac{N_{\text{PCA}}}{N_{\text{SF}}}. \tag{12.36}$$

The normalized saturated throughput for the ith priority PCA class, denoted by S_i, is given by

$$\begin{aligned} S_i &= \frac{E(\text{payload transmission time in a time slot for class } i)}{E(\text{length of a time slot})} F_{\text{PCA}} \\ &= \frac{p_{s,i} T_{E(L)}}{(1 - p_b)\delta + p_s T_s + (p_b - p_s) T_c} \frac{N_{\text{PCA}}}{N_{\text{SF}}}. \end{aligned} \tag{12.37}$$

Let T_H, L, $T_{E(L)}$, L^*, $T_{E(L^*)}$, T_{SIFS}, T_{ACK}, T_{RTS}, T_{CTS}, and $T_{\text{AIFS}(i)}$ denote, respectively, the time to transmit the header [including PLCP preamble and PLCP header consisting of MAC header, physical layer header, header check sequence (HCS), Reed–Solomon parity bits and tail bits], the length of the payload including frame check sum (FCS) and pad bits, the time to transmit a payload with length $E(L)$, the length of the longest frame in a collision, the time to transmit a payload with length $E(L^*)$, the SIFS time, the time to transmit an acknowledgment, the time to transmit a request-to-send (RTS) frame, the time to transmit a clear-to-send (CTS) frame, and an arbitration interframe space (AIFS) time for a class i PCA device. For the basic access, T_s and T_c are, respectively, given by

$$T_s = T_H + T_{E(L)} + T_{\text{SIFS}} + T_{\text{ACK}} + T_{\text{AIFS}(i)}, \tag{12.38}$$

$$T_c = T_H + T_{E(L^*)} + T_{\text{SIFS}} + T_{\text{ACK}} + \max(T_{\text{AIFS}(h)}), \qquad h = 0,1,\ldots,N-1. \tag{12.39}$$

For the RTS/CTS access, T_s and T_c are given, respectively, by

$$T_s = T_{\text{RTS}} + T_{\text{CTS}} + T_H + T_{E(L)} + 3T_{\text{SIFS}} + T_{\text{ACK}} + T_{\text{AIFS}(i)}, \tag{12.40}$$

$$T_c = T_{\text{RTS}} + T_{\text{SIFS}} + T_{\text{CTS}} + \max(T_{\text{AIFS}(h)}), \qquad h = 0,1,\ldots,N-1. \tag{12.41}$$

Let X_i denote the total number of backoff slots that a packet encounters without the case when the backoff counter freezes due to collisions in the PCA period or due to DRP transmissions or BP, for the ith priority class. The expected value of X_i, denoted by $E(X_i)$, is given by [similar to [7]]

$$E(X_i) = \sum_{j=0}^{L_{i,\text{retry}}} \frac{p_i^j(1-p_i)}{1-p_i^{L_{i,\text{retry}}+1}} \sum_{h=0}^{j} \frac{W_{i,h}}{2}. \tag{12.42}$$

Using (12.42) and the identities,

$$\sum_{j=0}^{n} jp^j = p\left[\frac{1-p^n}{(1-p)^2} - \frac{np^n}{1-p}\right] \tag{12.43}$$

and

$$\sum_{j=m}^{n} jp^j = \frac{p^{m+1}-p^{n+2}}{(1-p)^2} + \frac{mp^m-(n+1)p^{n+1}}{1-p}, \tag{12.44}$$

we have

$$E(X_i) = \begin{cases} \dfrac{(W_i+1)(1-p_i)\left[1-(2p_i)^{L_{i,\text{retry}}+1}\right]}{(1-2p_i)\left(1-p_i^{L_{i,\text{retry}}+1}\right)} - \left(\dfrac{W_i}{2}+1\right) \\ \quad - \dfrac{p_i\left(1-p_i^{L_{i,\text{retry}}}\right)}{2(1-p_i)\left(1-p_i^{L_{i,\text{retry}}+1}\right)} + \dfrac{L_{i,\text{retry}}\, p_i^{L_{i,\text{retry}}+1}}{2\left(1-p_i^{L_{i,\text{retry}}+1}\right)}, & m_i \geq L_{i,\text{retry}} \\ \dfrac{(W_i+1)(1-p_i)\left[1-(2p_i)^{m_i+1}\right]}{(1-2p_i)\left(1-p_i^{L_{i,\text{retry}}+1}\right)} - \left(\dfrac{W_i}{2}+1\right) \\ \quad - \dfrac{p_i(1-p_i^{m_i})}{2(1-p_i)\left(1-p_i^{L_{i,\text{retry}}+1}\right)} + \dfrac{m_i\, p_i^{m_i+1}}{2\left(1-p_i^{L_{i,\text{retry}}+1}\right)} \\ \quad + \dfrac{\left[(W_i+1)2^{m_i} - \left(\dfrac{W_i+1}{2}\right)2^{m_i} m_i\right] p_i^{m_i+1}\left(1-p_i^{L_{i,\text{retry}}-m_i}\right)}{1-p_i^{L_{i,\text{retry}}+1}} \\ \quad + \dfrac{[(W_i+1)2^{m_i}-1]\, p_i\left[p_i^{m_i+1} - p_i^{L_{i,\text{retry}}+1}\right]}{2(1-p_i)\left(1-p_i^{L_{i,\text{retry}}+1}\right)} \\ \quad + \dfrac{[(W_i+1)2^{m_i}-1]\, p_i\left[(m_i+1)p_i^{m_i} - \left(L_{i,\text{retry}}+1\right)p_i^{L_{i,\text{retry}}}\right]}{2\left(1-p_i^{L_{i,\text{retry}}+1}\right)}, & m < L_{i,\text{retry}}. \end{cases} \tag{12.45}$$

Only successful transmissions are considered above.

Let B_i denote the total number of slots when the backoff counter freezes, which a packet encounters, for the ith priority class. The expected value of B_i, denoted by $E(B_i)$, is given by

$$E(B_i) = \frac{E(X_i)}{1 - p_i P_{\text{PCA}} - P_{\text{BP}} - P_{\text{DRP}}}\left(p_i P_{\text{PCA}} + P_{\text{BP}} + P_{\text{DRP}}\right). \tag{12.46}$$

$E(B_i)$ and $E(X_i)$ can be treated as the total number of idle and busy slots that a packet encounters during backoff stages, respectively. Let $E(N_{i,\text{retry}})$ denote the average number of retries for the ith priority class. It is given by

$$\begin{aligned} E(N_{i,\text{retry}}) &= \sum_{j=0}^{L_{i,\text{retry}}} \frac{j p_i^j (1-p_i)}{1-p_i^{L_{i,\text{retry}}+1}} \\ &= \frac{p_i\left(1-p_i^{L_{i,\text{retry}}}\right)}{(1-p_i)\left(1-p_i^{L_{i,\text{retry}}+1}\right)} - \frac{L_{i,\text{retry}}\, p_i^{L_{i,\text{retry}}+1}}{1-p_i^{L_{i,\text{retry}}+1}}. \end{aligned} \tag{12.47}$$

Let D_i denote the class i packet delay. The average delay is given by

$$E(D_i) = E(X_i)\left\{P_{\text{PCA}}\delta + \overline{Y_{nz}}\delta\right\} + E(B_i)\left\{P_{\text{PCA}}\left[\frac{p_s}{p_b}T_s + \frac{p_b - p_s}{P_b}T_c\right] + \overline{Y_{nz}}\delta\right\}$$
$$+ E(N_{i,\text{retry}})\left\{r(T_c + T_o) + \overline{Y_z}\delta\right\} + T_s, \qquad (12.48)$$

where

$$\overline{Y_{nz}} = \sum_{y=1}^{n} P_{D_y}D_y + P_{\text{BP}}D_{\text{BP}}, \qquad (12.49)$$

$$\overline{Y_z} = \sum_{y=1}^{n} P_{Q+D_y}(Q+D_y) + P_{Q+\text{BP}}D_{Q+\text{BP}}, \qquad (12.50)$$

$$T_o = \text{SIFS} + T_{\text{ACK timeout}} \quad \text{(for basic access)}, \qquad (12.51)$$

$$T_o = \text{SIFS} + T_{\text{CTS timeout}} \quad \text{(for RTS/CTS access)}. \qquad (12.52)$$

In what follows we present results for the saturated throughput with one DRP class and two PCA classes ($N = 2$) as an illustrative example. The parameter values used in the numerical examples for the WiMedia MAC are tabulated in Table 12.5. The RTS/CTS access mode is assumed.

The saturated throughputs of two classes of PCA devices and one class of hard DRP devices with a data rate, R, of 480 Mbps are shown in Figure 12.10. The saturated throughput of class 0 PCA devices is higher than that of class 1 PCA devices, as class 0 PCA devices have higher priority than class 1 PCA devices. As the number of hard

TABLE 12.5 Parameter Values Used for WiMedia MAC

Symbol	Value	Symbol	Value
N_{SF}	256	$W_{0,j}$	{7, 15, 31, 63, 127, 255, 511}
N_{BSig}	2	$W_{1,j}$	{15, 31, 63, 127, 255, 511, 1023}
N_{BExt}	8	T_G	12 μs
N_{BS}	3	T_H	9.375 (PLCP preamble) + 5.075 (PLCP header) = 14.45 μs
N_{D}	96	L	4095 (payload) + 5 (FCS, tail bits, pad bit) = 4100 bytes
T_{MAS}	256 μs	Data rate, R	53.3, 200, 480 Mbps
W_0	7	$T_{E(L)} = T_{E(L^*)}$	615.385, 164.0, 68.33 μs
W_1	15	T_{SIFS}	10 μs
m_1	6	T_{ACK}	15.5, 14.73, 14.567 μs for one frame
m_2	6	T_{RTS}	14.45 μs
$L_{0,\text{retry}}$	6	T_{CTS}	14.45 μs
$L_{1,\text{retry}}$	6	$T_{\text{AIFS(0)}}$	46 μs
δ	9 μs	$T_{\text{AIFS(1)}}$	46 μs

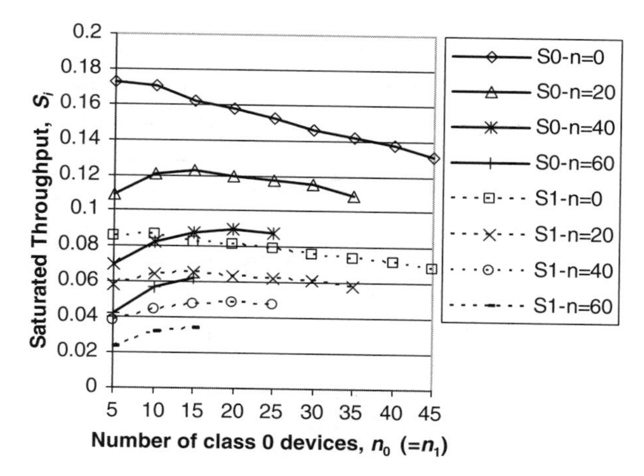

FIGURE 12.10 Saturated throughput for two PCA classes in the presence of DRP transmissions and BP with 480 Mbps.

DRP devices, n, increases, the saturated throughputs of the PCA devices decreases, as there are fewer time periods for PCA transmissions. Optimum points exist for each curve, which is also observed in IEEE 802.11 [5] and IEEE 802.11e [6]. Note that the maximum number of devices that can be supported in WiMedia is 96. We have rounded this number to 90 devices in our numerical examples. Thus, the sum of the number of hard DRP devices and the sum of the two classes of PCA devices is 90. This is the reason why as the number of hard DRP devices increases, the maximum number of PCA devices decreases. Figure 12.11 shows the saturated throughputs per

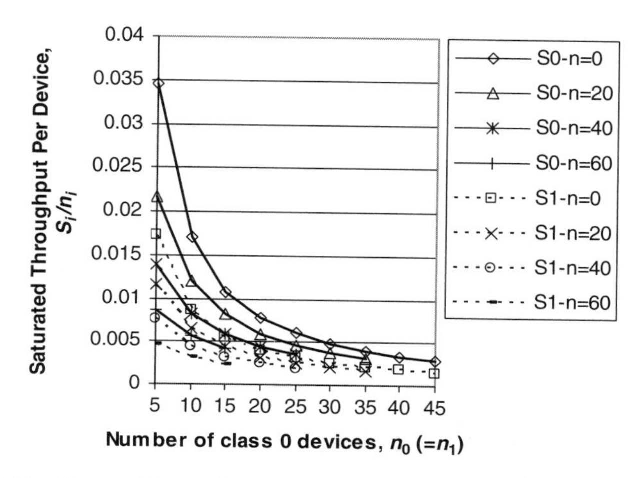

FIGURE 12.11 Saturated throughput per device for two PCA classes in the presence of DRP transmissions and BP with 480 Mbps.

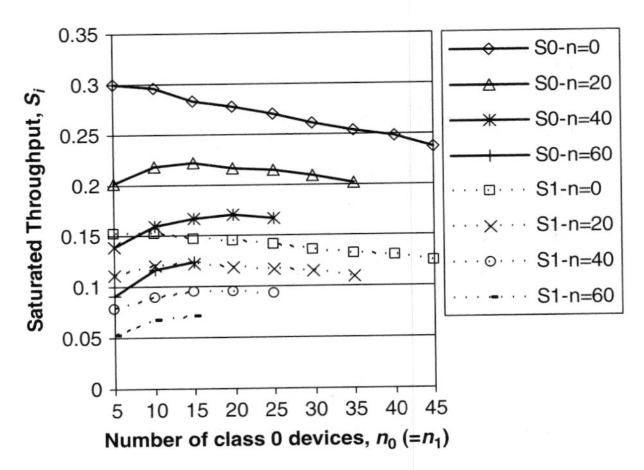

FIGURE 12.12 Saturated throughput for two PCA classes in the presence of DRP transmissions and BP with 200 Mbps.

device for two classes of PCA devices and one class of hard DRP devices with a data rate, R, of 480 Mbps. These throughputs per device decrease as more devices are accessing the PCA channel as shown in the figure. This figure also shows that the saturated throughputs per device decrease as the number of hard DRP devices increases, since this decreases the time periods for PCA transmissions.

The saturated throughputs of two classes of PCA devices and one class of hard DRP devices with data rates, R, of 200 and 53.3 Mbps are shown in Figures 12.12 and 12.13. From Figures 12.10, 12.12, and 12.13 it can be seen that the saturated throughput increases as the data rate decreases. The reason for this trend is that the

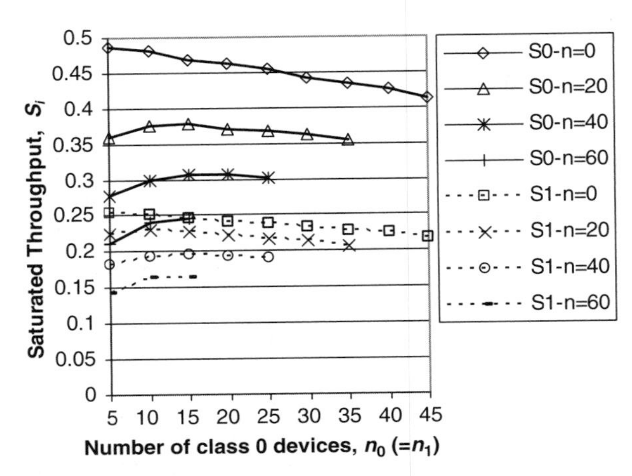

FIGURE 12.13 Saturated throughput for two PCA classes in the presence of DRP transmissions and BP with 53.3 Mbps.

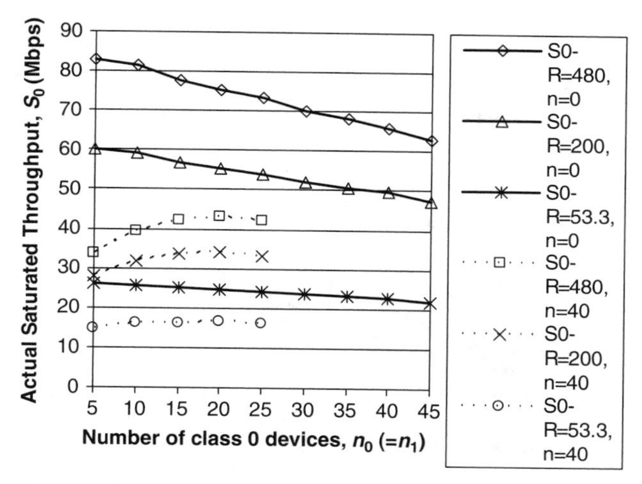

FIGURE 12.14 Actual saturated throughput for class 0 PCA in the presence of DRP transmissions and BP with $n \in \{0,40\}$.

payload transmission time increases as the data rate decreases. Note that the saturated throughputs are normalized in Figures 12.10, 12.12, and 12.13. Thus, it does not mean that the actual saturated throughput at the lower data rate is higher than that at a higher data rate. To compare the actual saturated throughputs at these data rates, the normalized saturated throughputs must be multiplied by their corresponding data rates. The actual saturated throughputs for class 0 and class 1 with $n \in \{0,40\}$ are shown in Figures 12.14 and 12.15, respectively. From these figures it can be seen that

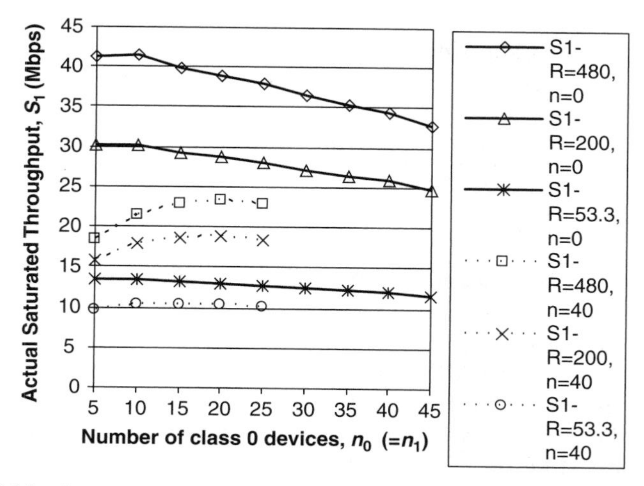

FIGURE 12.15 Actual saturated throughput for class 1 PCA in the presence of DRP transmissions and BP with $n \in \{0,40\}$.

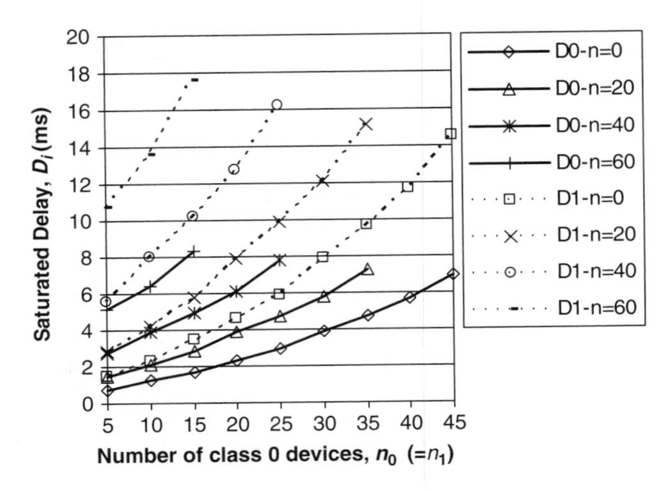

FIGURE 12.16 Saturated delay for two PCA classes in the presence of DRP transmissions and BP with 480 Mbps.

the actual saturated throughputs at a higher data rate are correspondingly higher that those at a lower data rate.

The saturated delays of two classes of PCA devices and one class of hard DRP devices with a data rate, R, of 480 Mbps are shown in Figure 12.16. The saturated delay of class 0 PCA devices, denoted by solid lines in the figure, is lower than that of class 1 PCA devices, denoted by dashed lines in the figure, as class 0 PCA devices have higher priority than class 1 PCA devices. As the number of hard DRP devices, n, increases, the saturated delays of both types of PCA devices increases, as there are fewer time periods for PCA transmissions, due to greater use of the bandwidth by DRP transmissions.

The saturated delays of two classes of PCA devices and one class of hard DRP devices with data rates, R, of 200 and 53.3 Mbps are shown in Figures 12.17 and 12.18, respectively. In general, the corresponding delays increase with the decrease in data rate as the payload transmission time increases.

12.4.8 Wireless Universal Serial Bus

Wireless USB (WUSB) uses another type of reservation in WiMedia MAC, known as private DRP. In the reserved slots owned by the WUSB using private DRP, micro-scheduled management commands (MMCs) are transmitted to control the WUSB channel by a host. These MMC packets are commands used to allocate and control the channel time between the host and the devices in communication. Each MMC controls its following reserved time, which is divided into micro-scheduled channel time allocations (MS-CTAs). Data communications between the host and devices can be done using these MS-CTAs as specified by the MMC.

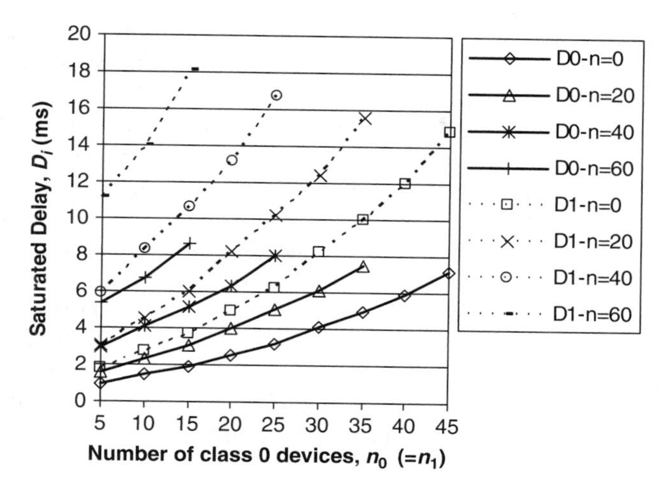

FIGURE 12.17 Saturated delay for two PCA classes in the presence of DRP transmissions and BP with 200 Mbps.

12.5 MOBILITY RESOURCE MANAGEMENT

WiMedia MAC is designed to handle mobility. When two groups of devices with different beacon periods (BPs) come together, there is a protocol in WiMedia MAC to merge the two groups of devices together. Two scenarios are possible. One of these scenarios is the case where the beacon periods do not overlap, while another scenario is the case where the beacon periods overlapped.

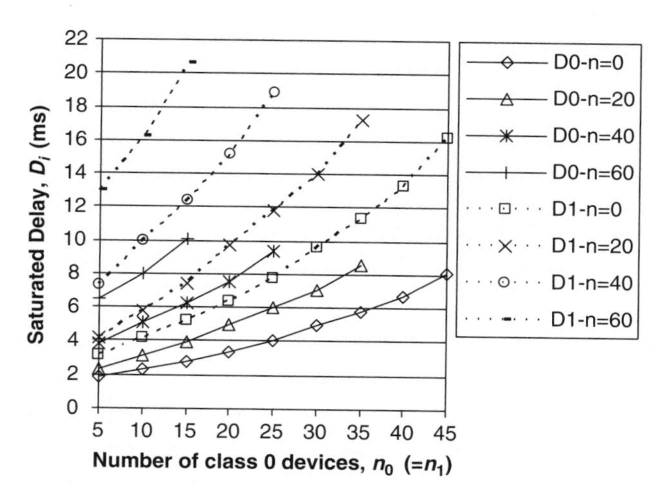

FIGURE 12.18 Saturated delay for two PCA classes in the presence of DRP transmissions and BP with 53.3 Mbps.

If the beacon periods do not overlap, the alien BPs will first be protected with a DRP information element (IE) in each group. The devices in one of the groups will initiate the switch to the alien BP if a beacon in that alien BP does not include a BP switch IE; that is, the group of devices in the alien BP does not intend to switch to another alien BP. The devices in the switching group will relocate its beacons to the alien BP within a certain time, depending on whether the alien BP falls within the first or second half of the superframe. The duration of the latter time is one and half times the duration of the former time. After merging, the BP switch IE and the alien BP protection will be removed and there will be only one BP in the superframe.

If the beacon periods overlapped, the devices whose beacon period start time (BPST) falls within an alien BP will move its BPST and beacons to the alien BP. This block of device beacons will be moved to the locations after the last location of the device beacon in the alien BP. The devices will not send any beacons in the old BP. After merging, there is only one BP in the superframe.

12.6 ROUTING

WiMedia specifications [1] do not specify any routing algorithm. WUSB uses a type of reservation in WiMedia MAC with a star topology. Thus, a routing algorithm is not necessary. However, WiMedia MAC is designed as a distributed MAC and can have a mesh topology in future applications. Thus, some of the routing algorithms described in Chapter 7 can be applied in these future applications.

12.7 QUALITY OF SERVICE

The qualities of service (QoSs) in WiMedia MAC are provided by its channel access methods. For real-time traffic, DRP can be used. The types of DRP include hard, soft, and private DRPs. For less stringent QoS traffic, PCA can be used. There are four types of access categories using PCA. This gives rise to relative QoS, depending on the minimum and maximum contention window sizes and the arbitration interframe space (AIFS). In general, the smaller the minimum and maximum contention window sizes and the AIFS, the higher the priority of that access category. As WiMedia is also rate adaptive in the PCA mode, a low-rate device will bring down the performance of other high-rate WiMedia PCA devices. Note that WiMedia has eight data rates.

12.8 APPLICATIONS

There can be many applications using WiMedia. One of these applications is WUSB, as described in Section 12.4.3. For example, a personal computer (PC) or laptop can be configured to be connected to a printer, a digital camera, a camcorder, a MP3/4 player, a storage device, a mobile phone, and a television set using WiMedia. Data, voice, video, pictures, and files can be transmitted through WiMedia WPAN architecture. No requirement for any cable with a high data rate of up to 480 Mbps is an advantage of WiMedia over existing low-rate systems.

SUMMARY

We have focused mainly on the physical layer and the medium access control in WiMedia as well as WUSB. Some related issues on network architecture, mobility radio management, routing, quality of service, and applications are also discussed and explained. A standard report on WiMedia specification may be found in [2].

Bluetooth is widely available in the market. The next high-data-rate WPAN to keep an eye on is the development of WiMedia that can support a raw data rate of up to 480 Mbps. A convergence architecture with Bluetooth and IP seating on top of WiMedia is also on the table for discussion. Detect and avoid (DAA) mechanisms for WiMedia/WiMax coexistence are also being standardized. Other WPAN standards to watch in the future are ECMA TC48 mmWave and IEEE 802.15.3c, which both use millimeter waves and operate at 60 GHz. ECMA TC48 mmWave can support a raw data rate of 6.478 Gbps, while IEEE 802.15.3c can support a raw data rate of up to 6 Gbps for single-carrier mode and up to 7.35 Gbps for OFDM mode. Furthermore, ECMA mmWave and IEEE 802.15.3c both use directional-antenna MACs. ECMA TC48 mmWave uses a MAC that is an extension from the WiMedia MAC, while IEEE 802.15.3c uses a MAC that is an extension from the IEEE 802.15.3b MAC. With such high raw data rates, ECMA TC48 mmWave and IEEE 802.15.3c WPANs are certainly the future winners in the race for higher data rate demand in WPANs.

REFERENCES

[1] "High rate ultra wideband PHY and MAC standard," *Standard ECMA-368*, ECMA International, Geneva, Dec. 2005.

[2] J. D. P. Pavon, S. N. Shankar, V. Gaddam, K. Challapali, and P. C. T. Chou, "The MBOA-WiMedia specification for ultra wideband distributed networks," *IEEE Commun. Mag.*, pp. 128–134, June 2006.

[3] Y.-C. Liang, S. Sun, X. Peng, and F. Chin, "Tutorial on emerging wireless standards on WRAN, WiFi, WiMedia and ZigBee," *10th IEEE International Conference on Communications Systems*, Singapore, Oct. 30, 2006.

[4] *IEEE Standard 802.11e*, Sept. 2005.

[5] G. Bianchi, "Performance analysis of the IEEE 802.11 distributed coordination function," *IEEE J. Sel. Areas Commun.*, vol. 18, no. 3, pp. 535–547, Mar. 2000.

[6] Z. N. Kong, D. H. K. Tsang, and B. Bensaou, "Performance analysis of IEEE 802.11e contention-based channel access," *J. Sel. Areas Commun.*, vol. 22, no. 10, pp. 2095–2106, Dec. 2004.

[7] Y. Xiao, "Performance analysis of priority schemes for IEEE 802.11 and IEEE 802.11e wireless LANs," *IEEE Trans. Wireless Commun.*, vol. 4, no. 4, pp. 1506–1515, July 2005.

[8] D. T. C. Wong, F. P. S. Chin, M. R. Shajan, and Y. H. Chew, "Performance analysis of saturated throughput of PCA in the presence of hard DRPs in WiMedia MAC," *IEEE Wireless Communications and Networking Conference 2007, Conference CD-ROM*, Hong Kong, Mar. 11–15, 2007.

[9] D. T. C. Wong, F. P. S. Chin, M. R. Shajan, and Y. H. Chew, "Saturated delay of PCA with hard DRPs in WiMedia MAC," IEEE PIMRC 2007, Conference CD-ROM, Athens, Greece, Sept. 4–6, 2007.

CHAPTER 13

CONVERGENCE OF NETWORKS

13.1 INTRODUCTION

In today's technological market, there are many types of networks. These networks include wireless personal area networks (WPANs), wireless local area networks (WLANs), wireless metropolitan area networks (WMANs), and cellular networks. In general, the range coverage of WPANs is smaller than that of WLANs, and that of WLANs is smaller than that of cellular networks. Between WMANs and cellular, their range coverage depends on the frequency band in which they are operating. In each type of network, there are also different categories of networks. For example, there are ZigBee, Bluetooth, WiMedia, and IEEE 8022.15.3c networks in WPANs. IEEE 802.11 WLANs can be further classified as IEEE 802.11a, IEEE 802.11b, IEEE 802.11g, and draft IEEE 802.11n. These WLANs differ in the data rates that they can deliver. Worldwide interoperability for microwave access (WiMax) is an example of WMAN. On the other hand, cellular networks can be divided into different generations. The first generation (1G) cellular networks are mostly no longer in use. A commonly used second-generation (2G) cellular network today is the global system for mobile communications (GSM). An enhancement of the GSM is the 2.5G general packet radio service (GPRS) cellular network. A third-generation (3G) cellular network in universal telecommunications systems (UMTSs) is the wideband code-division multiple access (WCDMA) network. Enhancement of the WCDMA is the 3.5G high-speed packet access (HSPA). In North America, there are 2G IS-95

Wireless Broadband Networks, By David Tung Chong Wong, Peng-Yong Kong, Ying-Chang Liang, Kee Chaing Chua, and Jon W. Mark

cellular networks and 3G CDMA2000 cellular systems. Another cellular network being standardized is the 3.9G long-term evolution (LTE).

Each of these networks is designed to deliver specific services. These services cannot be migrated to other types of networks. However, there are some bodies that address the needs whereby services can roam from one network to another network based on the best connection available. For example, a local area network (LAN) may be the best connection to the Internet in a fixed office environment. However, a WLAN may be a better option when users need to move from their office to a meeting room. If a user needs to be out of the office most of the time and is often "on the go," he or she may stay connected through a mobile phone or personal digital assistant (PDA). Thus, it would be very convenient for a user if the services she is using can be continued in any environment, whether she is in the office, a meeting room, or out of the office. The mobile device should have a multiradio that automatically connects the user to the best available network without loss of the service's session during vertical handoff and without intervention. Vertical handoff is the handling of connection from one network to another network that supports a different transmission rate. Thus, interworking is needed between different access networks. A few of these bodies that address the interworking mechanisms are the third-generation partnership project (3GPP), IEEE 802.11u, and IEEE 802.21. 3GPP evolved from the GSM and GPRS cellular networks. 3GPP identifies six scenarios for 3GPP/WLAN interworking, while IEEE 802.11u is to interwork IEEE 802.11 WLAN with external networks. IEEE 802.21 introduces a 2.5 layer between the second and third layers in the OSI/ISO model. The 2.5 layer introduces the media-independent handoff. The aim of this standard is to enable seamless handoff and interoperability between heterogeneous network types, which include both IEEE 802 and non-IEEE 802 networks.

A vision of a future convergence of networks envisaged for WPANs, WLANs, WiMax, and cellular networks are presented in this chapter. Issues arising from the interworkings of these networks are also discussed.

In Section 13.2 we present six 3GPP/WLAN interworking scenarios. In Section 13.3 we briefly describe the IEEE 802.11u for interworking with external networks. IEEE 802.21 media-independent handoff is described in Section 13.4. In Section 13.5 we take a critical look at an envisaged future cellular/WPANs/WLANs/WMAN with some interworking issues. A simple analytical model for cellular/WLAN interworking is presented in Section 13.6. The salient aspects of the chapter are summarized at the end of the chapter.

13.2 3GPP/WLAN INTERWORKING

Six scenarios for 3GPP/WLAN interworking are categorized in [1]. In this form of interworking, 3GPP refers to the third-generation partnership project, which evolved from 2G GSM and 2.5G GPRS cellular system; WLAN refers to IEEE 802.11a/b/g wireless local area networks. Both cellular systems and WLANs are widely deployed in the world today. To provide a seamless experience for a mobile device, interworking mechanisms between cellular systems and WLANs are needed to provide smooth vertical handoff in more complex interworking scenarios. The six scenarios are based on the simplest to the incrementally more complex interworking mechanisms. The six scenarios are as follows:

- *Scenario 1:* common billing and customer care
- *Scenario 2:* 3GPP system–based access control and charging
- *Scenario 3:* access to 3GPP system packet-switched (PS)-based services
- *Scenario 4:* service continuity
- *Scenario 5:* seamless service
- *Scenario 6:* access to 3GPP system circuit-switched (CS)-based services with seamless mobility

Each scenario is described briefly below.

13.2.1 Scenario 1

This scenario for common billing and customer care is the simplest of the six scenarios for 3GPP/WLAN interworking. There is only a single customer relationship for using both the 3GPP network and the WLAN. With this relationship, the customer will receive only one bill from the mobile system operator for services used in both networks. This also allows for simplified integrated customer care for both the customer and the mobile system operator.

Security may be independent for both networks. Thus, no new requirements are needed for the 3GPP specifications in this scenario. Therefore, there is no real interworking mechanism between the 3GPP network and the WLAN. Because of this, scenario 1 does not require special standardization activities.

13.2.2 Scenario 2

In this scenario for 3GPP system–based access control and charging, authentication, authorization, and accounting (AAA) are provided by the 3GPP cellular network. Thus, subscribers to the WLAN are subjected to the same AAA as in the 3GPP cellular network. Therefore, the operator can charge access to both networks in a consistent manner. Using the access method based on the 3GPP network has benefits for both the subscriber and the operator. First, the operator can easily allow existing 3GPP network subscribers to access the WLAN. The other benefit is that maintenance of the subscribers may also be simplified.

Only IP connectivity over the WLAN is needed. However, it is not necessary to provide a specific set of services in the WLAN.

13.2.3 Scenario 3

In this scenario for access to 3GPP system PS-based services, the goal is to extend to the WLAN the 3GPP cellular network PS-based services. These services include access point names (APNs), IP multimedia services (IMS), location-based services, instant messaging, presence-based services, multimedia broadcast multicast service (MBMS), and any services that are built based on a combination of these services.

Scenario 3 does not require service continuity between the 3GPP cellular network and the WLAN. That is, a service is not required to continue its session after the subscriber has handed off the 3GPP cellular network to the WLAN, and vice versa.

13.2.4 Scenario 4

In this scenario for access to 3GPP system PS-based services with service continuity, the goal is to allow the services supported in scenario 3 to survive a handoff between the 3GPP cellular network and the WLAN. The vertical handoff may be noticeable to the subscriber. However, there is no need for user intervention or for the mobile device or user equipment (UE) to reestablish sessions for the services. Due to the differences in the access technologies for the 3GPP cellular network and the WLAN, there may be a change in the quality of service (QoS) as a result of vertical handoff between these two access technologies. Change in the QoS after a vertical handoff is due to the varying capabilities and characteristics of radio access technologies. It is also possible that some services may be terminated, as the target network may not support services from the source network.

The criteria and decision mechanism for vertical handoff between the 3GPP cellular system and the WLAN are still under investigation [1].

13.2.5 Scenario 5

In this scenario for access to 3GPP system PS-based services with seamless services, the goal is to allow the services supported in scenario 3 not only to survive a handoff between the 3GPP cellular network and the WLAN but also to provide seamless service continuity. Seamless service continuity means that the data loss and break time during handoff between the access technologies are minimized.

The subscriber should not notice any significant changes in service after a vertical handoff between the 3GPP cellular network and the WLAN, or vice versa.

13.2.6 Scenario 6

In this scenario for access to 3GPP system CS-based services, the goal is to have 3GPP system CS-based services over the WLAN. The vertical handoff between these networks should be both seamless and transparent to the subscriber for these services.

13.2.7 Scenarios and Their Operational Capabilities

Table 13.1 shows the scenarios for 3GPP/WLAN interworking and their operational capabilities.

TABLE 13.1 Scenarios and Their Operational Capabilities

	Scenario 1	Scenario 2	Scenario 3	Scenario 4	Scenario 5	Scenario 6
Common billing	×	×	×	×	×	×
Common customer care	×	×	×	×	×	×
3GPP system-based access control		×	×	×	×	×
3GPP system-based access charging		×	×	×	×	×
Access to 3GPP system PS-based services from WLAN			×	×	×	×
Service continuity				×	×	×
Seamless service continuity					×	×
Access to 3GPP system CS-based services with seamless mobility						×

13.3 IEEE 802.11U INTERWORKING WITH EXTERNAL NETWORKS

The IEEE 802.11u standard is currently under standardization. The aim of this standard is to amend IEEE 802.11 to add features that will improve interworking with external networks. The interworking between IEEE 802.11 WLAN and external networks is achieved through medium access control (MAC) enhancements.

IEEE 802.11u addresses such areas as enrollment, network selection, emergency call support, user traffic segmentation, and service advertisement.

13.4 LAN/WLAN/WIMAX/3G INTERWORKING BASED ON IEEE 802.21 MEDIA-INDEPENDENT HANDOFF

The IEEE 802.21 standard is being standardized. The aim of this standard is to enable seamless handoff and interoperability between heterogeneous network types. These heterogeneous network types include both IEEE 802 networks and non–IEEE 802 networks. The IEEE 802 networks are the IEEE 802.3 local area network (LAN), IEEE 802.11 wireless LAN (WLAN), and IEEE 802.16 wireless metropolitan area network (WMAN). The LAN is Ethernet, while the WLAN is Wi-Fi. On the other hand, the WMAN is WiMax. The non-IEEE 802 networks are cellular networks from 3GPP and 3GPP2. 3GPP supports a 3G UMTS, and 3GPP2 supports 3G CDMA2000. UMTS evolved from 2G GSM and 2.5G GPRS; CDMA2000 evolved from the North America Standard IS-95. The fundamental assumption in the standard is that the mobile devices can support multiple radio interfaces, both wired and wireless, such as Ethernet, Wi-Fi, Bluetooth, WiMax, GSM, GPRS, UMTS, and CDMA2000.

The seamless handoff and interoperability between heterogeneous network types is achieved by introducing a conceptual layer 2.5. In the traditional OSI/ISO model, layer 1 is the physical layer, and layer 2 is the link layer, consisting of the data link sublayer and the medium access control (MAC) sublayer. Layer 3 is the network layer. This conceptual layer 2.5 is specified by the media independent handoff (MIH) function. Figure 13.1 shows IEEE 802.21 architecture, specifically the interactions between the IEEE 802.21 MIH function and other protocol stack elements.

In IEEE 802.21, one of the main ideas is to provide a common interface for managing events and control messages exchanged between network devices [2]. The outcome of the standard is to provide a general framework applicable from the management and control paradigms of each specific technology and a common interface to the upper layers [2]. These upper layers include the network layer up to the application layer. Besides these functionalities, the standard also introduces an information system (IS) for the storage, management, and communication of system-wide network information [2].

The IEEE 802.21 MIH function defines three services [3]:

1. Media-independent event service (MIES)
2. Media-independent command service (MICS)
3. Media-independent information service (MIIS)

The MIES provides services to the upper layers of a mobile device by notifying both local and remote events [3]. Local events occur within the local stack of the mobile device, while remote events occur in the IEEE 802.21 MIH function of another mobile device in the network. The event is based on a subscription and notification

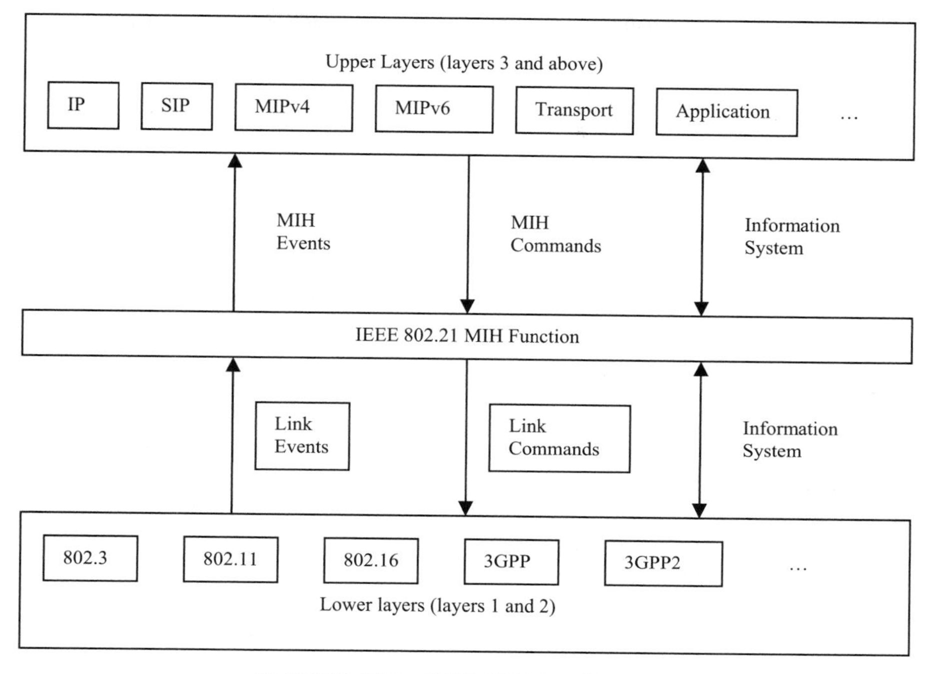

FIGURE 13.1 IEEE 802.21 architecture.

procedure. The upper layer protocols (layer 3 and above) register a certain set of events to the lower layers (layers 1 and 2) through the IEEE 802.21 MIH function, and they will be notified as these events take place [3]. Information on local events is passed from the lower layers to the higher layers through the IEEE 802.21 MIH function. Information on remote events can be passed to and from the IEEE 802.21 MIH function or layer 3 (L3) mobility protocol in a remote stack. Some common events that are provided by the IEEE 802.21 MIH function are as follows:

- Link up
- Link down
- Link parameters change
- Link going down
- Layer 2 (L2) handoff imminent

The MICS provides primitives to the higher layers to control the functions of the lower layers [3]. MICS commands are used to gather information about the status of the connected links. These commands are also used to execute higher layer mobility and connectivity decisions to the lower layers. Furthermore, these commands can be local and remote. Moreover, these commands can be from the higher layers to the IEEE 802.21 MIH function or from the IEEE 802.21 MIH function to the lower layers. Some examples of these commands that are incorporated in the IEEE 802.21 MIH function are as follows:

- MIH poll
- MIH scan
- MIH configure
- MIH switch

These commands instruct an IEEE 802.21 MIH mobile device to poll connected links to learn their up-to-date status, to scan for newly discovered links, to configure new links, and to switch between available links [3].

The MIIS defines information elements and corresponding query–response mechanisms. These elements and mechanisms allow the IEEE 802.21 MIH function to discover and obtain information and distribute within a geographical area network information relating to nearby networks [2,3]. The role of MIIS is to provide as much information as possible to the mobile devices on the networks available and the services that they can provide [2]. Furthermore, the MIIS provides the link to access information that is helpful to handoff decisions [3]:

- Network type
- Roaming partners
- Service providers of the neighboring networks
- Channel information
- MAC addresses
- Security information
- Other information on the higher layers

This information can be made available through the upper and lower layers. The MIIS uses a standard, platform-independent description language to represent the information [2]. External markup language (XML) and type length value (TLV) are two such examples. The information gathered by MIIS can be static or dynamic [2]. Examples of *static information* are the names and providers of the mobile device's neighboring network; examples of *dynamic information* are the information on the channel, the security, and the MAC addresses. In some scenarios, some layer 2 information may not be available, or not enough is available to make intelligent handoff decisions. In these cases, upper layers' services may be needed to help in making the decision.

There are two advantages of MIIS: that the information it provides can help significantly in the definition of high-level handoff decisions and policies [2], and that specific access-dependent discovery methods for automatic detection of neighboring networks are avoided.

The IEEE 802.21 standard does not specify rules or policies for handoff decisions and does not dictate that the handoff be a network or mobile-controlled handoff [4].

Concrete rules and policies are beyond the scope of the IEEE 802.21 standard, and their definition and specification are up to the wireless service provider [4]. A case study on WLAN/3G handoffs based on IEEE 802.21 may be found in [4].

13.5 FUTURE CELLULAR/WIMAX/WLAN/WPAN INTERWORKING

3GPP is considering cellular networks and WLANs interworking, while IEEE 802.21 is considering a media-independent handoff between LAN, WLAN, WiMax in IEEE, and cellular networks in 3GPP and 3GPP2. We envisage that other technologies for WPANs, such as Zigbee, Bluetooth, and WiMedia, will also be integrated together with LAN, WLANs, WiMax, GSM, GPRS, UMTS, IS-95, CDMA2000, HSPA, and LTE cellular networks, as shown in Figure 13.2.

ZigBee is a WPAN that targets applications that require a low data rate, long battery life, and secure networking. ZigBee uses IEEE 802.15.4 specifications. Bluetooth, which is also a WPAN, is used to connect such devices as mobile phones, computers, digital cameras, videogame consoles, and printers over a secure short-range unlicensed band. WiMedia is also a WPAN but can support a much higher data

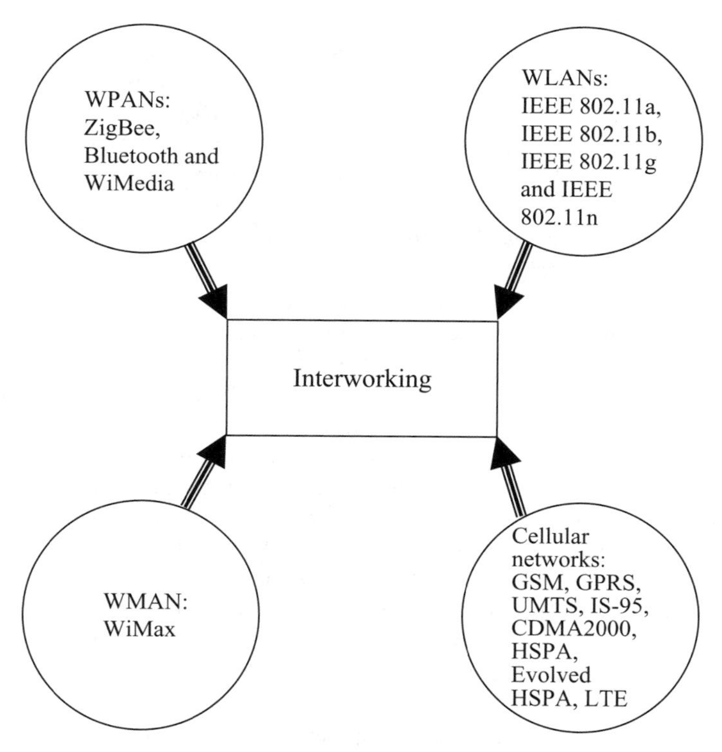

FIGURE 13.2 Future cellular/WiMax/WLAN/WPAN interworking.

rate, up to 480 Mbps. More details on WiMedia are provided in Chapter 12. The IEEE 802.15.3c draft and ECMA TC48 mmWave draft both operate in the 60-GHz range, and both use directional medium access control (MAC). The IEEE 802.15.3c draft MAC is an extension of the IEEE 802.15.3 MAC, and the ECMA TC48 mmWave draft MAC is an extension of the WiMedia ECMA-368 MAC.

IEEE 802.11 is a commonly used WLAN. IEEE 802.11b supports a data rate of up to 11 Mbps, while IEEE 802.11a and IEEE 802.11g support a data rate of up to 54 Mbps. The IEEE 802.11n draft standard supports a data rate of up to 600 Mbps. More details on WLAN are given in Chapter 11.

WiMax is a WMAN based on IEEE 802.16. It is used as a last-mile wireless broadband access alternative to cable and digital subscriber line (DSL). More details on WiMax are provided in Chapter 10.

GSM is a second-generation (2G) cellular network; GPRS is a 2.5G cellular network. UMTS is a third-generation (3G) cellular network that uses wideband code-division multiple access (WCDMA) as the multiple-access technology. IS-95 is a 2G code-division multiple access (CDMA)–based cellular network, and CDMA2000 is a 2.5G/3G CDMA standard. HPSA is a 3.5G cellular network designed on top of WCDMA. LTE is a 3.9G cellular network. More details on LTE are given in Chapter 9.

The convergence of networks will give rise to many technical challenges. Some of these challenges include quality-of-service issues, traffic class mappings among different access technologies, vertical handoff mechanisms between access technologies, protocol stack designs, AAA, service discovery mechanism, and routing. Issues in 3GPP/WLAN interworking and IEEE 802.21 media-independent handoff will certainly arise in the convergence of networks that is envisaged .

One of the fundamental factors in the access technologies is data rate. Table 13.2 shows the data rates for various access technologies. Due to the different data rates in the access technologies, some services may be terminated during vertical handoffs between access technologies. Rate-adaptive services are also needed when services move from one access technology to another.

Mappings of traffic classes between access technologies are dependent on the traffic classes available in each technology. ZigBee has two traffic classes: one based on a contention period (CP), the other based on a contention-free period (CFP) or guaranteed time slot (GTS). The former is based on CSMA/CA and is for non-real-time traffic; the latter is based on reservation and is for real-time traffic. Bluetooth can support voice and data traffic based on polling. WiMedia has four access categories in its prioritized channel access (PCA) mode, and real-time traffic can use its distributed reservation protocol (DRP) mode. The four access categories in PCA are background, best effort, video, and voice traffic. Similarly, IEEE 802.11e EDCA has four traffic access categories: background, best effort, video, and voice. WiMedia PCA is derived from IEEE 802.11e EDCA. Cellular networks 2G and above can support at least voice and data. UMTS can support four traffic classes: background, interactive, streaming, and conversational. WiMax can support at least four scheduling services: best effort, non-real-time polling, real-time polling, and unsolicited grant. Thus, it may not be possible to have one-to-one mapping of traffic classes across all the access technologies.

TABLE 13.2 Data Rates for Various Access Technologies

Access Technology	Data Rate
ZigBee	Up to 250 kbps
Bluetooth	Up to 1 Mbps
WiMedia	Up to 480 Mbps
802.15.3c	Up to 6 or 7.35 Gbps
ECMA TC48 mmWave	Up to 6.478 Gbps
WLAN	
802.11b	Up to 11 Mbps
802.11a	Up to 54 Mbps
802.11g	Up to 54 Mbps
802.11n	Up to 600 Mbps
GSM	Up to 9.6 kbps
IS-95	Up to 9.6 kbps (mandatory support)
GPRS	Up to 117 kbps
UMTS	Up to 2 Mbps
CDMA2000	Up to 2 Mbps
WiMax (fixed broadband wireless access)	In excess of 120 Mbps
HSPA	Up to 14.4 Mbps downlink/5.76 Mbps uplink
Evolved HSPA	Up to 42 Mbps downlink
LTE	Up to 100 Mbps downlink/50 Mbps uplink

The types of handoffs in different access technologies are also different. IEEE 802.11 WLAN uses hard handoff; that is, a connection is broken before a new connection is made: *break before make* handoff. GSM and GPRS cellular networks also use hard handoff, while 2G and 3G CDMA-based cellular networks such as IS-95, CDMA2000, and WCDMA in UMTS use soft handoff. A new connection is made before breaking the old connection in soft handoff, also known as *make before break*. On the other hand, LTE cellular network uses hard handoff. Thus, different horizontal handoff within each type of access technologies will make the design of vertical handoff between heterogeneous networks challenging.

New protocol stack design such as IEEE 802.21 media-independent handoff will certainly help to alleviate the vertical handoff issues. IEEE 802.21 media-independent handoff should be expanded to include access technologies that are not being considered at the moment as well as future access technologies.

New AAA scenarios also need to be studied to address security, charging, and access. Different access technologies also have different security procedures before a mobile device is granted access. The security architecture and protocol in IEEE 802.11 is called wired equivalent privacy (WEP). WEP is responsible for providing authentication, confidentiality, and data integrity. Authentication is the process of verifying that a mobile station or user requesting access is, in fact, a legitimate user. Confidentiality is achieved by sharing a secret key on how to encrypt and decrypt messages. The secret key is the method or algorithm or cipher that the mobile station and the access point use to decrypt messages. Integrity ensures that the message sent by the source to the destination has not been modified or tampered with. GSM uses a 128-bit preshared secret key for securing the interface between the mobile device and the base station. This key is also stored in the authentication center (AuC), which is a database that stores the secret keys of all subscribers. The secret key stored in the SIM card of the mobile device and the AuC forms the foundation for securing the GSM access interface.

GPRS uses the wireless application protocol (WAP). WAP is an open specification that offers a standard method to access Internet-based services and contents. An end-to-end security is achieved by wireless transport layer security (WTLS) in the WAP stack. There is also no key establishment in UMTS. UMTS also uses a 128-bit preshared secret key between the USIM card of the mobile device and the AuC. Bluetooth uses a large number of keys in its security process. The key hierarchy depends on whether it is unicast or broadcast communication. WiMedia uses a four-way handshaking procedure for two devices to establish pairwise temporal keys (PTKs) and a secure relationship [5]. A device may solicit or distribute group temporal keys (GTKs) within a secure relationship [5]. The security mechanisms in WiMedia control the security operation of devices by setting appropriate security modes [5]. They allow devices to authenticate each other, to derive PTKs, and to establish secure relationships [5]. They also enable devices to solicit or distribute GTKs within established secure relationships [5]. WiMax supports two encryption standards: data encryption standard 3 (DES3) and advanced encryption standard (AES). Thus, different access technologies use different types of security.

Charging or billing is certainly of paramount importance to both the operator and the mobile user. Any simplification in the bill would be most welcome by the mobile user. Moreover, the operator may also offer an overall cheaper plan to the subscriber using the heterogeneous networks due to new revenues generated in the new services introduced across the access technologies. Service discovery mechanisms are also needed for a mobile device to discover new services and networks within its region. A mobile device can find more about the available access technologies and their services in its vicinity. These will enable it to connect to the *available best connection* (ABC). Effective routing protocols in the WPANs are also needed to support all possible services and their QoS constraints. Enabling services with QoS constraints to be met in multihop WPANs is certainly an important technical challenge.

Many challenging technical issues need to be solved before the convergence of networks envisaged in this section can be realized. However, once it is achieved, mobile users can enjoy the fruits of this labor.

13.6 ANALYTICAL MODEL FOR CELLULAR/WLAN INTERWORKING

In this section an analytical model for cellular/WLAN interworking is presented. The analysis follows that in [6,7]. In an integrated cellular/WLAN system, one or more WLANs are assumed to be operating within each cell of the cellular system. There are two coverage areas in cellular/WLAN interworking: the cellular-only coverage area and the dual cellular/WLAN coverage area. Cellular horizontal handoffs occur between adjacent cells of the cellular network, while WLAN horizontal handoffs occur between coverage areas of adjacent WLANs. Vertical handoffs occur between moving from a cellular-only coverage area to a dual cellular/WLAN coverage area, and vice versa. The following system-level parameters are used throughout this section:

M^c	number of cells in the cellular system
A_i^c	set of cells adjacent to cell i
W_i^c	set of WLANs within the coverage of cell i
A_k^c	set of WLANs adjacent to WLAN k
D_k^w	set containing the overlaying cell of WLAN k in a dual cellular/WLAN coverage
C_i^c	capacity of each cell i of the cellular system
C_k^w	capacity of each WLAN k

The connections-level parameters used are as follows:

B_{ni}^c	new connection blocking probability in cell i
B_{hi}^c	handoff connection blocking probability in cell i
B_{nk}^w	new connection blocking probability in WLAN k
B_{hk}^w	handoff connection blocking probability in WLAN k
n_i	number of connections in progress in cell i
m_k	number of connections in progress in WLAN k
λ_i^c	arrival rate of new connections in cell i
λ_k^w	arrival rate of new connections in WLAN k
$1/\mu_i^c$	mean connection holding time of a connection in cell i
$1/\mu_k^w$	mean connection holding time of a connection in WLAN k

An assumption in this analysis is that each connection uses only one channel or unit of bandwidth.

Let q_{iT}^c be the probability that a connection in cell i of the cellular system terminates and leaves the system at the end of a channel holding time. Thus, the probability that it moves within the system and continues in an adjacent cell i or WLAN k is $1 - q_{iT}^c$. Let q_{ij}^c and q_{ik}^c be, respectively, the probability of attempting a horizontal handoff to adjacent cell j and the probability of attempting a vertical handoff to WLAN k within cell i. With these definitions, we have

$$1 - q_{iT}^c = \sum_{j \in A_i^c} q_{ij}^c + \sum_{k \in W_i^c} q_{ik}^c. \tag{13.1}$$

Let q_{kT}^w be the probability that a connection in WLAN k terminates and leaves the system at the end of a channel holding time. Thus, the probability that it continues and moves to the overlaying cell or an adjacent WLAN of WLAN k is $1 - q_{kT}^w$. Let q_{kl}^w and q_{ki}^w be, respectively, the probability of attempting a horizontal handoff to adjacent WLAN l and the probability of attempting a vertical handoff to overlaying

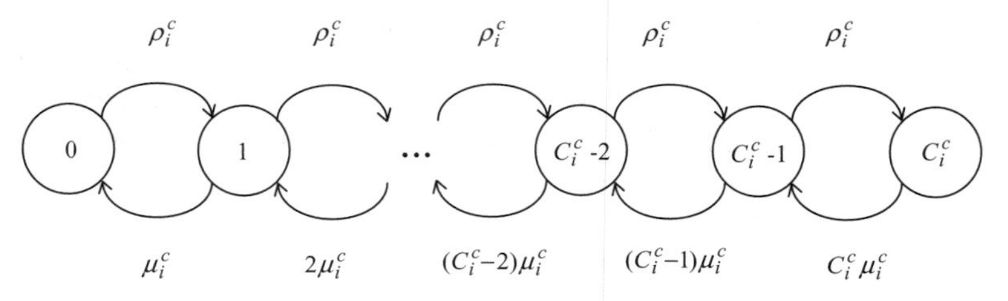

FIGURE 13.3 Markov chain for cell i in cellular/WLAN interworking.

cell i. With these definitions, we have

$$1 - q_{kT}^w = \sum_{l \in A_k^w} q_{kl}^w + \sum_{i \in D_k^w} q_{ki}^w. \qquad (13.2)$$

Let us next consider the traffic equations in the cellular network. The occupancy of a cell evolves according to a one-dimensional Markov chain, as shown in Figure 13.3. The birth rate is ρ_i^c, while the death rate in state n_i of cell i is $n_i \mu_i^c$. Let υ_{ji}^c and υ_{ki}^w be, respectively, the horizontal handoff rate of cell j offered to adjacent cell i and the vertical handoff rate of WLAN k offered to overlay cell i. The total traffic offered to cell i is given by

$$\rho_i^c = \lambda_i^c + \sum_{j \in A_i^c} \upsilon_{ji}^c + \sum_{k \in W_i^c} \upsilon_{ki}^w, \qquad (13.3)$$

where

$$\upsilon_{ji}^c = \lambda_j^c (1 - B_{nj}^c) q_{ji}^c + \sum_{x \in A_j^c} \upsilon_{xj}^c (1 - B_{hj}^c) q_{ji}^c + \sum_{y \in W_j^c} \upsilon_{yj}^w (1 - B_{hj}^c) q_{ji}^c, \quad (13.4)$$

$$\upsilon_{ki}^w = \lambda_k^w (1 - B_{nk}^w) q_{ki}^w + \sum_{x \in A_k^w} \upsilon_{xk}^w (1 - B_{hk}^w) q_{ki}^w + \sum_{y \in D_k^w} \upsilon_{yk}^w (1 - B_{hk}^w) q_{ki}^w. \quad (13.5)$$

From the birth–death process (see the Appendix), the steady-state probability of state n_i in cell i is given by

$$p_{n_i}^c = \frac{\left(\frac{\rho_i^c}{\mu_i^c}\right)^k \frac{1}{n_i!}}{\sum_{r=0}^{C_i^c} \left(\frac{\rho_i^c}{\mu_i^c}\right)^r \frac{1}{r!}}, \qquad 0 \le n_i \le C_i^c. \qquad (13.6)$$

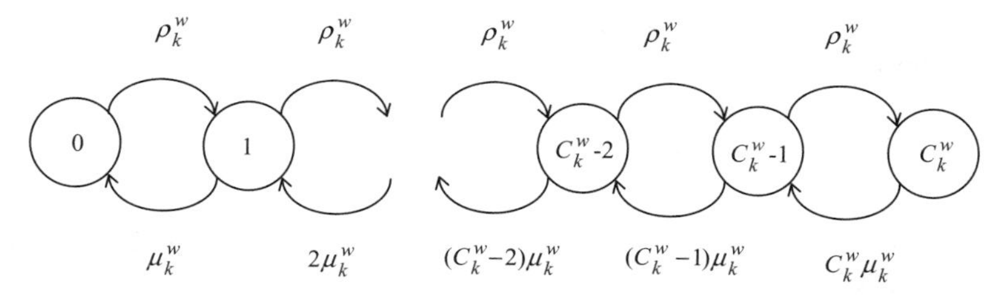

FIGURE 13.4 Markov chain for WLAN k in cellular/WLAN interworking.

Let us next consider the traffic equations in the WLAN. The occupancy of WLAN k also evolves according to a one-dimensional Markov chain, as shown in Figure 13.4. The birth rate is ρ_k^w, while the death rate in state m_k of WLAN k is $m_k\mu_k^w$. Let υ_{lk}^w and υ_{jk}^w be, respectively, the horizontal handoff rate of WLAN l offered to adjacent WLAN k and the vertical handoff rate of cell j offered to WLAN k for overlaying cells j and k. The total traffic offered to WLAN k is given by

$$\rho_k^w = \lambda_k^w + \sum_{l\in A_k^w} \upsilon_{lk}^w + \sum_{j\in D_k^w} \upsilon_{jk}^w, \tag{13.7}$$

where

$$\upsilon_{lk}^w = \lambda_l^w(1-B_{nl}^w)q_{lk}^w + \sum_{x\in A_l^w} \upsilon_{xl}^w(1-B_{hl}^w)q_{lk}^w + \sum_{y\in D_l^w} \upsilon_{yl}^w(1-B_{hl}^w)q_{lk}^w, \tag{13.8}$$

$$\begin{aligned}\upsilon_{jk}^w &= \lambda_j^c(1-B_{nj}^c)R_{jk}q_{jk}^c + \sum_{x\in A_j^c} \upsilon_{xj}^c(1-B_{hj}^c)R_{jk}q_{jk}^c \\ &\quad + \sum_{y\in W_j^c} \upsilon_{yj}^w(1-B_{hj}^c)R_{ik}q_{jk}^c.\end{aligned} \tag{13.9}$$

In (13.9), R_{ik} is the coverage factor between WLAN k and overlay cell j and $0 < R_{ik} \le 1$. This factor considers the coverage ratio between the radio coverage area of WLAN k and the radio coverage area of cell j.

Similarly, from the birth–death process (see the Appendix), the steady-state probability of state m_k in WLAN k is given by

$$p_{m_k}^w = \frac{\left(\dfrac{\rho_k^w}{\mu_k^w}\right)^k \dfrac{1}{m_k!}}{\displaystyle\sum_{r=0}^{C_k^w}\left(\dfrac{\rho_k^w}{\mu_k^w}\right)^r \dfrac{1}{r!}}, \qquad 0 \le m_k \le C_k^w. \tag{13.10}$$

With no guard channels, the new and handoff connection blocking probabilities in cell i are the same and are given by

$$B_{ni}^{c} = B_{hi}^{c} = P_{C_i^c}^{c}. \tag{13.11}$$

Similarly, with no guard channels, the new and handoff connection blocking probabilities in WLAN k are given by

$$B_{nk}^{w} = B_{hk}^{w} = P_{C_k^w}^{w}. \tag{13.12}$$

The traffic equations of the horizontal and vertical handoff rates can be solved using an iterative algorithm on repeated substitutions until convergence is achieved [6]. This is similar to those algorithms for finding the handoff rates in Chapter 5.

SUMMARY

The six 3GPP/WLAN interworking scenarios and their operational capabilities are presented in Section 3.2, while a brief description of IEEE 802.11u is presented in Section 3.3. IEEE 802.21 media-independent handoff is presented in Section 3.4.

3GPP/WLAN interworking between cellular technologies and WLANs, IEEE 802.11u interworking of IEEE 802.11 WLANs with external networks, and IEEE 802.21 media-independent handoff for heterogeneous access technologies such as IEEE 802.3, IEEE 802.11, IEEE 802.16, 3GPP cellular networks, and 3GPP2 cellular networks are certainly the specifications and standards to look for in the convergence of networks today.

The future envisaged for cellular/WiMax/WLAN/WPAN interworking in a convergence of networks, and some issues relating to the interworkings of various access technologies, are discussed in this chapter. Once these interworkings between technologies are achieved, together with new rate adaptive services across them, users can enjoy truly seamless and user-transparent connectivity based on the best connection available. Finally, an analytical model for cellular/WLAN interworking is presented. Other analytical models for cellular/WLAN interworking may be found in [6–9] and references therein.

REFERENCES

[1] "3rd Generation Partnership Project: technical specification group services and system aspects; feasibility study on 3GPP system to wireless local area network (WLAN) internetworking, (Release 7)," *3GPP TR 22.934 V7.0.0 (2007-06)*, 2007.

[2] F. Cacace and L. Vollero, "Managing mobility and adaptation in upcoming 82.21-enabled devices," *WMASH 2006,* Los Angeles, CA, Sept. 29, 2006.

[3] A. Dutta, S. Das, D. Famolari, Y. Ohba, K. Taniuchi, V. Fajardi, R. M. Lopez, T. Kodama, and H. Schulzrinne, "Seamless proactive handoff across heterogeneous access networks," *Wireless Personal Commun.*, vol. 43, pp. 837–855, 2007.

[4] A. de la Olivia, T. Melia, A. Vidal, C. J. Bernardos, I. Soto, and A. Banchs, "A case study: IEEE 802.21 enabled mobile terminals for optimized WLAN/3G handovers," *Mobile Comput. Commun. Rev.*, vol. 11, no. 2, Apr. 2007.

[5] "High rate ultra wideband PHY and MAC standard," *Standard ECMA-368*, ECMA International, Geneva, Dec. 2005.

[6] E. Stevens-Navarros, C. Sun, and V. W. S. Wong, "A survey of analytical modeling for cellular/WLAN interworking," in *Emerging Technologies in Wireless LANs: Theory, Design and Deployment*, Cambridge University Press, New York, 2007.

[7] E. Stevens-Navarros and V. W. S. Wong, "Resource sharing in an integrated wireless cellular/WLAN system," *Canadian Conference on Electrical and Computer Engineering 2007,* Vancouver, British Columbia, Canada, Apr. 2007.

[8] W. Song, H. Jiang, and W. Zhuang, "Performance analysis of the WLAN-first scheme in cellular/WLAN interworking," *IEEE Trans. Wireless Commun.*, vol. 6, no. 5, pp. 1932–1943, May 2007.

[9] W. Song and W. Zhuang, "Multi-service load sharing for resource management in the cellular/WLAN integrated network," *IEEE Trans. Wireless Commun.*, to appear.

APPENDIX

BASICS OF PROBABILITY, RANDOM VARIABLES, RANDOM PROCESSES, AND QUEUEING SYSTEMS

A.1 INTRODUCTION

In this appendix we review the basics of probability, random variables, exponential random process, birth–death processes, and queueing systems.

A.2 PROBABILITY

A.2.1 Set Operations

The following are basic set operations:

1. $A \cap B = B \cap A$
2. $A \cap (B \cup C) = (A \cap B) \cup (A \cap C)$
3. $(A \cap B) \cap C = A \cap (B \cap C) = A \cap B \cap C$
4. $A \cup B = B \cup A$
5. $A \cup (B \cap C) = (A \cup B) \cap (A \cup C)$
6. $(A \cup B) \cup C = A \cup (B \cup C) = A \cup B \cup C$
7. $(\overline{A \cup B}) = \overline{A} \cap \overline{B}$
8. $(\overline{A \cap B}) = \overline{A} \cup \overline{B}$

Wireless Broadband Networks, By David Tung Chong Wong, Peng-Yong Kong, Ying-Chang Liang, Kee Chaing Chua, and Jon W. Mark

A.2.2 Elements of Probability

Let us first define a trial, a sample space, and an event before introducing probability through set operations.

- A *trial* is a single performance of an experiment for which there is an outcome.
- A *sample space*, S, is the set of all possible outcomes in any given experiment.
- An *event*, A, is a subset of the sample space. If two events have no common outcomes, they are *mutually exclusive*.

Axioms

1. $P(A) \geq 0$
2. $P(S) = 1$
3. $P\left(\bigcup_{n=1}^{N} A_n\right) = \sum_{n=1}^{N} P(A_n)$ if$(A_m \cap A_n) = \emptyset$, where $\emptyset$ is the null set.

Joint Probability

1. $P(A \cap B) = P(A) + P(B) - P(A \cup B)$
2. $P(A \cup B) = P(A) + P(B) - P(A \cap B) \leq P(A) + P(B)$

For $A \cap B = \emptyset,\ P(A \cap B) = P(\emptyset) = 0$ (mutually exclusive).

Conditional Probability The conditional probability of an event A, given B, with $P(B) > 0$, is given by

$$P(A|B) = \frac{P(A \cap B)}{P(B)}. \tag{A.1}$$

If A and B are mutually exclusive ($A \cap B = \emptyset$), we have

$$P(A|B) = 0. \tag{A.2}$$

If $A \cap C = \emptyset$, we have

$$P[(A \cup C)|B] = P(A|B) + P(C|B). \tag{A.3}$$

Total Probability If $B_m \cap B_n = \emptyset,\ m \neq n = 1, 2, \ldots, N, \bigcup_{n=1}^{N} B_n = S,$ we have

$$P(A) = \sum_{n=1}^{N} P(A|B_n)P(B_n). \tag{A.4}$$

Bayes' Theorem

$$P(B_n|A) = \frac{P(B_n \cap A)}{P(A)}, \qquad P(A) \neq 0, \tag{A.5}$$

$$P(A|B_n) = \frac{P(A \cap B_n)}{P(B_n)}, \qquad P(B_n) \neq 0, \tag{A.6}$$

$$P(B_n|A) = \frac{P(A|B_n)P(B_n)}{P(A)} = \frac{P(A|B_n)P(B_n)}{\sum_{n=1}^{N} P(A|B_n)P(B_n)}. \tag{A.7}$$

Independent Events Given that events A, B, and C are independent, we have

$$P(A|B) = P(A), \tag{A.8}$$

$$P(B|A) = P(B), \tag{A.9}$$

$$P(A \cap B) = P(A) \cdot P(B), \tag{A.10}$$

$$P(A \cap B \cap C) = P(A) \cdot P(B) \cdot P(C). \tag{A.11}$$

A.3 RANDOM VARIABLES

A random variable is a real function of the elements of a sample space, S.

A.3.1 Conditions

1. Set $\{X \leq x\}$ is an event for any real number x.
2. $P(X = -\infty) = 0$.
3. $P(X = \infty) = 0$.

A.3.2 Discrete Random Variables

A discrete random variable has only discrete values. The sample space can be discrete, continuous, or a mixture.

Probability Mass Function

$$f_X(x) = P(X = x), \qquad x = 0, 1, 2, \ldots \quad \text{(discrete sample space)} \tag{A.12}$$

Total Probability

$$\sum_{x=0}^{\infty} f_X(x) = 1 \tag{A.13}$$

Cumulative Distribution Function

$$F_X(x) = \sum_{i=0}^{x} f_X(i) \tag{A.14}$$

Expected or Mean Value

$$E[X] = \sum_{x=0}^{\infty} x f_X(x) \tag{A.15}$$

or

$$E[X] = \sum_{x=0}^{\infty} (1 - F_X(x)) \qquad \text{if } x > 0 \tag{A.16}$$

Variance

$$\text{Var}[X] = E[(X - E[X])^2] = E[X^2] - (E[X])^2, \qquad \text{where } E[X^2] = \sum_{x=0}^{\infty} x^2 f_X(x). \tag{A.17}$$

A few common discrete random variables, with their mean, second moments, and variance, are shown in Table A.1.

TABLE A.1 Common Discrete Random Variables

Discrete Random Variable	$f_X(x)$	$E[X]$ or $\overline{X}$	$E[X^2]$ or $\overline{X^2}$	Var$[X]$
Bernoulli	$\begin{cases} 1-p, x=0 \\ p, x=1 \end{cases}$	p	p	$p(1-p)$
Binomial	$\binom{n}{x} p^x (1-p)^{n-x}, \quad x = 0,1,2,\ldots,n$ $\binom{n}{x} = \dfrac{n!}{x!\,(n-x)!}$	np	$n^2p^2 +$ $np(1-p)$	$np(1-p)$
Geometric	$p(1-p)^{x-1}, \quad x = 1,2,\ldots$	$\dfrac{1}{p}$	$\dfrac{2-p}{p^2}$	$\dfrac{1-p}{p^2}$
Poisson	$\dfrac{\lambda^x e^{-\lambda}}{x!}, \quad x = 0,1,2,\ldots$	λ	$\lambda^2 + \lambda$	λ

A.3.3 Continuous Random Variables

A continuous random variable has a continuous range of values. The sample space is continuous.

Probability Density Function

$$f_X(x) = P(X = x), x \in (-\infty, \infty) \tag{A.18}$$

Total Probability

$$\int_{-\infty}^{\infty} f_X(x)\,dx = 1. \tag{A.19}$$

Cumulative Distribution Function

$$F_X(x) = \int_{-\infty}^{x} f_X(x)\,dx \tag{A.20}$$

Expected or Mean Value

$$E[X] = \int_{-\infty}^{\infty} x f_X(x)\,dx \tag{A.21}$$

or

$$E[X] = \int_{0}^{\infty} [1 - F_X(x)]\,dx, \qquad \text{if } x \geq 0 \tag{A.22}$$

Variance

$$\begin{aligned}\mathrm{Var}[X] &= E[(X - E[X])^2] = E[X^2] - (E[X])^2, \qquad \text{where} \quad E[X^2]\\ &= \int_{-\infty}^{\infty} x^2 f_X(x)\,dx\end{aligned} \tag{A.23}$$

A few common continuous random variables, with their mean, second moments, and variance, are shown in Table A.2.

TABLE A.2 Common Continuous Random Variables

Continuous Random Variables	$f_X(x)$	$E[X]$ or $\overline{X}$	$E[X^2]$ or $\overline{X^2}$	$\mathrm{Var}[X]$
Uniform	$\frac{1}{b-a}, x \in [a, b]$	$\frac{a+b}{2}$	$\frac{a^2+b^2+ab}{3}$	$\frac{(b-a)^2}{12}$
Exponential	$\lambda e^{-\lambda x}, \quad x \in [0, \infty)$	$\frac{1}{\lambda}$	$\frac{2}{\lambda^2}$	$\frac{1}{\lambda^2}$
Gaussian (normal)	$\frac{1}{\sqrt{2\pi}\sigma} \exp\left[-\frac{1}{2}\left(\frac{x-\mu}{\sigma}\right)^2\right]$	μ	$\mu^2+\sigma^2$	σ^2

Memoryless Property of Exponential Distribution

Exponential pdf

$$f_X(x) = \lambda e^{-\lambda x}, \qquad x \geq 0, \quad \lambda > 0 \tag{A.24}$$

Exponential CDF

$$F_X(x) = 1 - e^{-\lambda x} \tag{A.25}$$

Memoryless Property

$$\begin{aligned} P[X > s + t | X > t] &= \frac{P[X > s + t, X > t]}{P[X > t]} \\ &= \frac{P[X > s + t]}{P[X > t]} \\ &= \frac{P[X > s]P[X > t]}{P[X > t]} \\ &= P[X > s] \\ &= e^{-\lambda s} \quad \text{(independent of } t\text{)} \end{aligned} \tag{A.26}$$

A.4 POISSON RANDOM PROCESS

Let $X(t)$ be the number of Poisson points in $[0, t)$ for $t \geq 0$. Given t, $X(t)$ is a Poisson random variable with parameter λt.

$$P[X(t) = x] = \frac{(\lambda t)^x}{x!} e^{-\lambda t}, \tag{A.27}$$

$$E[X(t)] = \lambda t, \tag{A.28}$$

$$E[X^2(t)] = \lambda t + \lambda^2 t^2, \tag{A.29}$$

$$\text{Var}[X(t)] = \lambda t. \tag{A.30}$$

A.4.1 Interarrival Times of a Poisson Process

The interarrival times of a Poisson process are independent, identically distributed exponential random variables with mean $1/\lambda$.

A.4.2 Decomposition of a Poisson Process

$X(t),\ t \geq 0$ is a Poisson random process with rate λ. Event i is class i with probability p_i such that

$$\sum_{i=1}^{n} p_i = 1. \tag{A.31}$$

Let $X_i(t)$ = number of class i arrivals in [0, t) such that

$$X(t) = \sum_{i=1}^{n} X_i(t). \tag{A.32}$$

$X_i(t)$ is a Poisson random process with rate $p_i\lambda$, and the $X_i(t)$'s are independent. That is, the decomposition of a Poisson random process will result in n other Poisson random processes.

A.4.3 Superposition of Poisson Processes

If the $X_i(t)$'s are Poisson random processes with rate λ_i such that

$$X(t) = \sum_{i=1}^{n} X_i(t), \tag{A.33}$$

then $X(t),\ t \geq 0$ is a Poisson random process with rate λ such that

$$\lambda = \sum_{i=1}^{n} \lambda_i. \tag{A.34}$$

That is, the superposition of Poisson random processes results in a Poisson random process with a rate equal to the sum of the individual rates.

A.5 BIRTH–DEATH PROCESSES

A birth–death process is a continuous-time Markov chain with discrete state space in which only transitions to neighboring states are permitted. The state transition diagram is shown in Figure A.1. $K(t) = k$ is the number in the system (customers, packets, calls, channels, etc.). λ_k is the arrival rate when $K(t) = k$. μ_k is the service rate when $K(t) = k$. $\{K(t) : t \geq 0\}$ is a continuous-time Markov chain. The steady-state (equilibrium) distribution of the probability of the number in the system

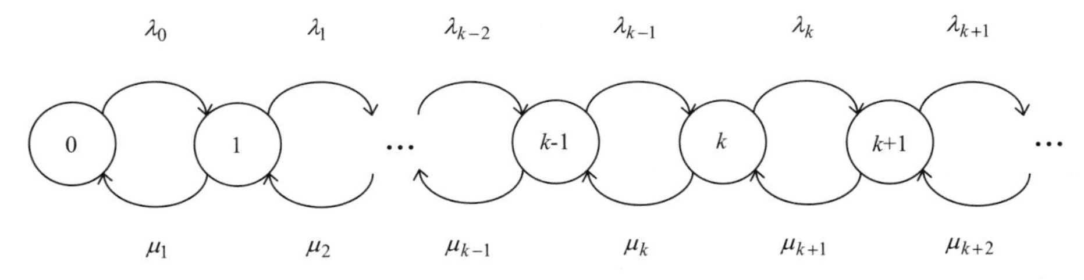

FIGURE A.1 State transition diagram of a birth–death process.

(state k), p_k, is given by

$$p_k = \lim_{t \to \infty} P[K(t) = k]. \tag{A.35}$$

By global balance equations:

$$\begin{aligned} \lambda_0 p_0 &= \mu_1 p_1, \qquad k = 0, \\ p_1 &= \frac{\lambda_0}{\mu_1} p_0, \end{aligned} \tag{A.36}$$

$$(\lambda_k + \mu_k) p_k = \lambda_{k-1} p_{k-1} + \mu_{k+1} p_{k+1}, \qquad k = 1, 2, \ldots. \tag{A.37}$$

By local balance equations (across a boundary), we have

$$\begin{aligned} \lambda_{k-1} p_{k-1} &= \mu_k p_k, \qquad k = 1, 2, \ldots, \\ p_k &= \frac{\lambda_{k-1}}{\mu_k} p_{k-1} \\ &= p_0 \prod_{i=1}^{k} \frac{\lambda_{i-1}}{\mu_i}. \end{aligned} \tag{A.38}$$

By the total probability property, we have

$$\begin{aligned} &\sum_{k=0}^{\infty} p_k = 1, \\ &p_0 + p_1 + p_2 + \cdots = 1, \\ &p_0 \left[1 + \sum_{k=1}^{\infty} \prod_{i=1}^{k} \frac{\lambda_{i-1}}{\mu_i} \right] = 1, \\ &p_0 = \frac{1}{1 + \sum_{k=1}^{\infty} \prod_{i=1}^{k} \frac{\lambda_{i-1}}{\mu_i}]}. \end{aligned} \tag{A.39}$$

The steady-state probability of being in state k, p_k, is given by

$$p_k = \frac{\prod_{i=1}^{k} \frac{\lambda_{i-1}}{\mu_i}}{1 + \sum_{k=1}^{\infty} \prod_{i=1}^{k} \frac{\lambda_{i-1}}{\mu_i}]}. \tag{A.40}$$

The mean number of customers in the system, $\overline{N}$, is given by

$$\overline{N} = \sum_{k=0}^{\infty} k p_k. \tag{A.41}$$

A.6 BASIC QUEUEING SYSTEMS

A.6.1 Kendall's Notation

An $A/B/C/K/m/Z$ queueing system means:

A	arrival process
B	service process
C	number of servers
K	maximum capacity (buffer size)
m	population of users
Z	service discipline

For arrival or service processes:

M	exponential distribution
D	deterministic (constant) distribution
G	general distribution

For principal service disciplines:

FIFO	first in, first out
LCFS	last come, first served
FIRO	first in, random out

When the last three elements of Kendall's notation are not specified, it is understood that $Z = \text{FIFO}$, $m = +\infty$, and $K = +\infty$.

A.6.2 *M/M/1*

The following are the assumptions for an *M/M/1* queue:

- Poisson arrival process (exponential interarrival times)
- Exponential service time
- One server
- Infinite storage spaces

The state transition diagram for an *M/M/1* queue is shown in Figure A.2. From this diagram, we have

$$\lambda_k = \lambda \qquad \forall k, \tag{A.42}$$

$$\mu_k = \mu \qquad \forall k, \tag{A.43}$$

$$p_0 = 1 - \rho, \qquad \rho = \frac{\lambda}{\mu} < 1. \tag{A.44}$$

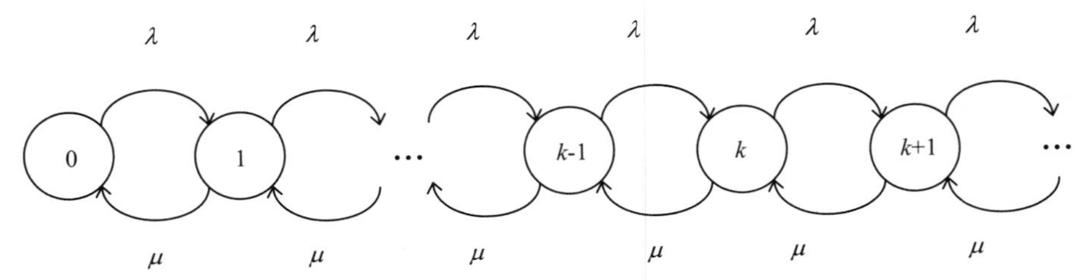

FIGURE A.2 State transition diagram of an *M*/*M*/1 queue.

The steady-state probability of being in state k, p_k, is given by

$$p_k = (1-\rho)\rho^k \quad \text{(geometric)}. \tag{A.45}$$

The mean number of customers in the system, $\overline{N}$, is given by

$$\overline{N} = \sum_{k=0}^{\infty} k p_k = \frac{\rho}{1-\rho}. \tag{A.46}$$

Figure A.3 shows an *M*/*M*/1 queue. From Little's law,

$$\overline{N} = \lambda \overline{T}, \tag{A.47}$$

where λ is the mean arrival rate for the system and $\overline{T}$ is the mean time spent in the system. The mean time spent in the system, $\overline{T}$, is given by

$$\overline{T} = \frac{\overline{N}}{\lambda} = \frac{\frac{1}{\mu}}{1-\rho} = \frac{1}{\mu-\lambda}, \qquad \rho = \frac{\lambda}{\mu}. \tag{A.48}$$

The mean queueing delay, $\overline{W}$, is given by

$$\begin{aligned}\overline{W} &= \overline{T} - \frac{1}{\mu}, \qquad \frac{1}{\mu} \text{ is the mean service time} \\ &= \frac{\mu - (\mu - \lambda)}{\mu(\mu - \lambda)} \\ &= \frac{1}{\mu}\frac{\rho}{1-\rho}.\end{aligned} \tag{A.49}$$

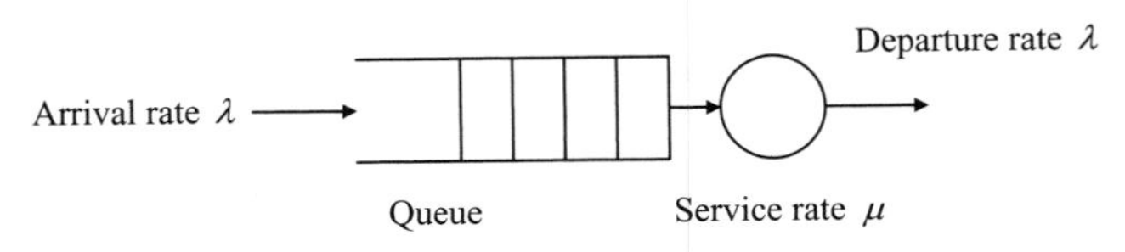

FIGURE A.3 *M*/*M*/1 queue.

Using Little's law, the mean number of customers in the queue, $\overline{N}_q$, is given by

$$\overline{N}_q = \lambda\overline{W} = \frac{\rho^2}{1-\rho}. \tag{A.50}$$

Using Little's law, the mean number of customers in service, $\overline{N}_s$, is given by

$$\overline{N}_s = \lambda\frac{1}{\mu} = \rho. \tag{A.51}$$

A.6.3 *M/M/1/K* (Finite Storage)

The following are the assumptions for an *M/M/1/K* queue:

- Poisson arrival process (exponential interarrival times)
- Exponential service time
- One server
- K storage spaces

The state transition diagram for an *M/M/1/K* queue is shown in Figure A.4. From this diagram we have

$$\lambda_k = \begin{cases} \lambda, & k < K \\ 0, & k = K, \end{cases} \tag{A.52}$$

$$\mu_k = \mu \qquad \forall k, \tag{A.53}$$

$$p_k = \rho p_{k-1}, \qquad \rho = \frac{\lambda}{\mu}, \quad k \le K \tag{A.54}$$

$$= \rho^k p_0$$

$$= 0, \qquad k > K. \tag{A.55}$$

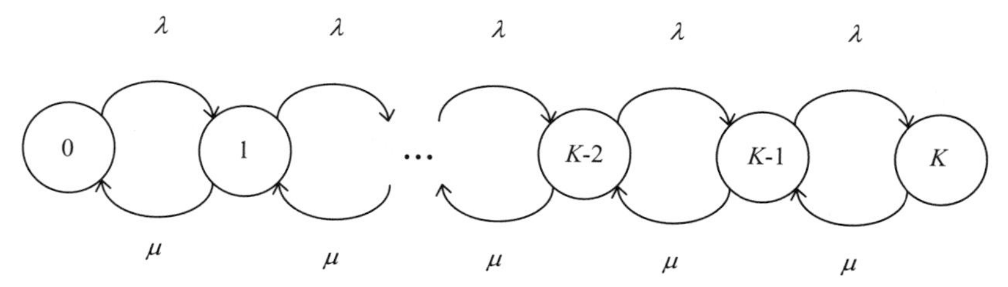

FIGURE A.4 State transition diagram of an *M/M/1/K* queue.

By the total probability property,

$$\begin{aligned}
&\sum_{k=0}^{K} p_k = 1,\\
&p_0[1+\rho+\rho^2+\cdots+\rho^K] = 1,\\
&p_0 = \frac{1-\rho}{1-\rho^{K+1}}.
\end{aligned} \tag{A.56}$$

The steady-state probability of being in state k, p_k, is given by

$$p_k = \begin{cases} \dfrac{(1-\rho)\rho^k}{1-\rho^{K+1}}, & k \leq K \\ 0, & k > K. \end{cases} \tag{A.57}$$

The blocking probability, P_B, is given by

$$\begin{aligned}
P_B &= P[\text{an arrival sees the system full}]\\
&= p_K\\
&= \frac{(1-\rho)\rho^K}{1-\rho^{K+1}}.
\end{aligned} \tag{A.58}$$

The mean number in the system, $\overline{N}$, is given by

$$\overline{N} = \sum_{k=0}^{\infty} k p_k = \frac{\rho}{1-\rho} - \frac{(K+1)\rho^{K+1}}{1-\rho^{K+1}}. \tag{A.59}$$

Figure A.5 shows an *M/M/1/K* queue. The actual arrival rate to the system, λ_a, is given by

$$\lambda_a = (1 - P_B)\lambda. \tag{A.60}$$

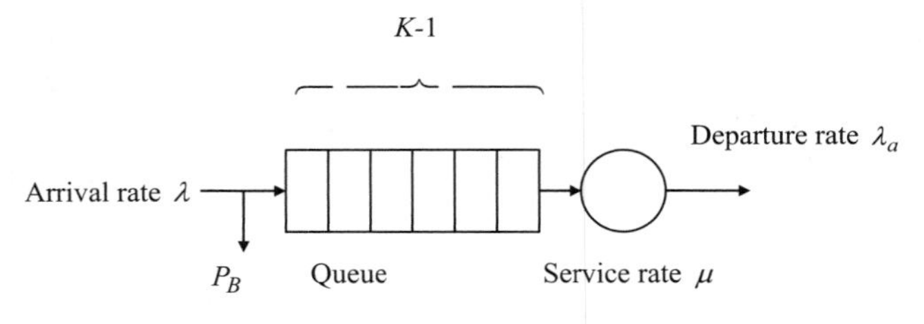

FIGURE A.5 *M/M/1/K* queue.

The mean time spent in the system, $\overline{T}$, is given by

$$\overline{T} = \frac{\overline{N}}{\lambda_a}. \tag{A.61}$$

The mean number of customers in service, $\overline{N}_s$, is given by

$$\overline{N}_s = 1 - p_0. \tag{A.62}$$

The mean number of customers in the queue, $\overline{N}_q$, is given by

$$\overline{N}_q = \overline{N} - \overline{N}_s. \tag{A.63}$$

The mean queueing delay, $\overline{W}$, is given by

$$\overline{W} = \frac{\overline{N}_q}{\lambda_a}. \tag{A.64}$$

A.6.4 *M*/*M*/*m* (*m* Servers System)

The following are the assumptions for an *M*/*M*/*m* queue:

- Poisson arrival process (exponential interarrival times)
- Exponential service time
- *m* servers
- Infinite storage spaces

The state transition diagram for an *M*/*M*/*m* queue is shown in Figure A.6. From this diagram we have

$$\lambda_k = \lambda \qquad \forall k, \tag{A.65}$$

$$\mu_k = \begin{cases} k\mu, & k \leq m \\ m\mu, & k > m, \end{cases} \tag{A.66}$$

$$p_k = p_0 \prod_{i=0}^{k-1} \frac{\lambda}{(i+1)\mu}, \qquad k \leq m$$

$$= p_0 \left(\frac{\lambda}{\mu}\right)^k \frac{1}{k!}, \tag{A.67}$$

$$p_k = p_0 \prod_{i=0}^{m-1} \frac{\lambda}{(i+1)\mu} \prod_{j=m}^{k-1} \frac{\lambda}{m\mu}, \qquad k > m$$

$$= p_0 \left(\frac{\lambda}{\mu}\right)^k \frac{1}{m!\, m^{k-m}}. \tag{A.68}$$

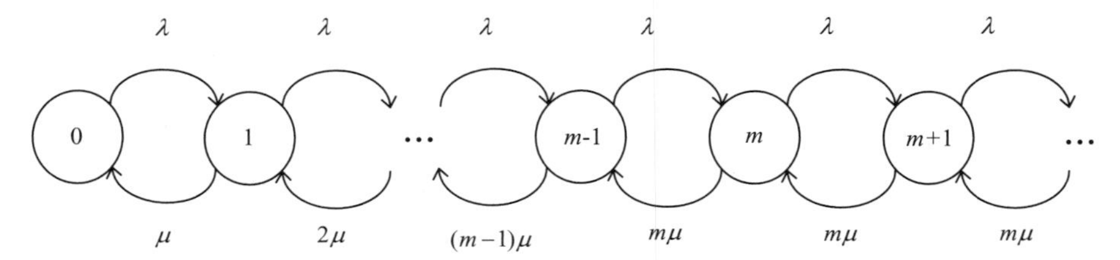

FIGURE A.6 State transition diagram of an *M/M/m* queue.

The steady-state probability of being in state k, p_k, is given by

$$p_k = \begin{cases} p_0 \dfrac{(m\rho)^k}{k!}, & k \leq m, \quad \rho = \dfrac{\lambda}{m\mu} < 1 \\ p_0 \dfrac{\rho^k m^m}{m!}, & k > m. \end{cases} \tag{A.69}$$

By the total probability property,

$$\begin{aligned} \sum_{k=0}^{\infty} p_k &= 1, \\ p_0 \left[1 + \sum_{k=1}^{m-1} \frac{(m\rho)^k}{k!} + \frac{m^m}{m!} \sum_{k=m}^{\infty} \rho^k \right] &= 1, \\ p_0 \left[\sum_{k=0}^{m-1} \frac{(m\rho)^k}{k!} + \frac{(m\rho)^m}{m!} \sum_{k=0}^{\infty} \rho^k \right] &= 1, \\ p_0 = \left[\sum_{k=0}^{m-1} \frac{(m\rho)^k}{k!} + \frac{(m\rho)^m}{m!} \frac{1}{1-\rho} \right]^{-1}&. \end{aligned} \tag{A.70}$$

Figure A.7 shows an *M/M/m* queue. The probability that an arrival will find all servers busy and will be forced to wait in queue is an important performance measure of the *M/M/m* system. Since an arriving customer finds the system in a "typical" state, we have

$$\begin{aligned} P[\text{queueing}] &= \sum_{k=m}^{\infty} p_k \\ &= p_0 \frac{(m\rho)^m}{m!\,(1-\rho)}, \end{aligned} \tag{A.71}$$

known as the *Erlang C formula*. This formula is often used in telephony (and more generally, in circuit-switching systems) to estimate the probability of a call request finding all the m circuits of a transmission line busy. In an *M/M/m* model it is assumed

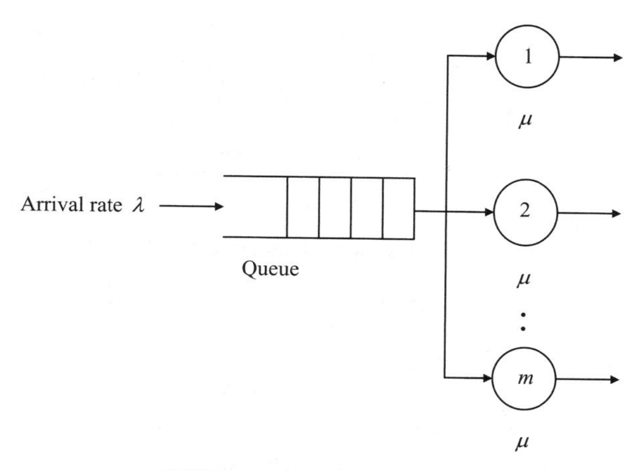

FIGURE A.7 *M/M/m* queue.

that such a call request "remains in queue," that is, attempts continuously to find a free circuit. The mean number of customers in the queue, $\overline{N}_q$, is given by

$$\overline{N}_q = \sum_{k=m}^{\infty} (k-m)p_k = \frac{(m\rho)^m \rho}{m!\,(1-\rho)^2} p_0. \tag{A.72}$$

The mean queueing delay, $\overline{W}$, is given by

$$\overline{W} = \frac{\overline{N}_q}{\lambda}. \tag{A.73}$$

The mean time spent in the system, $\overline{T}$, is given by

$$\overline{T} = \overline{W} + \frac{1}{\mu}. \tag{A.74}$$

The number of customers in the system, $\overline{N}$, is given by

$$\overline{N} = \lambda \overline{T}. \tag{A.75}$$

A.6.5 *M/M/m/m* (*m* Servers Loss System)

The following are the assumptions for an *M/M/m/m* queue:

- Poisson arrival process (exponential interarrival times)
- Exponential service time
- m servers
- m storage spaces

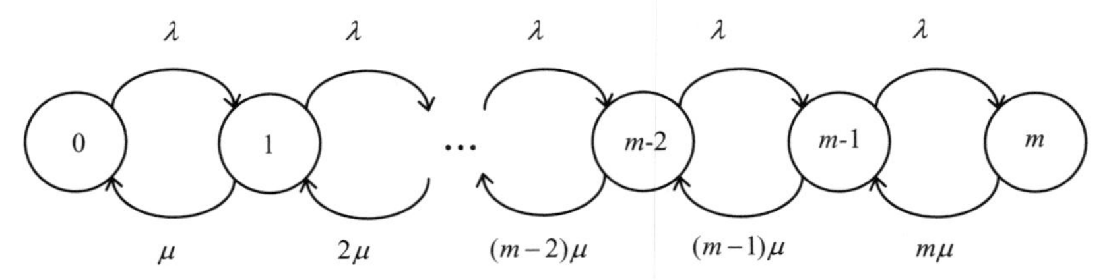

FIGURE A.8 State transition diagram of an *M/M/m/m* queue.

The state transition diagram for an *M/M/m/m* queue is shown in Figure A.8. From this diagram we have

$$\lambda_k = \begin{cases} \lambda, & 0 \leq k < m \\ 0, & k = m, \end{cases} \tag{A.76}$$

$$\mu_k = k\mu, \qquad 0 \leq k \leq m, \tag{A.77}$$

$$\begin{aligned} p_k &= p_0 \prod_{i=1}^{k} \frac{\lambda_{i-1}}{\mu_i}, \qquad 0 \leq k \leq m \\ &= p_0 \left(\frac{\lambda}{\mu}\right)^k \frac{1}{k!}. \end{aligned} \tag{A.78}$$

By the total probability property,

$$\begin{aligned} \sum_{k=0}^{m} p_k &= 1, \\ p_0 \left[\sum_{k=0}^{m} \left(\frac{\lambda}{\mu}\right)^k \frac{1}{k!}\right] &= 1, \\ p_0 &= \left[\sum_{k=0}^{m} \left(\frac{\lambda}{\mu}\right)^k \frac{1}{k!}\right]^{-1}. \end{aligned} \tag{A.79}$$

The steady-state probability of being in state k, p_k, is given by

$$p_k = \frac{\left(\frac{\lambda}{\mu}\right)^k \frac{1}{k!}}{\sum_{k=0}^{m} \left(\frac{\lambda}{\mu}\right)^k \frac{1}{k!}}. \tag{A.80}$$

Figure A.9 shows an *M/M/m/m* queue. The probability that an arrival will find all servers busy and will therefore be lost is an important performance measure of the *M/M/m/m* system. The blocking probability, P_B, is given by

$$P_B = p_m = \frac{\left(\frac{\lambda}{\mu}\right)^m \frac{1}{m!}}{\sum_{k=0}^{m} \left(\frac{\lambda}{\mu}\right)^k \frac{1}{k!}}, \tag{A.81}$$

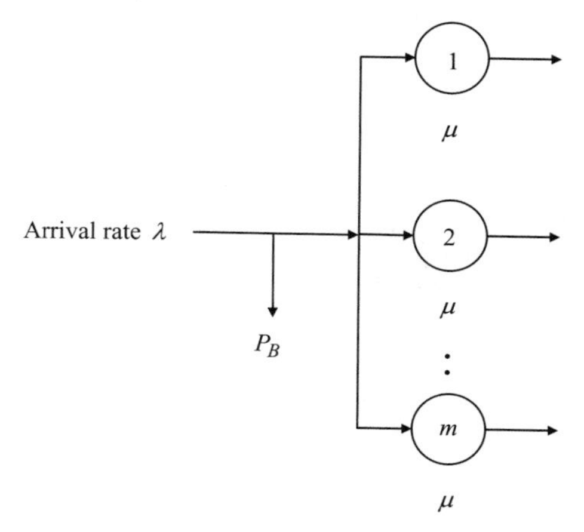

FIGURE A.9 *M/M/m/m* queue.

known as the *Erlang B formula*. This formula is widely used to evaluate the blocking probability of telephone systems. The actual arrival rate at the system, λ_a, is given by

$$\lambda_a = (1 - P_B)\lambda. \tag{A.82}$$

The mean time spent in the system, $\overline{T}$, is given by

$$\overline{T} = \frac{1}{\mu}. \tag{A.83}$$

The mean number of customers in the system, $\overline{N}$, is given by

$$\overline{N} = \lambda_a \overline{T} = \frac{\lambda}{\mu}(1 - P_B). \tag{A.84}$$

Note that there is no waiting room or queue. Therefore, there is no queueing delay and each customer is either served or lost.

A.6.6 *M/M/∞* (∞ Servers)

The following are the assumptions for an *M/M/∞* queue:

- Poisson arrival process (exponential interarrival times)
- Exponential service time
- Infinite servers
- Infinite storage spaces

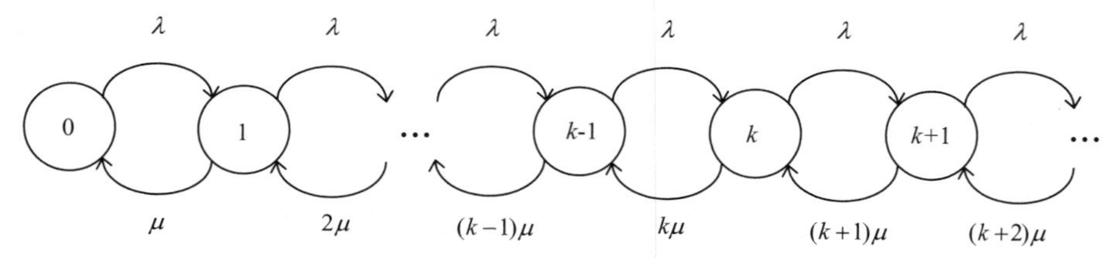

FIGURE A.10 State transition diagram of an $M/M/\infty$ queue.

The state transition diagram for an $M/M/\infty$ queue is shown in Figure A.10. From this diagram we have

$$\lambda_k = \lambda, \tag{A.85}$$

$$\mu_k = k\mu, \tag{A.86}$$

$$\begin{aligned} p_k &= \frac{1}{k!}\frac{\lambda}{\mu}p_{k-1}, \qquad k = 1, 2, 3, \ldots \\ &= \frac{1}{k!}\left(\frac{\lambda}{\mu}\right)^k p_0. \end{aligned} \tag{A.87}$$

By the total probability property,

$$\begin{aligned} &\sum_{k=0}^{\infty} p_k = 1, \\ &p_0 \sum_{k=0}^{\infty} \frac{1}{k!}\left(\frac{\lambda}{\mu}\right)^k = 1, \\ &p_0 e^{\lambda/\mu} = 1, \\ &p_0 = e^{-\lambda/\mu}. \end{aligned} \tag{A.88}$$

The steady-state probability of being in state k, p_k, is given by

$$p_k = \frac{\left(\frac{\lambda}{\mu}\right)^k e^{-\lambda/\mu}}{k!} \quad \text{(Poisson)} \tag{A.89}$$

It can be shown that the number in the system is Poisson distributed even if the service time is not exponential (i.e., $M/G/\infty$). The mean time spent in the system, $\overline{T}$, is given by

$$\overline{T} = \frac{1}{\mu}. \tag{A.90}$$

The mean number in the system, $\overline{N}$, is given by

$$\overline{N} = \frac{\lambda}{\mu}. \tag{A.91}$$

There is no waiting in the queue.

A.6.7 *M/G/*1 Queueing System

The following are the assumptions for an $M/G/1$ queue:

- Poisson arrival process (exponential interarrival times)
- General service time, Y (packet service time which depends on the distribution of the packet length)
- One server
- Infinite storage spaces (infinite buffer)

The mean service time, $\overline{Y}$, is given by

$$\overline{Y} = \frac{1}{\mu}. \tag{A.92}$$

Let $\overline{Y^2}$ denote the second moment of service time. From queueing theory, the mean queueing delay, $\overline{W}$, is given by

$$\overline{W} = \frac{\lambda\overline{Y^2}}{2(1-\rho)} = \frac{\rho}{2(1-\rho)}\frac{\overline{Y^2}}{\overline{Y}}, \qquad \text{where } \rho = \frac{\lambda}{\mu} = \lambda\overline{Y}. \tag{A.93}$$

The total time spent in the system, in queue and in service, is

$$\overline{T} = \overline{Y} + \frac{\lambda\overline{Y^2}}{2(1-\rho)}. \tag{A.94}$$

Applying Little's law, the mean number of customers in the queue, $\overline{N}_q$, and the mean number of customers in the system, $\overline{N}$, are given by

$$\overline{N}_q = \frac{\lambda^2\overline{Y^2}}{2(1-\rho)} = \lambda\overline{W}, \tag{A.95}$$

$$\overline{N} = \rho + \frac{\lambda^2\overline{Y^2}}{2(1-\rho)}. \tag{A.96}$$

When the service times are *exponentially distributed* ($M/M/1$), we have $\overline{Y^2} = 2/\mu^2 = 2(\overline{Y})^2$, and the mean queueing delay reduces to

$$\overline{W} = \frac{1}{\mu}\frac{\rho}{(1-\rho)} = \frac{\rho}{1-\rho}\overline{Y} \qquad (M/M/1). \tag{A.97}$$

When the service times are *deterministic or identical for all customers* ($M/D/1$), we have $\overline{Y^2} = 1/\mu^2 = (\overline{Y})^2$, and the mean queueing delay reduces to

$$\overline{W} = \frac{1}{\mu}\frac{\rho}{2(1-\rho)} = \frac{\rho}{2(1-\rho)}\overline{Y} \qquad (M/D/1) \tag{A.98}$$

A.6.8 *M*/*G*/1 with Vacation

The following are the assumptions for an $M/G/1$ queue with vacation:

- Poisson arrival process (exponential interarrival times)
- General service time, Y (packet service time which depends on the distribution of the packet length)
- One server
- Infinite storage spaces (infinite buffer)
- Server goes on vacation with duration V

From queueing theory, the mean queueing delay, $\overline{W}$, is

$$\begin{aligned}\overline{W} &= \frac{\lambda\overline{Y^2}}{2(1-\rho)} + \frac{\overline{V^2}}{2\overline{V}}, \qquad \text{where } \rho = \frac{\lambda}{\mu} = \lambda\overline{Y} \\ &= \frac{\rho}{2(1-\rho)}\left(\frac{\overline{Y^2}}{\overline{Y}}\right) + \frac{\overline{V^2}}{2\overline{V}}.\end{aligned} \tag{A.99}$$

The total time spent in the system, in queue and in service, is

$$\overline{T} = \overline{Y} + \overline{W}. \tag{A.100}$$

A.6.9 *M*/*G*/1 with Vacations and with *M* Users

The following are the assumptions for an $M/G/1$ queue with vacation and M users:

- Poisson arrival process (exponential interarrival times)
- General service time, Y (packet service time which depends on the distribution of the packet length)
- One server

- Infinite storage spaces (infinite buffer)
- Server goes on vacation with duration V
- M users

From queueing theory, the mean queueing delay, $\overline{W}$, is

$$\overline{W} = \frac{\lambda \overline{Y^2}}{2(1-\rho)} + \frac{(M-\rho)\overline{V}}{2(1-\rho)} + \frac{\text{Var}[V]}{2\overline{V}}, \qquad \text{where } \rho = \lambda \overline{Y}$$

$$= \frac{\rho}{2(1-\rho)} \frac{\overline{Y^2}}{\overline{Y}} + \frac{(M-\rho)\overline{V}}{2(1-\rho)} + \frac{\text{Var}[V]}{2\overline{V}}. \tag{A.101}$$

The total time spent in the system, in queue and in service, is

$$\overline{T} = \overline{Y} + \overline{W}. \tag{A.102}$$

REFERENCES

[1] S. M. Ross, *Introduction to Probability Models*, 5th ed., Academic Press, San Diego, CA, 1993.

[2] A. Leon-Garcia, *Probability and Random Processes for Electrical Engineering*, 2nd ed., Addison-Wesley, Reading, MA, 1994.

[3] L. Kleinrock, *Queueing Systems*, Vol. 1, *Theory*, Wiley, New York, 1975.

INDEX

Wireless Broadband Networks, By David Tung Chong Wong, Peng-Yong Kong, Ying-Chang Liang, Kee Chaing Chua, and Jon W. Mark